Reviews of Accelerator Science and Technology

Volume 9

Reviews of Accelerator Science and Technology

Edited by Alexander W. Chao and Weiren Chou

Published

Vol. 1 (2008)	Overview of Accelerator Science and Technology and the Community
Vol. 2 (2009)	Medical Applications of Accelerators
Vol. 3 (2010)	Accelerators as Photon Sources
Vol. 4 (2011)	Accelerator Applications in Industry and the Environment
Vol. 5 (2012)	Applications of Superconducting Technology to Accelerators
Vol. 6 (2013)	Accelerators for High Intensity Beams
Vol. 7 (2014)	Colliders
Vol. 8 (2015)	Accelerator Applications in Energy and Security
Vol. 9 (2016)	Technology and Applications of Advanced Accelerator Concepts

Reviews of Accelerator Science and Technology

Volume 9: Technology and Applications of Advanced Accelerator Concepts

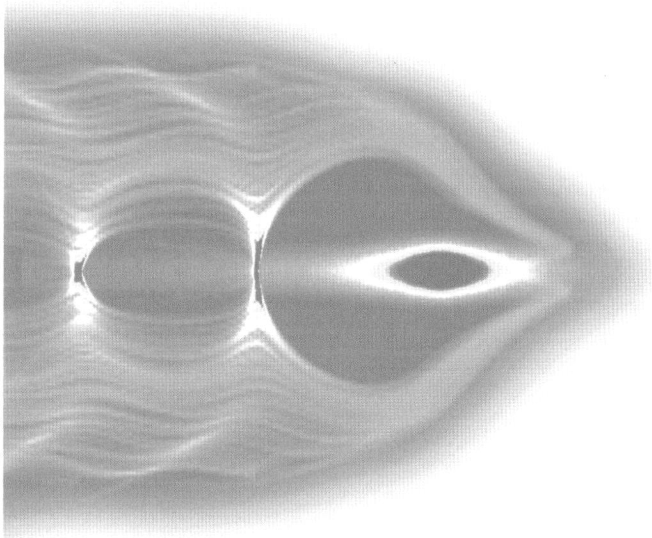

Editors

Alexander W. Chao
SLAC National Accelerator Laboratory, USA

Weiren Chou
Fermi National Accelerator Laboratory, USA

World Scientific

NEW JERSEY · LONDON · SINGAPORE · BEIJING · SHANGHAI · HONG KONG · TAIPEI · CHENNAI · TOKYO

Published by

World Scientific Publishing Co. Pte. Ltd.
5 Toh Tuck Link, Singapore 596224
USA office: 27 Warren Street, Suite 401-402, Hackensack, NJ 07601
UK office: 57 Shelton Street, Covent Garden, London WC2H 9HE

British Library Cataloguing-in-Publication Data
A catalogue record for this book is available from the British Library.

REVIEWS OF ACCELERATOR SCIENCE AND TECHNOLOGY
Volume 9: Technology and Applications of Advanced Accelerator Concepts

ISBN 978-981-3209-57-2

Printed in Singapore

Reviews of Accelerator Science and Technology
Vol. 9 (2016) v–vi

DOI: 10.1142/S179362681602001X

World Scientific
www.worldscientific.com

Editorial Preface

The theme of *Reviews of Accelerator Science and Technology* (RAST) Volume 9 is Technology and Applications of Advanced Accelerator Concepts. Since its invention in the 1920s, particle accelerators have made tremendous progress in accelerator science, technology and applications. However, the fundamental acceleration principle, namely, to apply an external radiofrequency (RF) electric field to accelerate charged particles, remains unchanged. As this method (either room temperature RF or superconducting RF) is approaching its intrinsic limitation in acceleration gradient (measured in MeV/m), it becomes apparent that new methods with much higher acceleration gradient (measured in GeV/m) must be found for future very high energy accelerators as well as future compact (table-top or room-size) accelerators. This volume introduces a number of advanced accelerator concepts (AAC) — their principles, technologies and potential applications. For the time being, none of them stands out as a definitive direction in which to go. But these novel ideas are in hot pursuit and look promising. Furthermore, some AAC requires a high power laser system. This has the implication of bringing two different communities — accelerator and laser — to join forces and work together. It will have profound impact on the future of our field.

This volume begins with an article on the US road map for AAC by Eric. R. Colby and L. K. Len. Following this overview, five articles discuss various advanced accelerator concepts that have been proposed: laser-driven plasma acceleration by Kazuhisa Nakajima, electron and positron beam driven plasma acceleration by Mark J. Hogan, proton beam driven plasma acceleration by Erik Adli and Patric Muggli, dielectric laser acceleration by K. P. Wootton, J. McNeur and K. J. Leedle, and dielectric wakefield acceleration by Chunguang Jing.

Following the introduction of these novel concepts, there is a review article by Jonathan Wheeler, Gerard Mourou and Toshiki Tajima, which addresses the laser technology that plays a critical role in further development of acceleration beyond the TeV range. Another technology article by Jean-Luc Vay and Remi Lehe reviews numerical simulations for plasma and laser acceleration.

Several articles discuss various aspects of envisioned applications of these advanced accelerator concepts: luminosity limitations in linear colliders based on plasma acceleration by Valeri Lebedev, Alexey Burov and Sergei Nagaitsev, application for colliders by Daniel Schulte, and medical applications by Mitsuru Uesaka and Kazuyoshi Koyama.

In this volume, William A. Barletta has written a review of an important book to our community, *Engines of Discovery: A Century of Particle Accelerators* by Andrew Sessler and Edmund Wilson.

As in the past volumes, we include two special articles, one on the accelerator community, another on "person-of-the-issue". The community article is particle accelerators in China by Chuang Zhang and Shouxian Fang, which gives a comprehensive overview of the rapidly growing accelerator community in China. The person-of-the-issue is Jerome Lewis Duggan, who was a pioneer and founder of a huge community of industrial and medical accelerators in the US. This article is written by Floyd Del McDaniel and Barney L. Doyle.

We wish to dedicate this volume to Jialin Xie, a member of the RAST Editorial Board, a great accelerator scientist and a founding father of the Chinese accelerator community.

Alexander W. Chao
SLAC National Accelerator Laboratory, USA
achao@slac.stanford.edu

Weiren Chou
Fermi National Accelerator Laboratory, USA
chou@fnal.gov

Contents

Reviews of Accelerator Science and Technology
Vol. 9 (2016) 1–18

DOI: 10.1142/S1793626816300012

World Scientific
www.worldscientific.com

Roadmap to the Future

Eric R. Colby* and L. K. Len†

Office of High Energy Physics,
US Department of Energy,
Washington DC, 20585, USA
**Eric.Colby@science.doe.gov*
†LK.Len@science.doe.gov

Most particle accelerators today are expensive devices found only in the largest laboratories, industries, and hospitals. Using techniques developed nearly a century ago, the limiting performance of these accelerators is often traceable to material limitations, power source capabilities, and the cost tolerance of the application. Advanced accelerator concepts[a] aim to increase the gradient of accelerators by orders of magnitude, using new power sources (e.g. lasers and relativistic beams) and new materials (e.g. dielectrics, metamaterials, and plasmas). Worldwide, research in this area has grown steadily in intensity since the 1980s, resulting in demonstrations of accelerating gradients that are orders of magnitude higher than for conventional techniques. While research is still in the early stages, these techniques have begun to demonstrate the potential to radically change accelerators, making them much more compact, and extending the reach of these tools of science into the angstrom and attosecond realms. Maturation of these techniques into robust, engineered devices will require sustained interdisciplinary, collaborative R&D and coherent use of test infrastructure worldwide. The outcome can potentially transform how accelerators are used.

Keywords: Advanced accelerator; plasma; laser; wakefield; dielectric.

1. Introduction

In the nearly 150 years since the invention of Crookes' tube, particle accelerators have been reinvented numerous times, moving far beyond their original induction coil and rectifier forms. The needs of scientific, medical, and industrial applications have shaped the accelerators in use today, demanding higher energy, beam power, and brightness, while simultaneously demanding lower energy consumption and cost.

Sustained accelerator research has produced a succession of new techniques that have advanced the performance of accelerators by eight orders of magnitude in energy and more than six orders of magnitude in power, with significant reductions in size and cost occurring as well.

There is both the capability and the need to continue both trends by several orders of magnitude. Increasing the high-energy capability of accelerators is a central aim of R&D on advanced accelerator concepts, along with reducing size, cost, and power consumption. We will discuss the scientific potential, R&D questions, and other challenges that lie ahead for advanced accelerator concepts. Before we do so, it is worth discussing the scientific and funding context within which this R&D thrust has emerged.

2. Acceleration Techniques

All particle accelerators rely on energy transfer from electric fields to charged particles, whether particles traverse these fields once (linear accelerators) or many times (circular accelerators).

In the case of linear accelerators, machine size is driven by the achievable accelerating gradient, a property that depends as much on the fundamental physical limitations of the accelerator as on the cost-effectiveness of the power source used to drive it. The majority of conventional linear electron

[a] "Advanced accelerator concepts" here refers specifically to novel high-gradient acceleration techniques. It should be noted that this term is often used in a broader context that includes high-power accelerators (e.g. superconducting radio frequency accelerators), high-field superconducting magnets, high-power radio frequency sources, new sources of particles, new methods for high-intensity beams (e.g. integrable optics or advanced feedback and control), and high-gradient RF, among other areas.

accelerators operate at gradients of ~20–30 MV/m, a limit determined by tolerable breakdown rates and the economics of the application.

The largest currently conceived 500 GeV linear accelerator is the International Linear Collider, which requires roughly 31 km to achieve its center-of-mass energy [1]. Decades of R&D has gone into understanding how to reliably and economically produce the machine's specified 31.5 MV/m gradient. Clearly, developing robust, cost-effective acceleration techniques capable of significantly higher accelerating gradients would impact both the economics and the scientific capability of such a machine.

Setting aside dc accelerators and induction accelerators, which are effective sources of low-voltage and very-high-current beams, respectively, the vast majority of accelerators rely on electromagnetic (EM) fields to accelerate charged particles.

Free space EM fields have an electric field that is predominantly transverse to the propagation direction, requiring some physical arrangement to be made to achieve efficient energy transfer between the fields and the particle beam.[b] Typically, this is accomplished by bending either the particle trajectories or the fields (i.e. diffraction). To achieve sustained energy transfer, some provision must also be made to maintain synchronism between the propagating fields and the particle beam. Achieving both conditions in a low-cost device has proven to be a challenge that is still being addressed today.

EM wave diffraction and confinement is by far the most widely used approach, and requires materials (conductors, typically) to be placed in the very near field of the beam to achieve strong coupling. Periodic waveguide structures operating in the microwave frequency range are the most common geometry used, as it is possible to arrange for strong coupling and phase synchronism, with considerable latitude left over to optimize other properties, such as peak field stress, surface heating, beam focusing, and wakefield characteristics. Accelerators of this kind are gradient-limited by material breakdown and the capabilities and cost of the power source.

[b]The discussion here is limited to processes that are first-order in the electric field strength. Higher-order energy exchange is possible at highly relativistic field strengths, and does not have such phasing restrictions. The peak intensities required, however, are at the edge of what is possible with today's laser technology.

Resonant transverse motion can directly extract energy from transverse EM fields (as in inverse free electron lasers) even in the absence of an EM guiding structure. This latter point proved an advantage in the early demonstrations of laser-driven accelerators. Such accelerators are limited by synchrotron radiation losses, which increase rapidly with energy.

3. Advanced Accelerator Concepts

Advanced accelerator concepts, as they have come to be called, aim to achieve much higher accelerating gradients by overcoming the power source and material limitations of their conventional counterparts.

Experiments have shown that shorter exposure to high-field stresses reduces both the breakdown rate and the long-term material damage [2]. Therefore, power sources capable of delivering very high peak powers in very short pulses (such as nanoseconds for microwave frequencies, or femtoseconds for optical frequencies) have an advantage in producing very high gradients. Lasers and charged particle beams can both achieve this, although the coupling mechanisms are different in each case.

New material choices are also being explored for the coupling structure, including dielectrics, metamaterials (e.g. photonic band gap structures), and plasmas. Dielectrics offer potential cost and performance advantages owing either to the simplification of machined geometries (in the micro- and mm-wave ranges) or to lithographic fabrication (in the terahertz and optical ranges). They can also have significantly lower loss and higher damage tolerance in the terahertz and optical frequency ranges. Metamaterials (MM) offer engineered permittivity and permeability properties at the expense of fabrication complexity, and photonic band gap structures offer the ability to ameliorate long-range wakefield effects. Each of these solid state material choices brings a gradient limit, either through breakdown or through Kerr effect dephasing. Experiments to date on waveguide geometries suggest that this gradient limitation may be well in excess of 1 GV/m for picosecond exposure times [3].

In principle, a structure that is reparable could exceed material damage limitations, and it is this property that makes plasmas interesting as accelerator "structures." Cavitation of the electrons in a plasma can produce enormous field gradients through the separation of charges. Provided that such cavities can be made to move synchronously with the accelerating beam, these fields can be used to transfer energy from the plasma to the beam. Such relativistic plasma cavities have been produced using electron, positron, and laser beams (and, soon, proton beams) to drive gradients well beyond 100 GV/m for deeply nonlinear plasma acceleration [4]. Achieving high-quality, energy-efficient acceleration in a robust, engineered system may reduce the loaded gradients closer to 10 GV/m.

Research in plasma wakefield acceleration is scientifically rich, as the accelerator structure (the plasma cavity) is both dynamic and modified by the accelerated beam. A high accelerating gradient requires high plasma densities $O(10^{17}/\text{cm}^3)$, which brings strong transverse forces and a characteristic scale length for the dynamics which is $O(100\,\text{microns})$, three orders of magnitude smaller than for conventional RF accelerators. These highly coupled and often strongly nonlinear systems demand time-intensive computer simulations to gain physical insight. It is precisely these characteristics that give rise to the extraordinarily rich physics of these systems, and have led to a wealth of science beyond the simple use as accelerators.

Progress in demonstrating and understanding the physics of these advanced accelerator techniques began in the 1980s with the inception of an accelerator science program at the US Department of Energy's Office of High Energy Physics. The program featured a dedicated workshop series (originally called the "Laser Acceleration of Particles" workshop [5]) to share important results in the nascent field, and this workshop series has continued to this day as the Advanced Accelerator Concepts Workshop [6]. Activity worldwide has grown significantly in this area, and has recently spawned European and Asian regional workshops on advanced accelerator concepts.

4. Accelerator R&D at the DOE's Office of High Energy Physics[c]

Research in high-energy physics and in nuclear physics have long been the primary driving forces of accelerator technology development. More recently, synchrotrons and free electron lasers have emerged as powerful tools of the basic energy sciences.

Among these fundamental sciences, high energy physics (HEP) has required accelerators of exponentially increasing energy. This has driven high-risk, high-payoff research to understand the ultimate limits of what can be done in a laboratory, driving R&D in advanced accelerator concepts. Of course, higher gradients would also lead to more compact, lower-energy accelerators, and this has been the driving force behind a wealth of new applications.

Aspects of advanced accelerator concepts R&D are supported by several US federal sponsors, including the Department of Energy, the Department of Defense, and the National Science Foundation. The DOE remains the dominant sponsor of long-term accelerator R&D, with the Office of High Energy Physics (OHEP) being the primary sponsor of advanced concepts.

Within the OHEP, accelerator research is funded through two portfolios: the General Accelerator R&D portfolio (GARD) and the Accelerator Stewardship (Stewardship) portfolio. Additional R&D is supported in the DOE Small Business Innovative Research and Small Business Technology Transfer program under the HEP technical topics.

4.1. *The GARD program*

The GARD program focuses on accelerator R&D that is expected to benefit HEP [7]. R&D in this area generally is chosen to meet both medium-term (e.g. significant accelerator upgrade) and long-term (e.g. new accelerator techniques) objectives.

Funding occurs in five thrust areas:

(i) Accelerator and Beam Physics
(ii) Advanced Accelerator Concepts
(iii) Particle Sources and High-Power Targetry
(iv) RF Acceleration Technology
(v) Superconducting Magnets & Materials

Advanced accelerator concepts R&D, the focus of this article, accounts for approximately 2.3% of the overall OHEP portfolio and approximately 26% of the GARD portfolio. Support includes three dedicated national laboratory accelerator facilities — the Argonne Wakefield Accelerator (AWA), the Berkeley Laboratory Laser Accelerator (BELLA), and

[c]DOE refers to the US Department of Energy. The Office of High Energy Physics is part of the DOE's Office of Science, the largest supporter of physical science research in the US.

the Facility for Advanced Accelerator Experimental Tests (FACET) at SLAC — for advanced concepts R&D. The primary goal of this R&D is to develop technologies that may eventually provide e^+e^- linear colliders (LCs) well into the TeV center-of-mass energy range.

The three accelerator facilities are designed to study different aspects of wakefield-based acceleration, with AWA and FACET testing beam-driven waveguide and plasma wakefield concepts, and BELLA testing laser-driven plasma concepts.

Within the GARD program, R&D is also conducted in accelerator and beam physics to advance the theory and simulation tools used to design accelerators, particle sources and high-power beam targets, RF acceleration technology that spans from conventional RF sources to normal-conducting and superconducting RF, high field bending and focusing superconducting magnets and materials, diagnostics, stabilizing feedback systems, and accelerator controls.

4.2. *Accelerator Stewardship*

The Stewardship program focuses on accelerator R&D which is expected to primarily benefit stakeholders *beyond* HEP while bringing a synergistic value to HEP [8]. The portfolio supports medium-term use-inspired basic R&D, long-term basic R&D, and the operation of a dedicated accelerator facility (the Accelerator Test Facility at Brookhaven) which is used to support, in part, advanced accelerator concepts R&D. R&D topics on the applied R&D track are identified by the community through basic research needs workshops [9]. R&D topics on the long-term basic R&D track are identified by individual investigators, but awards are made on the basis of scientific merit *and* the specific recommendations of the intended stewardship customer.

Stewardship accounts for approximately 1.3% of the overall OHEP portfolio, and Advanced Accelerator Concepts R&D approximately 30% of the Stewardship portfolio.

High-repetition-rate ultrafast laser R&D is a key applied R&D thrust area in the Stewardship portfolio, with a near-term goal of developing kilowatt-class laser systems suitable for laser-driven accelerators (both plasma- and structure-based), and for a variety of other applications, such as driving new sources of terahertz and X-ray radiation.

5. Accelerator R&D Planning

Long-range planning for accelerator R&D for HEP has a number of steps, designed to connect the research with the long-term needs of the science. This process is often conducted on a regional basis (e.g. Europe, US, Asia) with an international view.

The most recent US planning process began in 2012–13 with the so-called "Snowmass" process, during which the HEP community took stock of the most important R&D questions and possible experiments that could address them. The Particle Physics Project Prioritization Panel ("P5") formulated a set of recommendations [10] to bring the most important science to fruition over the next decade. P5 made several recommendations regarding advanced technology R&D, including one to pursue accelerator R&D with a focus on outcomes and capabilities that will dramatically improve cost-effectiveness for mid-term and far-term accelerators.

An accelerator R&D subpanel was then convened by HEPAP to further develop the P5 recommendations into a set of recommendations specific to this area [11]. Following the report from the "GARD Subpanel," a series of community-led workshops in the GARD thrust areas has been commissioned to fully elaborate the P5 plan into technical goals and roadmaps to guide R&D investments over the next decade. To date, Roadmap Workshops have covered the Superconducting Magnets & Materials and Advanced Accelerator Concepts thrusts.

6. Roadmap for Advanced Accelerator Concepts

On February 2–3, 2016, the OHEP convened a workshop in Gaithersburg, Maryland to develop an advanced accelerator concepts research roadmap. The workshop was in response to a recommendation by the HEPAP Accelerator R&D Subpanel to "convene the university and laboratory proponents of advanced acceleration concepts to develop R&D roadmaps with a series of milestones and common down selection criteria towards the goal for constructing a multi-TeV e^+e^- collider." During the workshop, key proponents of laser-driven plasma wakefield acceleration (LWFA), particle-beam-driven plasma wakefield acceleration (PWFA), and dielectric wakefield acceleration (DWFA), along with invited university, laboratory, and international experts, presented and critically discussed individual concept roadmaps. This roadmap workshop followed several preparatory, community-organized workshops, which provided significant background and input to the roadmap process.

Individual R&D roadmaps for LWFA, PWFA and DWFA were presented which outlined the beam issues and challenges related to creating a multi-TeV collider. Synergies between the domestic and global efforts, potential early applications, and diagnostic and computer simulation needs were also identified.

Roadmaps were developed with two key timescales and milestones in mind. The primary, long-term goal is a multi-TeV collider, with R&D culminating in completion of a technical design report (TDR) sometime in the 2035–40 time period. The secondary, nearer-term goal is the completion of a TDR for a potential early application in the 2025–30 time period. Participants developed both broad-view roadmaps to an e^+e^-collider operating by about 2040, and more detailed roadmaps covering research milestones over the next decade, which included preparing a TDR for a near-term application. Most often, the near-term goal was an XFEL or a compact gamma-ray source.

To accomplish these goals, the next ten years of AAC research should focus on addressing common challenges identified during the workshop:

(i) Higher energy staging of electron acceleration with independent drive beams, yielding equal energy gains in each stage, and with $\geq$90% beam transmission through both stages;
(ii) Understanding mechanisms for emittance growth and developing methods for achieving emittances that are needed to build a collider;
(iii) Demonstration of a single electron acceleration stage at higher energy;
(iv) Demonstration and understanding of positron acceleration;
(v) Continuous, joint development of a comprehensive and realistic operational parameter set for a multi-TeV collider, to guide operating specifications for AAC.

The phrase "higher energy" in the first and third common challenges encompasses the different capabilities of the concepts and facilities, and is taken to be in the multi-GeV range for LWFA and PWFA and in the multiple-100 MeV range for DWFA.

What follows are the roadmaps for LWFA, PWFA and DWFA that have been developed to address the above common challenges within the two timescales mentioned above. An additional roadmap is included for laser development in support of LWFA as advances in this driving technology are needed to support both the R&D and the possible near-term applications of this concept.

6.1. *Laser wakefield acceleration R&D*

A research program toward completion of a TDR for a multi-TeV e^+e^- linear collider based on LWFA technology is presented in Fig. 1.

In preparation for development of a conceptual linear collider design, R&D over the next ten years would continue the ongoing invention and discovery phase, where the main goals would be understanding of the key physics concepts and innovation of LWFA collider concepts. This phase would include experiments demonstrating significant progress toward the collider goal (see Fig. 2). To identify and solve outstanding physics and technology issues, several key experiments are required.

Presently LWFA can accelerate tens of pC of charge to several GeV in a single stage [12]. The ten-year R&D goal is to accelerate 100 pC of charge to 10 GeV in a single LWFA stage. Accomplishing this requires development of techniques for matched guiding of the laser pulse in the plasma. Other R&D would be focused on improvements in stability, reproducibility, and tunability, which would enable near-term applications.

With the completion of a 10 GeV electron LWFA stage, the 10 GeV beam may be employed for electron–positron pair creation and subsequent positron beam capture and LWFA. Development of an LWFA-based positron source would enable compact experiments on LWFA and focusing of positron beams.

Critical to the collider application is demonstration of multi-GeV LWFA staging with independent, equal-energy, drive beams. Following the successful demonstration of LWFA staging at the 100-MeV-energy level [13], a multi-GeV (e.g. 5 GeV and 5 GeV) staging experiment is required. The goal of this experiment is to demonstrate high beam capture efficiency (>90%), while maintaining a large average/geometric accelerating gradient (>5 GV/m) and preserving the beam emittance between LWFA stages. This experiment will allow study and understanding of emittance growth mechanisms, as well as testing of growth mitigation techniques.

Plasma target development is required to enable the key experiments. Shaping and precise control of plasma target profiles is required for the collider application. In particular, development of longitudinally tapered and near-hollow plasma channels, extending tens of centimeters, requires R&D.

Optimizing the energy transfer efficiency between laser and particle beam will be critical to realizing the potential of LWFA. Novel methods for extracting energy from plasma wakes via particle bunch shape (or current pulse) tailoring must be developed; techniques to reduce the remaining energy in linear wakes (hence also reducing the power loading on the structures) by exploring techniques to recycle residual wake energy. Methods for bunch shape tailoring and wake energy extraction would also benefit the beam-driven plasma systems.

Contemporaneously with the demonstration of key experiments, novel diagnostics for LWFA beams and plasma targets must be invented and high-fidelity and high-speed simulation tools must be developed. Modeling of plasma targets will require 3D magneto-hydrodynamic (MHD) codes to be developed, with the proper low-temperature physics and chemistry included. The development of the MHD codes will benefit from collaborations with Livermore and Sandia National Laboratories, leveraging National Nuclear Security Administration investments. Capabilities for rapid modeling of multi-GeV LWFA stages (laser and beam plasma interaction) are required for parameter exploration and start-to-end modeling of LWFA-based colliders. This requires a sustained community effort in the development of open source code suites that integrate all the recent algorithmic advances (e.g. boosted frame, Maxwell solvers with azimuthal Fourier decomposition, spectral solvers, control of numerical Cherenkov instability, laser envelope solvers, adaptive mesh refinement) for the plasma-based accelerator modules, and include advanced beam dynamics modules for transport through conventional transport sections. Integration of these innovations, together with porting of the codes to massively parallel many core and GPU architectures, will enable speedups by orders of magnitude, from days-to-weeks at present for 3D simulations of one multi-GeV stage, to minutes-to-hours on exascale-capable supercomputers.

These simulation tools must be validated with experimental measurements. Diagnostics are required to resolve femtosecond slice properties

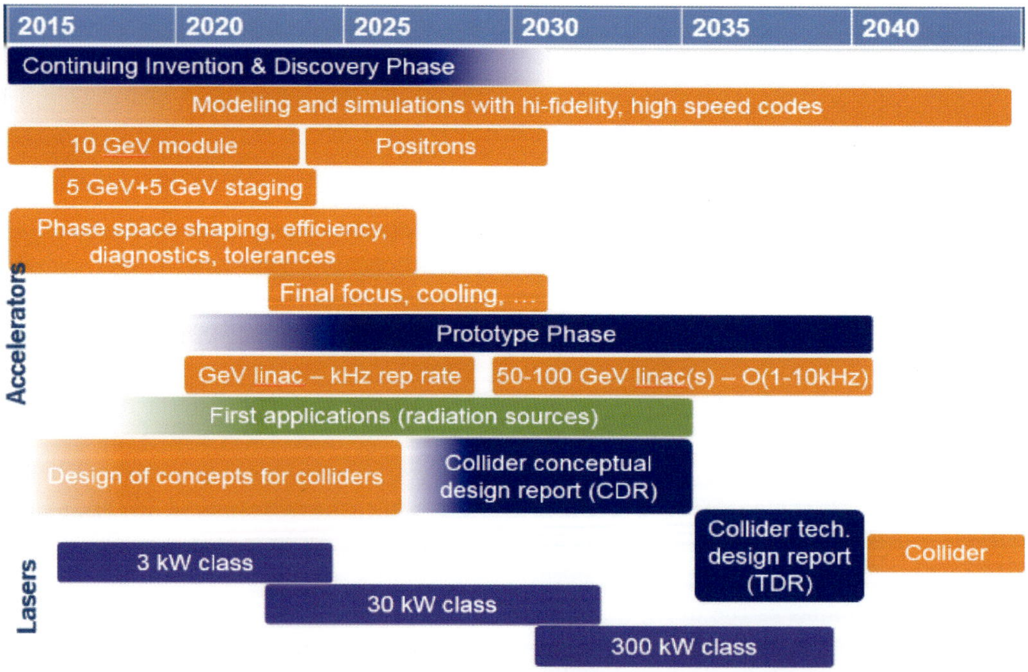

Fig. 1. High-level R&D roadmap for a laser wakefield accelerator–based collider.

of the LWFA beam, and techniques for noninvasive beam phase space measurements (for example, with resolution on the order of tens of nanometers normalized transverse emittance) are needed. Methods to measure the evolution of the 3D plasma profile are also required.

In addition, innovative LWFA collider concepts must be explored over the next ten years. For example, compact final focus methods with large momentum acceptance, rapid cooling methods for positron beams that are compatible with short (tens of microns) beams, and laser plasma methods for

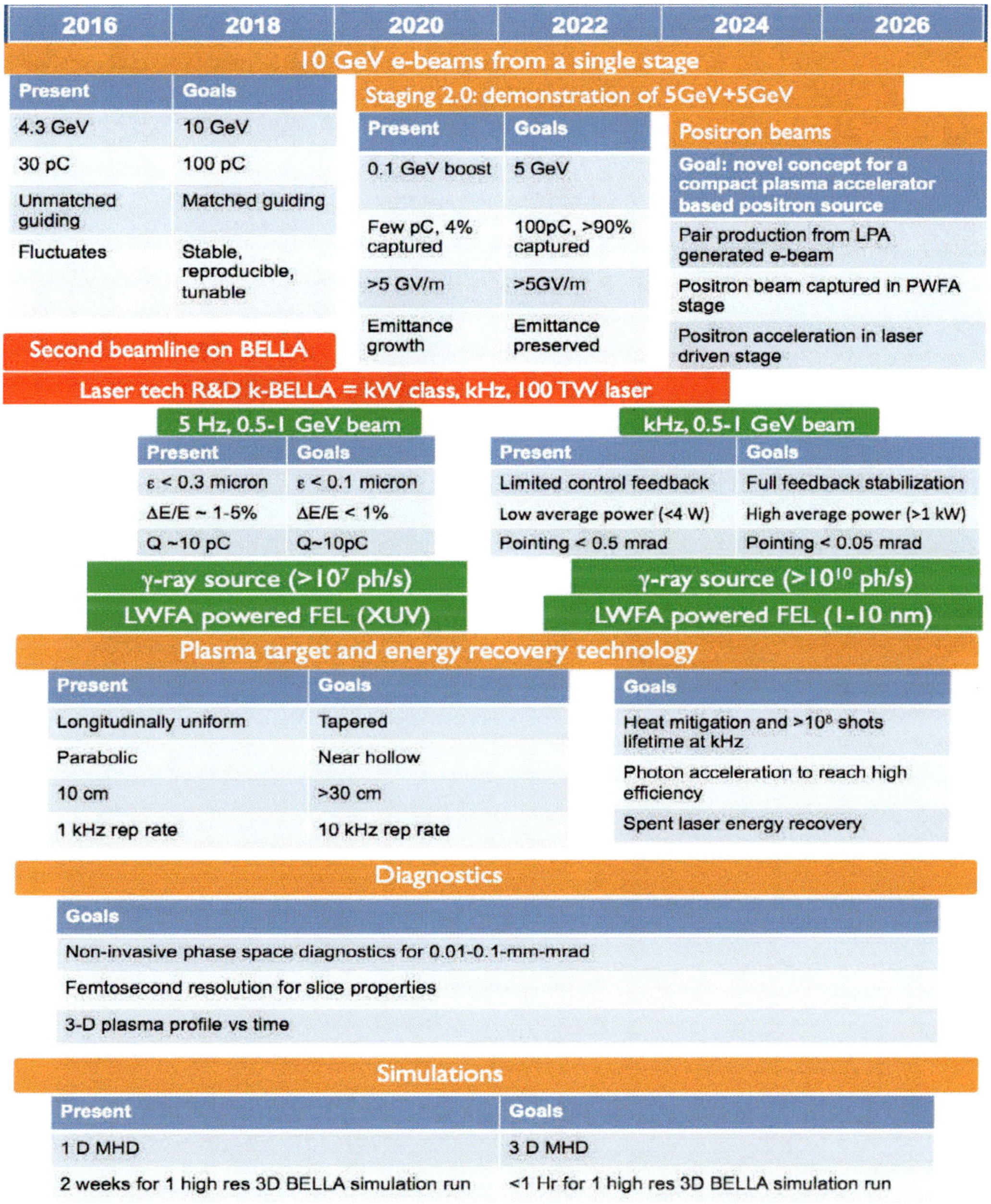

Fig. 2. LWFA ten-year roadmap and milestones. The orange boxes denote activities, with the present status and goals listed below. The red boxes denote facility investments that must occur to enable staging at multi-GeV energies (BELLA, 2nd line) and a high-average-power demonstration facility (k-BELLA). The green boxes denote early applications that are enabled by the development of the LWFAs. Applications on the left and the right hand side will be possible with today's lasers and with high-repetition-rate systems, respectively.

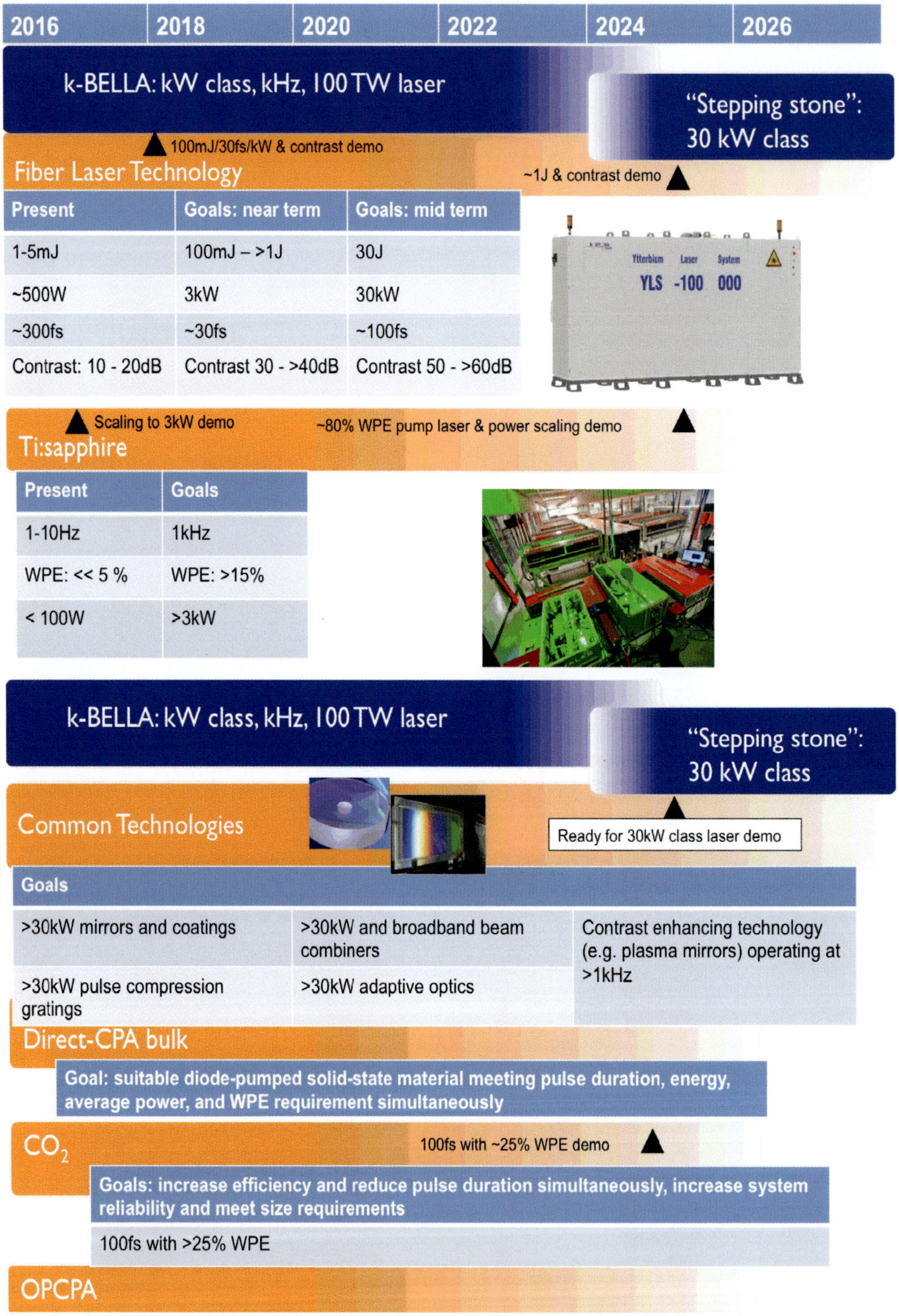

Fig. 3. Ten-year roadmap and milestones for laser technology development. In the next decade, technology would be developed and implemented for a high-average-power demonstration facility (k-BELLA) at the few-kW level, followed by scaling suitable technology to the tens-of-kW level. The different technologies that are available today have different levels of maturity, and all require R&D. R&D is also needed for common technologies such as mirror coatings, pulse compression and clean-up techniques (in both space and time).

generation of ultracold, polarized electron beams require investigation.

6.1.1. *Near-term applications*

During the next ten years, the development of non-HEP applications of LWFA technology will be a critical component of the collider R&D roadmap. Stable, reproducible, and tunable LWFA enables application of several-hundred-MeV to multi-GeV electron beams for light source applications. The first applications to be considered over the next decade are a free electron laser (FEL) powered by an LWFA (initially operating at XUV wavelengths for ultrafast science applications) and a Thomson scattering gamma ray source producing MeV photons for nuclear detection. These light sources would initially operate at the few-Hz repetition rate. Following successful demonstrations, and the development of kW-class laser systems, the FEL could be extended to the soft X-ray regime (few nm) at kHz rep rates, and the gamma ray source could be capable of producing 10^{10} photons/second or more.

6.1.2. *Associated technology R&D*

In parallel with the accelerator development, laser development is also required to realize an LWFA collider. In the next few years, R&D on kW average power laser systems will enable a Joule-class kHz-rate laser system (e.g. "k-BELLA") for LWFA R&D. At kHz repetition rates, studies of feedback and stabilization techniques are possible, enabling improved LWFA beam quality, as well as increased average beam power. A kW laser system allows testing of high-average-power beam issues, such as heat mitigation and lifetime, at kHz rep rates. To achieve the ~10 kHz rep rates and high average power needed for a collider requires further laser technology development. Approaches based on coherent combining of multiple fiber lasers, or on solid state lasers, CO_2 lasers, and on optical parametric chirped-pulse amplifiers, should be considered (see Fig. 3). Each of these approaches requires a different R&D level: some have a clear technology development pathway toward LWFA drivers, while others still need to address fundamental challenges.

Associated laser technologies include development of high-power optics technology (mirrors, diffraction gratings, beam combiners) to withstand hundreds of kW of optical power, as well as techniques for achieving high pulse contrast at ~10 kHz. In addition, methods for improved efficiency should be investigated, including spent laser energy recovery and photon acceleration to remove the energy remaining in the plasma wakefield.

6.2. *Particle-beam-driven plasma wakefield acceleration R&D*

A research program toward completion of a TDR for a multi-TeV e^+e^- linear collider based on PWFA technology is presented in Fig. 4.

The two areas of beam plasma physics considered most pressing for research in the next decade are emittance preservation and positron acceleration. Additional priorities include beam loading, higher transformer ratios, beam dynamics and tolerances, plasma source development, staging, and first applications. A detailed roadmap for beam-driven plasma wakefield accelerator R&D for the next decade is summarized in Fig. 5.

Operating in the nonlinear blowout regime, PWFA has demonstrated acceleration of ~50 pC of electrons with gradients ~7 GeV/m for a maximum energy gain of 9 GeV, a total energy spread of 4% and with an efficiency >10% [14, 15]. In addition, a new regime for accelerating positrons in a plasma wakefield was discovered and shown to accelerate more than 100 pC of positrons at a rate of more than 3 GeV/m for an energy gain of more than 4 GeV with an energy spread of less than 2% and an efficiency greater than 30% [16].

Emittance preservation during acceleration will be critical for nearly any application of PWFA, and especially so for colliders where low-emittance beams are needed to produce the required luminosity for physics at the TeV scale. Experiments will aim to characterize and limit the mechanisms for emittance growth while accelerating ~100 pC of electrons at high gradients (>GeV/m) and strong beam loading (10% energy spread and >10% efficiency). Characterization of PWFA beams will require diagnostics to be developed with micron spatial resolution, femtosecond time resolution, and the ability to discriminate between the drive beam and the accelerated trailing beam.

An electron–positron collider is envisioned as a tool for precision physics at the TeV scale. Experiments will aim to benchmark simulation codes, and

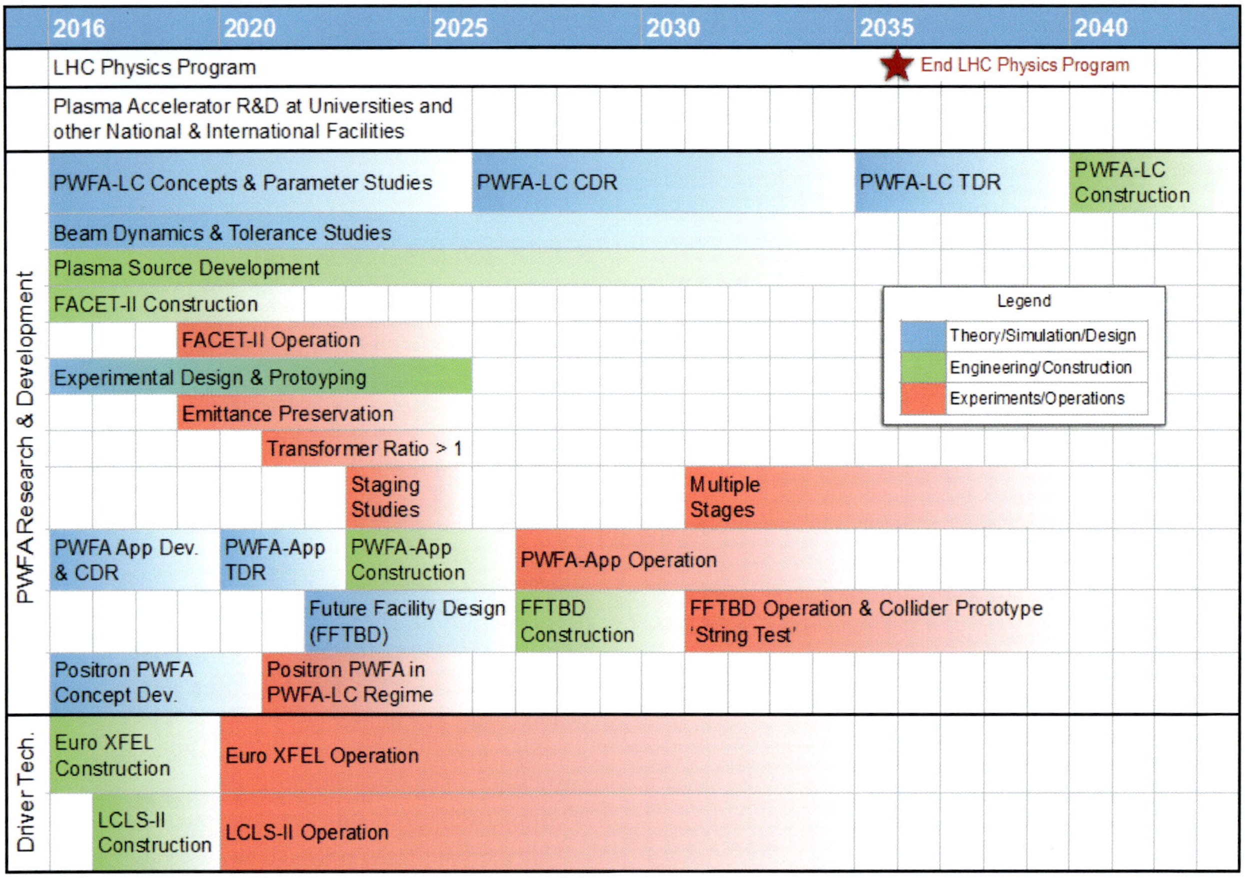

Fig. 4. High-level R&D roadmap for a particle-beam-driven plasma wakefield accelerator–based collider.

compare and contrast the three candidate regimes for positron acceleration in plasmas: quasi-nonlinear, hollow channel, and the new regime discovered at FACET in 2015. In parallel with the electron acceleration experiments, these efforts will also work to characterize and limit the mechanisms for emittance growth while accelerating ~100 pC of positrons at >GeV/m, with more than 1 GeV of energy gain to identify the optimal configuration for plasma acceleration of positrons for collider applications.

Staging multiple plasma cells together to reach very high energies will likely be necessary in any collider application. At FACET-II, an independent witness beam injector will be used to understand the tolerances for merging and extracting beams with different energies into and out of a plasma stage. The knowledge gained from these experiments will pave the way for designing systems with multiple plasma stages to reach progressively higher energy.

Correctly shaped beam current profiles are predicted to produce higher efficiency and beam loading with correspondingly higher electrical efficiency, smaller final energy spread, and a larger transformer ratio (larger energy boost per stage and efficiency). Current experiments work with Gaussian current profiles and transformer ratios of 1 or less. The flexibility of independent drive and witness beams will allow independently shaping the current profiles of the different beams. Experiments will provide an understanding of how much greater than 1 the practical transformer ratio can be while still preserving beam quality.

The transport and focusing of particle beams into and out of the plasma stage must be carefully managed to preserve the beam emittance. A recent concept design accomplishes this by adjusting the plasma density transitions at the entrance and exit of the plasma cell [17]. Development of suitable plasma sources will be essential to the demonstration of emittance preservation in plasma accelerators. Hollow or nearly hollow channel plasmas have been proposed as a candidate regime for

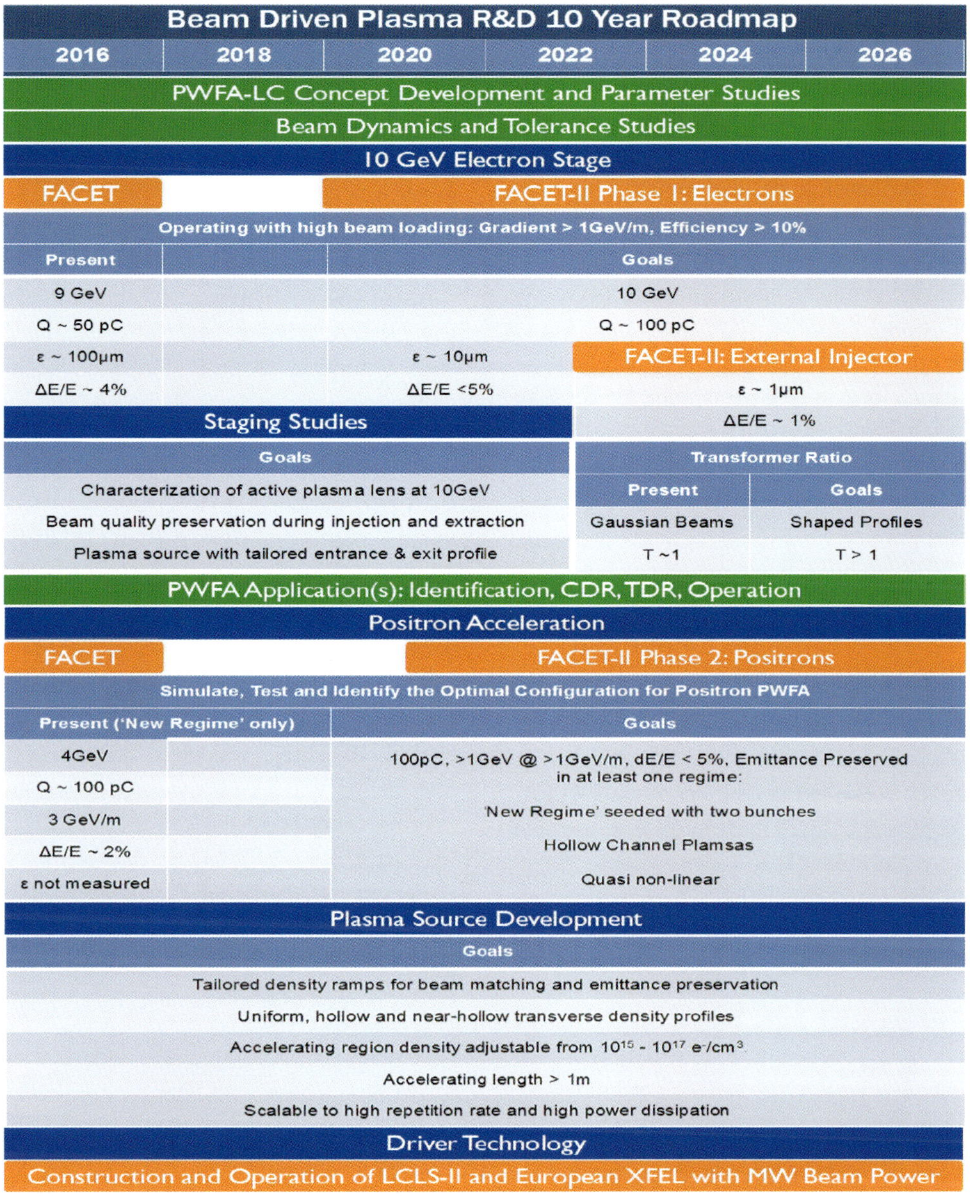

Fig. 5. Particle-driven PWFA: ten-year R&D roadmap and milestones.

positron acceleration. Some recent experiments have demonstrated hollow channels free of plasma on axis [18]. Future experiments will require an even greater degree of control, and plasma source development must proceed hand in hand with every experimental program.

In addition, the footprint of a collider and the geometric (effective) accelerating gradient will depend on the distance between successive plasma accelerating stages. So-called active plasma lenses [19] have demonstrated strong focusing that may aid matching high-energy beams between plasma cells and reduce the interstage distance [13]. Characterization of such lenses with high-peak-current, low-emittance 10 GeV beams will enable a much greater understanding of their utility in collider applications.

6.2.1. *Near-term applications*

The roadmap in Fig. 5 indicates the first applications for PWFA to be explored within the next decade. The exact nature of the first application is still to be determined, but candidates ranging from high-intensity gamma ray beams for nuclear physics to a new generation of X-ray free-electron lasers have been discussed. The development of any PWFA application will drive tolerance and systems engineering studies that will feed into future collider designs.

Near-term application design efforts will drive development of enhanced simulation capabilities. Hardware and software developments will reduce the timeframe for detailed simulations from weeks to hours or even minutes. By providing rapid turnaround time and full physics modeling, simulations will advance from *a posteriori* analytic tools to predictive design tools. Integration with traditional accelerator codes will also greatly enhance the utility of plasma codes for numerical design and prototyping of collider systems.

6.3. *Dielectric wakefield acceleration R&D*

A research program toward completion of a TDR for a multi-TeV e^+e^- linear collider based on DWFA technology is presented in Fig. 6.

The primary technical challenges for a DWFA-based linear collider (LC) are more similar to those of conventional accelerators than to the plasma wakefield concepts discussed previously. They are: developing polarized electron and positron sources, main beam acceleration, an efficient drive beam power source, staging of DWFA sections, and optimizing the beam delivery and final collider parameters.

The DWFA LC concept uses a short-RF-pulse, dielectric two-beam accelerator (TBA) as the main linac for both electron and positron acceleration. With a target gradient of 300 MV/m (effective gradient 200 MV/m), lower-cost dielectric structures, and a drive-beam-based power source, bunch shaping to raise the efficiency, and high-efficiency klystrons, the DWFA scheme has the potential for a significant reduction in cost compared with current LC technology.

As shown in Fig. 6, in the first five-year period the DWFA R&D will focus on extraction of GW-level RF power from the drive beam to support gradients of up to 300 MV/m in a single TBA module. A demonstration of staged acceleration of two TBA modules, including a beam kicker and RF delay lines, is also planned.

In the following five-year period, the DWFA R&D will focus on building a high-fidelity, integrated, dielectric wakefield TBA test facility. This test facility will simultaneously deliver high-gradient, staging, and acceleration of a linear collider quality beam. Source R&D to develop an e^- source to demonstrate operation at beam currents relevant to an operational LC is also planned. The technical goals are summarized in Fig. 6.

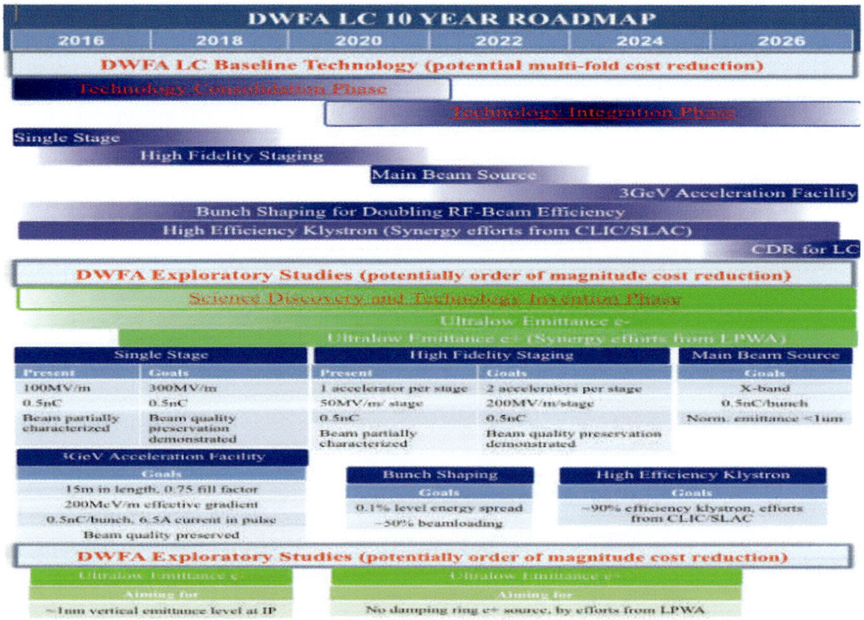

Fig. 6. High-level technical goals for DWFA.

6.3.1. *Technical challenges*

The technical challenges for achieving the DWFA roadmap milestones are significant and can broadly be divided into three technological areas: RF structures, drive beam generation and transport, and RF power transport. DWFA LC is based on a modular design and its most fundamental building blocks are the structures: the dielectric power extractors and the dielectric accelerators.

DWFA structure development is one of the primary challenges, with near-term R&D focusing on developing full-featured, dielectric structures that integrate dielectric tuning, wakefield damping, RF breakdown mitigation, broadband impedance matching, and multipactor suppression. Gradient and lifetime testing at higher gradient and power will be key to demonstrating the technology.

Drive beam production and transport are critical technologies needed for a DWFA-based collider. This includes developing electron sources capable of producing high-bunch-charge (e.g. 16 bunches of 40 nC each) drive beams at sensible cost. R&D to improve the stability of high-charge operation and the uniformity of the drive bunch train is needed to demonstrate the TBA scheme at higher power and gradient. Challenging beam line optics must be developed to deliver the drive beam through multiple stages of dielectric power extractors — a challenge, since the beam energy changes rapidly as energy is extracted.

Gigawatt-level RF power is needed to establish a >300 MV/m gradient in DWFA accelerators. To support this level of power transport, broadband RF components capable of transporting GW-level pulses with low loss must be developed, along with techniques for RF phase control.

A high wall plug efficiency (>10%) is highly desirable for a TeV-class LC, in order to reduce site power. Instead of using the usual Gaussian e^- and e^+ distribution, a longitudinally shaped bunch can be used to increase the beam loading, which in turn can double the RF-to-beam efficiency. Coupled with the recent research into high-efficiency klystrons, the site power could be reduced significantly.

The dependence of luminosity on beam power and emittance shows that for the same luminosity the beam power can be reduced by a factor of 3 to 5, if the beam's vertical emittance is reduced from the current level (tens of nanometers) down to 1 nm. Investigations to reduce e^- emittance based on various phase space manipulation efforts (e.g. emittance exchange) and reduction of source emittance are planned. The ultralow-emittance e^- beam, if coupled with an ultralow-emittance e^+ beam, has the potential to lower the overall cost of the LC.

6.3.2. *Near-term applications*

A ten-year roadmap for DWFA applications can be found in Fig. 7. The technology offers a promising

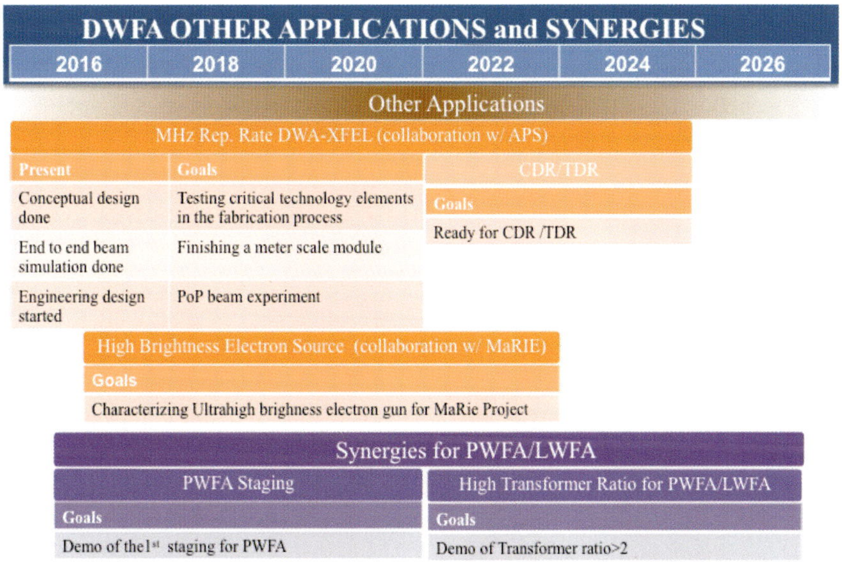

Fig. 7. DWFA application development and synergies: ten-year roadmap.

approach for a next-generation XFEL light source. The AWA team has collaborated with the Advanced Photon Source and Euclid Techlabs to develop a multiline, multiuser soft X-ray FEL, completing theoretical studies and end-to-end numerical simulations. Demonstration of a meter- scale DWFA module will be key to showing viability for such applications.

In 2016, the Matter–Radiation Interactions in Extremes (MaRIE) project by Los Alamos National Laboratory started collaboration with the AWA group to develop an ultrahigh-brightness electron source for the project. The joint efforts will benefit the DWFA LC exploratory effort in its development of an ultralow-emittance e^- source.

7. Dielectric Laser Acceleration R&D

Dielectric laser acceleration (DLA) refers to laser-powered solid state accelerators made partially or completely with dielectrics. Like their microwave counterparts, dielectric accelerators are typically composed of near-field structures that guide and diffract laser radiation to produce sustained high-gradient coupling. This technology trades difficult beam transport and alignment issues (apertures must be $\sim\lambda$ for high efficiency) for extreme compactness, mass production capability (for example by lithography), and the possibility of running CW.

This R&D area was rated a low priority by the GARD Subpanel and was not considered by the Advanced Accelerator Concepts Roadmap Workshop, but remains an area of modest worldwide R&D, and so is discussed here briefly.

Of the advanced technologies, DLA is perhaps the least developed, owing to the very significant challenges associated with working at wavelengths 10,000 times shorter than for conventional linacs. Demonstrations of practical structures that produce high gradient (690 MeV/m) [20] and staging demonstrations at 10 microns [21] and 0.8 microns [22] have begun to show the potential, but many issues remain to be addressed.

As with the other advanced concepts, the primary challenges are to (1) demonstrate the practical gradient limit (believed to be >1 GeV/m), (2) demonstrate operation at a very high repetition rate and discover the practical bunch charge limits of such devices (required to produce useful beam power for most applications), (3) demonstrate practical technologies for accelerating structures, beam focusing, bending, and diagnosis, (4) develop techniques for achieving the submicron-class alignment required to preserve emittance, (5) develop electron and positron sources matched to the unique phase space requirements, and (6) develop fully compatible, high-efficiency laser sources. Of these activities, (1) and (3) have received the most attention to date. For this technology to move forward to near-term applications, (2), (5) and (6) must be addressed.

The challenges associated with developing DLA structures have led to interesting conceptual connections with solid state physics, principally through the exploration of natural and engineered materials with tailored optical properties. Photonic band gap materials, photonic crystal fibers, negative permeability, permittivity, and index materials are each being investigated, along with suitable geometries to build accelerators. The fabrication of such metamaterials is challenging, but the technology continues to advance. Demonstrations of the fabrication costs and robustness of such materials will be critical steps toward feasibility.

As with the other advanced concepts, beam breakup must be thoroughly explored, as the very strong coupling impedances of most DLA concepts mean that the transverse wakefield effects are also strong. Some combination of strategic lattice design and BNS damping will be needed to push average beam currents into a realm where interesting applications can be enabled. Such effects will set very tight tolerances [e.g. $O(\lambda/100)$] on the alignment of accelerating and focusing elements in a DLA-based LC.

While integrating thousands of such microscopic structures together to produce an LC is a very significant challenge, what can be achieved within the confines of a single silicon wafer already opens up new applications. Within a 6-inch wafer, for example, beam energies of tens of MeV appear achievable at gradients already demonstrated. This realization has driven thoughts about new applications ranging from endoscopically mounted electron sources to directly irradiate tumors, to ultracompact oncology linacs to compact sources of X-rays and gamma rays. The intrinsic attosecond- and micron-scale operation of the DLA could potentially offer synchronized attosecond sources of electrons and radiation (e.g. THz, visible, X-rays) that provide the tools for directly monitoring and *controlling* chemical reactions.

8. Physics and Technical Challenges

All acceleration techniques share common challenges that must be addressed before the technology can be developed into a robust, well-engineered driver for a science program. These challenges are embodied in the five common high-level challenges set out in the chart for the roadmap workshop (see the introduction to Sec. 6). In this section, we further elaborate on these challenges, as they must be addressed before these advanced concepts can be useful for widespread use.

The first of these goals — higher-energy staging of electron acceleration with independent drive beams, equal energy gain, and 90% beam transmission — has only been partially met with preliminary demonstrations by some of the advanced concepts. Further work is needed to demonstrate independently driven stages with high capture efficiency and emittance-preserving matching and transport. For plasma concepts in particular, the internal and external optical functions typically differ by three orders of magnitude. Staging demonstrations are a first step to achieving higher integrated energy and to allowing trades to be made between single-accelerator-section energy gain and engineering considerations such as power efficiency and component costs.

Emittance preservation is a critical issue for linear collider applications, as luminosity is directly tied to beam emittance. Emittance is potentially damaged by several different physical processes: (1) variation in the strength of the accelerating field in space and time, (2) higher-order aberrations of the beam transport lattice, (3) phase space distortion by space charge fields, (4) interaction of the bunch with structure wakefields, and (5) collisions between the beam and background gas or plasma. Of these effects, (1) through (4) lead to correlated emittance growth which potentially can be reduced by enforcing appropriate symmetries of the accelerator, but which can easily be converted to irreversible emittance growth if mishandled. Effect (5) is stochastic in nature and cannot be corrected by linear means. Lattice design studies, including tolerance analysis, will be required to explore compatible solutions and develop strategies for preserving the high quality of the unusual beams needed by advanced concepts.

To achieve high-energy extraction efficiency, it is tempting to raise the single bunch charge to load fields heavily on a bunch-by-bunch basis. This has the effect of raising the (longitudinally varying) transverse kick to the bunch, requiring stronger focusing to keep the bunch in the accelerator, and diluting the emittance. Stronger transverse fields in turn tighten the tolerances on component misalignment and higher-order multipole content of focusing elements. Careful examination is required for the tradeoffs between single-bunch charge, multipulse structure, tolerances, and overall energy efficiency that accurately capture the challenges of controlling beam breakup instabilities and preserving the emittance.

Demonstrating positron acceleration is a key step for advanced concepts that are not trivially similar for bunches of the opposite charge. Vacuum concepts need only change the operating phase of the RF or laser pulse, but the plasma cavity formed in plasma wakefield accelerators reacts very differently, depending on the sign of the charge of the accelerated bunch. Demonstrating high-quality acceleration of positron bunches is therefore critical for LC applications of the technology.

Finally, joint development of multi-TeV LC conceptual designs based upon each of the advanced concepts is an important step in connecting the early-stage R&D with the application, guiding the direction of this very-long-term R&D. The notional machine parameters necessarily depend on a large number of assumptions about the eventual physical and technical limitations of each concept, and therefore have very large uncertainties. The quantitative elaboration of such conceptual LC designs requires a systems-level approach to thinking about the technology, however, and tends to make clear the highest R&D priorities, even if the end result may be decades away. Care should be exercised when discussing such designs to correctly report the large risk and uncertainty that they entail, and to not overstate the technical readiness of any of these concepts.

As the technical demonstrations progress in sophistication, the community which developed the present LC proposals has begun to engage substantively in discussions about the challenges that remain. It will be beneficial for this trend to continue.

At present, there is no demonstrated solution for building a multi-TeV lepton collider that is both scientifically interesting and economically viable. A number of proposals have been made which either suffer from a lack of adequate engineering demonstration or are economically untenable to build and operate, or both. Exploring new options is essential.

Sustained R&D on Advanced Accelerator Concepts has shown that orders-of-magnitude gains in the accelerating gradient are possible, as are efficient energy transfer and emittance preservation, but many questions remain about achieving all of these performance characteristics simultaneously and economically, as must be done for a large-scale LC.

9. The Need for Global Collaboration

The lofty goals of HEP will continue to demand exponential growth in the capabilities of accelerators in energy and intensity.

As accelerator technologies grow more complex, the required intellectual breadth, invested time, and cost of R&D also grow. This, together with flat funding, poses a distinct challenge that is changing how the field works.

Fewer accelerator R&D facilities can be built and operated, and those that are built must be configured with a national, strategic view in mind. Moreover, the increasing complexity of the science and technology frequently require broader multi-institutional teams to be formed in order to successfully carry out experiments. International collaboration must increasingly occur, so as to make the most effective use of worldwide investments.

Europe and Asia are aggressively investing in Advanced Accelerator Concepts R&D, particularly in laser-driven techniques. In Europe, umbrella organizations such as European Coordination for Accelerator R&D (EuCARD-2, renewing as ARIES), the European Network on Novel Accelerators (EuroNNAc2), and the European Committee for Future Accelerators (ECFA) provide international direction and organization of advanced accelerator concepts R&D, while targeted networks such as LASERLAB–Europe and LA^{3}NET, and projects such as EuPRAXIA, have brought considerable international organization to R&D on lasers and plasma wakefield accelerators.

In Asia, the historical strength in laser-driven accelerator concepts in Japan at ILE (Osaka) and KPSI has been joined by CoReLS in South Korea, and several >100 TW-class facilities in China (e.g. AAP, CLAPA, IOP, LLP, L2PA, SIOM, XG-III). Many of the higher-power facilities study high-energy-density physics (especially fusion) and radiation generation, conducting R&D on laser-driven particle acceleration on the side. Some programs, however, are specifically dedicated to studying techniques for generating high-quality beams of electrons, photons, ions, and neutrons. Networks, such as the Asian Intense Laser Network (AILN), coordinate and organize regional R&D through activities such as the ASILS and ASSS symposia on Intense Laser Science and Laser-Plasma Acceleration and Radiation.

It would be a beneficial step for US topical networks to develop that promote stronger domestic collaboration, organize international collaboration, and take a synoptic, community-optimized view of the research direction of advanced accelerator concepts.

More than ten petawatt-class facilities are expected to be operational outside the US by late 2018, including the three ELI centers ELI-Beamlines, ELI-ALPS, and ELI-NP, which will be dedicated to advanced particle and radiation sources, attosecond science, and nuclear physics, respectively.

International R&D facilities for advanced accelerator concepts are also being built or are operating, either as standalone facilities (e.g. SINBAD at DESY) or as additions to scientific facilities (e.g. AWAKE at CERN, FLASHForward at DESY).

Many of these centers represent hundreds of millions of dollars of capital investment, and rapid growth of expertise in beams, lasers, and their applications. Many possess unique capabilities that will not be duplicated in the US. It would be a lost opportunity if US scientists did not actively collaborate with these international centers.

10. Conclusion

After 150 years of ever-expanding development and use of accelerators, there is still remarkable progress occurring that promises to reduce the size and cost of this technology by orders of magnitude, and extend its reach well into the attosecond and angstrom domains.

Producing robust, engineered technology that can serve the needs of science and other applications requires R&D on interesting questions of plasma science, extreme field interactions, and beam dynamics on distance and time scales that are orders of magnitude shorter than previously considered. New sources of particles and new power sources must be developed to meet the requirements of these new technologies. Beam physics must advance, and computational tools must grow in sophistication to handle ever more strongly coupled systems. Interesting engineering questions must also be answered on how materials react to very high radiation and particle fluxes, and how to achieve stable control on time and distance scales that are orders of magnitude faster and smaller than for current accelerators. Understanding the underlying science and engineering of these advanced accelerator techniques will be key to developing solid, reliable technology.

With the significant reduction in cost and size of accelerators that these technologies promise will come a variety of new tools of science, and a wealth of new applications.

Acknowledgments

The authors acknowledge the extensive community engagement in developing the R&D roadmap for advanced accelerator concepts, first through a set of topical community workshops, then in the AAC Roadmap Workshop itself. The AAC Roadmap Workshop participants were: T. Antonsen, I. Ben-Zvi, G. Blazey (Workshop Chair), Y. Cai, W. Chou, M. Downer, W. Gai, M. Hogan, C. Jing, C. Joshi, W. Leemans, M. Litos, S. Nagaitsev, J. Rosenzweig, C. Schroeder, A. Seryi, and B. Weng. The content of Secs. 6.1–6.3 was written by the workshop participants, led by W. Leemans (Sec. 6.1), M. Hogan (Sec. 6.2), and W. Gai (Sec. 6.3), integrated into the workshop report by G. Blazey, and extensively edited for inclusion here by E.C. and L.K. The full text may be found in the Roadmap Workshop Report [23].

References

[1] T. Behnke, J. Brau, B. Foster, J. Fuster, M. Harrison, J. Patterson, M. Peskin, M. Stanitski, N. Walker and H. Yamamoto (eds.), ILC Technical Design Report (2013).

[2] M. Dal Forno, V. Dolgashev, G. Bowden, C. Clarke, M. Hogan, D. McCormick, A. Novokhatski, B. Spataro, S. Weathersby and S. Tantawi, *Phys. Rev. Accel. Beams* **19**, 011301 (2016).

[3] M. C. Thompson, H. Badakov, A. M. Cook, J. B. Rosenzweig, R. Tikhoplav, G. Travish, I. Blumenfeld, M. J. Hogan, R. Ischebeck, N. Kirby, R. Siemann, D. Walz, P. Muggli, A. Scott and R. B. Yoder, *Phys. Rev. Lett.* **100**, 214801 (2008).

[4] E. Esarey, C. Schroeder and W. P. Leemans, *Rev. Mod. Phys.* **81**, 1229–1285 (2009).

[5] P. Channell (ed.), Laser Acceleration of Particles Workshop (19–23 Feb. 1982, Los Alamos, NM), *AIP Conf. Proc.* **91**, 1 (1982).

[6] The Advanced Accelerator Concepts Workshop is approaching its 17th conference as of this writing; http://aac2016.org/.

[7] For more information see the Office of Science webpage: http://science.energy.gov/hep/research/advanced-technology-rd/.

[8] For more information see the Office of Science webpage: http://science.energy.gov/hep/research/accelerator-stewardship/.

[9] Basic Research Needs Workshops for the Accelerator Stewardship program may be found at http://science.energy.gov/hep/research/accelerator-stewardship/workshop-reports/.

[10] The Particle Physics Project Prioritization Panel report "Building for Discovery: Strategic Plan for US Particle Physics in the Global Context" (May 2014) may be found at http://science.energy.gov/~/media/hep/hepap/pdf/May-2014/FINAL_P5_Report_Interactive_060214.pdf.

[11] The GARD Subpanel report "Accelerating Discovery: A Strategic Plan for Accelerator R&D in the US" (Apr. 2015) may be found at http://science.energy.gov/~/media/hep/hepap/pdf/Reports/Accelerator_RD_Subpanel_Report.pdf.

[12] W. P. Leemans, A. J. Gonsalves, H.-S. Mao, K. Nakamura, C. Benedetti, C. B. Schroeder, Cs. Tóth, J. Daniels, D. E. Mittelberger, S. S. Bulanov, J.-L. Vay, C. G. R. Geddes and E. Esarey, *Phys. Rev. Lett.* **113**, 245002 (2014).

[13] S. Steinke, J. van Tilborg, C. Benedetti, C. G. R. Geddes, C. B. Schroeder, J. Daniels, K. K. Swanson, A. J. Gonsalves, K. Nakamura, N. H. Matlis, B. H. Shaw, E. Esarey and W. P. Leemans, *Nature* **530**, 190–193 (2016).

[14] M. Litos, E. Adli, W. An, C. I. Clarke, C. E. Clayton, S. Corde, J. P. Delahaye, R. J. England, A. S. Fisher, J. Frederico, S. Gessner, S. Z. Green, M. J. Hogan, C. Joshi, W. Lu, K. A. Marsh, W. B. Mori, P. Muggli, N. Vafaei-Najafabadi, D. Walz, G. White, Z. Wu, V. Yakimenko and G. Yocky, *Nature* **515**, 92–95 (2014).

[15] M. Litos, E. Adli, J. M. Allen, W. An, C. I. Clarke, S. Corde, C. E. Clayton, J. Frederico, S. J. Gessner and S. Z. Green, *Plasma Phys. Control. Fus.* **58**(3) (2016).

[16] S. Corde, E. Adli, J. M. Allen, W. An, C. I. Clarke, C. E. Clayton, J. P. Delahaye, J. Frederico, S. Gessner, S. Z. Green M. J. Hogan, C. Joshi, N. Lipkowitz, M. Litos, W. Lu, K. A. Marsh, W. B. Mori, M. Schmeltz, N. Vafaei-Najafabadi, D. Walz,

V. Yakimenko and G. Yocky, *Nature* **524**, 442–445 (2015).
[17] X. L. Xu, Y. P. Wu, C. J. Zhang, F. Li, Y. Wan, J. F. Hua, C.-H. Pai, W. Lu, P. Yu, W. An, W. B. Mori, M. J. Hogan, C. Joshi, *et. al.* arXiv: 1411.4386v2.
[18] S. Gessner, E. Adli, J. M. Allen, W. An, C. I. Clarke, C. E. Clayton, S. Corde, J. P. Delahaye, J. Frederico, S. Z. Green, C. Hast, M. J. Hogan, C. Joshi, C. A. Lindstrøm, N. Lipkowitz, M. Litos, W. Lu, K. A. Marsh, W. B. Mori, B. O'Shea, N. Vafaei-Najafabadi, D. Walz, V. Yakimenko and G. Yocky, *Nat. Commun.* **7**, Article No. 11785 (2016).
[19] J. van Tilborg, S. Steinke, C. G. R. Geddes, N. H. Matlis, B. H. Shaw, A. J. Gonsalves, J. V. Huijts, K. Nakamura, J. Daniels, C. B. Schroeder, C. Benedetti, E. Esarey, S. S. Bulanov, N. A. Bobrova, P. V. Sasorov and W. P. Leemans, *Phys. Rev. Lett.* **115**, 184802 (2015).
[20] K. P. Wootton, Z. Wu, B. M. Cowan, A. Hanuka, I. V. Makasyuk, E. A. Peralta, K. Soong, R. L. Byer and R. J. England, *Opt. Lett.* **41**(12), 2696–2699 (2016).
[21] W. D. Kimura, A. van Steenbergen, M. Babzien, I. Ben-Zvi, L. P. Campbell, D. B. Cline, C. E. Dilley, J. C. Gallardo, S. C. Gottschalk, P. He, K. P. Kusche, Y. Liu, R. H. Pantell, I. V. Pogorelsky, D. C. Quimby, J. Skaritka, L. C. Steinhauer and V. Yakimenko, *Phys. Rev. Lett.* **86**, 4041 (2001).
[22] C. Sears, E. Colby, R. England, R. Ischebeck, C. McGuinness, J. Nelson, R. Noble, R. Siemann, J. Spencer, D. Walz, T. Plettner and R. Byer, *Phys. Rev. ST Accel. Beams* **11**, 101301 (2008).
[23] Advanced Accelerator Development Strategy Report, Report of the Roadmap Workshop held Feb. 2–3 2016, G. Blazey (Chair); posted at http://science.energy.gov/hep/community-resources/reports/.

Eric R. Colby is the Senior Technical Advisor to the Department of Energy's Office of High Energy Physics, and the program manager for the Office of Science Accelerator Stewardship program. Before joining the DOE, he was the Director of Accelerator Research at SLAC National Accelerator Laboratory, working in the fields of RF high-power generation, electron source design, and laser acceleration, and leading the construction of three electron accelerator facilities at three national labs. He has held leadership roles on numerous advisory panels and in the American Physical Society, and has published over 75 articles in accelerator physics. He received his Ph.D. in Physics from UCLA in 1997.

L. K. Len is the program manager for the General Accelerator Research and Development program in the Office of High Energy Physics in the Department of Energy. Prior to that, he was the vice president and senior scientist of FM Technologies, Inc., where he worked on various R&D projects related to particle beam and radiation generation, and their applications.

Reviews of Accelerator Science and Technology
Vol. 9 (2016) 19–61
© World Scientific Publishing Company
DOI: 10.1142/S1793626816300024

Laser-Driven Plasma Electron Acceleration and Radiation

Kazuhisa Nakajima
Center for Relativistic Laser Science,
Institute for Basic Science (IBS), Gwangju 61005, South Korea
naka115@dia-net.ne.jp

Laser-driven plasma acceleration of electron beams is reviewed from the viewpoint of the underlying physics and recent progress in the experimental research. Betatron radiation cogenerated from laser plasma accelerators is mentioned in terms of electron beam dynamics and the radiation spectrum. At the end, future perspectives of possible applications are presented.

Keywords: Laser plasma accelerators; laser wakefield acceleration; betatron radiation; laser-driven FEL; high-energy frontier collider.

1. Introduction

Laser-driven plasma accelerators are one of the prevailing advanced accelerator concepts that provide us with compact particle and radiation sources for a wide range of sciences. In contrast to structure-based accelerators, including conventional radio frequency (RF) accelerators and advanced accelerator concepts, plasma-based accelerators sustain an ultrahigh-gradient accelerating field of the order of 100 GV/m in plasma at an electron density of $\sim 10^{18}\,\mathrm{cm}^{-3}$, free from the material breakdown and surface heating that limits the maximum accelerating field of structure-based accelerators. Since such ultrahigh fields are generated in a form of large-amplitude plasma wake as a result of electron and ion charge separation produced by an ultrafast laser pulse propagating in plasma nearly at the speed of light, energetic electrons are dominantly trapped and accelerated, while ions remain at rest during the interaction with wakefields. In that scenario, this article concentrates on electron acceleration and radiation based on laser wakefield accelerators. From the viewpoints of particle acceleration physics and a broad range of potential applications, laser-driven ion plasma accelerators are currently also vibrant topics for advanced accelerator research, as referenced in Ref. 1.

Historically, laser-driven plasma acceleration has evolved from a groundbreaking concept by Tajima and Dawson [2] into a potential technology for the next-generation particle accelerators, which was motivated by a long-standing consciousness that the state-of-the-art high-energy accelerators had become too large and costly, and possibly had approached the end of the road [3]. To date, high-energy accelerators are based on high-power RF technologies that accelerate charged particles with electric fields up to 100 MV/m at most, which is a limit stably produced in metallic, electromagnetic cavities due to electrical surface heating and breakdown. As illustrated with the Large Hadron Collider (LHC) [4] and the future linear collider [5] beyond the TeV energy range, the overall accelerator complex will range over several tens of kilometers in size and require enormous expenditure to be built. A remarkable scaling-down of high-energy accelerators may be simply brought about by using much higher accelerating fields than the present limits. In this context, various novel concepts of charged particle accelerators that utilize superhigh fields of lasers and plasmas have been proposed at the early time of the laser, since the 1960s [6].

In the meantime, experimental progress in laser wakefield acceleration of electron beams was initiated by demonstrating ultrahigh-gradient acceleration of the order of $\sim$100 GeV/m, using chirped pulse amplification lasers with 10 TW class peak power and 1 ps pulse duration, focused onto a gas jet [7, 8]. Such experiments were characterized in terms of the self-modulated wakefield regime [9], where laser power should be higher than the critical power for relativistic self-focusing and a laser pulse duration is longer than a plasma period. In this regime, the laser pulse undergoes temporal intensity modulation and self-guiding through nonlinear interactions with plasma so that large-amplitude plasma waves can be resonantly excited. Ultimately, their wave breaking occurs, generating relativistic electrons to be continuously trapped and accelerated by wakefields throughout an acceleration distance. Therefore, electron beams produced from single-stage experiments show energy spectra with 100% energy spread, as characterized by a Maxwellian distribution with the highest-energy tail reaching at most 100 MeV [10]. The energy gain of accelerated electrons should be determined by the acceleration distance, which is restricted due to dephasing of electrons with respect to the correct acceleration phase of the wakefield and due to depletion of the laser pulse energy. For most experiments using a supersonic gas jet, the acceleration distance extends only to a few mm, so that the energy gain has been limited to the order of 200 MeV [11].

For many practical applications of electron beams, the quality, stability and controllability in the beam performance, such as energy, energy spread, emittance and charge, are indispensable in addition to compact and robust features of accelerators. In this context, breakthrough experiments [12–14] have succeeded in producing high-quality electron beams, so-called quasi-monoenergetic beams, with ultrashort pulse, small energy spread and low emittance. Quasi-monoenergetic electron beams have been obtained from the use of ultrashort laser pulses with duration of the order of several tens of femtoseconds and controlling plasma density precisely to make the dephasing length long enough to exceed the acceleration distance. Under these conditions, provided that plasma electrons expelled by the ponderomotive force (radiation pressure) of the laser pulse form a plasma cavity called a "bubble" [15], some of them are injected into a wake by a restoring force exerted from an ion channel remaining unshielded behind the laser pulse. As a result of the beam loading of trapped electrons, the nonlinear wakefield amplitude inside the bubble decreases below the trapping threshold. Consequently, electrons trapped in the wakefield undergo the processes of acceleration and bunching toward the wave crest to increase their energy and brightness unless the acceleration distance exceeds the dephasing length. This is a scenario of quasi-monoenergetic electron beam acceleration, based on the self-injection mechanism in the bubble regime, which is visually shown by multidimensional particle-in-cell (PIC) simulations [16, 17].

Since self-injection is an inartificial method relying on self-focusing and self-compression of the laser pulse and expansion of the bubble [18], which occur during the propagation of relativistic laser pulses, initially heated (accelerated) electrons with large transverse momentum are injected into nonlinear wakefields which excite betatron oscillation of accelerated electrons due to a strong focusing field. In contrast to the self-injection,various controlled injection schemes have been proposed for producing high-quality electron beams even in the quasilinear wakefield regime driven by the laser pulse with a moderate intensity; for instance, colliding optical injection [19, 20], density transition injection [21], density down-ramp [22] and ionization-induced injection [23–25]. These injection schemes provide us with high-quality electron beam injectors for a front end of multistage high-energy accelerators in the conventional paradigm. As the simplest case, two-stage laser plasma acceleration has been successfully demonstrated in combination with ionization-induced injection [26, 27].

For laser plasma acceleration reaching GeV-level energies, it is essential to propagate intense laser pulses over a centimeter-scale distance in underdense plasma. For this purpose, a preformed plasma density channel with a parabolic radial distribution [28] has been developed for guiding a laser beam over many Rayleigh lengths without diffraction which limits the acceleration distance to a few mm in a uniform plasma. Plasma density channels stabilize propagation of relativistically intense laser pulses under the matched condition, preventing laser plasma nonlinear instabilities, such as filamentation and hosing, which often occur in the self-guiding [29, 30]. Employing a centimeter-scale plasma waveguide, laser plasma acceleration experiments on GeV-level electron acceleration have been carried out with a gas-filled or ablative discharge capillary [31–33] to demonstrate quasi-monoenergetic electron beams. Recently, experimental progress in laser plasma acceleration beyond 1 GeV has been achieved by using a sub-PW class laser and a cm-scale gas cell relying on the self-guiding [34–36] or capillary discharge plasma waveguide [37, 38]. For almost the last two decades, a number of laser plasma accelerator experiments have been carried out under various conditions. Comparing these experimental results with theoretical laser wakefield acceleration models may be useful for finding a correct scaling law capable of predicting energy gain, accelerated electron charge and the required laser and plasma conditions.

In laser plasma acceleration, electrons injected into plasma wakefields with initially transverse momentum undergo betatron oscillation due to strong focusing forces, leading to emission of synchrotron radiation, referred to as betatron radiation, which affects energy loss and transverse emittance of the electron beam via the radiation reaction force, as well as a promising broadband radiation source of X-rays and/or gamma rays capable of providing the peak brightness comparable to the third-generation synchrotron radiation light sources.

Significant advances in laser plasma acceleration are being made with potential applications of resultant electron and photon beams with ultrafast and brilliant properties as well as compact monolithic sources, often referred to as an all-optical device. Such devices include an ultrafast electron diffractometer (UED) using kHz laser-wakefield-generated MeV electron beams [39], a very-high-energy (VHE) electron radiation therapy using ~200 MeV laser-driven electron beams [40], femtosecond betatron X-rays for X-ray radiography [41, 42] and X-ray absorption spectroscopy for probing the warm dense matter [43], X-ray free electron laser ranging from ultraviolet to hard X-ray coherent radiation [44–48] and gamma ray sources via betatron radiation [49, 50] and inverse Compton scattering [48, 51–54] from the interaction of laser photons with laser-plasma-accelerated electrons, in addition to applications to high-energy accelerators such as high-energy frontier electron–positron colliders [55, 56].

The remainder of this article is organized as follows. Section 2 describes the underlying physics of laser plasma acceleration on excitation of plasma wakefields, injection of electrons and propagation of laser pulses. Section 3 presents electron beam dynamics and betatron radiation in laser wakefield accelerators. Section 4 provides perspectives on some specific applications based on laser plasma acceleration. At the end, Sec. 5 gives a conclusion.

2. The Physics of Laser Plasma Acceleration

2.1. *Laser wakefield acceleration in the linear regime*

Plasmas can sustain ultrahigh electric fields, and optically guide laser pulses as well as transport particle beams under appropriate conditions. In the non-relativistic limit, the acceleration gradient is limited to the wave-breaking field [57], where electron motion reaches the phase velocity of the plasma wave in the limit of $v_p = \omega_p/k_p \to c$, given by

$$E_0[\mathrm{V/cm}] = \frac{m_e c \omega_p}{e} \simeq 0.96(n_e[\mathrm{cm}^{-3}])^{1/2}, \qquad (1)$$

where $\omega_p = ck_p = (4\pi n_e e^2/m_e)^{1/2} = c(4\pi r_e n_e)^{1/2}$ is the electron plasma frequency with the electron mass m_e, the plasma electron density n_e and the classical electron radius $r_e = e^2/m_e c^2$. This means that the plasma density of $n_e = 1 \times 10^{18}\,\mathrm{cm}^{-3}$ can sustain an acceleration gradient of 100 GeV/m. When an intense laser pulse propagates in underdense plasma, for which the electron density is smaller than the critical plasma density $n_c = m_e\omega_0^2/4\pi e^2 = \pi/r_e\lambda_0^2$ for laser frequency ω_0 and wavelength $\lambda_L = 2\pi c/\omega_0$, plasma electrons are expelled from the region of a laser pulse due to the ponderomotive force $\mathbf{F}_p = -m_e c^2 \nabla(a_0^2/2)$, defined by averaging the nonlinear force over $2\pi/\omega_0$ exerted on plasma electrons by the laser field, with the normalized vector potential a_0 defined as $a_0 \equiv eA_0/m_e c^2$ in terms of the peak amplitude of the vector potential A_0 normalized to the electron rest energy $m_e c^2$ as

$$a_0 = \left[\frac{2e^2\lambda_L^2 I_L}{\pi m_e^2 c^5}\right]^{1/2} \simeq 8.55 \times 10^{-10} I_L^{1/2}\left[\frac{\mathrm{W}}{\mathrm{cm}^2}\right]\lambda_L[\mu\mathrm{m}], \qquad (2)$$

where I_L is the peak intensity for the linearly polarized laser. Physically a_0 is equal to the normalized momentum of the electron quiver motion in the laser field. This effect excites a large-amplitude plasma wave (wakefield) with the phase velocity v_p approximately equal to the group velocity v_g of the laser pulse, i.e. $v_p \simeq v_g = c(1 - \omega_p^2/\omega_0^2)^{1/2}$. In the cold fluid limit of a plasma, the momentum equation of electron fluid is

$$\frac{\partial \mathbf{p}}{\partial t} + \mathbf{v} \cdot \nabla \mathbf{p} = -e\left(\mathbf{E} + \frac{\mathbf{v}}{c} \times \mathbf{B}\right), \tag{3}$$

where the electric $\mathbf{E}$ and magnetic $\mathbf{B}$ fields of the laser can be written as

$$\mathbf{E} = -\frac{\partial \mathbf{A}}{c\partial t}, \quad \mathbf{B} = \nabla \times \mathbf{A}, \tag{4}$$

where $\mathbf{A}$ is the vector potential of the laser. For linearly polarized laser, the vector potential is written as $\mathbf{A} = A_0 \cos(k_0 z - \omega_0 t)\mathbf{e}_x$, where $\mathbf{e}_x$ is the unit vector of the polarization direction. In the linear limit $|\mathbf{a}| = e|\mathbf{A}|/(m_e c^2) \ll 1$, letting $\mathbf{v} = \mathbf{v}_0 + \delta \mathbf{v}$, the leading order equation of motion $m_e \partial \mathbf{v}_0/\partial t = -e\mathbf{E} = e\partial \mathbf{A}/c\partial t$ gives the quiver velocity $\mathbf{v}_0 = e\mathbf{A}/m_e c = c\mathbf{a}$. The second-order motion is given by

$$\begin{aligned} m_e\left(\frac{\partial \delta \mathbf{v}}{\partial t}\right) &= -m_e[(\mathbf{v}_0 \cdot \nabla)\mathbf{v}_0 + c\mathbf{v}_0 \times (\nabla \times \mathbf{a})] \\ &= -m_e c^2[(\mathbf{a} \cdot \nabla)\mathbf{a} + \mathbf{a} \times (\nabla \times \mathbf{a})] \\ &= -m_e c^2 \nabla\left(\frac{a^2}{2}\right) \equiv \mathbf{F}_p. \end{aligned} \tag{5}$$

The linearized Poisson and the continuity equations

$$\nabla \cdot \mathbf{E} = -4\pi e \delta n, \quad \frac{\partial \delta n}{\partial t} + n_0 \nabla \cdot \mathbf{v} = 0, \tag{6}$$

where n_0 is the ambient electron density and δn the density perturbation, lead to equations for excitation of the density oscillation (plasma wave) and electrostatic potential ϕ defined by $\mathbf{E}(\mathbf{r},t) = -\nabla\phi(\mathbf{r},t)$:

$$\left(\frac{\partial^2}{\partial t^2} + \omega_p^2\right)\left(\frac{\delta n}{n_0}\right) = \frac{c^2 \nabla^2 a^2(\mathbf{r},t)}{2}, \tag{7}$$

$$\left(\frac{\partial^2}{\partial t^2} + \omega_p^2\right)\phi(\mathbf{r},t) = \frac{\omega_p^2 m_e c^2 a^2(\mathbf{r},t)}{2e}, \tag{8}$$

where $a^2(\mathbf{r},t) \ll 1$ is the normalized intensity of a drive laser pulse and $\delta n/n_0 \ll 1$ is the density perturbation of the plasma wave. Using the Green function of the harmonic oscillation, the solutions for the density oscillation, electrostatic potential and electric field excited by the laser field $\mathbf{A}(\mathbf{r},t)$ are given by

$$\begin{aligned} \frac{\delta n(\mathbf{r},t)}{n_0} &= \frac{1}{\omega_p}\int_0^t dt' \sin\omega_p(t-t') \times \left[\frac{c^2\nabla^2 a^2(\mathbf{r},t')}{2}\right], \\ \phi(\mathbf{r},t) &= \left(\frac{m_e c^2 \omega_p}{e}\right)\int_0^t dt' \sin\omega_p(t-t') \times \left[\frac{a^2(\mathbf{r},t')}{2}\right], \\ \mathbf{E}(\mathbf{r},t) &= -\frac{m_e c^2 \omega_p}{e}\int_0^t dt' \sin\omega_p(t-t') \times \left[\frac{\nabla^2 a^2(\mathbf{r},t')}{2}\right]. \end{aligned} \tag{9}$$

Assuming that all of the axial and time dependencies can be expressed as a function of a single variable $\zeta = z - v_p t$, the wake potential $\Phi = e\phi$ is derived from a simple harmonic equation,

$$\frac{\partial^2 \Phi}{\partial \zeta^2} + k_p^2 \Phi = \frac{k_p^2 m_e c^2 a^2(r,\zeta)}{2}, \tag{10}$$

where $k_p = \omega_p/v_p$ and $a(r,\zeta) = eA(r,\zeta)/m_e c^2$ is the normalized vector potential of the laser pulse. The wake potential is obtained as

$$\Phi(r,\zeta) = -\frac{m_e c^2 k_p}{2}\int_\zeta^\infty d\zeta' \sin k_p(\zeta-\zeta')a^2(r,\zeta'). \tag{11}$$

Thus, the axial and radial electric fields are obtained from $eE_z = -\partial\Phi/\partial z$ and $eE_r = -\partial\Phi/\partial r$, respectively. Considering laser wakefields driven by a linearly polarized laser pulse with a ponderomotive potential of the Gaussian mode given by $a^2(r,\zeta) = (a_0^2/2)\exp(-2r^2/r_L^2 - \zeta^2/\sigma_L^2)$, where σ_L is a temporal $1/e$ half-width and r_L is a $1/e^2$ spot radius, the axial and radial wakefields [47] are

$$\frac{E_z(r,\zeta)}{E_0} = \left(\frac{\sqrt{\pi}}{8}\right)a_0^2 k_p \sigma_L e^{-2r^2/r_L^2 - k_p^2\sigma_L^2/4} \times [C(\zeta)\cos k_p\zeta - S(\zeta)\sin k_p\zeta], \tag{12}$$

$$\frac{E_r(r,\zeta)}{E_0} = -\left(\frac{\sqrt{\pi}}{2}\right)a_0^2\left(\frac{r\sigma_L}{r_L^2}\right)e^{-2r^2/r_L^2 - k_p^2\sigma_L^2/4} \times [C(\zeta)\sin k_p\zeta + S(\zeta)\cos k_p\zeta], \tag{13}$$

respectively, where $C(\zeta) = 1 - \mathrm{Re}[\mathrm{erf}(\zeta/\sigma_L - ik_p\sigma_L/2)]$, $S(\zeta) = \mathrm{Im}[\mathrm{erf}(\zeta/\sigma_L - ik_p\sigma_L/2)]$, and $\mathrm{erf(z)} = (2/\sqrt{\pi})\int_0^z e^{-z'^2}dz'$ is the complex error function. Behind the laser pulse at $\zeta \ll -\sigma_L$, taking into account $C(\zeta) \to 2$ and $S(\zeta) \to 0$, Eqs. (12) and (13) result in $E_z/E_0 \propto 2\cos k_p\zeta$ and $E_r/E_0 \propto 2\sin k_p\zeta$, respectively. Hence the radial wakefield is zero along the axis and its phase is shifted by $\pi/2$ from that of the axial wakefield. An electron displaced from the axis will experience simultaneous axial accelerating and radial focusing forces in a phase region of $k_p\Delta\zeta = \pi/4$. The maximum accelerating wakefield driven by a Gaussian laser pulse with $1/e$ half-width σ_L is given by

$$eE_M \simeq \left(\frac{\sqrt{\pi}}{4}\right) m_e c^2 a_0^2 (k_p^2 \sigma_L) \exp\left(\frac{-k_p^2\sigma_L^2}{4}\right). \quad (14)$$

For a given plasma density, the maximum field is $E_M \simeq 0.38a_0^2 E_0$ at $k_p\sigma_L = \sqrt{2}$, while for a given pulse duration the maximum field is $eE_M \simeq 0.65 m_e c^2 a_0^2/\sigma_L$ at $k_p\sigma_L = 2$.

In the linear laser wakefield with the accelerating field $E_z = E_M \cos\Psi$, equations of the longitudinal motion of an electron with the normalized velocity $\beta_z = v_z/c \approx 1$ and energy $\gamma = E/m_e c^2$ are given by [58]

$$\frac{d\gamma}{dz} = k_p \frac{E_M}{E_0}\cos\Psi, \quad \frac{d\Psi}{dz} = k_p\left(1 - \frac{\beta_p}{\beta_z}\right) \approx \frac{k_p}{2\gamma_g^2}, \quad (15)$$

where $\Psi = k_p(z - v_p t) + \Psi_0$ is a phase of the plasma wave, $\beta_p = v_p/c \approx v_g/c = \beta_g$ and $\gamma_g = (1-\beta_g^2)^{-1/2} \gg 1$. Integrating Eqs. (15), the energy and phase of the electron can be calculated as

$$\gamma(z) = \gamma_0 + 2\gamma_g^2\left(\frac{E_M}{E_0}\right)[\sin\Psi(z) - \sin\Psi_0],$$
$$\Psi(z) \approx \frac{k_p z}{2\gamma_g^2} + \Psi_0. \quad (16)$$

Setting the initial electron phase $\Psi_0 = 0$ at $z = 0$, the maximum energy gain is given by

$$\Delta\gamma_{\max} = \gamma_{\max} - \gamma_0 = 2\gamma_g^2\left(\frac{E_M}{E_0}\right), \quad (17)$$

at $z = \lambda_p\gamma_g^2/2$. As shown from Eqs. (16), setting $\Psi_0 = -\pi/2$, the maximum energy gain reaches $\Delta\gamma_{\max} = 4\gamma_g^2 E_M/E_0$ at $z = \lambda_p\gamma_g^2$. However, electrons undergo both acceleration and focusing only for $0 \le \Psi \le \pi/2$. Hence, one can define the dephasing length as $L_{\rm dp} = \lambda_p\gamma_g^2/2$.

Typical longitudinal, transverse and on-axis accelerating wakefields $E_z(r,\zeta)/E_0$, $E_r(r,\zeta)/E_0$ and $E_z(0,\zeta)/E_0$ in the quasilinear regime are calculated by two-dimensional particle-in-cell (2D PIC) simulation, as shown in Figs. 1(a), 1(b) and 1(c), respectively, for $a_0 = 1.4, k_p\sigma_L = 1, k_p r_L = 3$ and $n_e = 1\times 10^{18}\,\mathrm{cm}^{-3}$. The peak amplitudes of linear wakefields described by Eqs. (12) and (13) are verified by comparison with the scaled 2D PIC simulations in the

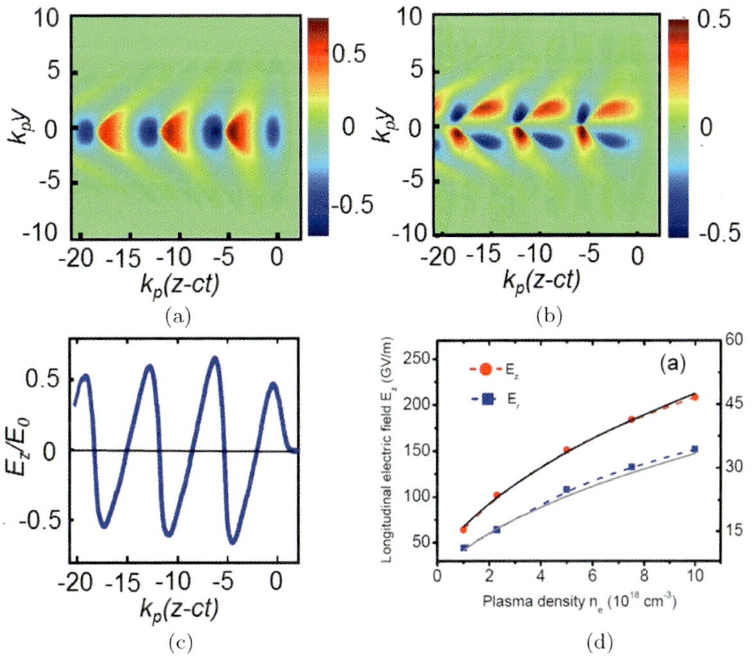

Fig. 1. Laser wakefields obtained from 2D PIC simulation for $a_0 = 1.4, k_p\sigma_L = 1, k_p r_L = 3$ and $n_e = 1\times10^{18}\,\mathrm{cm}^{-3}$: (a) longitudinal electric field $E_z(k_p y, k_p\zeta)/E_0$, (b) transverse electric field $E_r(k_p y, k_p\zeta)/E_0$, (c) on-axis accelerating field $E_r(0, k_p\zeta)/E_0$, where (d) peak amplitudes of longitudinal E_z and transverse wakefield E_r as a function of plasma density n_e. The solid lines show the longitudinal and transverse wakefields calculated from Eqs. (12) and (13), respectively.

quasilinear regime for $a_0 \simeq 1.4, k_p\sigma_L = 1, k_p r_L = 3$ and $n_e = 10^{18}$–$10^{19}\,\text{cm}^{-3}$, as shown in Fig. 1(d) [56].

2.2. *1D nonlinear wakefield*

In most cases of laser wakefield accelerator experiments, efficient electron acceleration requires the relativistic laser intensity with $a_0 \geq 1$, for which the linearized equation on wake potential Φ given by Eq. (10) is invalid because of the quiver velocity of plasma electron $v_\perp = ca/\gamma \to c$ for $a \gg 1$, where $\gamma = \gamma_\perp\gamma_z = [(1+a^2)/(1-\beta_z^2)]^{1/2} \to a$. Hence, the nonlinear wakefield is obtained from the 1D cold relativistic fluid model described as a set of equations: the electron momentum equation is

$$\partial_t \mathbf{p} + (\mathbf{v}\cdot\nabla)\mathbf{p} = -e(\mathbf{E} + \boldsymbol{\beta}\times\mathbf{B}), \tag{18}$$

where $\mathbf{p} = m_e\gamma\mathbf{v}, \gamma = (1 + p^2/m_e^2c^2)^{1/2}, \mathbf{E} = -(1/c)\partial_t\mathbf{A} - \mathbf{e}_z\partial_z\Phi, \boldsymbol{\beta} = \mathbf{v}/c$ and $\mathbf{B} = \nabla\times\mathbf{A}$ for the vector potential of the laser pulse $\mathbf{A}$ and the electrostatic potential in plasma Φ. Since conservation of the canonical momentum gives $\mathbf{p}_\perp/m_ec = e\mathbf{A}/m_ec^2 = \mathbf{a}(z,t)$,

$$\gamma = \left[1 + \left(\frac{p_\perp}{m_ec}\right)^2 + \left(\frac{p_z}{m_ec}\right)^2\right]^{1/2} = \gamma_\perp\gamma_z, \tag{19}$$

where $\gamma_\perp = (1+a^2)^{1/2}$, $\gamma_z = (1-\beta_z^2)^{-1/2}$ and $\beta_z = v_z/c$. The electrostatic potential Φ in plasma is obtained from the continuity equation

$$\left(\frac{1}{c}\right)\partial_t n + \nabla\cdot(n\boldsymbol{\beta}) = 0, \tag{20}$$

and the Poisson equation

$$\nabla^2\Phi = 4\pi e[n(z,t) - n_0], \tag{21}$$

where $n(z,t)$ is the plasma electron density and n_0 is the ambient plasma density. The vector potential $\mathbf{A}$ is obtained from the wave equation

$$\nabla^2\mathbf{A} - \left(\frac{1}{c^2}\right)\partial_t^2\mathbf{A} = -\left(\frac{4\pi}{c}\right)\mathbf{J} + \left(\frac{1}{c}\right)\nabla\partial_t\Phi, \tag{22}$$

using the Coulomb gauge $\nabla\cdot\mathbf{A} = 0$, where $\mathbf{J} = -en\mathbf{v}$ is the current density. Using Eq. (19) and $\varphi = e\Phi/m_ec^2$, the momentum equation (18) is rewritten as

$$\left(\frac{1}{c}\right)\partial_t\left(\gamma_\perp\sqrt{\gamma_z^2-1}\right) + \partial_z(\gamma_\perp\gamma_z) = \partial_z\varphi, \tag{23}$$

the continuity equation (20)

$$\left(\frac{1}{c}\right)\partial_t n + \partial_z\left(\frac{n\sqrt{\gamma_z^2-1}}{\gamma_z}\right) = 0, \tag{24}$$

the Poisson equation (21)

$$\nabla^2\varphi = k_p^2\left(\frac{n}{n_0} - 1\right) \tag{25}$$

and the wave equation (22)

$$\nabla^2\mathbf{a} - \left(\frac{1}{c^2}\right)\partial_t^2\mathbf{a} = \frac{k_p^2\left(\dfrac{n}{n_0}\right)\mathbf{a}}{\gamma_\perp\gamma_z}, \tag{26}$$

taking into account Eqs. (20) and (21). Assuming a laser pulse form as $\mathbf{a}(z,t) = (1/2)\mathbf{a}_0(\zeta,\tau)\exp(-i\theta) +$ c.c., where $\zeta = z - v_gt, \theta = \omega_0t - k_0z, v_g = \partial\omega_0/\partial k_0$ is the group velocity of the laser pulse and τ is a slow timescale such that $\partial^2 a_0/\partial\tau^2 \ll a_0\omega_0^2$. In these equations, changing the independent variables (z,t) into (ζ,τ), one has two integrals as a function of ζ from Eqs. (23) and (24):

$$\begin{aligned} \gamma_\perp\left(\gamma_z - \beta_g\sqrt{\gamma_z^2-1}\right) - \varphi &= 1, \\ n\left(\beta_g\gamma_z - \sqrt{\gamma_z^2-1}\right) &= n_e\beta_g\gamma_z, \end{aligned} \tag{27}$$

respectively, with $\beta_g = v_g/c$, assuming the initial equilibrium prior to the laser pulse, i.e. $n = n_0$, $\gamma_z = 1$, $\varphi = a_0 = 0$ at $\zeta = 0$. As a result, one has two coupled equations describing the evolution of electrostatic and vector potentials from Eqs. (25) and (26):

$$\partial_\zeta^2\varphi = \frac{k_p^2\sqrt{\gamma_z^2-1}}{\beta_g\gamma_z - \sqrt{\gamma_z^2-1}}, \tag{28}$$

$$2i\omega_0\partial_\tau\mathbf{a}_0 + 2c\beta_g\partial_\tau\partial_\zeta\mathbf{a}_0 + c^2\left(\frac{\omega_p^2}{\omega_0^2}\right)\partial_\zeta^2\mathbf{a}_0 - \partial_\tau^2\mathbf{a}_0 = -\omega_p^2\left[1 - \frac{\beta_g}{\gamma_\perp\left(\beta_g\gamma_z - \sqrt{\gamma_z^2-1}\right)}\right]\mathbf{a}_0, \tag{29}$$

respectively, using $\beta_g^2 = 1 - \omega_p^2/\omega_0^2$ and Eq. (27). Using Eq. (27), one rewrites Eqs. (28) and (29) as

$$\partial_\zeta^2\varphi = k_p^2\gamma_g^2\left[\beta_g\left(1 - \frac{1+\bar{a}^2}{\gamma_g^2(1+\varphi)^2}\right)^{-1/2} - 1\right], \tag{30}$$

$$\frac{2}{c}\partial_\tau\left[i\left(\frac{k_0}{\beta_g}\right)\mathbf{a}_0+\beta_g\partial_\zeta\mathbf{a}_0\right] + (1-\beta_g^2)\partial_\zeta^2\mathbf{a}_0-\left(\frac{1}{c^2}\right)\partial_\tau^2\mathbf{a}_0 = -k_p^2\left[1-\frac{\beta_g}{1+\varphi}\left(1-\frac{1+\bar{a}^2}{\gamma_g^2(1+\varphi)^2}\right)^{-1/2}\right]\mathbf{a}_0, \tag{31}$$

where $\bar{a}^2 = a_0^2$ for circular polarization and $\bar{a}^2 = a_0^2/2$ for linear polarization. For $\beta_g \simeq 1$, i.e. $\omega_p^2 \ll \omega_0^2$, the wake potential and the vector potential are simply obtained from the coupled equations:

$$\partial_\zeta^2\varphi = \frac{k_p^2}{2}\left[\frac{1+\bar{a}^2}{(1+\varphi)^2}-1\right], \tag{32}$$

$$\frac{2}{c}\partial_\tau(ik_0\mathbf{a}_0+\partial_\zeta\mathbf{a}_0) = -k_p^2\frac{\varphi}{1+\varphi}\mathbf{a}_0. \tag{33}$$

For $\bar{a}^2 \ll 1$ and $\varphi \ll 1$, Eq. (32) becomes Eq. (10) for the wake potential in the linear regime. For a short pulse $c\tau_L < \lambda_p$, neglecting the intensity modulation of the laser pulse due to plasma wave excitation, the nonlinear equation (32) of the wake potential can be solved independently. As a result, the longitudinal wakefield and the electron density oscillation are calculated as $\varepsilon \equiv eE_z/mc\omega_p = -(1/k_p)\partial_\zeta\varphi$ and $n/n_0 = (1/k_p^2)\partial_\zeta^2\varphi + 1$, respectively.

2.3. *Self-modulated laser wakefield acceleration*

For a long laser pulse with a length $c\tau_L \geq \lambda_p$, one obtains the following equation on $a_0^2(\zeta)$ from Eq. (33):

$$\partial_\zeta^2 a_0^2 + \frac{2\varphi}{3(1+\varphi)}k_p^2 a_0^2 = 0. \tag{34}$$

A solution to Eq. (34) becomes $a_0^2(\zeta) \propto |\cos(\alpha k_p\zeta)|$, where $\alpha = [(2/3)\varphi/(1+\varphi)]^{1/2}$. As a result, a temporal profile of the laser pulse undergoes the periodic intensity modulation with period $\lambda_p/2\alpha$ due to density oscillation of plasma waves. This effect resonantly enhances excitation of plasma waves, as shown in Fig. 2. While an initially Gaussian laser pulse with FWHM length $c\tau_L \simeq 2.1\lambda_p$ excites the wake potential with a maximum amplitude $\varphi_{\max} \approx (1+\bar{a}^2)^{1/2} - 1 \simeq 0.4$ and a maximum longitudinal wakefield with $|E_{z\,\max}|/E_0 \approx 0.04$ only within the pulse duration, the modulated pulse with period $\lambda_p/(2\alpha) \approx 1.1\lambda_p$ resonantly excites the wakefield

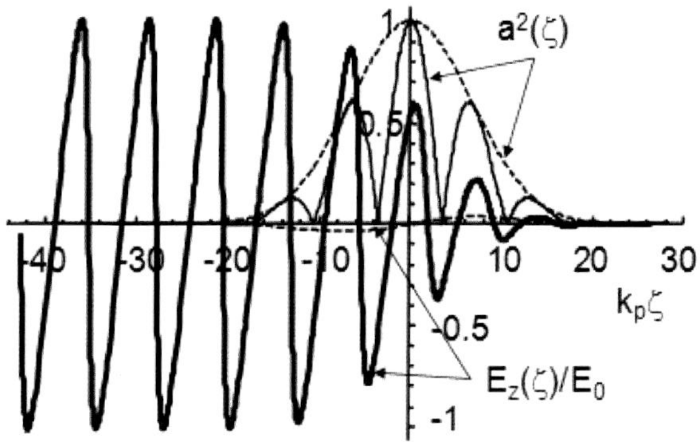

Fig. 2. Longitudinal wakefields $E_z(\zeta)/E_0$ excited by a modulated laser pulse with $a^2(\zeta) = (a_0^2/2)\exp[-(k_p\zeta)^2/(k_p\sigma_L)^2] \times |\cos(0.44k_p\zeta)|$ (solid curves) and a Gaussian laser pulse with $a^2(\zeta) = (a_0^2/2)\exp[-(k_p\zeta)^2/(k_p\sigma_L)^2]$, where $a_0 = 1.4$ and $k_p\sigma_L = 3\pi$ (dashed curves).

with $\varphi_{\max} \approx 1.6$ and $|E_{z\,\max}|/E_0 \approx 1$ behind the pulse duration.

Self-modulation of a laser pulse with the pulse duration τ_0 much greater than the inverse plasma frequency ω_p^{-1} is considered as the nonlinear interaction with plasma waves according to the process similar to the stimulated forward Raman scattering (FRS), which is characterized by observation of multiple sidebands of the transmitted laser spectrum with the mth harmonic frequency $\omega_0 \pm m\omega_p$ [10]. However, increasing the growth of the plasma wave by increasing the FRS growth rate cannot lead to an indefinitely increasing plasma wave amplitude, due to the wave-breaking limit in the relativistic regime, i.e. $a_0 > 1$, where saturation of the FRS growth rate and phase mismatching between the modulated intensity profile and the excited nonlinear wakefield occur.

Most of the initial experiments on laser wakefield acceleration are carried out in the self-modulated wakefield regime, using 10–100 J class chirped pulse amplification (CPA) lasers with 1 ps pulse duration. Figure 3(a) shows the electron energy spectra observed from experiments on electron acceleration in the self-modulated wakefield excited by a 50 J, 1 ps laser pulse with intensity $I_L > 10^{19}\,\text{cm}^{-3}$ ($a_0 \sim 3$) for plasma densities $n_e = 0.4$–$1.5\times10^{19}\,\text{cm}^{-3}$, when focused at the front edge of a helium gas jet produced by a supersonic nozzle with a 4 mm diameter by an off-axis parabolic mirror in $f/4$ geometry [10, 59]. The gas jet should have a very homogeneous density profile and a sharp edge [11]. As the plasma density

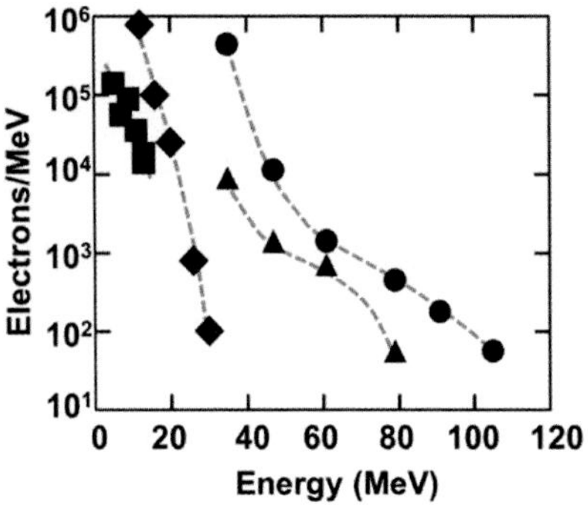

(a)

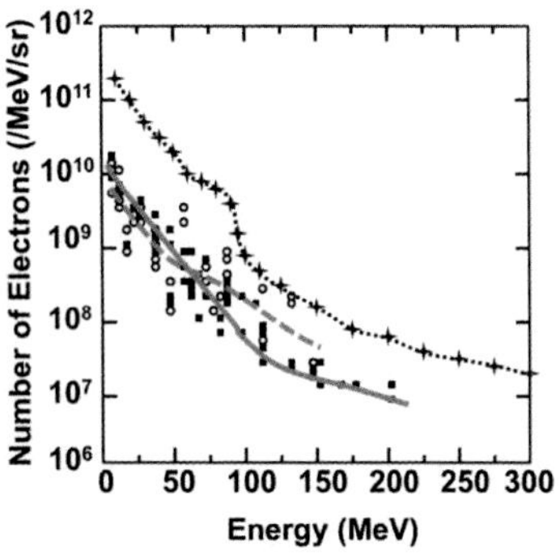

(b)

Fig. 3. (a) Energy spectra of electrons accelerated by self-modulated wakefields at $n_e = 4 \times 10^{18}\,\text{cm}^{-3}$(■), $7 \times 10^{18}\,\text{cm}^{-3}$(♦), $1.1 \times 10^{19}\,\text{cm}^{-3}$(▲) and $1.5 \times 10^{19}\,\text{cm}^{-3}$(•) for intensities $I_L > 10^{19}\,\text{W/cm}^2 (a_0 \sim 3)$, when a 50 J, 1 ps laser pulse is focused on a 10-μm-diameter spot (FWHM) [10]. (b) Energy spectra of electrons accelerated by forced wakefields driven by a 1 J, 30 fs laser pulse with $a_0 \sim 1.2$ at $n_e = 2.5 \times 10^{19}\,\text{cm}^{-3}$(■) and $6 \times 10^{19}\,\text{cm}^{-3}$(○) [11]. The electron energy spectrum observed at $n_e = 7.7 \times 10^{18}\,\text{cm}^{-3}$ (+) is shown for a 650 fs (FWHM), 160 J laser pulse focused on a 6-μm-diameter spot ($I_L \sim 10^{20}\,\text{W/cm}^2$; $a_0 \simeq 15$) [60].

increases, the number and maximum energy of electrons increase remarkably, up to more than 100 MeV in a high-energy tail. Though electron acceleration is associated with FRS, the observation of the transmitted laser spectra results in the broadening of the sidebands, showing loss of the plasma wave coherence due to the onset of wave breaking [10].

On the contrary, the nonlinear interaction of an ultrashort laser pulse with duration $\tau_L \sim 2\pi\omega_p^{-1}$ is capable of drastically efficient production of accelerated electrons with a high energy and a large current, as shown in Fig. 3(b). In the experiment carried out by using a 30 fs, 1 J laser, one obtains a large current of accelerated electrons with the total charge >5 nC and the maximum energy beyond 200 MeV, when the laser pulse is focused on the edge of a gas jet of density $n_e \sim 2.5 \times 10^{19}\,\text{cm}^{-3}$ by a parabola with a 1 m focal length that gives $a_0 \sim 1.2$ [10, 11]. Though the wave breaking of a relativistic plasma wave is responsible for such a current of energetic electrons, the spectrum of the transmitted laser shows no satellites as seen in the self-modulated wakefield, except for a broadening of the spectrum, evidently implying that the pulse is not modulated at the plasma frequency. Since the wave breaking requires $a_0 > 2$ in the 1D PIC simulation, the 3D effect, such as self-focusing is correlated with production of energetic electrons. The nonlinear wakefield excited by the ultrashort pulse without self-modulation, called a "forced laser wakefield," produces a well-collimated electron beam with divergence <2° for >50 MeV electrons, compared to the fact that the emitted beam has a relatively high divergence (for example, a divergence of >12° FWHM for 20 MeV electrons), which is comparable to the initial laser divergence, in the self-modulated wakefield regime, where a direct resonant interaction with the laser field and self-generated fields due to the high current bunch of electrons occur. Hence, in the forced wakefield regime, the reduced beam divergence demonstrates that electrons are purely accelerated by the plasma wave and the effect of direct laser acceleration is minimal.

Apart from the nonlinear wakefield acceleration of electrons driven by the ultrashort pulse, the interaction of an extremely intense long laser pulse ($\tau_L \gg 2\pi/\omega_p$) with underdense plasma at intensities $I_L > 10^{20}\,\text{W/cm}^2$ reveals phenomena of electron acceleration different from those in the self-modulated wakefield and forced wakefield regimes. In Fig. 3(b), shown is a result of the experiment performed using a Petawatt-class laser with 160 J in a duration of $\tau_L =$ 650 fs (FWHM), when the laser is focused on a 6-μm-diameter spot at the edge of a supersonic helium gas jet plasma with a 2 mm diameter by using an $f/3$ off-axis parabolic mirror that produces peak intensities in excess of $3 \times 10^{20}\,\text{W/cm}^2$ in vacuum, corresponding to $a_0 \simeq 15$ [60]. The electron energy spectrum measured at the plasma density $n_e = 7.7 \times 10^{18}\,\text{cm}^{-3}$ shows the highest electron energies, up to 300 MeV, where the signal descends into the background. The beam divergence is approximately 50 mrad (<3°) for electrons above 1.5 MeV, which is rather smaller than that in the forced wakefield regime. In contrast to the self-modulated wakefield regime, transmitted laser spectra measured simultaneously with the electron spectra show much lower sideband intensities than those at lower intensity and lack of correlation between laser pulse modulation in the transmitted spectra and the effective temperature of the electron beam. Furthermore, a large broadening of the satellites indicates that plasma waves are being driven to wave breaking, due to which an order-of-magnitude larger number of electrons are observed compared to those in the forced wakefield regime. These observations suggest that the principal mechanism may not be plasma wave acceleration. The results of the 2D PIC simulations reveal that plasma waves are not the dominant acceleration mechanism, that the majority of fast electrons exhibit stochastic acceleration in the nonlinear focusing field of the ponderomotively driven channel; and that the highest-energy electrons exhibit betatron-like resonance motion, of which the transverse and longitudinal momenta phase space, i.e. $p_y - p_x$ of electrons shows a distinct parabolic shape reminiscent of the well-known $p_x = (1/2)p_y^2$ trajectory of an electron in a plane wave and "filled-in" parabola, indicating a diffusion in phase space corresponding to a stochastic acceleration.

2.4. *Nonlinear laser wakefield in the bubble regime*

Most of the laser plasma acceleration experiments that successfully demonstrated the production of quasi-monoenergetic electron beams with narrow energy spread have been elucidated in terms of the self-injection and acceleration mechanism in the bubble regime [15–17]. In these experiments, electrons are self-injected into a nonlinear wake, often referred to as a bubble, i.e. a cavity void of plasma electrons consisting of a spherical ion column surrounded by a narrow electron sheath, formed behind the laser pulse, instead of a periodic plasma wave in the linear regime. A phenomenological theory of laser plasma interaction in the bubble regime is described as follows, based on the results of the multidimensional PIC simulations. Plasma electrons radially expelled by the radiation pressure of the laser form a sheath with a thickness of the order of the plasma skin depth c/ω_p outside an ion sphere, remaining unshielded behind the laser pulse. Consider the fields inside a spherical electron cavity moving in plasma with the relativistic velocity $v_B \simeq c$ along x axis, where v_B is the bubble velocity, assuming that the ion dynamics is neglected, i.e. $R_B \ll c/\omega_{pi}$, where $\omega_{pi} = (4\pi Z^2 e^2 n_0/m_i)^{1/2}$ is the ion plasma frequency with ion charge state Z, mass m_i and density n_0. Inside the cavity with the electron density $n = 0$ and the normalized laser vector potential $\mathbf{a}_\perp = e\mathbf{A}_\perp/m_e c^2 = 0$, the normalized wakefield potential $\Phi = a_x - e\varphi/m_e c^2$ is given by the spherical symmetric solution to the equation [15]

$$\Delta\Phi = \frac{1}{r^2}\frac{\partial}{\partial r}\left(r^2\frac{\partial\Phi}{\partial r}\right) = \frac{3}{2}k_p^2, \tag{35}$$

where $\Phi = 1 - k_p^2(R_B^2 - r^2)/4, a_x = -e\varphi/m_e c^2 = \Phi/2$ with $\Phi(R_B) = 1, r^2 = \xi^2 + y^2 + z^2$ and $\xi = x - v_B t$. Accordingly, the electromagnetic fields inside the relativistic cavity are given by $\mathbf{E} = -\nabla\varphi - (1/c)\partial_t\mathbf{A}$ and $\mathbf{B} = \nabla \times \mathbf{A}$ as

$$\begin{aligned} \frac{eE_x}{m_e c\omega_p} &= \partial_\xi\Phi = \frac{\xi}{2}, \quad \frac{eB_x}{m_e c\omega_p} = 0, \\ \frac{eE_y}{m_e c\omega_p} &= \frac{-eB_z}{m_e c\omega_p} = \frac{\partial_y\Phi}{2} = \frac{k_p y}{4}, \\ \frac{eE_z}{m_e c\omega_p} &= \frac{eB_y}{m_e c\omega_p} = \frac{\partial_z\Phi}{2} = \frac{k_p z}{4}. \end{aligned} \tag{36}$$

The Lorentz force acting on a relativistic electron with $v_x = c$ inside the cavity is given by $\mathbf{F} = -e\mathbf{E} - (e/c)\mathbf{v} \times \mathbf{B}$ as

$$\begin{aligned} F_x &= -eE_x = -m_e c^2 k_p \partial_\xi\Phi = \frac{-m_e c^2 k_p^2 \xi}{2}, \\ F_y &= -e(E_y - B_z) = -m_e c^2 k_p \partial_y\Phi \\ &= \frac{-m_e c^2 k_p^2 y}{2}, \\ F_z &= -e(E_z + B_z) = -m_e c^2 k_p \partial_z\Phi \\ &= \frac{-m_e c^2 k_p^2 z}{2}. \end{aligned} \tag{37}$$

The Lorentz force peaks for an electron with $v_x = v_B = c$, while it is zero for the electron with $v_x = -c$, because of the relativistic compensation of the electrostatic force by the self-magnetic force.

At the beginning of the interaction, as there is no bunch, the cavity shape is determined by balancing the Lorentz force of the ion sphere exerted on the electron sheath containing a return current carried by weakly relativistic electrons with the ponderomotive force of the laser pulse. The transverse radius R_B of the cavity at the center of the laser pulse with circular polarization is approximately estimated from the equation

$$\frac{k_p R_B}{4} \approx -\frac{1}{k_p}\frac{\partial}{\partial R_B}\sqrt{1+a^2(R_B)}, \tag{38}$$

where $a^2(r) = a_0^2 \exp(-2r^2/r_L^2)$ and $a_0^2 \to a_0^2/2$ for a laser pulse with linear polarization. In the case where the laser spot radius is equal to the cavity radius, i.e. $R_B \approx r_L$, Eq. (38) gives the cavity radius $k_p R_B \approx k_p r_L \approx 1.72\sqrt{a_0}(1+7.4/a_0^2)^{-1/4}$. For $a_0^2 \gg 1$, the PIC simulations [61] result in

$$k_p R_B \sim k_p r_L \approx 2\sqrt{a_0}. \tag{39}$$

From the accelerating field Eq. (36) and the cavity radius Eq. (39), the maximum accelerating field E_B becomes

$$\frac{E_B}{E_0} = \left(\frac{1}{2}\right)\alpha_c k_p R_B \approx \alpha_c\sqrt{a_0}, \tag{40}$$

where $E_0 = m_e c\omega_p/e$ is given by Eq. (1) and α_c represents a factor taking into account the difference between the simulation and the theoretical estimation. Figure 4 shows the laser wakefield simulation in the bubble regime for $a_0 = 4$, illustrating that the bubble size and the maximum accelerating field are in good agreement with the theoretical estimation Eqs. (39) and (40) with $\alpha_c = 0.9$, respectively.

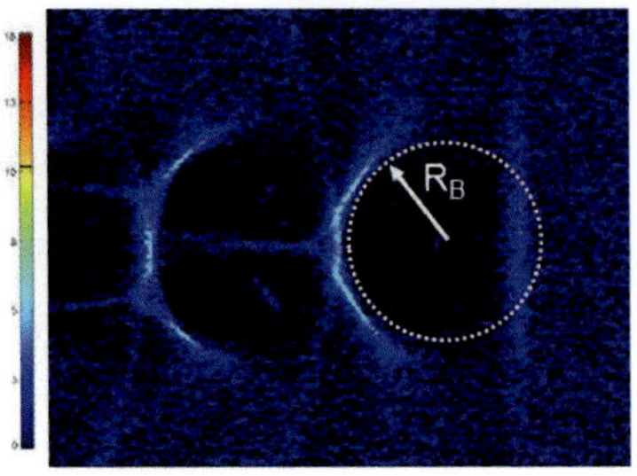

(a)

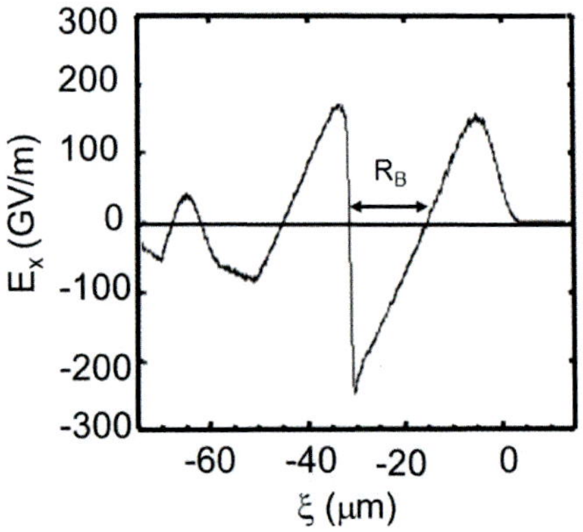

(b)

Fig. 4. Laser wake in the bubble regime calculated by the 2D PIC simulation for $a_0 = 4, n_e = 2\times10^{18}\,\text{cm}^{-3}, \lambda_p = 23.6\,\mu\text{m}$ and $r_L = 20\,\mu\text{m}(k_p r_L = 5.3)$: (a) electron density distribution, where the cavity with radius $R_B = 15\,\mu\text{m}(k_p R_B = 2a_0^{1/2} = 4)$ is shown by a dotted circle; (b) the accelerating wakefield with the maximum electric field $E_x \simeq 250\,\text{GV/m} \approx 0.9a_0^{1/2}E_0$, where $E_0 \simeq 136\,\text{GV/m}$.

As shown in the PIC simulation Fig. 4(a), electron density in the sheath strongly peaks and a spiky accelerating field emerges at the back of the plasma cavity so that electrons at that portion are accelerated by the strong longitudinal wakefield and get trapped into the plasma cavity when the electron velocity exceeds the bubble velocity, i.e. $v_x \geq v_B$. Once an electron bunch is trapped in the bubble, loading of trapped electrons reduces the wakefield amplitude to below the trapping threshold and stops further injection. Consequently, trapped electrons undergo acceleration and the bunching process within a separatrix on the phase space of the wakefield. This is a simple scenario for producing quasi-monoenergetic electron beams in the bubble regime.

Using Eq. (40), the equations of longitudinal motion of a trapped electron are approximately written as [62]

$$\frac{dW}{dx} = \frac{1}{2}m_e c^2\alpha_c k_p^2(R_B - \xi), \qquad \frac{d\xi}{dx} = 1 - \frac{v_B}{c} \approx \frac{3}{2}\frac{\omega_p^2}{\omega_0^2}, \tag{41}$$

where $\xi = x - v_B t (0 \leq \xi \leq R_B)$ is the longitudinal coordinate of the bubble frame moving at the velocity v_B and $v_x = dx/dt$ is the longitudinal electron velocity. The bubble velocity is given by

$$v_B \approx v_g - v_{\text{etch}} \approx c\left[1 - \left(\frac{1}{2}+1\right)\frac{\omega_p^2}{\omega_0^2}\right], \tag{42}$$

where v_g is the group velocity of the laser pulse in uniform plasma and $v_{\text{etch}} \sim c\omega_p^2/\omega_0^2$ is the velocity at which the laser front etches back due to the local pump depletion [17]. Integrating Eqs. (41), the energy and phase of the electron can be calculated as

$$W(x) = W_0 + \frac{1}{3}m_e c^2\alpha_c\gamma_g^2 k_p^2 R_B\xi\left(1 - \frac{\xi}{2R_B}\right), \qquad \xi = \frac{3x}{2\gamma_g^2}, \tag{43}$$

where W_0 is the injection energy and $\gamma_g = (1 - v_g^2/c^2)^{-1/2} = \omega_0/\omega_p = (n_c/n_e)^{1/2}$ is the relativistic factor for the group velocity. Hence, the maximum energy gain is obtained at $\xi = R_B$ as

$$\Delta W_{\max} = W_{\max} - W_0 \approx \frac{1}{6} m_e c^2 \alpha_c \gamma_g^2 k_p^2 R_B^2 \approx \frac{2}{3} m_e c^2 \alpha_c a_0 \frac{n_c}{n_e}. \quad (44)$$

The dephasing length $L_{\rm dp}$ turns out to be

$$k_p L_{dp} \approx \frac{2}{3} k_p R_B \gamma_g^2 = \frac{4}{3}\sqrt{a_0}\frac{n_c}{n_e}. \quad (45)$$

In order to reach the maximum energy gain given by Eq. (44) for a high-quality beam with small energy spread, the dephasing length should be less than the pump depletion length due to pulse front erosion, i.e. $L_{pd} \sim c\tau_L n_c/n_e \geq L_{dp}$. Therefore, the pulse length is set to be $c\tau_L \geq (2/3\pi)a_0^{1/2}\lambda_p$.

Quasi-monoenergetic electron beams with a-few-percent energy spread have been first observed in the laser wakefield acceleration experiments using 10 TW class laser pulses with 30–50 fs duration, focused onto a helium gas jet by a large f-number ($\sim f/20$) off-axis parabola [12, 14] or onto a laser-induced density plasma channel by a small f-number ($\sim f/4$) parabola [13], giving relativistic intensities with $a_0 > 1$, as shown in Fig. 5. In these experiments, plasma density is scanned to find the optimum density at which monoenergetic spectra are generated. At higher density than the optimum, the electron energy spectra show a Maxwellian profile with continuous energy spread. At the lower densities above the threshold where plasma wave breaking occurs, successive plasma periods accelerate trapped electron bunches to different energies, producing the multiple spikes in the spectrum. As a result of careful density control, the quasi-monoenergetic structure in the spectrum is produced at the optimum density, where the plasma length is equal to the dephasing length [12].

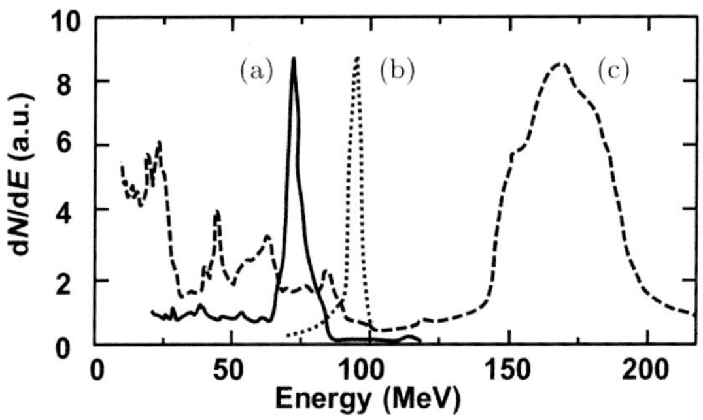

Fig. 5. Measured energy spectra of electrons at the experiments on laser wakefield acceleration using a gas jet, carried out for laser power P, pulse duration τ, intensity I (a_0) and plasma density n_e: (a) $P = 11$ TW, $\tau = 45$ fs, $I = 2.5 \times 10^{18}$ W/cm^2 ($a_0 \simeq 1.1$), $n_e = 2\times10^{19}$ cm^{-3} [12]; (b) $P = 9$ TW, $\tau = 55$ fs, $I = 1.1 \times 10^{19}$ W/cm^2 ($a_0 \simeq 2.3$), $n_e = 1.9 \times 10^{19}$ cm^{-3} [13]; and (c) $P = 30$ TW, $\tau = 33$ fs, $I = 3.2 \times 10^{18}$ W/cm^2($a_0 \simeq 1.3$), $n_e = 6 \times 10^{18}$ cm^{-3} [14]. The quasi-monoenergetic beam has energy E, FWHM relative energy spread: (a) $E = 78$ MeV, $\Delta E/E = 6\%$; (b) $E = 86$ MeV, $\Delta E/E = 4\%$; (c) $E = 170$ MeV, $\Delta E/E = 24\%$.

Most cases of simulations of production of monoenergetic electron beams show that high-energy electrons are injected and accelerated within the first period of the plasma wave. This fact has been experimentally observed by measuring the transverse profile of the electron beam accelerated in laser wakefields [63].

When the laser pulse length is comparable to the plasma wavelength, i.e. $c\tau_L \simeq \lambda_p$, the electron beam profile becomes elliptical due to an increase in the beam emittance in the plane of laser polarization, caused by the interaction of the electrons with the laser electric field. When the pulse length is made shorter to $c\tau_L \simeq \lambda_p/2$, the degree of ellipticity is decreased and the beam profile tends to be circular, because the electrons only extend into the bubble by a half plasma wavelength that corresponds to the accelerating phase.

2.5. *Electron injection into the laser wakefield*

In contrast with conventional RF accelerators operating in the GHz frequency range with a wavelength of the order of $\sim$10 cm, the transverse and longitudinal phase space of the laser wakefield have only a tiny size characterized by a plasma wavelength of the order of $\sim$10 μm at the plasma density of 10^{19} cm^{-3}. Hence, efficient electron injections from an external injector such as an RF linac and a photocathode RF gun have not always been successful for generating intense high-quality electron beams because of the difficulty in spatial and temporal phase matching of RF-accelerator-produced electron beams with plasma waves [64, 65]. In this context, besides the self-injection in a bubble wakefield model, several injection schemes that inject plasma electrons into the accelerating wakefield are proposed and some of them can be demonstrated successfully as controlled injection; for instance, optical injection employing two colliding pulses, ionization-induced injection and density tailoring injection including density transition, and down-ramp and up-ramp injection.

2.5.1. *Trapping and single-particle dynamics of an electron in a plasma wave*

Consider the single-particle dynamics of an electron in a plasma wave described by the normalized electrostatic potential $\varphi = e\Phi/m_e c^2$, assuming that φ oscillates between the values $\varphi_{\min} \le \varphi \le \varphi_{\max}$ periodically with a plasma wavelength λ_p and it is a function of only the variable $\zeta = z - v_p t$, where v_p is the phase velocity of the plasma wave and $v_p \simeq v_g$ for the plasma wave driven by the laser pulse with the group velocity v_g. The 1D equations of motion describing the evolution of the electron energy in terms of the relativistic factor γ and the phase of the electron relative to that of the plasma wave $\psi = k_p\zeta$ are given by

$$\frac{d\gamma}{d\hat{z}} = \frac{\partial\varphi}{\partial\psi}, \quad \frac{d\psi}{d\hat{z}} = 1 - \frac{\beta_p}{\beta}, \tag{46}$$

where the normalized variables $\beta = v/c, \gamma = (1-\beta^2)^{-1/2}, \beta_p = v_p/c$ and $\hat{z} = k_p z$ are used. The electron motion in the plasma wave is described by the Hamiltonian [66]

$$H(\gamma, \psi) = \gamma(1 - \beta\beta_p) - \varphi(\psi), \tag{47}$$

i.e. $\partial H/\partial\psi = -d\gamma/d\hat{z}$ and $\partial H/\partial\gamma = d\psi/d\hat{z}$ and $H(\gamma, \psi) =$ const. along a given electron orbit. For a given Hamiltonian constant, the electron energy is given by

$$\gamma = \gamma_p^2(H + \varphi) \pm \gamma_p\beta_p\sqrt{\gamma_p^2(H+\varphi)^2 - 1}, \tag{48}$$

where the + sign corresponds to the higher-energy orbit and the − sign corresponds to the lower-energy orbit, respectively, in (γ, ψ) phase space of the electron motion.

The separatrix $\gamma_s(\psi)$, that separates the closed orbits (trapped electrons) from the open orbits (untrapped electrons), is given by

$$H(\gamma_s, \psi) = H(\gamma_p, \psi_{\min}) = \frac{1}{\gamma_p} - \varphi_{\min}, \tag{49}$$

where $\varphi_{\min} = \varphi(\psi_{\min})$ and the points at $\gamma = \gamma_p$, $\psi = \psi_{\min} \pm mk_p\lambda_p$ (m is an integer) in (γ, ψ) phase space are the unstable fixed points ($d\gamma/d\hat{z} = d\psi/d\hat{z} = 0$), while the points at $\gamma = \gamma_p, \psi = \psi_{\max} \pm mk_p\lambda_p, \psi_{\max} = \psi_{\min} + k_p\lambda_p/2$ are the stable fixed points, where $\varphi(\psi_{\max}) = \varphi_{\max}$. For the wake potential φ in the linear wakefield regime with

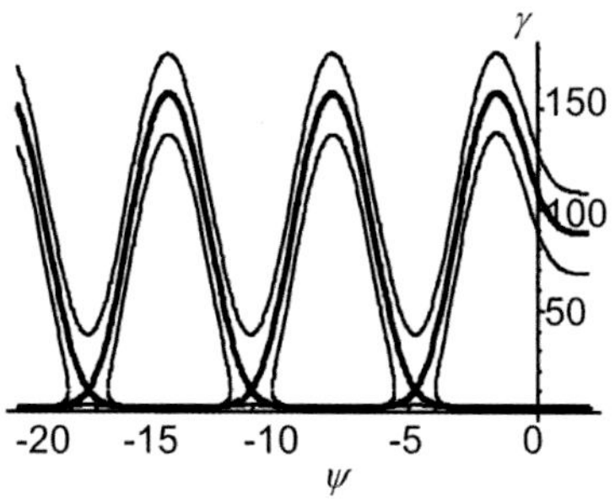

(a)

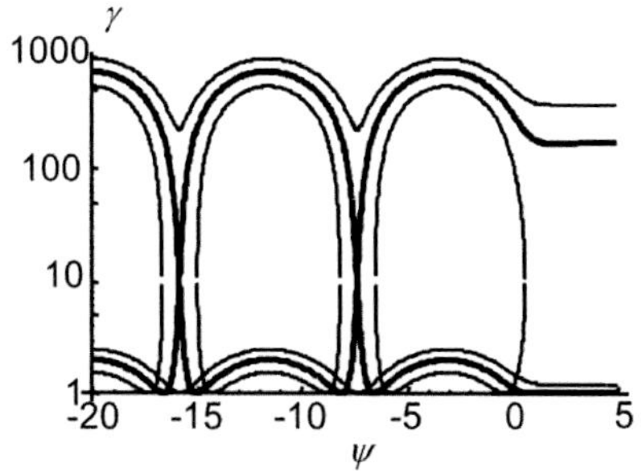

(b)

Fig. 6. Electron orbits (thin solid line) and separatrix (thick solid line) in (γ, ψ) phase space for (a) the linear wakefield with $a_0 = 1$ and (b) the 1D nonlinear wakefield with $a_0 = 3$. Both wakefields are driven by a Gaussian laser pulse with length $k_p\sigma_L = 1$ in the plasma density $n_e = 0.01n_c$ ($\gamma_p = 10$). The separatrix corresponds to the Hamiltonian constant $H_0 = 1/\gamma_p - \phi_{\min}$, i.e. $H_S = 0.445$ for case (a) and $H_S = 0.831$ for case (b), respectively. The closed orbit of a trapped electron in the separatrix is shown at the Hamiltonian $H_T = H_S - 0.1$ (1) for case (a) [(b)]. The open orbit for an untrapped electron is shown at the Hamiltonian $H_U = H_S + 0.1(1)$ for case (a) [(b)].

$a_0 = 1$ and $\gamma_p = 10$, the electron orbits and the separatrix in (γ, ψ) phase space are shown in Fig. 6(a). The maximum $\gamma_{\max}$ and minimum $\gamma_{\min}$ energy, represented by γ_m, for electrons that lie on the separatrix are given by $H(\gamma_m, \psi_{\max}) = H(\gamma_p, \psi_{\min})$ or $H(\gamma_m, \varphi_{\max}) = H(\gamma_p, \varphi_{\min})$, i.e.

$$\gamma_m = \gamma_p(1 + \gamma_p\Delta\varphi) \pm \gamma_p\beta_p[(1 + \gamma_p\Delta\varphi)^2 - 1]^{1/2}, \tag{50}$$

where $\Delta\varphi = \varphi_{\max} - \varphi_{\min}$ and the ± give $\gamma_{\max}$ and $\gamma_{\min}$, respectively. The minimum energy necessary for trapping an electron, $\gamma_{\min}$, is located at the phase $\psi = \psi_{\max}$ ($\varphi = \varphi_{\max}$). Although $\gamma_{\min} = 1$ ($v_{\min} = 0$) when the plasma wave amplitude satisfies $\Delta\varphi = (\gamma_p - 1)/\gamma_p$, the plasma wave does not trap the background electrons that are undergoing the plasma oscillations and at $\psi_{\min}$, flowing backward (i.e. in the direction opposite v_p) at the maximum fluid velocity. For the wave amplitudes $\Delta\varphi < (\gamma_p - 1)/\gamma_p, \gamma_{\min} > 1$ and the wave will trap electrons at $\psi_{\min}$ with velocities $v_{\min} = c(1 - \gamma_{\min}^{-2})^{1/2} > 0$. For the wave amplitudes $\Delta\varphi > (\gamma_p - 1)/\gamma_p, \gamma_{\min} > 1$ and the wave will trap electrons at $\psi_{\min}$ with the negative velocities $v_{\min} = -c(1 - \gamma_{\min}^{-2})^{1/2} < 0$. The wave breaking occurs when the wave amplitude is sufficiently large so as to trap the background electrons. At wave breaking, one has $\gamma_{\min} = \gamma_p$ and $v_{\min} = -v_p$ at the phase $\psi_{\max}$. From Eq. (50), wave breaking occurs at $\Delta\varphi = 2(\gamma_p^2 - 1)/\gamma_p$.

Consider the 1D nonlinear wakefield driven by a short intense laser pulse, assuming that the plasma wave and laser intensity profile are a function of only $\zeta = z - v_p t$, where $v_p \simeq v_g$. The quasistatic, cold fluid equation, Eq. (30), is rewritten as [66].

$$\frac{d^2 X}{d\psi^2} = \gamma_p^2 \left[\beta_p X \left(X^2 - \frac{\gamma_\perp^2}{\gamma_p^2}\right)^{-1/2} - 1\right], \quad (51)$$

where $X = 1 + \varphi, \psi = k_p\zeta$ and $\gamma_\perp = (1 + \bar{a}^2)^{1/2}$. Multiplying $dX/d\psi$ on both sides, the integral of Eq. (51) is given by

$$X'^2 = 2\gamma_p^2 \left\{X_0 - X + \beta_p \left[\left(X^2 - \frac{\gamma_\perp^2}{\gamma_p^2}\right)^{1/2} - \left(X_0^2 - \frac{\gamma_\perp^2}{\gamma_p^2}\right)^{1/2}\right]\right\}, \quad (52)$$

where the initial condition $X = X_0$ at $X' = 0$, and $X' = d\varphi/d\psi$ is the normalized axial electric field of the plasma wave, i.e. $X' = \varepsilon = E_z/E_0$ with the cold, nonrelativistic wave-breaking field $E_0 = m_e c^2 k_p/e$. In the region behind the driver (i.e. $\mathbf{a}^2 = 0$), the maximum amplitude of the nonlinear plasma wave, $X_{\max} = 1 + \varphi_{\max}$, is described by Eq. (52) with $\gamma_\perp = 1$ and $X_0 = X_{\max}$. The maximum electric field amplitude, $X'_{\max} = \varepsilon_{\max}$, occurs at $X = 1$, as indicated by Eq. (51), and is given by Eq. (52) when $\gamma_\perp = 1, X_0 = X_{\max}$ and $X = 1$. The minimum potential $X_{\min}$ is given by setting Eq. (52) equal to zero with $\gamma_\perp = 1$ and $X_0 = X_{\max}$. The maximum $X_{\max}$ and the minimum $X_{\min}$, denoted by X_m, can be written as

$$X_m = 1 + \frac{\varepsilon_{\max}^2}{2} \pm \beta_p \left[\left(1 + \frac{\varepsilon_{\max}^2}{2}\right)^2 - 1\right]^{1/2}, \quad (53)$$

where the $\pm$ give $X_{\max}$ and $X_{\min}$, respectively. Accordingly, one has

$$\Delta\varphi = \varphi_{\max} - \varphi_{\min} = 2\beta_p \left[\left(1 + \frac{\varepsilon_{\max}^2}{2}\right)^2 - 1\right]^{1/2}, \quad (54)$$

and $\Delta\varphi \simeq \beta_p \varepsilon_{\max}^2$ for $\varepsilon_{\max}^2 \gg 2$ and $\Delta\varphi \simeq 2\beta_p\varepsilon_{\max}$ for $\varepsilon_{\max}^2 \ll 2$. Hence, for a given peak electric field amplitude $\varepsilon_{\max}$, inserting Eq. (54) into Eq. (50) obtains the maximum energy gain of a trapped electron in the nonlinear plasma wave. The wave breaking of a nonlinear plasma wave occurs when the electron fluid velocity of the plasma wave becomes equal to the phase velocity v_p. At this point the plasma density becomes singular when $X_{\min} \to 1/\gamma_p$, as indicated by Eq. (51). This corresponds to $X_{\max} \to (2\gamma_p^2 - 1)/\gamma_p$ and $\varepsilon_{\max}^2 \to 2(\gamma_p - 1)$, i.e. $\Delta\phi = 2(\gamma_p^2 - 1)/\gamma_p, \varepsilon_{\max} \simeq \varepsilon_{WB} \simeq [2(\gamma_p - 1)]^{1/2}$ at the wave-breaking. Substituting this into Eq. (50), one obtains the minimum trapping energy, $\gamma_{\min} = \gamma_p$, and the maximum energy of the trapped electron, $\gamma_{\max} \to 4\gamma_p^3 - 3\gamma_p$, at the wave-breaking limit. For a 1D nonlinear wakefield with $a_0 = 3$ and $\gamma_p = 10$, described by Eq. (51), the electron orbits and the separatrix are shown in Fig. 6(b). Note that when $\Delta\phi > (\gamma_p - 1)/\gamma_p$ and $\gamma_{\min} > 1$ at $\psi_{\max}$, there are two points on the separatrix $\psi_{\max} \pm \delta\psi$ for which $\gamma = 1$.

2.5.2. *Self-injection of electrons into a plasma bubble*

An intense laser pulse propagating in plasma expels electrons due to its ponderomotive force and forms a narrow electron sheath surrounding a cavity of plasma electrons behind the pulse, referred to as a bubble. Here, we consider self-injection of electrons into a nonlinear wakefield in the bubble (the first period of the wake).

The self-injection of electrons in a bubble wake is analytically investigated by solving the equations of the electron motion inside the spherical bubble with radius R_B [67]:

$$\frac{dp_x}{dt} = f(r)\left[\frac{yp_y}{4\gamma} - \frac{\xi(1 + v_B)}{4}\right],$$
$$\frac{dp_y}{dt} = -f(r)\left(1 + \frac{p_x}{\gamma}\right)\frac{y}{4}, \quad (55)$$
$$\frac{d\xi}{dt} = \frac{p_x}{\gamma} - v_B, \quad \frac{dy}{dt} = \frac{p_y}{\gamma},$$

where $\xi = x - v_B t, \mathbf{v}_B = \mathbf{e}_x v_B \simeq \mathbf{e}_x$ is the bubble velocity, $\gamma^2 = 1 + p_x^2 + p_y^2, r^2 = \xi^2 + y^2 + z^2$, and $v_B = (1 - 1/\gamma_B^2)^{-1/2}$ with a Lorentz factor of the bubble motion $\gamma_B \simeq \gamma_p$. The screen effect of an electron sheath around the bubble cavity is taken into account by $f(r) = (1/2)\{1 + \tanh[(R_B - r)/d]\}$, where d is the width of the electron sheath. Here all variables are denoted by dimensionless units, normalizing the time to ω_p^{-1}, the length to $c/\omega_p = 1/k_p$, the velocity to c, the electromagnetic fields to $m_e c\omega_p/e$, and the electron density n to n_e, which is the background plasma density. The trajectories of the electrons can be shown by the numerical solution of the coupled equations (55) for the initially static ($\mathbf{p} = 0$) electron, located at the bubble border ($y \approx R_B$) at $t = 0$, as shown in Fig. 7.

The analysis of the condition for electron capture in the bubble results in $R_B \geq \sqrt{2}\gamma_B$, corresponding to $\varepsilon_{\max} \geq \gamma_B/\sqrt{2}$. Assuming the bubble radius given by Eq. (39), the electron trapping threshold is obtained as $a_{\text{th}} \geq \gamma_B^2/2$. In the numerical solutions shown by Fig. 7, the case (a) satisfies the trapping condition $R_B = 8 > \sqrt{2}\gamma_B \simeq 5.7$ for the sheath thickness $d = 0.3$ and the initial position $y = 5$, while in the case (b) the trapping occurs at $R_B = 5 \ll \sqrt{2}\gamma_B \simeq 14.1$ with the sheath thickness $d = 2$ and the initial position $y = 5.361$. However, the bubble model assuming a uniform sheath thickness may give an overestimate of the trapping threshold.

The experimental observations of an energy density threshold for the stable production of monoenergetic electron beams reveal that the stable injection occurs in the range of plasma densities $1.0 - 1.6 \times 10^{19}\,\text{cm}^{-3}$ when there is an intensity corresponding to the normalized vector potential $a_0 \geq 3$ for the matched spot given by Eq. (39) as a result of the compression of the laser pulse in time and space to dimensions on the order of the plasma wavelength λ_p [68]. For lower densities such as $n_e < 5 \times 10^{18}\,\text{cm}^{-3}$, the self-injection threshold is measured in terms of P/P_c, where P is the laser power and P_c is the critical power for self-focusing given by $P_c = 17\omega_0^2/\omega_p^2$ [GW], using a 200 TW, 60 fs laser, focused into a vacuum spot radius $r_0 = 15\,\mu$m at the $1/e^2$ intensity in the 5 mm and 8 mm helium gas jets [69]. The total charge is self-injected and accelerated to high energies when $P/P_c \geq 3$. Using $P/P_c = k_p^2 r_0^2 a_0^2/32$, the self-injection threshold corresponds to $a_{\text{th}} \geq 2$ (1.6) at density $n_e = 3\,(5) \times 10^{18}\,\text{cm}^{-3}$.

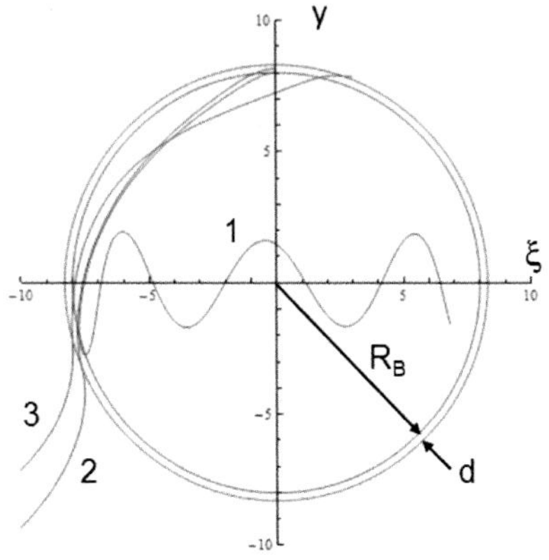

(a)

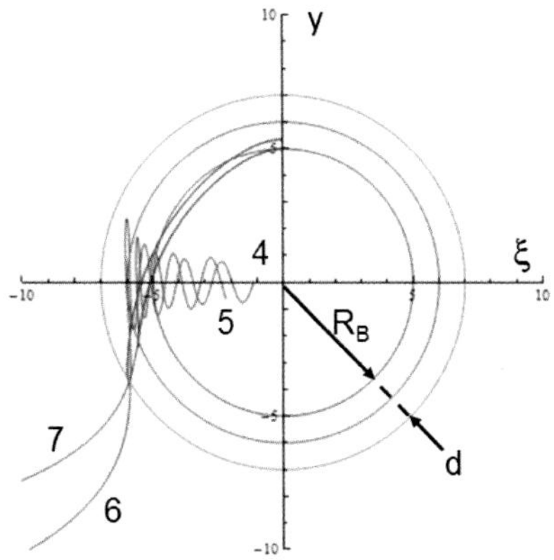

(b)

Fig. 7. Trajectories obtained from the numerical solution of the coupled equations (55) for (a) $\gamma_B = 4, R_B = 8, d = 0.3$, and (b) $\gamma_B = 10, R_B = 5, d = 1$ (track 7) and $d = 2$ (tracks 4, 5, 6). All trajectories have the same initial momentum, $\mathbf{p} = 0$. The initial positions (ξ, y) are $(0, 8)$ for track 1, $(0, 8.15)$ for track 2, $(3, 7.85)$ for track 3, $(0, 5)$ for track 4, $(0, 5.361055)$ for track 5, $(0, 5.361056)$ for track 6, and $(0, 5)$ for track 7.

2.5.3. *Optical injection with colliding pulses*

Electron injection to the trapped phase space of wakefields with the order of plasma wavelength λ_p can be manipulated by ultrashort optical pulses with precision of the pulse length $c\tau_L \leq \lambda_p$. The injection methods, referred to as optical injection, are transverse injection [70, 71], employing two crossing laser pulses, both of which require high-intensity pulses, and colliding pulse injection [72, 73], which employs a drive pulse for wakefield excitation and counterpropagating two injection pulses with different frequencies to produce a beat wave for trapping the electrons in plasma. For experimental simplicity, the colliding pulse injection scheme is modified into a two-pulse system with the same frequency and polarization, consisting of a wake drive pulse and a counterpropagating [19] or crossing injection pulse [20] which generate a standing wave for accelerating plasma electrons inside the separatrix in the phase space of plasma waves.

Consider colliding two pulses with a temporally Gaussian profile [72],

$$\begin{aligned}\mathbf{a}_0 &= a_0\mathbf{e}_x \exp\left[\frac{-(z-\beta_g ct)^2}{4\sigma_L^2}\right]\cos(k_0 z - \omega_0 t),\\ \mathbf{a}_1 &= a_1\mathbf{e}_x \exp\left[\frac{-(z+\beta_g ct)^2}{4\sigma_L^2}\right]\cos(k_0 z + \omega_0 t),\end{aligned} \tag{56}$$

where the parameters indicating the pump pulse are denoted by subscript 0, those indicating the counterpropagating injection pulse are denoted by subscript 1, σ_L is the rms pulse length, $\beta_g c = c(1-\omega_p^2/\omega_0^2)^{1/2}$ is the group velocity of the laser pulse, $\mathbf{e}_x$ is the polarization vector of the laser, and ω_0 and k_0 are the frequency and wave number of the laser, respectively, assuming that both pump and injection laser pulses have the same frequency, wave number and polarization. In the absence of the wakefield, the electron energy equation is described as

$$\frac{d\gamma}{dt} = \left(\frac{\mathbf{u}}{\gamma}\right)\frac{\partial \mathbf{a}}{\partial t} = \left(\frac{1}{2\gamma}\right)\frac{\partial a^2}{\partial t}, \tag{57}$$

where $\mathbf{u} = \mathbf{p}/m_e c$ is the normalized electron momentum, $\gamma = (1+u^2)$ is the Lorentz factor and $a^2 = \mathbf{a}\cdot\mathbf{a} = (\mathbf{a}_0+\mathbf{a}_1)\cdot(\mathbf{a}_0+\mathbf{a}_1)$ is calculated by

$$\begin{aligned}a^2 &= a_0^2 \exp\left[\frac{-(z-\beta_g ct)}{2\sigma_L^2}\right]\cos^2(k_0 - \omega_0 t)\\ &\quad + a_1^2 \exp\left[\frac{-(z+\beta_g ct)}{2\sigma_L^2}\right]\cos^2(k_0 z + \omega_0 t)\\ &\quad + a_0 a_1 \exp\left[\frac{-(z^2+\beta_g^2 c^2 t^2)}{2\sigma_L^2}\right]\\ &\quad \times (\cos 2k_0 z + \cos 2\omega_0 t),\end{aligned} \tag{58}$$

with the first and second terms being the ponderomotive potentials for exciting the plasma wakefields. For the condition generating the standing wave, $z = \pm(\omega_0/k_0)t$ and $|z| \ll \sigma_L$, Eq. (57) is rewritten as

$$\frac{d\gamma}{dt} \simeq \left(\frac{2}{\gamma}\right) k_0 a_0 a_1 \sin(2k_0 z). \tag{59}$$

The maximum normalized energy is calculated as $\gamma_{\max} \simeq (1+2a_0a_1)^{1/2}$. Hence, when this energy exceeds the trapping threshold, plasma electrons are injected into the wakefield. The Hamiltonian describing the electron motion in the wakefield and the standing wave is given by [73]

$$H = (1+u_z^2)^{1/2} - \beta_p u_z - \varphi(\psi), \tag{60}$$

where $u_z m_e c$ is the electron axial momentum, $c\beta_p$ is the phase velocity of the wake, $\varphi = \varphi_0 \cos\psi$ is the wake potential after the laser pulse, and $\varphi_0 = \pi(a_0^2 - a_1^2)/4$. The potential φ_{inj} in the laser pulse is given by

$$\varphi_{\text{inj}}(\psi) = \left(\frac{1}{\pi}\right)\varphi_0\left[1+\sin\psi + \left(\frac{3\pi}{2} - \frac{\psi}{2}\right)\cos\psi\right]. \tag{61}$$

The momentum of the background plasma electron in an untrapped orbit u_{back} is obtained from Eq. (60) with $H = 1$. The momentum of an electron in a trapped orbit is given by Eq. (60) with $H = H(u_z = \gamma_p\beta_p, \psi = \pi) = \gamma_p^{-1} - \varphi_{\text{inj}}(\pi)$. The maximum (+) and minimum (−) momenta of the electron in the standing wave orbit are $u_{\text{st}} = \pm 2(a_0a_1)^{1/2}$, respectively. The conditions for trapping the background plasma electrons are $(u_{\text{st}})_{\max} \geq u_{\text{trap}}$ and $(u_{\text{st}})_{\min} \leq u_{\text{back}}$. The injection threshold of the standing wave is given by $a_{\text{th}} = (a_0a_1)_{\text{th}}^{1/2} = (1 - H_{\text{trap}})/4\beta_p$. The number of trapped electrons is estimated by integrating $(a_0a_1)^{1/2}/a_{\text{th}}$ over the injection phase volume. The optimum standing wave field is obtained as $(a_0a_1)^{1/2} \sim 0.7$ from the analytical estimation and the 1D PIC simulations [20]. At low a_1, the number of trapped electrons decreases due to the low injection field, while at high a_1 that also decreases due to the high threshold, since the accelerating field of the plasma wave is perturbed by the counterpropagating plasma wave excited by the injection pulse.

The colliding optical injection with two counterpropagating laser pulses has been demonstrated at the optimum condition $(a_0a_1)^{1/2} \simeq 0.7$, employing the pump pulse of $a_0 = 1.3$ and the injection pulse of $a_1 = 0.4$, each of which has a 30 fs duration and parallel linear polarization, focused at a 2 mm supersonic helium gas jet [19]. In this experiment, the electron self-injection was suppressed by operating the plasma density below the threshold of self-injection at $n_e = 7.5\times10^{18}\,\text{cm}^{-3}$. As a result, this method produced very stable electron beams with peak energy 117 ± 7 MeV, FWHM energy spread $11\pm2\%$, charge 19 ± 6.8 pC, FWHM beam divergence 5.8 ± 2 mrad, and beam pointing stability 0 ± 1.8 mrad, where each parameter shows the statistics of mean $\pm$ standard deviation over 20 consecutive shots. In addition to enhanced stability, tuning of the beam energy can be carried out by adjusting the colliding position in the gas jet, simply changing the delay between two laser pulses. When the lasers collide at the entrance of the gas jet, electrons are injected at an early stage and accelerated over the whole gas jet length to result in the high-energy beam with a small angular divergence, while injection at the exit of the gas jet limits the acceleration length and leads to a low-energy beam with a large angular divergence. In this experiment, an estimate of the average accelerating electric field was $E_z \simeq 2.7$ GV/cm with energy gain 190 MeV over 700 μm. The good stability and tunability of the electron beam energy indicate that the electrons are injected during the overlap of two laser pulses ($\sim$30 fs) and the trapped electron bunch originates from a single period of the plasma wave ($\lambda_p/c = 40$ fs ≥ 30 fs), in contrast with the self-injection mechanism induced by the nonlinear evolution of the laser pulse through strong self-focusing and self-compression.

In order to avoid technical problems such as damage to the laser system and complexity of the optics for applications and diagnosis of electron beams, associated with the counterpropagating colliding scheme, the countercrossing optical injection is carried out by employing two pulses consisting of the pump pulse with energy 212 mJ and duration 70 fs (3 TW), focused on the spot radius 12.5 μm at the intensity $I \sim 6.8 \times 10^{17}$ W/cm^2 ($a_0 = 0.6$), and the injection pulse with energy 10 mJ and duration 70 fs (0.14 TW), focused on the spot radius 15 μm at the intensity $I \sim 2 \times 10^{16}$ W/cm^2 ($a_0 = 0.1$) [20]. Both pulses are focused onto a supersonic helium gas jet with 1.3 mm width at the plasma density $n_e = 3.95 \times 10^{19}$ cm^3 below the self-injection threshold. The injection pulse is focused in the region at the beginning of a plasma channel formed by the pump pulse at the angle of 135° with respect to the propagation axis of the pump pulse. The collision of two pulses produced a quasi-monoenergetic electron bunch with peak energy 14.4 ± 0.7 MeV, energy spread 1.5 ± 0.2 MeV ($10.6 \pm 1.5\%$) and charge 21.8 ± 3.8 pC, where the beam parameters show the statistics of mean$\pm$standard deviation over 18 shots, produced with the reproducibility of 50%.

2.5.4. *Density gradient/transition injection*

In the 1D wakefield, an electron with a velocity equal to the phase velocity of the plasma wave, i.e. $v_z = v_p$, is trapped at the back of the wake period. In the bubble regime, the electron trapping into the plasma cavity with the radius R_B requires $k_p R_B \geq \sqrt{2}\gamma_B$ [67], where γ_B is the Lorentz factor at the back of the bubble, $\gamma_B \simeq \gamma_p = (1 - \beta_p^2)^{-1/2}$, and $\beta_p = v_p/c$. Accordingly, the electron injection can be enhanced by decreasing the phase velocity or by increasing the bubble size. Correspondingly, both effects are controlled by tailoring the plasma density $n_e(z)$. In the linear regime $a_0 \ll 1$, the wake phase velocity is given by [74]

$$\beta_p(z) = \frac{\beta_g}{1 - \left(\frac{|\zeta|}{\omega_p}\right)\frac{d\omega_p}{dz}} \simeq \frac{1 - n_e/2n_c}{1 - \left(\frac{|\zeta|}{2n_e}\right)\frac{dn_e}{dz}}, \tag{62}$$

where $\beta_g = (1 - \omega_p^2/\omega_0^2)^{1/2}$ is the laser group velocity, z the propagation coordinate and $|\zeta| = |z - \beta_g ct|$ the distance behind the front of the laser pulse. For a down-ramp density gradient $dn_e/dz < 0$, the phase velocity $\beta_p < \beta_g < 1$ and β_p decreases as the laser pulse propagates. The threshold γ_p is reduced at the back of the first wake period by a factor of $1 - (1/2)(\lambda_p/\Delta z)^2(\Delta n_e/n_e)^2$, where Δz is the characteristic length over which a density reduction is Δn_e. For controlling injection into the first wake period, where electrons are most likely trapped and accelerated, the characteristic length of the density gradient should be as short as the plasma wavelength, i.e. $\Delta z \sim \lambda_p$.

In the experiments, a down-ramp density gradient is produced by: a thin slit gas jet, downstream of which the laser is focused [75]; a downward density transition formed on a shock front generated by a knife edge in the gas jet [21]; laser machining of the gas jet [76]; and a gas jet nozzle embedded in a capillary discharge plasma waveguide [22].

Here, the electron injector employing a sharp downward density transition is described as an example of the density tailoring controlled injection scheme, which requires a single laser pulse and a tailored density gradient in the plasma. The density transition exploits the shock front formation in a supersonic helium gas jet prior to ionization by the driving laser pulse. The jet is generated by a pulsed de Laval nozzle with an exit diameter of $300\,\mu$m, where the shock is generated by introducing a knife edge laterally. The typical scale length of an abrupt change of the flow parameters in the supersonic flow is on the order of a few times the molecular mean free path, depending on the initial Mach number and shock angle. The ratio of gas densities in front of and behind the shock is given by $n_1/n_2 = 1-2/(\kappa+1)[1-1/(M_1^2 \sin^2 \alpha_{\rm sh})]$, where n_1 and n_2 are the gas densities before and after the shock, respectively, κ is the specific heat ratio of 5/3 for a monoatomic gas, M_1 is the initial Mach number of the gas flow, and $\alpha_{\rm sh}$ is the angle between the gas flow and the shock front, as shown in Fig. 8. In the experiment, the laser pulses with 65 mJ energy on target and an FWHM duration 8 fs are focused to an FWHM spot diameter of $12\,\mu$m, yielding a peak intensity of $2.5\times10^{18}\,\mathrm{W/cm^2}(a_0 \sim 1.1)$. As shown in Figs. 8(a) and 8(b), a shadowgraph measurement of the plasma shows $\alpha_{\rm sh} = 16\pm0.5°$ and $n_1/n_2 \simeq 1.6$, corresponding to a Mach number $M_1 = 5.1\pm0.2$. The width of the density transition is evaluated as $\sim5\,\mu\mathrm{m}(\approx \lambda_p)$ at the plasma density after transition $n_{e2} \sim 3.8\times10^{19}\,\mathrm{cm^{-3}}$. The results of electron beam parameters produced by density transition injection (self-injection) are energy $23.3^{+3.3}_{-3.0}$ $(26.0^{+8.2}_{-6.6})$ MeV, FWHM energy spread $9^{+6}_{-8}(12^{+9}_{-10})\%$, FWHM divergence $8.9^{+3.1}_{-3.3}\,(10.9^{+3.5}_{-3.7})$ mrad, and charge $3.3^{+2.0}_{-2.2}(3.7^{+2.9}_{-3.1})$ pC with rms deviations for higher-than-average and lower-than-average values. As shown in Figs. 8(c) and 8(d), it is evident that the density transition injection yields a cleaner background-free and more reproducible beam than that obtained with the self-injection.

2.5.5. *Ionization-induced injection*

A mechanism of the ionization-induced trapping is elucidated by the fact that inside a plasma bubble, most likely trapped are a number of electrons that are produced from impurity of gas with a large difference of the ionization potential between the outer shell electrons and the inner shell ones, such as nitrogen, of which two K-shell electrons are ionized by the optical field ionization over the barrier suppression ionization threshold, $I_{\rm BSI} = cU_i^4/(128\pi e^6 Z^2) \approx 1.04\times10^{19}\,(1.62\times10^{19})\,\mathrm{W/cm^2}$, for the nitrogen ionization state N^{6+} (N^{7+}), where Z is the charge state and the ionization potential $U_i = 552.057\,(667.029)\,\mathrm{eV}$, corresponding to $a_{\rm th} \approx 2.9$ (3.6). The outer shell electrons up to N^{5+} are ionized below the intensity of $I_{\rm BSI} \simeq 1.47\times10^{16}$ $(9\times10^{15})\,\mathrm{W/cm^2}$ for the N^{5+} (N^{4+}) ionization potential of $U_i = 97.888\,(77.472)\,\mathrm{eV}$ and can be considered preionized in the leading front of the laser

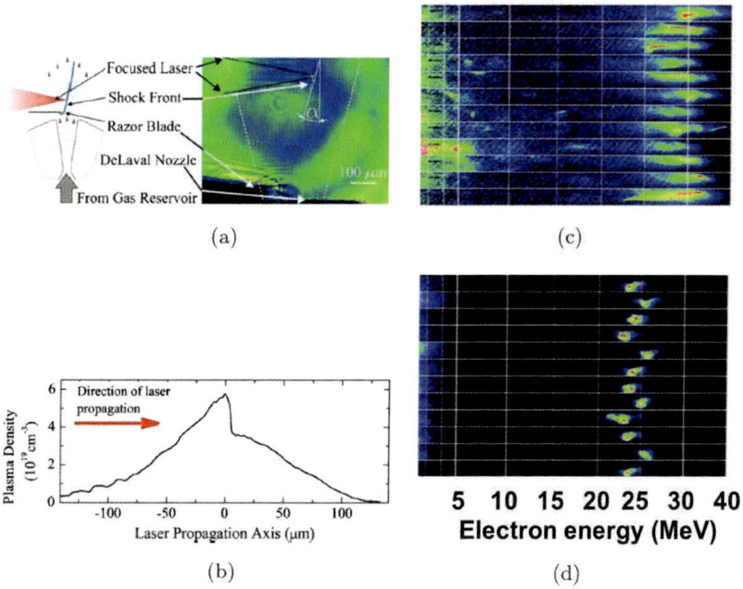

Fig. 8. The experiment of density transition injection on a shock front [21]. (a) A target setup and shadowgraph image. A razor blade is introduced laterally into a supersonic gas jet. Inside plasma, one can see a tilted shock front and a small, bright spot indicative of electron injection at the laser focus. (b) The measured electron density along the laser propagation axis: the electron energy spectra and divergence (c) for the self-injection and (d) for the density transition injection.

pulse before the bubble formation. Hence, the inner shell electrons are produced only near the peak intensity of the laser pulse, which is located near the bubble center on the propagation axis, where the wake potential is a maximum, and thereby the expelling ponderomotive force of the laser pulse is a minimum. Contrary to preionized free electrons, whose trajectories move along a narrow sheath with radius R_B outside the bubble, the ionized electrons emitted from the inner shell move close to the bubble axis toward the back of the bubble, where the wake potential is a minimum, and eventually trapped into the wakefield when electrons gain sufficient kinetic energy required for trapping. This mechanism occurs at an intensity as low as the optical field ionization threshold for the inner shell electrons of impurity gas and significantly increases the trapped charge. Since trapping occurs close to the bubble axis, the amplitudes of the betatron oscillation after trapping decrease compared to self-injection from the electron sheath.

According to theoretical considerations on ionization-induced injection [77], for trapping electrons ionized at the peak of the laser electric field, the minimum laser intensity is given by $1 - \gamma_p^{-1} \leq 0.64a_{\text{min}}^2$, where γ_p is the Lorentz factor given by $\gamma_p = (1 - \beta_p^2)^{-1/2}$ and β_p is the phase velocity of the plasma wave. For electrons to be trapped at or in front of the laser envelope, the intensity must be $a_0 \geq 1.7$ for $\gamma_p = 33$. The 1D PIC simulations show that the maximum number of trapped electrons is saturated to be approximately $N_{e\,\text{max}} \sim 5 \times 10^6\,\mu\text{m}^{-2}$ at the mixed gas length $L_{\text{mix}} \approx 1000\lambda_L$ for plasma density $n_e = 0.001n_c$ ($1.7 \times 10^{18}\,\text{cm}^{-3}$) with a nitrogen concentration of $\alpha_{\text{N}} = 1\%$ and laser parameters $a_0 = 2$ and $c\tau_L \approx 15\lambda_L$ due to the beam-loading effects and initially trapped particle loss from the separatrix in the phase space. The number of trapped electrons scales as $N_e[\mu\text{m}^{-2}] \sim 8 \times 10^7 \alpha_{\text{N}} k_p L_{\text{mix}} (n_e/n_c)^{1/2}$ for $\alpha_{\text{N}} k_p L_{\text{mix}} \leq 2$. The energy spread is also proportional to both the mixed gas length and the nitrogen concentration.

Recent experiments [23–27] support the ionization-induced trapping mechanism which reduces the self-injection threshold, which is estimated to be $a_{\text{th}} \sim 3.8$ as a result of an overview of many experiments, to $P/P_c \sim 1.4$ ($a_{\text{th}} \sim 1.6$) for a 9:1 He : N_2 gas mixture of $n_e \sim 1.4 \times 10^{19}\,\text{cm}^{-3}$ [23], increases 4–5 times the charge for a 1.2% N_2 98.8% He gas mixture of $n_e \sim 2 \times 10^{19}\,\text{cm}^{-3}$ with a 30 TW, 30 fs laser pulse, and produces significantly collimated electron beams [24]. If the laser intensity is well beyond the ionization threshold over a long distance, most produced electron beams have a large energy spread due to the continuous ionization and trapping of the inner shell electrons. In addition to the use of a shorter mixed gas length and a lower concentration of the high-Z gas for improving the electron beam quality, controlling self-focusing of the laser beam by adjusting the laser energy so as to shorten the distance over which the inner shell electrons are ionized and trapped in a plasma wake, the production of quasi-monoenergetic electron beams can be created by ionization-induced injection [25]. Note that the control of self-focusing is carried out by focusing a laser power well beyond the critical power for relativistic self-focusing, i.e. $P/P_c \sim 10$.

On the analogy of conventional accelerators, a staged laser plasma accelerator composed of an injector stage and an accelerator stage is conceivable for operating each stage under a different plasma condition to produce high-quality electron beams with small energy spread and low transverse emittance. In particular, for the ionization-induced injection scheme, it is essential to separate the injector operated at a high density of a mixed gas of the order of 1 mm length from the accelerator operated at a low density of a pure low-Z gas of the order of a cm scale for suppressing self-injection of background electrons. Figure 9 shows an experiment successfully demonstrating a two-stage laser wakefield accelerator with ionization-induced injection [26].

2.6. *Optical guiding in laser wakefield accelerators*

In addition to the injection into a correct accelerating phase of plasma waves, one of the critical issues in a laser plasma accelerator concept is optical guiding of relativistic laser pulses in underdense plasma over a long distance to the extent of a dephasing length, which scales as $L_{\text{dp}} \propto n_e^{-3/2}$, where n_e is the operating plasma density of laser plasma accelerators. To date, most experiments on laser plasma accelerators have been carried out by employing ultrashort pulse lasers with duration $\tau_0 = 30 - 80\,\text{fs}$, focused onto a short-scale plasma target, for instance a few-mm-long gas jet and a

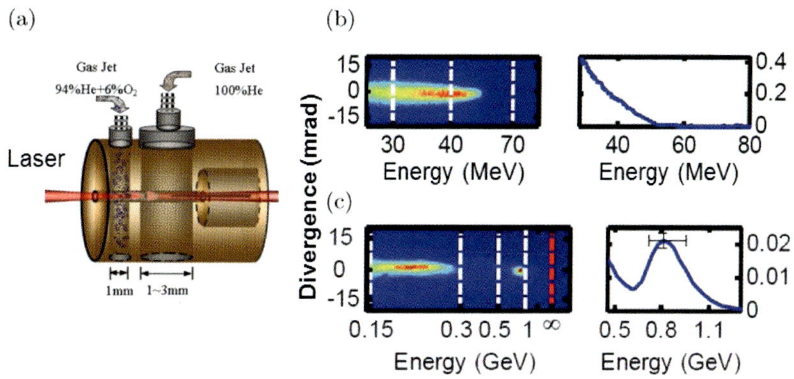

Fig. 9. The experiment on two-stage laser wakefield accelerators using ionization-induced injection, carried out by using 40–60 TW laser pulses with 40 fs duration ($\lambda_0 = 800$ nm), focused by an $f/20$ off-axis parabola onto a 16 μm spot radius at the $1/e^2$ intensity in gas cells [26]. (a) Two segments of gas cells composed of a 1-mm-thick injector cell filled with 6% O_2 gas and 94% He gas flow, and a 3-mm-thick accelerator cell filled with pure He gas flow. (b) Energy spectra of electrons produced from the injector cell at $n_e \simeq 5.7 \times 10^{18}\,\text{cm}^{-3}$, when a 60 TW laser power is focused at a 1.2 mm position from the cell entrance. Note that electron beams have a Maxwellian spectrum with the cutoff energy of 50 MeV. (c) Energy spectra of electron beams from the accelerator stage operated at $n_e \simeq 2.5 \times 10^{18}\,\text{cm}^{-3}$, where electron beams are seeded by the ionization-induced injector. At a 45 TW drive laser power, the electron beam has 0.8 GeV peak energy, 25% energy spread (FWHM), 2.6 mrad divergence, and 3.7 pC charge, associated with a 200 MeV broad peak containing 320 pC charge.

few-cm-long plasma channel at the density in the range of $n_e = 10^{18}$–$10^{19}\,\text{cm}^{-3}$, where a very large amplitude plasma wave of the order of 100 GV/m is excited and traps energetic electrons to be efficiently accelerated inside a wake to high energies of the order of 1 GeV. Here we consider the condition for propagation of an intense laser beam in plasma, and overview the experiments on laser wakefield acceleration from the methodological point of view for optical guiding, characterized as self-guiding and channel-guiding.

2.6.1. *Propagation of intense laser beams in plasma channels*

The propagation of a laser beam in plasma channels with a parabolic density channel of the form $n(r) = n_0 + \Delta n r^2/r_0^2$, where n_0 is the density at $r = 0, r_0$ the laser spot radius and Δn the channel depth at $r = r_0$. Self-guiding propagation is considered as $\Delta n = 0$. The wave equation for the normalized vector potential $\mathbf{a}(r, z, t)$ with the Coulomb gauge $\nabla \cdot \mathbf{a} = 0$ describing the three-dimensional evolution of a laser pulse with duration τ_L in a fully ionized plasma can be written as [78]

$$\nabla^2 \mathbf{a} - \left(\frac{1}{c^2}\right) \partial_t^2 \mathbf{a} = k^2(1 - \eta^2)\mathbf{a}, \tag{63}$$

where $k = \omega/c$ is the free-space wave number along the propagation direction. The (squared) refractive index for linearly polarized electromagnetic waves in the long pulse limit ($ck_p\tau_L \gg 1$) is given by [79]

$$\eta^2(r, z) = 1 - \left(\frac{k_p^2}{k^2}\right) \gamma_\perp^{-1} \left(1 + k_p^{-2} \nabla_\perp^2 \gamma_\perp + \frac{r^2}{r_{\text{ch}}^2}\right), \tag{64}$$

where $k_p = (4\pi r_e n_e)^{1/2}$ is the plasma wave number evaluated with the unperturbed on-axis density n_e and classical electron radius r_e, and $\gamma_\perp = (1 + a^2/2)^{1/2}$ is the relativistic factor of the laser intensity for the linear polarization. In Eq. (64), the first term represents free-space propagation, and three terms in the square brackets correspond to relativistic self-focusing, ponderomotive channeling and a preformed plasma channel with the channel radius $r_{\text{ch}} = r_0(\Delta n/n_0)^{-1/2}$ at which the plasma density doubles, respectively. Using $\zeta = z - \beta_g ct$ and $\mathbf{a}(r, \zeta) = a(r, \zeta)\exp(ik\zeta)\mathbf{e}_x + \text{c.c.}$, where $c\beta_g = c^2k/\omega$ is the linear group velocity, the wave equation can be reduced to the paraxial form,

$$(\nabla_\perp^2 + 2ik\partial_z)a = k^2\left[1 - \frac{1}{\beta_g^2} + \frac{k_p^2}{\gamma_\perp k^2}\left(1 + \frac{1}{k_p^2}\nabla_\perp^2\gamma_\perp + \frac{r^2}{r_{\text{ch}}^2}\right)\right]a, \tag{65}$$

where $\nabla_\perp^2 \equiv r^{-1}\partial_r(r\partial_r) + r^{-2}\partial_\phi^2$ in cylindrical coordinates (r, ϕ, z). For an axially symmetric case, a solution to Eq. (65) is the lowest-order Gaussian mode, $a(r, z) = a_0 \exp[i\theta - (1 - i\alpha)r^2/r_s^2]$, where $a_0(z), \theta(z), \alpha(z)$ and $r_s(z)$, which represent the amplitude, phase shift, curvature and spot radius of the laser field [80], respectively, satisfy

$$\partial_z(a_0 r_s) = 0, \quad \alpha = \left(\frac{kr_s}{2}\right)\partial_z r_s, \tag{66}$$

$$\begin{aligned}\frac{\partial^2 r_s}{\partial z^2} &- \frac{8k_p^2}{k^2 r_s a_0^2}\left(1-\gamma_{\perp 0}+2\ln\frac{\gamma_{\perp 0}+1}{2}\right)\\ &- \frac{16}{k^2 r_s^3 a_0^2}\ln\gamma_{\perp 0} + \frac{2k_p^2 r_s}{k^2 r_{\mathrm{ch}}^2}\\ &\times\left[\frac{4}{a_0^2}\left(1-\gamma_{\perp 0}+\ln\frac{\gamma_{\perp 0}+1}{2}\right)\right.\\ &\left.+H\left(\frac{-a_0^2}{2}\right)\right] = 0,\end{aligned} \tag{67}$$

$$\begin{aligned}\frac{\partial\theta}{\partial z} &= \frac{k}{2}\left(\frac{1}{\beta_g^2}-1\right) - \frac{8\ln\gamma_{\perp 0}}{kr_s^2 a_0^2} - \frac{4k_p^2}{ka_0^2}\ln\frac{\gamma_{\perp 0}+1}{2}\\ &+ \frac{k_p^2 r_s^2}{2kr_{\mathrm{ch}}^2}\left[\frac{8}{a_0^2}\left(1-\gamma_{\perp 0}+\ln\frac{\gamma_{\perp 0}+1}{2}\right)\right.\\ &\left.+H\left(\frac{-a_0^2}{2}\right)\right],\end{aligned} \tag{68}$$

where $\gamma_{\perp 0} = (1+a_0^2/2)^{1/2}$, $H(-a_0^2/2) = {}_4F_3(\frac{1}{2},1,1,1;2,2,2;-a_0^2/2)$ and ${}_pF_q$ denotes the generalized hypergeometric series of order q and class $q-p+1$. In the limit of $a_0^2 \ll 1$, the laser spot radius Eq. (67) and the phase shift Eq. (68) become [80, 81]

$$\frac{\partial^2 r_s}{\partial z^2} = \frac{4}{k^2 r_s^3}\left(1-\frac{r_s^4}{r_m^4}\right), \quad \frac{\partial\theta}{\partial z} = -\frac{2}{k}\left(\frac{1}{r_s^2}-\frac{1}{r_m^2}\right), \tag{69}$$

where $r_m = r_0(\Delta n_c/\Delta n)^{1/4} = r_{\mathrm{ch}}^{1/2}(\pi r_e n_0)^{-1/4}$ is the matched spot radius and $\Delta n_c = 1/\pi r_e r_0^2$ is the critical channel depth. The group velocity is given by $\beta_g^2 = 1 - k_p^2/k^2 - 4/k^2 r_m^2$. The spot radius and phase Eqs. (69) indicate that no variation of the spot size and phase occurs for the matched condition, i.e. $r_s = r_m = r_0$.

Using $X = r_s/\hat{a}\hat{r}$ and $a_0 = X^{-1}$, where $\hat{a}$ and $\hat{r}$ are the vacuum amplitude and the minimum spot size at the focus, respectively, the equation for the laser beam envelope is written as [78, 79 82, 83]

$$\frac{\partial^2 X}{\partial z^2} + \frac{1}{\hat{a}^4 Z_R^2}\frac{\partial V}{\partial X} = 0, \tag{70}$$

where V is defined by

$$\begin{aligned}\frac{\partial V}{\partial X} &= -64\frac{P}{P_c}X\left(1-\gamma_{\perp X}+2\ln\frac{\gamma_{\perp X}+1}{2}\right)\\ &- \frac{4\ln\gamma_{\perp X}}{X} + 2N_{\mathrm{ch}}X\\ &\times\left[4X^2\left(1-\gamma_{\perp X}+\ln\frac{\gamma_{\perp X}+1}{2}\right)\right.\\ &\left.+H\left(-\frac{X^{-2}}{2}\right)\right],\end{aligned} \tag{71}$$

$\gamma_{\perp X} = (1+X^{-2}/2)^{1/2}$, $Z_R = k\hat{r}^2/2$ is the Rayleigh length in vacuum, $P/P_c = k_p^2\hat{a}^2\hat{r}^2/32$ is the ratio of the laser beam power to the critical power for the relativistic self-focusing at the plasma density n_0, given by $P_c = 17n_c/n_0$ [GW], and $N_{\mathrm{ch}} = [16P/k_p r_{\mathrm{ch}} P_c]^2$ is a dimensionless focusing strength parameter that includes both relativistic and channel focusing. The phase equation (68) is rewritten as

$$\begin{aligned}\frac{\partial\theta}{\partial z} &= \frac{k}{2}\left(\frac{1}{\beta_g^2}-1\right)\\ &- \frac{1}{\hat{a}^2 Z_R}\left\{64\frac{P}{P_c}X^2\ln\frac{\gamma_{\perp X}+1}{2} + 4\ln\gamma_{\perp X}\right.\\ &- N_{\mathrm{ch}}X^2\left[8X^2\left(1-\gamma_{\perp X}+\ln\frac{\gamma_{\perp X}+1}{2}\right)\right.\\ &\left.\left.+H\left(-\frac{X^{-2}}{2}\right)\right]\right\}.\end{aligned} \tag{72}$$

Equations (70) and (72) describe the evolution of the scaled spot radius $X(z) = r_s(z)/\hat{a}\hat{r}$ and the phase with the initial conditions $r_s = r_i$, $(\partial r_s/\partial z)_i$, and θ_i at $z = 0$. The matched condition for the beam propagating with a constant spot size, i.e. the equilibrium solution $r_s = r_0 = \hat{r}$, is derived from $\partial V/\partial X = 0$ at $X = 1/a_0$ in Eq. (71). Thus, the matched spot radius $r_m = r_0$ is given by

$$\begin{aligned}k_p^2 r_m^2 &= 2\ln\gamma_{\perp 0}\left\{\gamma_{\perp 0} - 1 - 2\ln\frac{\gamma_{\perp 0}+1}{2}\right.\\ &+ \frac{\Delta n}{n_0}\left[1-\gamma_{\perp 0}+\ln\frac{\gamma_{\perp 0}+1}{2}\right.\\ &\left.\left.+\frac{a_0^2}{4}H\left(-\frac{a_0^2}{2}\right)\right]\right\}^{-1},\end{aligned} \tag{73}$$

and the matched power $P_m = (k_p^2 r_m^2 a_0^2/32)P_c$. For the matched condition, the laser phase velocity is derived from $\partial\theta/\partial z = 0$ at $X = 1/a_0$ in Eq. (72) as

$$\frac{\omega^2}{c^2k^2} = \frac{1}{\beta_g^2} = 1 + \frac{k_p^2}{k^2} \times \left[\frac{8}{a_0^2}\left(\gamma_{\perp 0} - 1 - \ln\frac{\gamma_{\perp 0}+1}{2}\right) + \frac{\Delta n}{n_0}H\left(-\frac{a_0^2}{2}\right)\right]. \tag{74}$$

In the limit of $a_0^2 \ll 1$, Eq. (74) leads to the standard dispersion relation $\omega^2 = c^2k^2 + \omega_p^2 + 4c^2/r_m^2$ for an electromagnetic wave with spot radius r_m in a homogeneous plasma with plasma frequency ω_p. Hence, the group velocity in plasma channels becomes

$$\beta_g^2 = \frac{c^2k^2}{\omega^2} \approx 1 - \frac{k_p^2}{\kappa_{\rm ch}k^2}, \tag{75}$$

where a correction factor of the group velocity is defined as

$$\kappa_{\rm ch} = \frac{a_0^2}{8}\left[\left(\gamma_{\perp 0} - 1 - \ln\frac{\gamma_{\perp 0}+1}{2}\right) + \frac{\Delta n}{n_0}\frac{a_0^2}{8}H\left(-\frac{a_0^2}{2}\right)\right]^{-1}. \tag{76}$$

For the matched condition due to self-guiding and ponderomotive channeling, the spot radius and the group velocity are given by Eqs. (73) and (75), respectively, where the channel depth is set to $\Delta n = 0$. Figure 10 shows the dimensionless matched spot radius k_pr_m and the correction factor of the group velocity $\kappa_{\rm ch}$ as a function of a_0 for the normalized channel depth $\Delta n/n_0$, including self-guiding $\Delta n/n_0 = 0$.

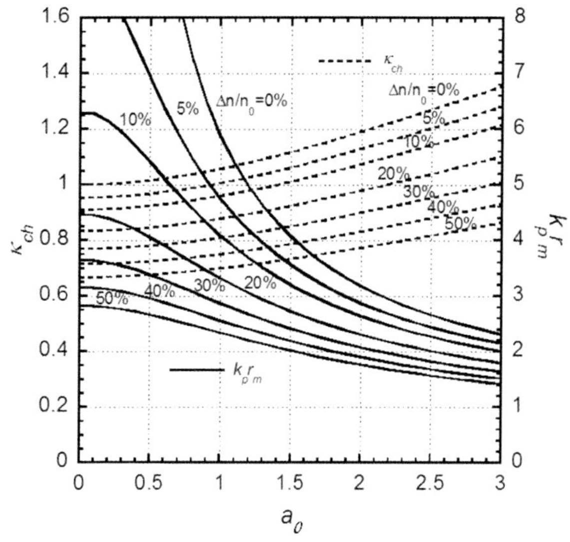

Fig. 10. The dimensionless matched spot radius k_pr_m and the correction factor of the group velocity $\kappa_{\rm ch}$ as a function of the normalized vector potential a_0 for the normalized channel depth $\Delta n/n_0$, including the self-guiding case $\Delta n/n_0 = 0$.

2.6.2. *Self-guided laser wakefield accelerators*

The self-guiding of relativistically intense ($a_0 \geq 1$) ultrashort ($c\tau_L \simeq \lambda_p$) laser pulses has been experimentally investigated in the blowout (bubble) regime. When such a laser pulse with power $P > P_c$ enters an underdense plasma ($\omega_p < \omega_0$), the plasma electrons at the head of the pulse are completely blown out radially during the rise time of the pulse in the first plasma period. Most of the laser pulse resides inside the electron density depression and can therefore be guided. However, due to the inertia of the electrons, the density or refractive index channel forms on a longitudinal scale length of the order of c/ω_p. Hence, the very front of the laser pulse continuously erodes away due to diffraction, with the degree of guiding of the remaining pulse varying along the laser pulse [84]. An estimate of the erosion rate is stated as c/ω_p per the Rayleigh length Z_R, which would limit the distance over which such an ultrashort pulse can be self-guided to a few Z_R. However, when the spot radius is matched to the bubble radius R_B so that Eq. (39) can be satisfied, in spite of diffractive erosion, self-guiding and wake excitation is possible over tens of Z_R in the blowout regime [61]. For a matched spot $r_m \simeq R_B$, this can be expressed as $k_pr_m = 2\sqrt{2}(P/P_c)^{1/6}$ [85]. For $P/P_c > 1$ and $r_0 > r_m$, the spot size will converge to (and remain at) r_m [85, 86], where diffraction at the head of the laser pulse is minimized. As r_0 is reduced from the matched spot size, diffraction loss tends to increase. As a matched ultrashort laser pulse with $a_0 \geq 2$ propagates through the plasma, the front of the laser pulse locally pump depletes as the wake is excited, causing the photons to shift to a longer wavelength. These frequency downshifted photons slip back due to dispersion with a velocity $v_{\rm etch} = c\omega_p^2/\omega_0^2$ into the wake, where the transverse density depression is sufficient for guiding them. This $v_{\rm etch}$ is the difference between the linear and the nonlinear group velocity of the laser pulse [17]. As a result, to the first order, the nonlinear pump depletion length is given by

$$L_{\rm pd} = \frac{(c\tau_L)c}{v_{\rm etch}} \simeq \frac{(c\tau_L)\omega_0^2}{\omega_p^2} = \frac{(c\tau_L)n_c}{n_e}. \tag{77}$$

For $2 < a_0 < 20$, this scaling is independent of a_0, and thereby independent of power as long as $P/P_c > 1$. Beyond the pump depletion limit, the pulse is so severely etched that it is no longer intense enough to excite a wake and therefore no longer guided. The nonlinear pump depletion length can be obtained from measurements of the length of the self-guided plasma channel as a function of a_0 by observing Thomson scattering images from the laser–plasma interaction in plasma channels [30, 87, 88]. In this experiment [87], a typical measurement shows that channeling starts before the focus and stretches over a 2.8 mm length, corresponding to 5 times the Rayleigh length, at the peak power of 109 TW ($a_0 \simeq 4.7$), which is 16 times larger than the self-focusing critical power of $P_c \simeq 7$ TW at the density $4.3 \times 10^{18}\,\text{cm}^{-3}$. At the same time, two quasi-monoenergetic electron bunches are measured, resulting in the peak energy 200/220 MeV, energy spread 29%/17% (FWHM) and divergence 20/20 mrad (FWHM), respectively.

The dephasing length $L_{\text{dp}} \simeq \gamma_p^2 \lambda_p$, with $\gamma_p \simeq \omega_0/\omega_p \gg 1$, is typically longer than the Rayleigh length Z_R, the length over which the laser defocuses in vacuum, i.e. $Z_R = \pi r_0^2/\lambda_L$ for a Gaussian pulse. When going to lower density, to obtain a higher final energy, the ratio $L_{\text{dp}}/Z_R \simeq \gamma_p/\pi$ increases for a laser spot diameter $2r_0 \simeq \lambda_p$. Therefore, some form of optical guiding is required to ensure acceleration over a dephasing length. For acceleration to higher energies with a self-guided pulse, lower densities are required to extend the dephasing length. This increases not only the power needed for self-focusing ($P_c \propto n_e^{-1}$) but also the laser power required to obtain a threshold a_0 for injection, since the bubble size also increases with decreasing density.

Experiments on laser wakefield acceleration driven by a self-guided laser pulse have been carried out using 200 TW class laser pulses [69, 89]. The dephasing-limited energy gain given by the expression $E[\text{MeV}] \simeq 115 P[\text{TW}]^{1/4} L_{\text{acc}}[\text{mm}]^{1/2}$ [17] is in agreement with the experimentally measured energies, if the acceleration length L_{acc} is the gas jet length minus some extent of length ($\simeq 2.5$ mm) over which the laser spot evolves to its matched spot size. A 720 MeV electron beam accelerated in the length $L_{\text{acc}} = 5$ mm indicates that the nearly self-matched laser spot is guided over 8 mm ($\sim 10 Z_R$). The divergence $\Delta\theta_{\text{FWHM}}$ of the 720 MeV beam is 2.85 mrad. Using the electron beam size of $\sigma_r \simeq 1\,\mu$m measured from the simulations, a corresponding normalized emittance, $\varepsilon_n = \sigma_r \gamma \Delta\theta_{\text{FWHM}}$, is estimated to be $\varepsilon_n \sim 5$ mm-mrad.

2.6.3. *Channel-guided laser wakefield accelerators*

Plasma waveguides for guiding ultraintense short laser pulses in plasmas are produced by a number of methods, including laser-induced hydrodynamic expansion [90–92], and pulsed discharges of an ablative capillary [28, 93–96] or a gas-filled capillary [97–102]. However, the length of such a plasma channel has been limited to about 10 cm and the plasma density has been created for $n_0 \geq 10^{17}\,\text{cm}^{-3}$. Laser-induced plasma channels need a fast ignition by a few ns prior to propagating a main laser pulse, while a slow pulsed discharge capillary needs a jitter-free trigger within a few ns. To guide ultraintense laser pulses, plasma channels must be produced in fully ionized gases with low atomic number Z, such as hydrogen or helium. Not fully ionized particles will be further ionized by the laser, disturbing the electron density profile.

In slow pulsed capillary discharges [103, 104], the duration of discharge is much longer than any microscopic processes in the plasma, which are collisions between all kinds of particles (atoms, ions and electrons), irradiation processes (recombination radiation, line plasma emission by ions and atoms, bremsstrahlung) and ionization. The plasma density is set to be high enough so the collisions will dominate radiation processes (local thermal equilibrium — LTE regime). If the capillary length is much longer than its diameter, then plasma quantities vary primarily as functions of time and radius only, excluding the small regions near the open ends of the capillary. The characteristic acoustic time for this system is $\tau_a = r_{\text{ch}}/c_s$, where r_{ch} is the capillary radius and c_s is the plasma sound velocity. For typical capillary discharges, $c_s \sim 10^6\,\text{cms}^{-1}$, $r_{\text{ch}} \sim 0.025$ cm and the acoustic time is therefore a few tens of nanoseconds. This time is much shorter than the discharge rise time τ_{dis}, which typically is about 100 ns. Thus, after a few acoustic times the plasma pressure along the radius of the capillary is nearly constant. This is valid for any time period, $t \gg \tau_a$. When the current of the main discharge flows through the initial plasma, the plasma temperature starts to rise due to ohmic (Joule) heating, and the plasma density also increases due to the ionization of the gas in the gas-filled capillary or ablation from the capillary walls in the ablative capillary. There is a period of time during which the plasma temperature is generally higher at the center of the capillary than at the edges. The plasma is almost fully ionized, so its conductivity σ scales as $T_e^{3/2}$ and is nearly independent of plasma density. Thus, the electrical current of the main discharge goes primarily through the central part of the capillary plasma. Since the plasma pressure is nearly independent of radial position r, a maximum in temperature at the center of the capillary and the density distribution leads to an on-axis density minimum.

The plasma channel has a limited lifetime and will disappear due to diffusion of the cold dense plasma from the capillary walls to the center. For plasma densities of 10^{17}–$10^{19}\,\mathrm{cm}^{-3}$, the diffusion time τ_{dif} varies between 0.1 to a few tens of microseconds. These timescales are much longer than any laser propagation timescale. The rise time τ_{dis} of the main discharge current is set to satisfy $\tau_a \ll \tau_{\mathrm{dis}} \ll \tau_{\mathrm{dif}}$. In this regime, the depth of the plasma channel is largest near the maximum of the current. This requirement limits the value of the current driven through the capillary. The thermal plasma pressure at the current maximum must also be higher than the pressure of the magnetic field generated by the main discharge current. Thus, the Ampere force can be neglected and the capillary plasma is confined radially, due mainly to the capillary walls.

The electron acceleration in the discharge capillary waveguide has been demonstrated by employing both a gas-filled capillary [31, 32, 105, 106] and an ablative discharge capillary [33, 37, 94, 107] with 1.5–4 cm lengths at an on-axis density ranging from $1.9 \times 10^{18}\,\mathrm{cm}^{-3}$ to $8.4 \times 10^{18}\,\mathrm{cm}^{-3}$. In most of these experiments, since the input normalized laser field is in the quasilinear laser wakefield regime, i.e. $a_0 \simeq 0.7$–1.7, the electron injection occurs as a result of the laser pulse evolution such as self-focusing and self-compression, leading to the intensity self-amplification for reaching the self-injection threshold ($a_{\mathrm{th}} \approx 3$–$4$) [68].

The 3D axisymmetric PIC simulations [106] show that after a short distance ($z \approx 200\lambda_p$), a spot radius of the laser pulse, initially with $a_0 \simeq 1$, is self-focused to a spot radius given by the matching condition $r_{\mathrm{sf}} \equiv \lambda_p\sqrt{a_0}/\pi$ [61] for the bubble regime, followed by maintaining a constant mean spot size r_{sf} in the plasma channel, whereas in a gas cell the pulse diffracts rapidly for the same laser and plasma conditions. The small spot radius maintained in the plasma channel allows a_0 to grow through temporal completion of the pulse and enables the threshold value for self-injection to be reached at $z \approx 1000\lambda_p$. This not only reduces the effective acceleration distance but also degrades the quality, stability and reproducibility of electron beams from the discharge capillary. Therefore, it is essential that channel-guided laser wakefield accelerators operated in the quasilinear wakefield regime build in a short plasma section (a gas jet or gas cell) as an injector to control the electron injection.

In the experiment by Lu *et al.* [37], the ablative capillary is composed of two segments, a half of which is made of acrylic resin (C : O : H = 4 : 2 : 7) and polyethylene (C : O : H = 1 : 0 : 2). With a 4-cm-long polyethylene capillary, no electrons are accelerated regardless of varying the drive power from 80 TW to 130 TW and scanning the delay over a wide range. This result is quite consistent with the experiment by McGuffey *et al.* [107], where no electrons are observed from a 3 cm polyethylene capillary driven by a 30 fs, 35 TW laser pulse. Hence, in an acrylic capillary containing oxygen atoms, electrons from the K-shell of oxygen are likely injected into the wakefield due to the ionization-induced injection of O^{7+} with the ionization threshold $a_{\mathrm{th}} = 2.9$ when the laser pulse with $a_0 \geq 3$ is focused.

2.7. *Scaling of laser wakefield accelerators*

For the last two decades, a number of laser plasma accelerator experiments have been carried out under various conditions, primarily for self-guided laser wakefield accelerators [26, 27, 34–36, 69, 89, 108, 109] and channel-guided laser wakefield aceelerators [31–33, 37, 38, 106]. Comparing these data with theoretical laser wakefield acceleration models, it may be useful to find a correct scaling law capable of predicting energy gain, accelerated electron charge and the required laser plasma conditions. Since the maximum energy gain scales as $\Delta W_{\max} \propto n_c \propto \lambda_L^{-2}$ for a given a_0, most of the previous experiments employed the

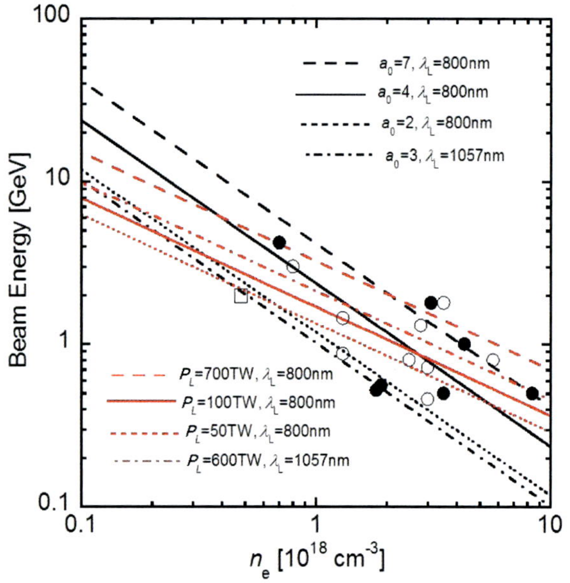

Fig. 11. Beam energy scaling of GeV-class laser wakefield accelerators: comparison of measured electron beam energies with the energy scaling formula (44) (black lines) for various values of a_0, assuming that $\alpha_c = 1$, and Eq. (78) (red lines) for various laser powers P_L as a function of the operating plasma density n_e for the self-guided case (open circle) at $\lambda_L = 800\,\text{nm}$, the channel-guided case (solid circle) at $\lambda_L = 800\,\text{nm}$, and the self-guided case (open square) at $\lambda_L = 1057\,\text{nm}$, respectively.

chirped pulse amplification lasers with wavelength $\lambda_L = 800\,\text{nm}$ and pulse duration $\tau_L \leq 80\,\text{fs}$, except for the case using a PW-class laser with wavelength $\lambda_L = 1057\,\text{nm}$ and $\tau_L \sim 150\,\text{fs}$ [35]. The validity of the energy scaling formulas based on the present analysis may be verified by comparison with these experimental results. Figure 11 shows the comparison of measured electron beam energies with the energy scaling formulas, which are given by Eq. (44) in terms of a_0 and n_e and the following formula [17] on the maximum energy gain E_b in terms of focused laser power P_L and n_e:

$$\Delta W_{\max}[\text{GeV}] \approx 1.7\left(\frac{P_L}{100\text{TW}}\right)^{1/3} \times \left(\frac{n_e}{10^{18}\,\text{cm}^{-3}}\right)^{-2/3}\left(\frac{\lambda_L}{0.8\,\mu\text{m}}\right)^{-4/3}, \tag{78}$$

assuming that $a_0 = (8P_L/P_c)^{1/3}$. As shown in the figure, the measured energy $\Delta W = 4.2\,\text{GeV}$ of Ref. 38 accelerated at $n_e = 7 \times 10^{17}\,\text{cm}^{-3}$ in a 9 cm gas-filled discharge capillary driven by a laser pulse with $\lambda_L = 800\,\text{nm}$ and $P_L = 233\,\text{TW}$ can be scaled by Eq. (44) for $a_0 \sim 5$ with $\alpha_c = 1$, while Eq. (78) predicts that $\Delta W_{\max} \sim 2.9\,\text{GeV}$, which is approximately 1.5 times lower than the measured energy.

3. Electron Beam Dynamics and Betatron Radiation in Laser Plasma Acceleration

3.1. *Betatron motion of an electron*

Beams that undergo strong transverse focusing forces $F_\perp = -m_e c^2 \kappa_F^2 x_\beta$ in plasma waves exhibit betatron oscillation, where x_β is the transverse amplitude of betatron oscillation and κ_F is a focusing constant exerted from transverse wakefields. The betatron motion of electrons emits synchrotron radiation, which affects energy loss and transverse emittance of the electron beam via the radiation reaction force. The motion of an electron traveling along the z axis in the accelerating force eE_z and the radial force $e(E_r - B_\theta)$ from the plasma wave evolves according to

$$\frac{du_x}{cdt} = -\kappa_F^2 x_\beta + \frac{F_x^{\text{RAD}}}{m_e c^2}, \quad \frac{du_z}{cdt} = k_p \frac{E_z}{E_0} + \frac{F_z^{\text{RAD}}}{m_e c^2}, \tag{79}$$

where $\mathbf{F}^{\text{RAD}}$ is the radiation reaction force and $\mathbf{u} = \mathbf{p}/mc$ is the normalized electron momentum. The classical radiation reaction force [110] is given by

$$\frac{\mathbf{F}^{\text{rad}}}{m_e c \tau_R} = \frac{d}{dt}\left(\gamma \frac{d\mathbf{u}}{dt}\right) + \gamma \mathbf{u}\left[\left(\frac{d\gamma}{dt}\right)^2 - \left(\frac{d\mathbf{u}}{dt}\right)^2\right], \tag{80}$$

where $\gamma = (1 + u^2)^{1/2}$ is the relativistic Lorentz factor of the electron and $\tau_R = 2r_e/3c \simeq 6.26 \times 10^{-24}\,\text{s}$. Since the scale length of the radiation reaction, i.e. $c\tau_R = 2r_e/3 \simeq 1.9\,\text{fm}$, is much smaller than that of the betatron motion, i.e. $\sim\lambda_p \gamma^{1/2}$, assumming that the radiation reaction force is a perturbation and $u_z \gg u_x$, the equations of motion (79) are approximately written as [111]

$$\frac{du_x}{dt} \simeq -c\kappa_F^2 x_\beta - \tau_R c^2 \kappa_F^2 u_x (1 + \kappa_F^2 \gamma x_\beta^2),$$
$$\frac{du_z}{dt} \simeq \omega_p \frac{E_z}{E_0} - \tau_R c^2 \kappa_F^4 \gamma^2 x_\beta^2, \quad \frac{dx_\beta}{dt} = \frac{cu_x}{\gamma} \simeq c\frac{u_x}{u_z}. \tag{81}$$

Finally, the particle dynamics is obtained from the coupled equations [62, 112, 113]

$$\frac{d^2x_\beta}{dt^2}+\left(\frac{\omega_p}{\gamma}\frac{E_z}{E_0}+\tau_R c^2\kappa_F^2\right)\frac{dx_\beta}{dt}+\frac{c^2\kappa_F^2}{\gamma}x_\beta=0,$$
$$\frac{d\gamma}{dt}=\omega_p\frac{E_z}{E_0}-\tau_R c^2\kappa_F^4\gamma^2x_\beta^2. \tag{82}$$

The radiated power in the classical limit is given by [110]

$$P_{\mathrm{RAD}}=\frac{2e^2\gamma^2}{3c}\left[\left(\frac{d\mathbf{u}}{dt}\right)^2-\left(\frac{d\gamma}{dt}\right)^2\right]$$
$$=\frac{2e^2\gamma^2}{3m_e^2c^3}[|\mathbf{F}_{\mathrm{ext}}|^2-|\mathbf{F}_{\mathrm{ext}}\cdot\mathbf{u}/\gamma|^2], \tag{83}$$

using $m_e c d\gamma/dt=\mathbf{F}_{\mathrm{ext}}\cdot\mathbf{u}/\gamma$, where $\mathbf{F}_{\mathrm{ext}}$ is the external force on the electron. As the force is transverse only, i.e. $\mathbf{F}_{\mathrm{ext}}=F_\perp\mathbf{e}_x$ and for a relativistic electron with $u_x^2\ll\gamma^2$, the radiated power can be written as [113]

$$P_{\mathrm{RAD}}=\frac{2e^2\gamma^2}{3m_e^2c^3}F_\perp^2=\frac{2}{3}ce^2\kappa_F^4\gamma^2x_\beta^2$$
$$=m_ec^4\tau_R\kappa_F^4\gamma^2x_\beta^2, \tag{84}$$

with $F_\perp=-m_ec^2\kappa_F^2x_\beta$.

In the bubble regime [16, 61], plasma electrons radially expelled by the radiation pressure of the laser pulse form a sheath with a thickness of the order of the plasma skin depth c/ω_p outside the ion sphere, which is remaining unshielded behind the laser pulse moving at relativistic velocity. As described in Sec. 2.4, the longitudinal field is given by $|E_z/E_0|=(1/2)\alpha_c k_p\xi$, where $0\le\xi\le R_B$. The transverse fields are composed of the electric fields from the ion sphere $E_{r\mathrm{ion}}/E_0=k_pr/2$, the radial plasma current $E_{r\mathrm{EM}}/E_0=-k_pr/4$ and the magnetic field from the radial plasma current $B_{\theta\mathrm{EM}}/E_0=-k_pr/4$. The total focusing force on a beam electron is $e(E_r-B_\theta)=(1/2)m_ec^2k_p^2r$. Hence, the accelerating field and the focusing constant are given by

$$\frac{E_z}{E_0}=\frac{1}{2}k_p(R_B-\xi)=\chi\left(1-\frac{\zeta}{\zeta_d}\right),$$
$$\kappa_F^2=\frac{1}{2}k_p^2, \tag{85}$$

where $\zeta=k_p\xi,\chi=a_0^{1/2},0\le\zeta\le\zeta_d$, and $\zeta_d=(4/3)\chi\gamma_g^2$ is the dimensionless dephasing length with $\gamma_g^2=n_c/n_e$. Using $d/cdt=d/dz$, the dimensionless variables $\Sigma_\beta=k_px_\beta,\zeta=k_pz,s=1-\zeta/\zeta_d$ and Eq. (85), the coupled equations (82) become

$$\gamma\frac{d^2\Sigma_\beta}{ds^2}-2\left[\gamma_ds+\frac{1}{4}\gamma\tau_R\omega_p\zeta_d\right]\frac{d\Sigma_\beta}{ds}+\frac{\zeta_d^2}{2}\Sigma_\beta=0,$$
$$\frac{d\gamma}{ds}=-2\gamma_ds+\frac{1}{2}\tau_R\omega_p\zeta_d\gamma^2\Sigma_\beta^2, \tag{86}$$

where $\gamma_d=(2/3)a_0\gamma_g^2=(1/2)\chi\zeta_d,\chi=2\gamma_d/\zeta_d$ and $\tau_R\omega_p=2k_pr_e/3\approx1.12\times10^{-10}(n_e/10^{17}\,\mathrm{cm}^{-3})^{1/2}$. Since the radiative term in Eq. (86) can be neglected for $\gamma_d\ll10^5$, i.e. the maximum energy at the dephasing length $W_d\ll50\,\mathrm{GeV}$ at the operating plasma density of $n_e\le10^{17}\,\mathrm{cm}^{-3}$, integrating the energy equation as $\gamma=\gamma_d(1-s^2)$, the equation of the betatron amplitude results in a Legendre differential equation,

$$(1-s^2)\frac{d^2\Sigma_\beta}{ds^2}-2s\frac{d\Sigma_\beta}{ds}+\nu(\nu+1)\Sigma_\beta=0, \tag{87}$$

where $\nu(\nu+1)=\zeta_d^2/2\gamma_d=(4/3)\gamma_g^2$. The solution to Eq. (87) is obtained from the Legendre functions of the first and second kinds, i.e. $P_\nu(s)$ and $Q_\nu(s)$, as

$$\Sigma_\beta(s)=AP_\nu(s)+BQ_\nu(s), \tag{88}$$

where A and B are determined by the initial condition $(\Sigma_{\beta i},\Sigma'_{\beta i})$. For $x_\beta(\zeta)=x_{\beta0}$ at $\zeta=0$, the betatron amplitude is simply given by

$$x_\beta(\zeta)=x_{\beta0}P_\nu\left(1-\frac{\zeta}{\zeta_d}\right). \tag{89}$$

For $\gamma_g\gg1$ and $\nu\approx2\gamma_g/\sqrt{3}\sim\gamma_g\gg1$, the asymptotic expressions of $P_\nu(s)$ and $Q_\nu(s)$ are given by

$$P_\nu(s)=\sqrt{\frac{2}{\nu\pi(1-s^2)^{1/2}}}\times\sin\left[\left(\nu+\frac{1}{2}\right)\cos^{-1}s+\frac{\pi}{4}\right],$$
$$Q_\nu(s)=\sqrt{\frac{2}{\nu\pi(1-s^2)^{1/2}}}\times\cos\left[\left(\nu+\frac{1}{2}\right)\cos^{-1}s+\frac{\pi}{4}\right]. \tag{90}$$

Thus, the solution for $\Sigma_\beta(s)$ can be written as

$$\Sigma_\beta(s)\approx\frac{\Sigma_0}{(1-s^2)^{1/4}}\cos(\gamma_g\cos^{-1}s+\varphi_0), \tag{91}$$

where Σ_0 and φ_0 are constants. This equation is rewritten as [114]

$$\Sigma_\beta(\zeta)\sim\frac{\Sigma_0'}{\gamma(\zeta)^{1/4}}\cos\left(\frac{\omega_p}{\omega_0}\int_0^\zeta\frac{d\zeta'}{[2\gamma(\zeta')]^{1/2}}+\varphi_0'\right). \tag{92}$$

Accordingly, the betatron amplitude x_β decreases as $\gamma^{-1/4}$ during acceleration, while the betatron frequency ω_β decreases as $\gamma^{-1/2}$.

3.2. *Radiative energy loss due to betatron radiation*

Assuming that the electron beam has a constant equilibrium radius defined by $\sigma_{\rm eq}^2 = \varepsilon_0\gamma^{1/2}/\kappa_F$, where $\varepsilon_0 = \varepsilon_{n0}/\gamma_i$ is the initial geometrical emittance at the injection energy γ_0 for the normalized emittance ε_{n0}, the radiative energy loss due to betatron radiation can be estimated from

$$\Delta\gamma_{\rm RAD} \approx 2\rho\int_{\gamma_0}^{\gamma_M} d\gamma(\gamma^2 + 2\rho\gamma^4), \tag{93}$$

where γ_M is the electron energy without radiation and $\rho \equiv \tau_R\omega_p k_p^2\sigma_{\rm eq}^2(E_0/E_z)(\kappa_F/k_p)^4$ is a coefficient of radiative loss, assuming that $\rho \ll 1$. Taking into account $\gamma_0 \ll \gamma_M$ and $\gamma_M = \gamma - \Delta\gamma_{\rm RAD}$, the relative radiative energy loss becomes [62]

$$\frac{\Delta\gamma_{\rm RAD}}{\gamma} \approx \frac{2}{3}\rho_0\gamma^{5/2}\frac{1+\left(\frac{6}{5}\right)\rho_0\gamma^{5/2}}{1-2\rho_0\gamma^{5/2}(1+2\rho_0\gamma^{5/2})},$$
$$\rho_0 = \left(\frac{8\pi}{3}\right)r_e^2 n_e\left(\frac{\varepsilon_{n0}}{\gamma_0}\right)\left(\frac{E_0}{E_z}\right)\left(\frac{\kappa_F}{k_p}\right)^3, \tag{94}$$

with the initial normalized emittance ε_{n0}. In the bubble regime wakefields driven by a_0, the relative radiation loss is approximately given by

$$\frac{\Delta\gamma_{\rm RAD}}{\gamma} \approx \frac{0.0136}{a_0^{1/2}}\left(\frac{\varepsilon_{n0}}{1\,\mu{\rm m}}\right)\left(\frac{n_e}{10^{17}\,{\rm cm}^{-3}}\right) \times\left(\frac{1\,{\rm GeV}}{W_i}\right)\left(\frac{W}{100\,{\rm GeV}}\right)^{5/2}, \tag{95}$$

where W_i is the injection energy. As an example, for $a_0 = 4$, $n_e = 1.4\times 10^{16}\,{\rm cm}^{-3}$ and $W_i = 0.1\,{\rm GeV}$, the relative energy loss at 100 GeV is $\Delta\gamma_{\rm RAD}/\gamma \approx 0.01\,(\varepsilon_{n0}/1\,\mu{\rm m})$.

3.3. *Betatron radiation spectrum*

Betatron radiation generated from laser-driven plasma accelerators provides us with a compact, all-optical source of high-energy photons with an ultrafast duration and broadband spectrum, which lead to a new promising tool for a wide range of sciences. For instance, betatron hard X-rays are useful for a high-resolution radiographic imaging [41], phase contrast single shot imaging [42, 115, 116] and probing of the warm dense matter for ultrafast time-resolved measurement [43]. Recent progress in laser wakefield acceleration toward multi-GeV electron beams [35, 36, 38] has extended the betatron radiation spectrum to the MeV gamma ray region [49, 50, 117, 118]. This capability of generating MeV betatron gamma rays provides a simpler alternative operated with only one laser to all-optical gamma ray sources consisting of two synchronized lasers for inverse Compton scattering off laser-wakefield-accelerated electrons [51, 52]. These devices facilitate new research fields, such as laser nuclear physics and nuclear photonics, to explore photonuclear spectroscopy [119] and the fundamental QED process [54].

The classical characteristics of the betatron radiation spectrum [114, 120] can be derived from calculation of the Liénard–Wiechert potentials [110] on a relativistic electron trajectory moving in a plasma wake, where electrons undergo betatron motion due to the strong focusing plasma wakefield, leading to emission of synchrotron radiation, often referred to as a plasma wiggler. The energy radiated per unit solid angle Ω per unit frequency ω is given by

$$\frac{d^2I}{d\omega d\Omega} = \frac{e^2}{4\pi^2 c}\left|\int_{-\infty}^{\infty}\frac{\mathbf{n}\times[(\mathbf{n}-\boldsymbol{\beta})\times\dot{\boldsymbol{\beta}}]}{(1-\boldsymbol{\beta}\cdot\mathbf{n})^2} \times\exp\left[i\omega\left(t-\frac{\mathbf{n}\cdot\mathbf{r}(t)}{c}\right)\right]dt\right|^2, \tag{96}$$

where $\mathbf{n}$ is the direction of observation, $\boldsymbol{\beta}$ the velocity of electrons normalized to c, $\dot{\boldsymbol{\beta}} = d\boldsymbol{\beta}/dt$, and $\mathbf{r}(t)$ the electron position at time t, respectively.

In contrast to synchrotron radiation in the conventional wiggler, where the betatron orbit of an electron is specified on the plane perpendicular to the magnetic field, the radiation spectrum emitted by electron beams accelerated in plasma wakefields is significantly distinct from the synchrotron radiation spectrum based on an energy constant betatron motion of a single electron due to effects of the acceleration process, finite energy spread and unspecified orbit plane. The spectrum of betatron radiation is characterized with the critical photon energy $E_c = 3\gamma^2\hbar\omega_\beta K_\beta$, where $K_\beta = \gamma\omega_\beta r_\beta/c$ is a parameter describing the strength of betatron oscillation of a single electron with energy $m_ec^2\gamma$ and radial amplitude r_β, and $\omega_\beta = \omega_p/\sqrt{2\gamma}$ is the betatron oscillation frequency with plasma frequency $\omega_p = (4\pi e^2 n_e/m_e)^{1/2}$. The photon numbers spectrum radiated by an electron per unit photon energy $E_\gamma = \hbar\omega$ and solid angle Ω observed at an angle θ and ϕ in the spherical coordinate system — $x = r\sin\theta\cos\phi, y = r\sin\theta\sin\phi, z = r\cos\theta$, assuming that an electron undergoes betatron oscillation on the (x, z) plane — can be given in the asymptotic limit $K_\beta \gg 1$, using the angle coordinates $\theta_x = \sin\theta\cos\phi$ and $\theta_y = \sin\theta\sin\phi$ [121],

$$\frac{d^2 I_\gamma(\theta_x, \theta_y)}{dE_\gamma d\Omega} \simeq \frac{6}{\pi^2}\alpha_f N_\beta \gamma^2 \frac{\zeta^2}{1+\gamma^2\theta_y^2} \times \left[K_{2/3}^2(\zeta) + \frac{\gamma^2\theta_y^2}{1+\gamma^2\theta_y^2}K_{1/3}^2(\zeta)\right], \quad (97)$$

where $\alpha_f = e^2/\hbar c \simeq 1/137$ is the fine structure constant, and $N_\beta = L_{\text{int}}/\lambda_\beta = L_{\text{int}}/\sqrt{2\gamma}\lambda_p$ is the number of betatron periods that the electron undergoes over the interaction length L_{int} with betatron and plasma wavelengths, $\lambda_\beta = 2\pi c/\omega_\beta$ and $\lambda_p = 2\pi c/\omega_p$, respectively. $K_{2/3}(\zeta)$ and $K_{1/3}(\zeta)$ are the modified Bessel functions of the second kind with $\zeta = \xi(1+\gamma_z^2\theta_y^2)^{3/2}/(1-\gamma_z^2\theta_x^2/K_\beta^2)^{1/2}$ and $\xi = E_\gamma/E_c$. Here the first term in the square brackets corresponds to radiation with polarization parallel to the plane of the electron orbit, and the second term to radiation with polarization perpendicular to that plane [110]. Equation (97) indicates that an angular dependence of the photon spectrum can be cast as a function of three dimensionless variables on photon energy ξ and two angles $\gamma\theta_x, \gamma\theta_y$.

Figures 12(a) and 12(b) show the spectral functions $d^2I_\gamma(\theta_x, 0)/dE_\gamma d\Omega$ and $d^2I_\gamma(0, \theta_y)/dE_\gamma d\Omega$ normalized by the peak fluence at $\theta_x = \theta_y = 0$, i.e. $d^2I_\gamma(0, 0)/dE_\gamma d\Omega = (6/\pi^2)\alpha_f N_\beta\gamma^2$, respectively [117]. As shown in Fig. 12, the emission angle θ_x in the electron orbit plane is confined to the electron deflection angle $\theta_{\max} = K_\beta/\gamma$, while the emission angle θ_y is confined within an angle $\sim 1/\gamma$ along the y axis perpendicular to the electron orbit plane. Averaging Eq. (97) over the angle coordinate θ_y yields [117]

$$\frac{d^2 I_\gamma}{dE_\gamma d\theta_x} = \int_{-1}^{1} d\theta_y \frac{d^2 I_\gamma}{dE_\gamma d\Omega} = \frac{\sqrt{3}}{2\pi}\alpha_f N_\beta \gamma S(\chi), \quad S(\chi) = \chi\int_\chi^\infty K_{5/3}(\eta)d\eta, \quad (98)$$

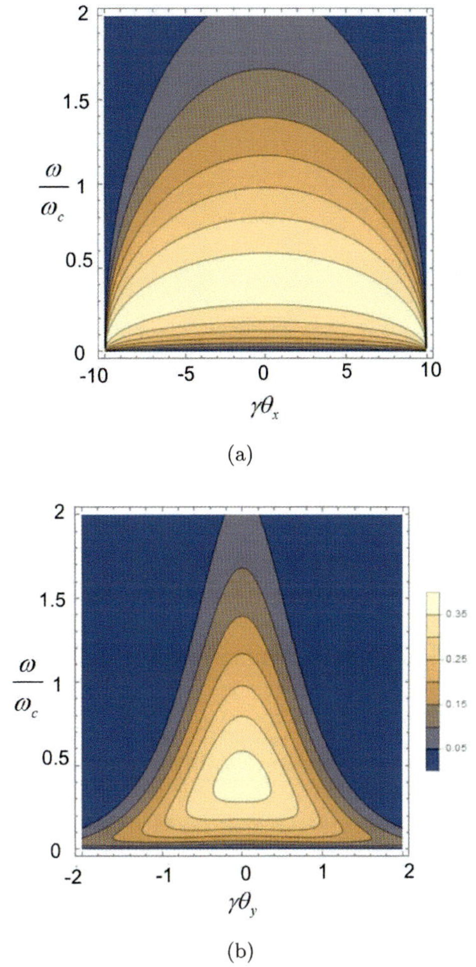

Fig. 12. Contour plot of the angle-dependent betatron radiation spectra along (a) the θ_x axis and (b) the θ_y axis emitted from an electron with energy $m_e c^2\gamma$ and betatron strength parameter $K_\beta = 10$ [117].

where $\chi = 2E_\gamma/[E_c(1 - \gamma^2\theta_x^2/K_\beta^2)^{1/2}]$. The spectral function $S(\chi)$ has a peak value at $\chi_p = 0.286$, i.e. $S(\chi_p) = 0.92$. Accordingly, the angular dependence of the peak photon energy is given by $E_{\gamma p}(\theta_x) = 0.143E_c[1 - (\theta_x/\theta_{\max})^2]^{1/2}$. Integrating Eq. (98) over photon energy E_γ, one can obtain the angular distribution of photon energy as [117]

$$\frac{dI_\gamma}{d\theta_x} = \frac{\alpha_f N_\beta \gamma E_c}{18}\sqrt{1 - \left(\frac{\theta_x}{\theta_{\max}}\right)^2}. \quad (99)$$

The betatron radiation spectrum given by Eq. (97) is deduced for radiation emitted by electrons with a constant energy. For betatron radiation emitted from electrons accelerated in laser plasma accelerators, the radiation spectrum can be significantly modified due to an energy increase and finite energy spread. During the electron accelerating process in the bubble regime of laser plasma accelerators with the dephasing length L_d, the electron energy $m_e c^2 \gamma$ at the position z is approximately given by [62]

$$\frac{d\gamma}{dz} = \frac{2\gamma_d}{L_d}\left(1 - \frac{z}{L_d}\right),$$
$$\gamma(z) = \gamma_0 + 2\gamma_d\left(\frac{z}{L_d} - \frac{1}{2}\frac{z^2}{L_d^2}\right), \quad (100)$$

where $m_e c^2 \gamma_d$ is the maximum electron energy at the dephasing length and $m_e c^2 \gamma_0$ the initial electron energy. Since electron energy increases during the acceleration process in plasma, the betatron frequency changes as $\omega_\beta(z) = \omega_p/\sqrt{2\gamma(z)}$ in plasma so that the number of betatron oscillation periods over the dephasing length is calculated as $N_\beta = \int_0^{L_d} dz/\lambda_\beta = \pi L_d/(2\sqrt{2\gamma_d}\lambda_p)$. Thus, the energy dependence of the critical photon energy E_c is written by $E_c(\gamma_d) = 3\gamma^2 \hbar\omega_\beta K_\beta = (3/2)\pi^3 m_e c^2 \lambda\!\!\!^{-}_e r_e r_\beta n_e \gamma_d^2$, where $\lambda\!\!\!^{-}_e = \hbar/m_e c \simeq 3.86 \times 10^{-11}$ cm is an electron Compton wavelength divided by 2π.

Consider the energy distribution of electrons inside a bunch with a parabolic density distribution $\rho_e(\gamma) = 3/(4\Delta\gamma)[1-(\gamma-\bar{\gamma})^2/\Delta\gamma^2]$ for $\bar{\gamma} - \Delta\gamma \leq \gamma \leq \bar{\gamma} + \Delta\gamma$, where $m_e c^2 \bar{\gamma}$ is the mean energy and $m_e c^2 \Delta\gamma$ a half energy spread; the critical photon energy of betatron radiation averaged over all bunch electrons results in [117]

$$\langle E_c \rangle = \int_0^\infty \rho_e(\gamma_d) E_c(\gamma_d) d\gamma_d$$
$$= \left(\frac{3}{2}\right)\pi^3 m_e c^2 \lambda\!\!\!^{-}_e r_e r_\beta n_e \bar{\gamma}_d^2 \left[1 + \left(\frac{\Delta\gamma^2}{5\bar{\gamma}_d^2}\right)\right]. \quad (101)$$

In consequence, the observable betatron radiation spectrum radiated by an electron beam with a finite energy spread in the acceleration process is characterized by the maximum mean energy $m_e c^2 \bar{\gamma}_d$ at the dephasing length L_d [114]. The effect of electron beam energy spread produces a broadened spectrum with a higher peak energy, compared to that of the monoenergetic case [117, 118].

3.4. *Measurement of betatron gamma rays from laser plasma accelerators*

Experiments on multi-GeV class laser plasma acceleration are carried out at the PW laser system (Fig. 13), capable of delivering 1.1 PW pulses of 28 fs FWHM duration at a repetition rate of 0.1 Hz [50]. The experimental setup is shown in Fig. 14(a). A 30 J laser pulse with 28–57 fs duration is focused by a spherical mirror with a 6 m focal length to a maximum intensity of 5.8×10^{19} W/cm^2 ($a_0 \simeq 5$) onto a gas cell with a variable length (2–3 cm), filled with

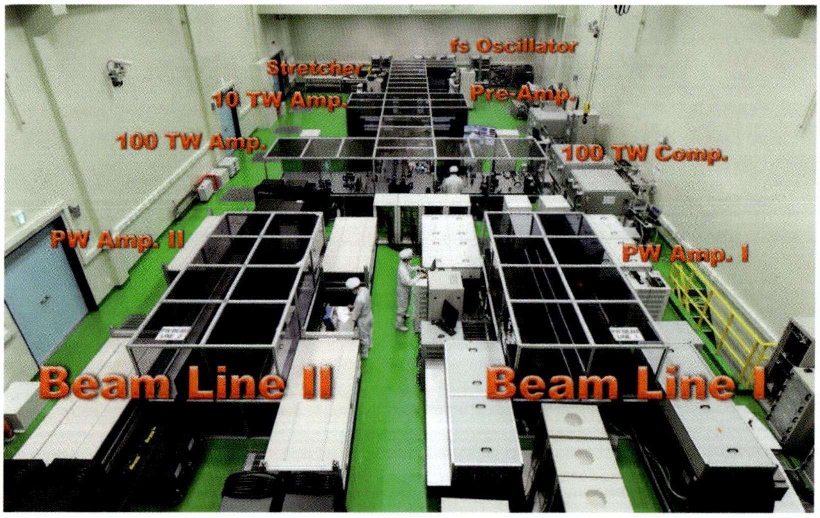

Fig. 13. The PW Ti:sapphire laser system, comprising Beam Line I delivering laser pulses with 30 fs duration and 1 PW peak power at 0.1 Hz and Beam Line II delivering laser pulses with 30 fs duration and 1.5 PW peak power at 0.1 Hz, in the Center for Relativistic Laser Science at the Institute for Basic Science (IBS-CoReLS) [122].

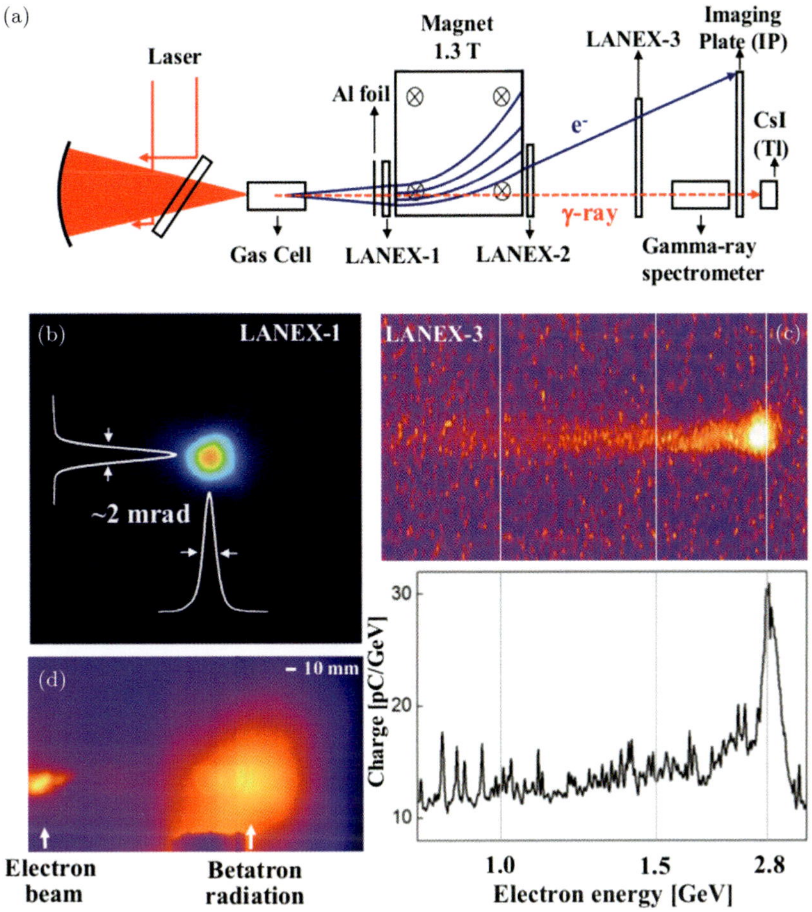

Fig. 14. (a) Schematic of the experimental setup, showing a spherical mirror having a focal length of 6 m onto a variable length gas cell of 2–3 cm, operated at the electron density of $1.2{-}1.5 \times 10^{18}\,\text{cm}^{-3}$. A typical electron beam from the gas cell is shown in (b) for a transverse profile recorded on LANEX-1 and (c) for a raw energy spectrum recorded on LANEX-3 and for the corresponding retrieved electron beam spectrum. (d) A typical image of electron beams and betatron radiation recorded on the IP located 2.5 m away from the magnet exit [50].

He gas at the electron density of $n_e = 1.2{-}1.5 \times 10^{18}\,\text{cm}^{-3}$.

As shown in Fig. 14(b), electron beams produced through the LWFA process are characterized with a 1.3 T and 30-cm-long magnetic spectrometer and three scintillator screens (Kodak LANEX) with a thickness of 300 μm: LANEX-1, located at the front of the magnet, images an electron beam profile; LANEX-2 and -3 are placed at the magnet exit and 0.53 m away from the magnet, respectively, to measure the accelerated electron beam spectrum using a two-screen technique. A typical retrieved electron beam spectrum is shown in Fig. 14(c). Betatron radiation from the gas cell is detected outside the interaction chamber through a 2-cm-thick acrylic window, while electron beams of interest are swept out from the laser propagation axis due to deflection in the magnet. Figure 14(d) shows both electron beam and betatron radiation images observed with the image plate (IP) placed 2.5 m away from the magnet exit, where two images are separated from each other at an ∼25 cm distance (∼100 mrad in angle). The differential filtering gamma ray spectrometer is aligned 3.4 m away from the gas cell on the laser propagation axis to measure betatron radiation spectra.

The spectrometer consists of a series of IPs (Fuji-BAS MS) alternatively stacked with 15 filters of increasing Z number from Al to Pb and then a thickness of Pb for differential filtering, as shown in Fig. 15(a). Each filter with a 15 mm diameter is sandwiched between 250-μm-thick Mylar sheets for minimizing the contribution of secondary electrons <150 keV and stainless steel frames with a 15 mm diameter hole, followed by the image plate disk with

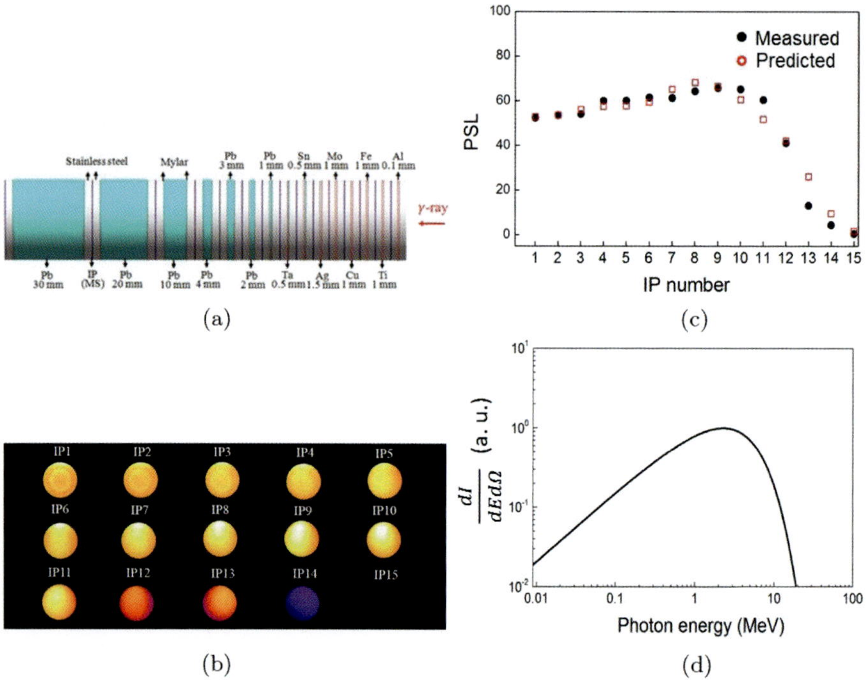

Fig. 15. (a) Schematic of the differential filtering gamma ray spectrometer, showing the configuration of metallic filters and their thickness. (b) A sample of scanned 15-IP images for a single shot. (c) The measured and predicted spectrometer responses. (d) The unfolded radiation spectrum [50].

a 30 mm diameter. The filters and image plates are stacked in a 1-cm-thick stainless steel housing case, which is shielded by a 10-cm-thick Pb wall after being set on the gamma ray beam line to reduce the background noises from the electron beam dump.

In the experiment, the spectrometer responses are measured with a 15-channel stack detector to increase the cutoff energies. Figure 15(b) shows a sample of the scanned images of 15 IPs taken for betatron radiation generated by an electron beam with the energy of 798 ± 67 MeV. The photostimulated luminescence (PSL) values of the Fuji BAS-MS IP are simulated by using the Monte Carlo simulation code GEANT4 [123] from the total energy deposited in a phosphor layer consisting of $BaFBr_{0.85}I_{0.15}$. This calibration factor from the deposited energy calculated by GEANT4 to the PSL values of the IP is $1/(13.63 \pm 1.71)$ [PSL/GeV]. The simulated responses are in good agreement with the measured and modeled sensitivities [124] for overall PSL values and the characteristic X-ray absorption energies at Br-K_α (13.5 keV), I-K_α (33 keV) and Ba-K_α (37.4 keV).

For broadband radiation, the total response Q_i of the ith IP to the photon spectral intensity $I(E_j)$ within the jth energy bin ($E_j \leq E_\gamma \leq E_{j+1}$) can be predicted by using the response matrix $R_i(E_j)$ as $Q_i = \sum_{j=1}^{M} R_i(E_j)I(E_j)$, where M is the total number of energy bins and the spectrometer comprises N channels. The response matrix can be calculated by GEANT4 as the response to the monoenergetic incident photons, i.e. $R_i(E_j) = Q_i/I(E_j)$. Thus, the photon spectrum can be obtained by minimizing the rms deviation of the predicted response Q_i from the measured PSL value P_i, $\sigma_{\text{dev}} = [(1/N)\sum_{i=1}^{N}(P_i - Q_i)^2]^{1/2}$, assuming a proper function of the photon spectrum as $I(E_\gamma) = \Phi(E_\gamma)$. The spectrum of the betatron radiation observed along the laser propagation axis $\theta = 0$ is approximated as

$$\frac{d^2I(0)}{dE_\gamma d\Omega} \propto \Phi_{\text{Beta}}(E_\gamma)$$

$$\equiv \Phi_0 \left(\frac{E_\gamma}{E_c}\right)^2 K_{2/3}^2\left(\frac{E_\gamma}{E_c}\right) \simeq AE_\gamma^\alpha \exp(-\beta E_\gamma), \tag{102}$$

where A, α and β are parameters for fitting to the theoretical function of the betatron radiation. This means that the betatron radiation spectrum can be approximately represented as a gamma distribution function by setting appropriate parameters to be $\Phi_0 = 2.71, A = 4.30/E_c^{\alpha}, \alpha = 0.768$ and $\beta = 1.88/E_c$, respectively. Figure 15(c) shows the measured and predicted spectrometer responses, and Fig. 15(d) shows the result of unfolding the spectrometer responses consisting of 15 IPs.

4. Future Perspectives

Nowadays, vibrant research is carried out on the laser plasma acceleration concept [2] in order to achieve high-energy, high-quality electron beams with multi-GeV energy in a cm-scale plasma [31, 34–38], 1%-level energy spread [33], a 1-mm-mrad-level transverse emittance [32], and a 1-fs-level bunch duration [125], ensuring that the stability of reproduction is as high as that of the present high-power, ultrashort-pulse lasers [88, 126]. Recently, two-staged laser plasma acceleration [26, 27, 127] has been successfully demonstrated. In this context, recent research on laser plasma acceleration moves on the embodiment of practical applications as laser-plasma-based particle and radiation sources.

4.1. *Design of laser plasma accelerators*

For applications of unique particle beam and radiation sources, one needs proper formulas for designing laser plasma accelerators to meet the requirement of electron beam parameters such as energy and charge, complying with the scaling law obtained from laser plasma accelerator experiments. Since, as shown in Sec. 2.7, most results of laser plasma acceleration experiments can be properly scaled by self-guided laser wakefield acceleration in the bubble regime, here we formulate underlying criteria for designing electron beam parameters for broad applications such as energy and charge.

For a given energy E_b GeV and charge Q_b pC, the parameters of self-guided laser wakefield accelerators can be designed as follows. In laser wakefield acceleration, an accelerated electron beam induces its own wakefield and cancels the laser-driven wakefield. Assuming the beam loading efficiency $\eta_b \equiv 1 - E_z^2/E_M^2$ defined by the fraction of the plasma wave energy absorbed by particles of the bunch with the rms radius σ_b, the beam-loaded field is given by $E_z = (1-\eta_b)^{1/2}E_M = \alpha_c E_M$, where E_M is the accelerating field without beam loading, given by $E_M \approx a_0^{1/2}E_0$ for the bubble regime $a_0 \geq 2$. Thus, a loaded charge is calculated as

$$Q_b \simeq \frac{e}{4k_L r_e}\frac{\eta_b k_p^2\sigma_b^2}{1-\eta_b}\frac{E_z}{E_0}\left(\frac{n_c}{n_e}\right)^{1/2} \approx 76[\text{pC}]\frac{1-\alpha_c^2}{\alpha_c^2}a_0^{1/2}k_p^2\sigma_b^2 n_{18}^{-1/2}, \quad (103)$$

denoting $n_{18} \equiv n_e/(10^{18}\,\text{cm}^{-3})$. The field reduction factor α_c is obtained from solving the equation $\alpha_c^2 + C\alpha_c^{3/2} - 1 = 0$, where the coefficient C is given by $C = (Q_b/123)\lambda_L^{-1}E_b^{-1/2}(k_p\sigma_b)^{-2}$. Then, the operating plasma density is determined from Eq. (44), taking into account relativistic correction of the group velocity of the laser pulse at wavelength λ_L μm, as

$$n_e[\text{cm}^{-3}] \approx 3.8\times 10^{17}\kappa_{\text{ch}}a_0\lambda_L^{-2}\left(\frac{E_b}{\alpha_c}\right)^{-1}, \quad (104)$$

where κ_{ch} is the relativistic correction factor of group velocity, given by Eq. (76). The accelerator length equal to the dephasing length, i.e. $L_{\text{acc}} = L_{\text{dp}}$, becomes

$$L_{\text{acc}}[\text{cm}] \approx 3.6a_0^{-1}\kappa_{\text{ch}}^{-1/2}\lambda_L\left(\frac{E_b}{\alpha_c}\right)^{3/2}, \quad (105)$$

while the pump depletion length due to pulse front erosion is given by $L_{\text{pd}} \approx c\tau_L n_c/n_e$. Since the dephasing length should be less than the pump depletion length, i.e. $L_{\text{pd}} \geq L_{\text{dp}}$, the pulse length is set to be

$$\tau_L[\text{fs}] \geq 38\kappa_{\text{ch}}^{1/2}\lambda_L\left(\frac{E_b}{\alpha_c}\right)^{1/2}. \quad (106)$$

For matched propagation of the laser pulse, the spot radius is set to be

$$r_L[\mu\text{m}] \approx 8.7R_m(a_0\kappa_{\text{ch}})^{-1}\lambda_L\left(\frac{E_b}{\alpha_c}\right)^{1/2}, \quad (107)$$

where $R_m = k_p r_m$ is the matched spot radius, given by Eq. (73). The matched peak power is given by

$$P_L[\text{TW}] \approx 1.6a_0\kappa_{\text{ch}}^{-1}R_m^2\left(\frac{E_b}{\alpha_c}\right), \quad (108)$$

and the required laser pulse energy becomes

$$U_L[\text{J}] = P_L\tau_L \geq 0.06a_0\kappa_{\text{ch}}^{-1/2}R_m^2\lambda_L\left(\frac{E_b}{\alpha_c}\right)^{3/2}. \quad (109)$$

4.2. *Applications to coherent X-ray radiation sources*

4.2.1. *SASE FEL for EUV light source*

Extreme ultraviolet (EUV) with wavelengths below about 50 nm and, more specifically, around and below 13.5 nm can be used in microlithography processes to enhance the resolution of optical systems that are limited by the diffraction limit of light accompanying miniaturization of semiconductor integrated circuits. This technology, called extreme ultraviolet lithography (EUVL), is capable of providing resolution below 30 nm, which has not been possible with conventional optical lithography utilizing deep ultraviolet (DUV) light sources with wavelengths of 248 nm or 193 nm.

The current technologies for generating high-power EUV radiation at 13.5 nm, referred to as laser-produced plasma (LPP), employ the deposition of laser energy into a source material such as molten tin (Sn) droplets, creating ionized gas microplasma at electron temperatures of several tens of electron volts. The current LPP radiation sources have a serious obstacle on the way to a high-volume manufacturing (HVM) source, such as small efficiency of the radiation source, a limited set of discrete wavelengths and the mitigation of the plasma debris required for the protection of the EUV optics. Free electron laser (FEL)-based radiation sources have evident advantages in wavelength tunability, high efficiency and high output power, compared to the current LPP radiation sources. The problem of debris mitigation does not exist at all. There is no need to use a multilayer-coated reflective collector, for which reflectivity is limited to about 70%.

A proposed FEL [128] producing a kW-level average output power of EUV radiation utilizes high-energy electron beams of the order of 1 GeV generated from a radio-frequency-based linear accelerator (RF linac). The RF linac may consist of a high-brightness electron injector, a several-stage magnetic bunch compressor system to compress bunch length, and a main linac composed of a series of room-temperature or superconducting RF cavities with the accelerating gradient of the order of 10 MV/m as well as a beam transport system, lastly followed by undulators with a total length of 30 m. The alternating magnetic fields of the undulator force relativistic electrons in a bunch to emit EUV radiation coherently on a sinusoidal trajectory due to the microbunching process, called a self-amplified spontaneous emission (SASE) FEL. The overall size of an RF-linac-driven, FEL-based EUV light source may require a 250-m-long facility for a linac-based light source or a 120-m-long, 60-m-wide area for a recirculator-based light source. The costs for construction and operation of such a facility may turn out to be so incredibly large as to hinder FEL-based EUV light sources from industrial realization of the next-generation lithography technology. A design of FEL-based EUV light source can be carried out by the one-dimensional FEL theory [129], as follows. The FEL amplification takes place in the undulator with the undulator period λ_u and the peak magnetic field B_u at the resonant wavelength λ_X, given by

$$\lambda_X = \lambda_u \left(1 + \frac{K_u^2}{2}\right) / 2\gamma^2, \tag{110}$$

where $\gamma = E_b/m_e c^2$ is the relativistic factor of the electron beam energy E_b, and $K_u = 0.934 B_u[\mathrm{T}]\lambda_u[\mathrm{cm}]$ is the undulator parameter. In the high-gain regime required for the operation of an SASE FEL, an important parameter is the Pierce parameter ρ_{FEL} given by

$$\rho_{\mathrm{FEL}} = (2\gamma)^{-1} \left(\frac{I_b}{I_A}\right)^{1/3} \left[\frac{\lambda_u K_u A_u}{2\pi\sigma_b}\right]^{2/3}, \tag{111}$$

where I_b is the beam current, $I_A = 17\,\mathrm{kA}$ is the Alfvén current, σ_b is the rms transverse size of the electron bunch, and the coupling factor is $A_u = 1$ for a helical undulator and $A_u = J_0(\Xi) - J_1(\Xi)$ for a planar undulator, where $\Xi = K_u^2/[4(1 + K_u^2/2)]$ and J_0 and J_1 are the Bessel functions of the first kind. Another important dimensionless parameter is the longitudinal velocity spread Λ of the beam normalized by the Pierce parameter,

$$\Lambda^2 = \rho_{\mathrm{FEL}}^{-2} \left\{ \left(\frac{\sigma_\gamma}{\gamma}\right)^2 + \left(\frac{\varepsilon_n^4}{\sigma_b^4}\right) \left[2\left(1 + \frac{K_u^2}{2}\right)\right]^{-2} \right\}, \tag{112}$$

where σ_γ/γ is the relativistic rms energy spread, and ε_n the normalized emittance. An e-folding gain length L_{gain} over which the power grows with propagation distance z exponentially according to $\exp(2z/L_{\mathrm{gain}})$ is given by

$$L_{\mathrm{gain}} = \frac{\lambda_u (1 + \Lambda^2)}{4\sqrt{3}\pi\rho_{\mathrm{FEL}}}. \tag{113}$$

In order to minimize the gain length, one needs a large Pierce parameter ρ_{FEL} and a normalized longitudinal velocity spread Λ sufficiently low compared to 1, which means a sufficiently small energy spread σ_γ/γ and $\varepsilon = \varepsilon_n/\gamma$. This expression applies to moderately small beam size σ_b such that the diffraction parameter $B \gg 1$, where B is defined as $B = 16\sqrt{2}\pi^2\gamma^{3/2}A_u K_u \sigma_b^2 (I_b/I_A)^{1/2}\lambda_u^{-2}(1+K_u^2/2)^{-3/2}$. A saturation length L_{sat} required to saturate the amplification can be expressed as

$$L_{\text{sat}} = L_{\text{gain}} \ln\left[\left(\frac{P_{\text{sat}}}{P_{\text{in}}}\right)\left(\Lambda^2 + \frac{3}{2}\right)\Big/\left(\Lambda^2 + \frac{1}{6}\right)\right], \tag{114}$$

where P_{in} and P_{sat} are an input and a saturated power, respectively. P_{in} and P_{sat} are related to an electron beam power P_b according to $P_b = \gamma I_b m_e c^2 = I_b E_b$, $P_{\text{sat}} \cong 1.37\rho_{\text{FEL}} P_b \exp(-0.82\Lambda^2)$, and $P_{\text{in}} \cong 3\sqrt{4\pi}\rho_{\text{FEL}}^2 P_b [N_{\lambda_X} \ln\,(N_{\lambda_X}/\rho_{\text{FEL}})]^{-1/2}$, where N_{λ_X} is the number of electrons per wavelength, given by $N_{\lambda_X} = I_b \lambda_X/ec$.

Table 1 summarizes design examples of main laser plasma accelerator and FEL parameters for the FEL EUV radiation source with the radiation power of 1 kW at the wavelengths of 13.5 nm (case A) and 6.7 nm (case B), using the undulator with the period of 15 mm and the gap of 3 mm, assuming an FWHM electron bunch duration of ~10 fs and a relative energy spread of $\Delta E/E_b \sim 1\%$ [53]. The laser plasma accelerator may be driven by a high-repetition-rate, high-average-power laser, such as the coherent combining fiber-based chirped pulse amplification (CPA) laser [130].

Table 1. Parameters for laser plasma accelerator-based EUV FEL light sources [53].

Case	A	B
Drive laser		
Wavelength [μm]	1	1
Average laser power [MW]	1.19	2.60
Repetition rate [MHz]	0.315	0.473
Laser energy per pulse [J]	3.79	5.51
Peak power [TW]	59	75
Pulse duration [fs]	65	73
Matched spot radius [μm]	27	30
Laser plasma accelerator		
Electron beam energy [MeV]	659	935
Plasma density [10^{17} cm^{-3}]	4.2	3.2
Accelerator length [mm]	51	74
Charge per bunch [nC]	0.5	0.5
Bunch duration [fs]	10	10
Energy spread [%]	~1.6	~1.1
Transverse beam size [μm]	25	25
Peak current [kA]	50	50
Average beam power [kW]	104	221
Free electron laser		
Undulator period (gap) [mm]	15 (3)	15 (3)
Radiation wavelength [nm]	13.5	6.7
Peak magnetic field [T]	1.425	1.425
Undulator parameter K_u	2.0	2.0
Pierce parameter [%]	1.60	1.125
Gain length [mm]	86	123
Saturation length [cm]	102	144
Number of periods	68	96
Spectral bandwidth [%]	1.5	1.0
Radiation cone angle [μrad]	82	48
Input power [MW]	5.3	5.3
Saturated power [GW]	317	317
Duration of EUV pulse [fs]	10	10
Average EUV power [kW]	1	1.5

The bunch duration of the electron beam in the injector stage at a plasma density of $n_e \approx 10^{18}$ cm^{-3} is assumed to be ~10 fs in FWHM, based on the measurement of electron bunch duration in the recent laser wakefield acceleration experiment [131]. The relative energy spread of an accelerated electron beam with the injection energy of $0.1E_b$, where E_b is the final beam energy in the accelerator stage, is assumed to be of the order of 10% in the injector stage. After acceleration up to 10-times-higher energy in the accelerator stage, the relative energy spread at the final beam energy is reduced to $\Delta E/E_b \sim 1\%$ due to adiabatic damping in the longitudinal beam dynamics. The transverse beam size is tuned by employing the beam-focusing system, for instance a set of Halbach-type permanent-magnet-based quadrupoles (PMQs) [47, 132].

For the application of laser plasma accelerators to FELs, it is essential to produce high-quality electron beams with a sufficiently small relative energy spread of the order of ρ_{FEL}, as well as small transverse emittance. In order to reduce the electron energy spread significantly, there has recently been proposed a complex laser plasma accelerator composed of the injection, bunch compression and acceleration stages [133]. Initially, an electron bunch with a negative chirp energy spread, i.e. higher energy at the head and lower energy at the tail, is injected into the second wave period at a sharp density down-ramp in the injector using density gradient injection (see Sec. 2.5.4), followed by a decreasing energy spread through acceleration in the wakefield with a positive dispersion. Then, the electron bunch is transferred to the zero-phase region at the center of the wakefield period in the bunch compressor with a lower plasma density. In this stage, bunch compression takes place as a result of the phase space rotation through two reversals of the energy chirp from an initially negative to a positive one, and finally back to a negative chirp before being seeded into the base of the first wave period in the accelerator, where the energy spread is minimized with the peak energy acceleration, providing a 0.1%-level energy spread and a sub-fs bunch [133].

4.2.2. *Laser plasma accelerator–driven hard X-ray FEL*

Consider the feasibility of a compact hard X-ray FEL capable of reaching a wavelength of $\lambda_X = 0.1\,\text{nm}$, which requires electron beam energy of multi-GeV range in the case of a modest undulator period of the order of a few centimeters. One of the prominent features of laser plasma accelerators is the ability to produce 1-fs-level bunch duration, which is unreachable by means of conventional accelerator technologies. The X-ray FELs rely on SASE, where the coherent radiation builds up in a single pass from the spontaneous (incoherent) undulator radiation. In an undulator the radiation field interacts with electrons snaking their way when overtaking them so that electrons are resonantly modulated into small groups (microbunches) separated by a radiation wavelength and emit coherent radiation with a wavelength equal to the microbunch period length. This process requires an extremely-high-current beam with small energy spread and emittance in addition to a long, precisely manufactured undulator. Therefore, the conventional accelerator-based FELs need a long section of the multistage bunch compressor called a "chicane," which compresses a bunch from an initial bunch length of a few picoseconds to the order of 100 fs to increase the current density of the electron beam up to the order of kiloamperes before being injected into the undulator, whereas the laser plasma accelerator-based FELs would have no need of any bunch compressor. Although the present laser plasma accelerators need further improvements in beam properties such as energy, current, qualities and operating stability, the beam current of 100 kA level (i.e. 100 pC electron charge within 1 fs bunch duration) allows a drastic reduction in the undulator length of several meters to reach the saturation of the FEL amplification. In addition to inherently compact laser and plasma accelerator, a whole FEL system will be operational on a table-top scale. The realization of laser-driven compact table-top X-ray FELs will benefit science and industry over a broad range by providing new tools enabling leading-edge research at small facilities, such as universities and hospitals [46].

An SASE FEL driven by an electron beam with energy $m_e c^2 \gamma$ requires transverse normalized emittance $\varepsilon_n < \gamma\lambda_X$, where λ_X is an FEL wavelength of radiation from the undulator with period λ_u, given by Eq. (110), and K_u is the undulator parameter with the magnetic field strength on the undulator axis B_u, i.e. $K_u = 0.934\lambda_u[\text{cm}]B_u[T]$. For lasing a hard X-ray region $\lambda_X \approx 0.1\,\text{nm}$ (photon energy $E_{\text{photon}} = 12.4\,\text{keV}$) from an undulator of $\lambda_u = 1.5\,\text{cm}$ with a magnetic field of $B_u = 1.425\,\text{T}$ ($K_u = 2.0$) at beam energy $E_b = 7.665\,\text{GeV}$ ($\gamma = 1.5 \times 10^4$), the normalized emittance should be $\varepsilon_n < 1.5\,\mu\text{m}\,\text{rad}$. In addition, it is essential for SASE FELs to inject electron beams with a very high peak current of the order of 100 kA. This requirement imposes a charge of ~200 pC on the laser plasma accelerator design in the case of an accelerated bunch length of 2 fs.

Here we present a design example of a 0.1 nm hard X-ray FEL driven by a 7.665 GeV electron beam delivered from a self-guided laser wakefield accelerator driven by a laser pulse with wavelength $\lambda_L = 0.8\,\mu\text{m}$ and a normalized vector potential of $a_0 = 2$, which has a group velocity correction factor of $\kappa_{\text{ch}} = 1.19$ and the matched spot radius $R_m = 3.2$. The wakefield reduction factor α_c due to a loaded charge $Q_b = 200\,\text{pC}$ is calculated from Eq. (103) for the electron beam radius $k_p\sigma_b = 1$, where the coefficients are $C = 0.8$ and $\alpha_c = 0.72$. The rms transverse size of the electron bunch is set to $\sigma_b = (\bar{\beta}_u\varepsilon_n/\gamma)^{1/2} = 10\,\mu\text{m}$ in the undulator, assuming the average beta function in the undulator $\bar{\beta}_u = 1\,\text{m}$ and the normalized emittance $\varepsilon_n = 1.5\,\mu\text{m}\,\text{rad}$. For a peak current $I_b = 100\,\text{kA}$ with the number of electrons per wavelength $N_{\lambda_X} \sim 2000$, the main laser plasma accelerator and FEL parameters are obtained from Eqs. (104)–(109) and Eqs. (111)–(114), respectively, as shown in Table 2.

Table 2. Parameters for laser plasma accelerator-based hard X-ray FEL light sources.

Drive laser	
Wavelength λ_L	$0.8\,\mu$m
Laser energy per pulse U_L	23 J
Peak power P_L	287 TW
Pulse duration τ_L	80 fs
Laser spot radius	$47\,\mu$m
Laser plasma accelerator	
Electron beam energy E_b	7.665 GeV
Plasma density n_c	$1.3 \times 10^{17}\,\text{cm}^{-3}$
Accelerator length L_{acc}	45 cm
Charge per bunch Q_b	0.2 nC
Bunch duration τ_b	2 fs
Energy spread (rms) σ_γ/γ	~0.3%
Transverse beam size σ_b	$10\,\mu$m
Peak beam current I_b	100 kA
Free electron laser	
Undulator period λ_u (gap g)	15 (3) mm
Radiation wavelength λ_X	0.1 nm
Peak magnetic field B_u	1.425 T
Undulator parameter K_u	2.0
Pierce parameter ρ_{FEL}	0.32%
Gain length L_{gain}	0.43 m
Saturation length L_{sat}	3.8 m
Number of periods	254
Spectral bandwidth $\Delta\lambda_X/\lambda_X$	0.4%
rms radiation cone angle θ_X	$3.6\,\mu$rad
Input power P_{in}	0.51 GW
Saturated power P_{sat}	1.47 TW
Duration of X-ray pulse τ_X	2 fs
Photon flux N_{photon}	$7.4 \times 10^{26}\,\text{s}^{-1}$
Peak brilliance B_{peak} photons/s/mm^2/mrad^{2}0.1% BW	4.7×10^{32}

In the saturation regime, the photons flux of X-ray radiation is

$$N_{\text{photon}} = \frac{P_{\text{sat}}}{eE_{\text{photon}}} \approx \frac{0.6\rho_{\text{FEL}}E_b I_b}{eE_{\text{photon}}} \approx 7.4 \times 10^{26}\,\text{s}^{-1}, \qquad (115)$$

for $E_b = 7.665\,\text{GeV}, E_{\text{photon}} = 12.4\,\text{keV}(\lambda_X = 0.1\,\text{nm}), \rho_{\text{FEL}} = 0.0032$ and $I_b = 100\,\text{kA}$.

$$B_{\text{peak}} = \frac{N_{\text{photon}}\gamma^2}{4\pi^2\varepsilon_n^2\left(\dfrac{10^3\Delta\lambda_X}{\lambda_X}\right)} \approx 4.7 \times 10^{32}\,\text{photons/s/mm}^2\text{/mrad}^2\text{/0.1\% BW}, \qquad (116)$$

where $\Delta\lambda_X/\lambda_X \sim 1/N_u \approx 0.004$ is the spectral bandwidth of X-ray radiation. This peak brilliance is comparable to large-scale X-ray FELs based on the conventional linacs [134].

4.3. *Feasibility study for 100 GeV laser plasma acceleration*

Recently, there has been a growing interest in rapid progress in laser-driven plasma-based accelerators by exploiting petawatt-class lasers, whereby high-quality electron beams can be accelerated to multi-GeV energies in a centimeter-scale plasma thanks to the laser wakefield acceleration mechanism, as reported so far, e.g. 1.8 GeV driven by 130 TW at SIOM [37], 2 GeV driven by 620 TW at TEXAS [35], 3 GeV driven by 210 TW at GIST [36] and 4.2 GeV driven by 230 TW at LBNL-BELLA [38]. Endeavors to accelerate further high-energy electron beams beyond 10 GeV are underway worldwide at large-scale laser and particle accelerator facilities. The BELLA (Berclay Lab Laser Accelerator) project [135] at LBNL is aimed at developing 10 GeV laser wakefield accelerators for high-quality electron beam production in the conventional accelerator paradigm, i.e. a staged accelerator comprising an injector and an accelerator driven by a 1.5 PW laser at 1 Hz. The FACET (Facility for Accelerator science and Experimental Test beams) project [136] at SLAC is aiming to accelerate 40 GeV electron beams by plasma wakefield acceleration driven by 20 GeV high-current electron bunches delivered from the 2 km SLAC linac. The AWAKE (proton-driven plasma wakefield accelerator) project [137] at CERN is planned to produce an approximately GeV-level energy gain of externally injected electron beams by means of a plasma wakefield generated by 450 GeV self-modulated proton bunches from the CERN-SPS proton synchrotron.

The state of the art of PW-class lasers allows us to study the feasibility of laser plasma accelerators toward 100 GeV in a full-scale experiment. The experiments are proposed for implementing the demonstration of a 100 GeV electron beam acceleration by means of a laser plasma accelerator driven with a multi-PW laser capable of delivering 3.5 kJ, 500 fs pulses [62]. This capability allows us to explore laser plasma acceleration operated in the entire laser wakefield regime, from the linear regime to the nonlinear bubble regime. Such a large-scale laser plasma accelerator comprises a gas jet or a short gas cell, which acts as an injector, followed by a long, uniform, low-density plasma or preformed plasma channel (plasma waveguide), which acts as the accelerating medium. Accelerated beams are detected and analyzed using diagnostic system composed of high-resolution spectrometers and beam-imaging detectors.

According to the design formulas (103)–(109), the parameters for the experiment on laser wakefield acceleration of electron beams up to 100 GeV can be designed by taking the normalized vector potential $a_0 = 3$ and the field reduction factor due to the beam loading $\alpha_c = 0.9$ for the multi-PW laser system with a wavelength of 1053 nm and a 500 fs pulse duration, as follows: an operating plasma density of $1.2 \times 10^{16}\,\mathrm{cm}^{-3}$, a dephasing-limited accelerator length of 12 m, a laser-focused spot radius of 110 μm, a laser peak power of 2.1 PW and the required pulse energy of 1 kJ. A design of the neutral gas-filled plasma waveguide for 100 GeV laser plasma accelerators has been presented in Ref. 138.

Injector. Electron beams can be produced and accelerated in the injector stage driven by the same laser pulse as that in the accelerator stage, relying on the self-injection mechanism, such as an expanding bubble self-injection mechanism [18], or the ionization-induced injection mechanism [23–25] with a short mixed gas cell.

Plasma waveguide. As described in Sec. 2.6, preformed plasma density channels for guiding ultraintense short laser pulses are produced by a number of methods, including laser-induced hydrodynamic expansion, and pulsed discharges of an ablative capillary or a gas-filled capillary. However, the length of such a plasma channel has been limited to less than 10 cm and the plasma density has been tuned for $>10^{17}\,\mathrm{cm}^{-3}$. An alternative method may be a hollow dielectric capillary tube filled with neutral gas for guiding intense short laser pulses over several meters. Since the laser is guided by Fresnel reflection at the inner capillary wall, this method relies on neither laser power nor plasma density. Adjusting the capillary tube radius with respect to the laser spot radius can propagate intense laser pulses with a peak intensity of the order of $10^{20}\,\mathrm{W/cm}^2$ over a 100 m scale.

UV-driven LWFA. In Eq. (44), it is obvious that higher a_0, i.e. focused laser intensity I_L and the shorter wavelength λ_L, produce a higher energy gain, which scales as $\propto a_0\lambda_L^{-2}$. For a given a_0, this is attributed to the longer self-guided length, corresponding to the longer dephasing length due to an increase in the critical density at the shorter wavelength. As a result, the operating plasma density at 351 nm can be increased up to $n_e = 1.2 \times 10^{17}\,\mathrm{cm}^{-3}$ for $a_0 = 4$ and $n_e = 2.4 \times 10^{17}\,\mathrm{cm}^{-3}$ for $a_0 = 8$, respectively, compared to $n_e = 1.4 \times 10^{16}\,\mathrm{cm}^{-3}$ for $a_0 = 4$ at 1053 nm. This increase indicates a decrease in the threshold of electron self-injection into laser wakefields, as well as a decrease in the critical power for self-guiding.

Table 3. Design parameters for 40 and 100 GeV self-guided LPAs in comparison with the results of 3D PIC simulations in Ref. 139.

Case	A	Ref. 139	B	C
E_b [GeV]	40	38	100	100
n_e [$10^{16}\,\mathrm{cm}^{-3}$]	3.2	2.2	1.2	51
L_{dp} [m]	4	5	12	1.2
λ_L [nm]	800	800	1053	351
a_0	2	2	3	8
r_0 [μm]	95	100	110	9.3
τ_L [fs]	224	160	500	225
P_L [PW]	1.2	1.4	2.1	0.95
U_L [kJ]	0.26	0.22	1.03	0.21
Q_b [pC]	127	300	250	64

Table 3 shows the design parameters of 40 and 100 GeV laser plasma accelerators in comparison with the results of 3D PIC simulations from the Lorentz-boosted frame OSIRIS code [139] at $\lambda_L = 800$ nm. The design parameters for 40 GeV laser plasma accelerators are approximately in good agreement with the Lorentz-boosted frame PIC simulation.

4.4. *Perspective on high-energy frontier colliders*

A novel concept of laser acceleration exploiting superhigh electromagnetic fields of lasers is motivated by a long-standing feeling that state-of-the-art high-energy accelerators become too large and costly, and possibly they approach the end of the road, as emblematically reminded by the demise of the Superconducting Super Collider (SSC) [140]. As a consequence, the application to high-energy frontier colliders may be one of the supreme goals for laser plasma accelerators, as illustrated with recent vibrant R&D projects, such as BELLA [135], FACET [136] and AWAKE [137], to be aimed at laser-plasma-based colliders in the TeV-range energies. In the conventional paradigm of high-energy particle accelerators, an entire accelerator structure of electron–positron colliders with the center-of-mass energy of TeV energies will constitute multistage laser wakefield accelerators, each of which is operated in the quasilinear wakefield regime rather than the nonlinear wakefield regime, i.e. the bubble regime operated at high normalized laser fields of $a_0 > 4$ [56], taking into account the capability of accelerating both electrons and positrons in a transverse focusing wakefield [55]. From the viewpoint of the critical requirements for the luminosity and operational power in the electron–positron collider, the operating plasma density n_e, which is the most crucial parameter of laser plasma accelerators, must be deliberately designed, based on the following considerations [56].

For the electron–positron linear collider, a critical requirement is the event rate, which is determined by the product of collision cross-section $\sigma(e^-e^+ \to e^-e^+) \propto E_b^{-2}$ and luminosity $L = f_c N_b^2/4\pi\sigma_x\sigma_y$, where f_c is the collision frequency, N_b is the number of particles per bunch, and σ_x and σ_y are the horizontal and vertical rms beam sizes at the collision point, respectively. The luminosity requirement for future TeV-range colliders with beam energy E_b TeV, collision beam area $\sigma_x\sigma_y\,\mathrm{nm}^2$ and number of particles per bunch $N_b\,[10^9]$ is approximately scaled as $L\,[10^{34}\,\mathrm{cm}^{-2}\mathrm{s}^{-1}] \approx 4E_b^2$ [55]. The required collision frequency is obtained as $f_c\,[\mathrm{kHz}] \simeq 5\sigma_x\sigma_y E_b^2/N_b^2$, and the beam power is given by $P_b\,[\mathrm{MW}] = f_c N_b E_b \approx 0.8E_b^3\sigma_x\sigma_y/N_b$. The number of particles per bunch is $N_b = Q_b/e \approx 1.49\eta_b(1-\eta_b)^{-1} \times (k_p\sigma_{x0})^2(E_z/E_0)n_{17}^{-1/2}$, where Q_b is a bunch charge, $\eta_b = 1-(E_z/E_M)^2$ the beam loading efficiency which is an energy fraction of the plasma wave with the beam-loaded field E_z and the maximum electric field E_M absorbed by particles of the bunch with a radius $r_b = \sqrt{2}\sigma_{x0}$, and $E_0[\mathrm{GV/m}] \simeq 30n_{17}^{1/2}$ the nonrelativistic wave-breaking amplitude of the plasma wave at $n_e = n_{17}\ [10^{17}\,\mathrm{cm}^{-3}]$. Thus, the required beam power is calculated as

$$P_b[\mathrm{MW}] \approx 0.54E_b^3\left(\frac{1-\eta_b}{\eta_b}\right)\frac{\sigma_x\sigma_y n_{17}^{1/2}}{k_p^2\sigma_{x0}^2}\left(\frac{E_0}{E_z}\right). \tag{117}$$

The average laser power per stage is $P_{\mathrm{avg}} = f_c U_L$ and the total wall plug power for the collider is $P_{\mathrm{wall}} = 2N_{\mathrm{stage}}P_{\mathrm{avg}}/\eta_L = 2f_c U_L N_{\mathrm{stage}}/\eta_L$, where U_L is the laser energy per stage, η_L the efficiency from the wall plug to the laser, and N_{stage} the number of stages per beam. For a typical quasilinear wakefield driven by a laser pulse with $a_0 = 1.4$ and pulse length $k_p\sigma_L = \sqrt{2}$, the average laser power per stage is given by

$$P_{\mathrm{avg}}[\mathrm{kW}] \approx 3.55E_b^2\left(\frac{1-\eta_b}{\eta_b}\right)^2\frac{(k_p r_L)^2}{(k_p\sigma_{x0})^4} \times \left(\frac{E_0}{E_z}\right)^2\frac{\sigma_x\sigma_y}{\lambda_L^2 n_{17}^{1/2}}, \tag{118}$$

where r_L is a spot radius of the laser with a wavelength of $\lambda_L\,\mu$m and the wall plug power yields

$$P_{\mathrm{wall}}[\mathrm{MW}] \approx \frac{0.78n_{17}^{1/2}E_b^3}{\eta_L} \times \left(\frac{1-\eta_b}{\eta_b}\right)^2\frac{\sigma_x\sigma_y(k_p r_L)^2}{(k_p\sigma_{x0})^4}\left(\frac{E_0}{E_z}\right)^3. \tag{119}$$

Since the matched beam radius scales as $\sigma_{x0} \propto n_e^{-1/2}$, the number of particles per bunch scales as $N_b \propto n_e^{-1/2}$ and the average laser power per stage scales as $P_{\mathrm{avg}} \propto n_e^{-1/2}$, the wall plug power results in $P_{\mathrm{wall}} \propto n_e^{1/2}$. The overall efficiency from the wall plug to the beam is given by

$$\eta_{\mathrm{overall}} = \frac{2P_b}{P_{\mathrm{wall}}} \approx 1.4\frac{\eta_L\eta_b}{1-\eta_b}\left(\frac{\sigma_{x0}}{r_L}\right)^2\left(\frac{E_z}{E_0}\right)^2. \tag{120}$$

Considering constraint on the operational cost of the future linear colliders that limits the wall plug power to a few 100 MW, the low operating plasma density in the range of 10^{15}–$10^{16}\,\mathrm{cm}^{-3}$ works in favor of the multi-TeV linear collider. Many of the underlying parameters for the design of laser plasma colliders

Table 4. Scaling dependence of LPA parameters on the operating plasma density.

Accelerating field E_z	$\propto n_e^{1/2}$
Stage length L_{stage}	$\propto n_e^{-3/2}$
Energy gain per stage W_{stage}	$\propto n_e^{-1}$
Number of stages N_{stage}	$\propto n_e$
Total linac length L_{total}	$\propto n_e^{-1/2}$
Number of particles per bunch N_b	$\propto n_e^{-1/2}$
Laser pulse duration τ_L	$\propto n_e^{-1/2}$
Laser peak power P_L	$\propto n_e^{-1}$
Laser energy per stage U_L	$\propto n_e^{-3/2}$
Synchrotron radiation loss $\Delta\gamma$	$\propto n_e^{1/2}$
Radiative energy spread σ_γ/γ	$\propto n_e^{1/2}$
Initial normalized emittance ε_{n0}	$\propto n_e^{-1/2}$
Collision frequency f_c	$\propto n_e$
Beam power P_b	$\propto n_e^{1/2}$
Average laser power P_{avg}	$\propto n_e^{-1/2}$
Wall plug power P_{wall}	$\propto n_e^{1/2}$

scale with respect to the operating plasma density, as shown in Table 4 [56].

5. Conclusion

Here we have reviewed electron acceleration from the viewpoint of the interaction between strong laser fields and electrons in underdense plasma, where underlying physics is often referred to as laser plasma acceleration. To date, a breakthrough concept of laser electron acceleration [2] has led to tremendous progress in laser-driven plasma-based accelerators combined with ultraintense lasers that have emerged from the chirped pulse amplification technique [141]. More specifically, in this decade, achievements in laser plasma accelerators delivering high-energy, high-quality electron beams and cogenerated betatron X/α-rays from plasma on a table-top scale have inspired various applications in the broad range of sciences and technologies, as well as an original concept for overcoming the limits of conventional accelerator technologies aimed at high-energy frontier colliders [3]. In addition to research on advanced accelerator technologies, laser plasma acceleration includes a fruitful physics in plasma wave generation, wave-wave/wave-particle interactions, particle dynamics and photon radiation, all phenomena of which intrinsically occur in the relativistic nonlinear regime of both plasma and electron dynamics. In this sense, it can be said that heuristic breakthroughs in the experimental research and/or computational research have taken a crucial role of making great advances in understanding the physics and developing the technologies of laser plasma acceleration. These findings will lead to the maturation of advanced accelerator technologies for practical applications and the emergence of a new scientific field in the near future.

Acknowledgments

The author would like to thank Prof. T. Tajima for his valuable suggestions and continual encouragement on all aspects of the author's work on laser plasma acceleration. He would also like to thank Prof. G. Mourou of IZEST, Ecole Polytechnique, for his support and encouragement regarding the 100 GeV ascent and EUV-FEL work. In addition, the author is grateful to Prof. C. H. Nam of the Center for Relativistic Laser Science, Institute for Basic Science, and Profs. R. X. Li, J. S. Liu and B. F. Shen of the Shanghai Institute of Optics and Fine Mechanics, CAS, for the collaboration on PW laser-driven electron acceleration. The author was supported by the Project Code (IBS-R012-D1).

References

[1] T. Tajima, D. Habs and X. Q. Yan, Laser acceleration of ions for radiation therapy, in *Review of Accelerator Science and Technology*, eds. A. W. Chao and W. Chou (World Scientific, Singapore, 2009), Vol. 2, pp. 201–208.

[2] T. Tajima and J. M. Dawson, *Phys. Rev. Lett.* **43**, 267 (1979).

[3] P. Channel, The brief report on the laser acceleration of particles workshop, *AIP Conf. Proc.* **91**, 1–27 (1982).

[4] S. Myers, *Int. J. Mod. Phys. A* **28**, 1330035 (2013).

[5] T. Behnke, J. E. Brau, B. Foster, J. Fuster, M. Harrison, J. M. Paterson, M. Peskin, M. Stanitzki, N. Walker and H. Yamamoto, The International Linear Collider Accelerator, in *The International Linear Collider Technical Design Report, Vol.* 1, *Executive Summary. ILC International Linear Collider*, pp. 9–26.

[6] K. Shimoda, *Appl. Opt.* **1**, 33 (1962).

[7] K. Nakajima, D. Fisher, T. Kawakubo, H. Nakanishi, A. Ogata, Y. Kato, Y. Kitagawa, R. Kodama, K. Mima, H. Shiraga, K. Suzuki, K. Yamakawa, T. Zhang, Y. Sakawa, T. Shoji, Y. Nishida, N. Yugami, M. Downer and T. Tajima, *Phys. Rev. Lett.* **74**, 4428 (1995).

[8] A. Modena, Z. Najmudin, A. E. Dangor, C. E. Clayton, K. A. Marsh, C. Joshi, V. Malka, C. B. Darrow, C. Danson, D. Neely and F. N. Walsh, *Nature* **377**, 606 (1995).

[9] N. E. Andreev, L. M. Gorbunov, V. I. Kirsanov, A. A. Pogosova and R. R. Ramazashvili, *Sov. Phys. JETP* **55**, 571 (1992).

[10] Z. Najmudin, K. Krushelnick, E. L. Clark, S. P. D. Mangles, B. Walton, A. E. Dangor, S. Fritzler, V. Malka, E. Lefebvre, D. Gordon, F. S. Tsung and C. Joshi, *Phys. Plas.* **10**, 2071 (2003).

[11] V. Malka, S. Fritzler, E. Lefebvre, M.-M. Aleonard, F. Burgy, J.-P. Chambaret, J.-F. Chemin, K. Krushelnick, G. Malka, S. P. D. Mangles, Z. Najmudin, M. Pittman, J.-P. Rousseau, J.-N. Scheurer, B. Walton and A. E. Dangor, *Science* **298**, 1596 (2002).

[12] S. P. D. Mangles, C. D. Murphy, Z. Najmudin, A. G. R. Thomas, J. L. Collier, A. E. Dangor, E. J. Divall, P. S. Foster, J. G. Gallacher, C. J. Hooker, D. A. Jaroszynski, A. J. Langley, W. B. Mori, P. A. Norreys, F. S. Tsung, R. B. Viskup, R. Walton and K. Krushelnick, *Nature* **431**, 535 (2004).

[13] C. G. R. Geddes, Cs. Toth, J. van Tilborg, E. Esarey, C. B. Schroeder, D. Bruhwiler, C. Nieter, J. Cary and W. P. Leemans, *Nature* **431**, 538 (2004).
[14] J. Faure, Y. Glinec, A. Pukhov, S. Kiselev, S. Gordienko, E. Lefebvre, F. Rousseau, J.-P. Burgy and V. Malka, *Nature* **431**, 541 (2004).
[15] A. Pukhov and J. Meyer-ter-vehn, *Appl. Phys. B* **74**, 355 (2002).
[16] I. Kostyukov, A. Pukhov and S. Kiselev, *Phys. Plas.* **11**, 5256 (2004).
[17] W. Lu, M. Tzoufras, C. Joshi, F. S. Tsung, W. B. Mori, J. Vieira, R. A. Fonseca and L. O. Silva, *Phys. Rev. ST Accel. Beams* **10**, 061301 (2007).
[18] S. Kalmykov, S. A. Yi, V. Khudik and G. Shvets, *Phys. Rev. Lett.* **103**, 135004 (2009).
[19] J. Faure, C. Rechatin, A. Norlin, A. Lifschitz, Y. Glinec and V. Malka, *Nature* **444**, 737 (2006).
[20] H. Kotaki, I. Daito, M. Kando, Y. Hayashi, K. Kawase, T. Kameshima, Y. Fukuda, T. Homma, J. Ma, L. M. Chen, T. Z. Esirkepov, A. S. Pirozhkov, J. K. Koga, A. Faenov, T. Pikuz, H. Kiriyama, H. Okada, T. Shimomura, Y. Nakai, M. Tanoue, H. Sasao, D. Wakai, H. Matsuura, S. Kondo, S. Kanazawa, A. Sugiyama, H. Daido and S. V. Bulanov, *Phys. Rev. Lett.* **103**, 194803 (2009).
[21] K. Schmid, A. Buck, C. M. S. Sears, J. M. Mikhailova, R. Tautz, D. Herrmann, M. Geissler, F. Krausz and L. Veisz, *Phys. Rev. ST Accel. Beams* **13**, 091301 (2010).
[22] A. J. Gonsalves, K. Nakamura, C. Lin, D. Panasenko, S. Shiraishi, T. Sokollik, C. Benedetti, C. B. Schroeder, C. G. R. Geddes, J. van Tilborg, J. Osterhoff, E. Esarey, C. Toth and W. P. Leemans, *Nat. Phys.* **7**, 862 (2011).
[23] A. Pak, K. A. Marsh, S. F. Martins, W. Lu, W. B. Mori and C. Joshi, *Phys. Rev. Lett.* **104**, 025003 (2010).
[24] C. McGuffey, A. G. R. Thomas, W. Schumaker, T. Matsuoka, V. Chvykov, F. J. Dollar, G. Kalintchenko, V. Yanovsky, A. Maksimchuk, K. Krushelnick, V. Y. Bychenkov, I. V. Glazyrin and A. V. Karpeev, *Phys. Rev. Lett.* **104**, 025004 (2010).
[25] C. Xia, J. Liu, W. Wang, H. Lu, W. Cheng, A. Deng, W. Li, H. Zhang, X. Liang, Y. Leng, X. Lu, C. Wang, J. Wang, K. Nakajima, R. Li and Z. Xu, *Phys. Plas.* **18**, 113101 (2011).
[26] J. S. Liu, C. Q. Xia, W. T. Wang, H. Y. Lu, C. Wang, A. H. Deng, W. T. Li, H. Zhang, X. Y. Liang, Y. X. Leng, X. M. Lu, C. Wang, J. Z. Wang, K. Nakajima, R. X. Li and Z. Z. Xu, *Phys. Rev. Lett.* **107**, 035001 (2011).
[27] B. B. Pollock, C. E. Clayton, J. E. Ralph, F. Albert, A. Davidson, L. Divol, C. Filip, S. H. Glenzer, K. Herpoldt, W. Lu, K. A. Marsh, J. Meinecke, W. B. Mori, A. Pak, T. C. Rensink, J. S. Ross, J. Shaw, G. R. Tynan, C. Joshi and D. H. Froula, *Phys. Rev. Lett.* **107**, 045001 (2011).
[28] Y. Ehrlich, C. Cohen, A. Zigler, J. Krall, P. Sprangle and E. Esarey, *Phys. Rev. Lett.* **77**, 4186 (1996).
[29] N. M. Naumova, J. Koga, K. Nakajima, T. Tajima, T. Zh. Esirkepov, S. V. Bulanov and F. Pegoraro, *Phys. Plas.* **8**, 4149 (2001).
[30] L. M. Chen, H. Kotaki, K. Nakajima, J. Koga, S. V. Bulanov, T. Tajima, Y. Q. Gu, H. S. Peng, X. X. Wang, T. S. Wen, H. J. Liu, C. Y. Jiao, C. G. Zhang, X. J. Huang, Y. Guo, K. N. Zhou, J. F. Hua, W. M. An, C. X. Tang and Y. Z. Lin, *Phys. Plas.* **14**, 040703 (2007).
[31] W. P. Leemans, B. Nagler, A. J. Gonsalves, C. Toth, K. Nakamura, C. G. R. Geddes, E. Esarey, C. B. Schroeder and S. M. Hooker, *Nat. Phys.* **2**, 696 (2006).
[32] S. Karsch, J. Osterhoff, A. Popp, T. P. Rowlands-Rees, Z. Major, M. Fuchs, B. Marx, R. Hörlein, K. Schmid, L. Veisz, S. Becker, U. Schramm, B. Hidding, G. Pretzler, D. Habs, F. Grüner, F. Krausz and S. M. Hooker, *New J. Phys.* **9**, 415 (2007).
[33] T. Kameshima, W. Hong, K. Sugiyama, X. Wen, Y. Wu, C. Tang, Q. Zhu, Y. Gu, B. Zhang, H. Peng, S.-I. Kurokawa, L. Chen, T. Tajima, T. Kumita and K. Nakajima, *Appl. Phys. Exp.* **1**, 066001 (2008).
[34] C. E. Clayton, J. E. Ralph, F. Albert, R. A. Fonseca, S. H. Glenzer, C. Joshi, W. Lu, K. A. Marsh, S. F. Martins, W. B. Mori, A. Pak, F. S. Tsung, B. B. Pollock, J. S. Ross, L. O. Silva and D. H. Froula, *Phys. Rev. Lett.* **105**, 105003 (2010).
[35] X. Wang, R. Zgadzaj, N. Fazel, Z. Li, S. A. Yi, X. Zhang, W. Henderson, Y. Y. Chang, R. Korzekwa, H. E. Tsai, C. H. Pai, H. Quevedo, G. Dyer, E. Gaul, M. Martinez, A. C. Bernstein, T. Borger, M. Spinks, M. Donovan, V. Khudik, G. Shvets, T. Ditmire and M. C. Downer, *Nat. Comms.* **4**, 1988 (2013).
[36] H. T. Kim, K. H. Pae, H. J. Cha, I. J. Kim, T. J. Yu, J. H. Sung, S. K. Lee, T. M. Jeong and J. Lee, *Phys. Rev. Lett.* **111**, 165002 (2013).
[37] H. Lu, M. Liu, W. Wang, C. Wang, J. Liu, A. Deng, J. Xu, C. Xia, W. Li, H. Zhang, X. Lu, C. Wang, J. Wang, X. Liang, Y. Leng, B. Shen, K. Nakajima, R. Li and Z. Xu, *Appl. Phys. Lett.* **99**, 091502 (2011).
[38] W. P. Leemans, A. J. Gonsalves, H.-S. Mao, K. Nakamura, C. Benedetti, C. B. Schroeder, Cs. Tóth, J. Daniels, D. E. Mittelberger, S. S. Bulanov, J.-L. Vay, C. G. R. Geddes and E. Esarey, *Phys. Rev. Lett.* **113**, 245002 (2014).
[39] Z.-H. He, A. G. R. Thomas, B. Beaurepaire, J. A. Nees, B. Hou, V. Malka, K. Krushelnick and J. Faure, *Appl. Phys. Lett.* **102**, 064104 (2013).
[40] K. Nakajima, J. J. Yuan, L. M. Chen and Z. M. Sheng, *Appl. Sci.* **5**, 1 (2015).

[41] S. Kneip, C. McGuffey, J. L. Martins, S. F. Martins, C. Bellei, V. Chvykov, F. Dollar, R. Fonseca, C. Huntington, G. Kalintchenko, A. Maksimchuk, S. P. D. Mangles, T. Matsuoka, S. R. Nagel, C. A. J. Palmer, J. Schreiber, K. Ta Phuoc, A. G. R. Thomas, V. Yanovsky, L. O. Silva, K. Krushelnick and Z. Najmudin, *Nat. Phys.* **6**, 980 (2010).

[42] S. Fourmaux, S. Corde, K. Ta Phuoc, P. Lassonde, G. Lebrun, S. Payeur, F. Martin, S. Sebban, V. Malka, A. Rousse and J. C. Kieffer, *Opt. Lett.* **36**, 2426 (2011).

[43] M. Z. Mo, Z. Chen, S. Fourmaux, A. Saraf, K. Otani, J. C. Kieffer, Y. Y. Tsui, A. Ng and R. Fedosejevs, *Rev. Sci. Instrum.* **84**, 123106 (2013).

[44] K. Nakajima, M. Kando, T. Kawakubo, T. Nakanishi and A. Ogata, *Nucl. Instrum. Methods A* **375**, 593 (1996).

[45] F. Grüner, S. Becker, Schramm, T. Eichner, M. Fuchs, R. Weingartner, D. Habs, J. Meyer-ter-Vehn, M. Geissler, M. Ferrario, L. Serafini, B. Van der Geer, H. Backe, W. Lauth and S. Reiche, *Appl. Phys. B* **86**, 431 (2007).

[46] K. Nakajima, *Nat. Phys.* **4**, 92 (2008).

[47] K. Nakajima, A. H. Deng, H. Yoshitama, N. A. M. Hafz, H. Y. Lu, B. F. Shen, J. S. Liu, R. X. Li and Z. Z. Xu, Laser-driven table-top X-ray FEL, in *Free Electron Laser*, ed. S. Varró (InTech, Rijeka, 2012), Chap. 5, pp. 119–150.

[48] K. Nakajima, *High Power Laser Sci. Eng.* **2**, e31 (2014).

[49] S. Cipiccia, M. R. Islam, B. Ersfeld, R. P. Shanks, E. Brunetti, G. Vieux, X. Yang, R. C. Issac, S. M. Wiggins, G. H. Welsh, M.-P. Anania, D. Maneuski, R. Montgomery, G. Smith, M. Hoek, D. J. Hamilton, N. R. C. Lemos, D. Symes, P. P. Rajeev, V. O. Shea, J. M. Dias and D. A. Jaroszynski, *Nat. Phys.* **7**, 867 (2011).

[50] J. H. Jeon, K. Nakajima, H. T. Kim, Y. J. Rhee, V. B. Pathak, M. H. Cho, J. H. Shin, B. J. Yoo, C. Hojbota, S. H. Jo, K. W. Shin, J. H. Sung, S. K. Lee, B. I. Cho, I. W. Choi and C. H. Nam, *Rev. Sci. Instrum.* **86**, 123116 (2015).

[51] S. Chen, N. D. Powers, I. Ghebregziabher, C. M. Maharjan, C. Liu, G. Golovin, S. Banerjee, J. Zhang, N. Cunningham, A. Moorti, S. Clarke, S. Pozzi and D. P. Umstadter, *Phys. Rev. Lett.* **110**, 155003 (2013).

[52] G. Sarri, D. J. Corvan, W. Schumaker, J. M. Cole, A. Di Piazza, H. Ahmed, C. Harvey, C. H. Keitel, K. Krushelnick, S. P. D. Mangles, Z. Najmudin, D. Symes, A. G. R. Thomas, M. Yeung, Z. Zhao and M. Zepf, *Phys. Rev. Lett.* **113**, 224801 (2014).

[53] K. Nakajima, *Proc. Jpn. Acad. Ser. B* **91**, 223 (2015).

[54] K. Homma, K. Matsuura and K. Nakajima, *Prog. Theor. Exp. Phys.* **2016**, 013C01 (2016).

[55] C. B. Schroeder, E. Esarey, C. G. R. Geddes, C. Benedetti and W. P. Leemans, *Phys. Rev. ST Accel. Beams* **13**, 101301 (2010).

[56] K. Nakajima, A. Deng, X. Zhang, B. Shen, J. Liu, R. Li, Z. Xu, T. Ostermayr, S. Petrovics, C. Klier, K. Iqbal, H. Ruhl and T. Tajima, *Phys. Rev. ST Accel. Beams* **14**, 091301 (2011).

[57] J. M. Dawson, *Phys. Rev.* **113**, 383 (1959).

[58] C. Chiu, S. Cheshkov and T. Tajima, *Phys. Rev. ST Accel. Beams* **3**, 101301 (2000).

[59] M. I. K. Santala, Z. Najmudin, E. L. Clark, M. Tatarakis, K. Krushelnick, A. E. Dangor, V. Malka, J. Faure, R. Allott and R. J. Clarke, *Phys. Rev. Lett.* **86**, 1227 (2001).

[60] S. P. D. Mangles, B. R. Walton, M. Tzoufras, Z. Najmudin, R. J. Clarke, A. E. Dangor, R. G. Evans, S. Fritzler, A. Gopal, C. Hernandez-Gomez, W. B. Mori, W. Rozmus, M. Tatarakis, A. G. R. Thomas, F. S. Tsung, M. S. Wei and K. Krushelnick, *Phys. Rev. Lett.* **94**, 245001 (2005).

[61] W. Lu, C. Huang, M. Zhou, W. B. Mori and T. Katsouleas, *Phys. Rev. Lett.* **96**, 165002 (2006).

[62] K. Nakajima, H. Y. Lu, X. Y. Zhao, B. F. Shen, R. X. Li and Z. Z. Xu, *Chin. Opt. Lett.* **11**, 013501 (2013).

[63] S. P. D. Mangles, A. G. R. Thomas, M. C. Kaluza, O. Lundh, F. Lindau, A. Persson, F. S. Tsung, Z. Najmudin, W. B. Mori, C.-G. Wahlström and K. Krushelnick, *Phys. Rev. Lett.* **96**, 215001 (2006).

[64] M. Kando, H. Ahn, H. Dewa, H. Kotaki, T. Ueda, M. Uesaka, T. Watanabe, H. Nakanishi, A. Ogata and K. Nakajima, *J. Appl. Phys.* **38**, L967 (1999).

[65] A. G. Khachatryan, F. A. vanGoor, K.-J. Boller, A. J. W. Reitsma and D. A. Jaroszynski, *Phys. Rev. ST Accel. Beams* **7**, 121301 (2004).

[66] E. Esarey and M. Pilloff, *Phys. Plas.* **2**, 1432 (1995).

[67] I. Kostyukov, E. Nerush, A. Pukhov and V. Seredov, *Phys. Rev. Lett.* **103**, 175003 (2009).

[68] S. P. D. Mangles, A. G. R. Thomas, O. Lundh, F. Lindau, M. C. Kaluzac, A. Persson, C.-G. Wahlström, K. Krushelnickd and Z. Najmudin, *Phys. Plas.* **14**, 056702 (2007).

[69] D. H. Froula, C. E. Clayton, T. Döppner, K. A. Marsh, C. P. J. Barty, L. Divol, R. A. Fonseca, S. H. Glenzer, C. Joshi, W. Lu, S. F. Martins, P. Michel, W. B. Mori, J. P. Palastro, B. B. Pollock, A. Pak, J. E. Ralph, J. S. Ross, C. W. Siders, L. O. Silva and T. Wang, *Phys. Rev. Lett.* **103**, 215006 (2009).

[70] D. Umstadter, J. K. Kim and E. Dodd, *Phys. Rev. Lett.* **76**, 2073 (1996).

[71] W.-M. Wang, Z.-M. Sheng and J. Zhang, *Appl. Phys. Lett.* **93**, 201502 (2008).

[72] H. Kotaki, S. Masuda, M. Kando, J. K. Koga and K. Nakajima, *Phys. Plas.* **11**, 3296 (2004).
[73] C. B. Schroeder, P. B. Lee, J. S. Wurtele, E. Esarey and W. P. Leemans, *Phys. Rev. E* **59**, 6037 (1999).
[74] F. Y. Li, Z. M. Sheng, Y. Liu, J. Meyer-ter-Vehn, W. B. Mori, W. Lu and J. Zhang, *Phys. Rev. Lett.* **110**, 135002 (2013).
[75] C. G. R. Geddes, K. Nakamura, G. R. Plateau, Cs. Toth, E. Cormier-Michel, E. Esarey, C. B. Schroeder, J. R. Cary and W. P. Leemans, *Phys. Rev. Lett.* **100**, 215004 (2008).
[76] J. Faure, C. Rechatin, O. Lundh, L. Ammoura and V. Malka, *Phys. Plas.* **17**, 083107 (2010).
[77] M. Chen, E. Esarey, C. B. Schroeder, C. G. R. Geddes and W. P. Leemans, *Phys. Plas.* **19**, 033101 (2012).
[78] B. Hafizi, A. Ting, P. Sprangle and R. F. Hubbard, *Phys. Rev. E* **62**, 4120 (2000).
[79] B. Hafizi, A. Ting, R. F. Hubbard, P. Sprangle and J. R. Penano, *Phys. Plas.* **10**, 1483 (2003).
[80] E. Esarey and W. P. Leemans, *Phys. Rev. E* **59**, 1082 (1999).
[81] C. B. Schroeder, C. Benedetti, E. Esarey, J. van Tilborg and W. P. Leemans, *Phys. Plas.* **18**, 083103 (2011).
[82] M. Liu, B. Zhou, Y. Yi, X. Liu and L. Tang, *Phys. Plas.* **14**, 103104 (2007).
[83] M. Liu, R. Li, Z. Xu and C.-J. Kim, *Phys. Lett. A* **373**, 363 (2009).
[84] P. Sprangle, E. Esarey and A. Ting, *Phys. Rev. Lett.* **64**, 2011 (1990).
[85] J. E. Ralph, K. A. Marsh, A. E. Pak, W. Lu, C. E. Clayton, F. Fang, W. B. Mori and C. Joshi, *Phys. Rev. Lett.* **102**, 175003 (2009).
[86] A. G. R. Thomas, Z. Najmudin, S. P. D. Mangles, C. D. Murphy, A. E. Dangor, C. Kamperidis, K. L. Lancaster, W. B. Mori, P. A. Norreys, W. Rozmus and K. Krushelnick, *Phys. Rev. Lett.* **98**, 095004 (2007).
[87] H. Yoshitama, T. Kameshima, Y.-Q. Gu, Y. Guo, C.-Y. Jiao, H.-J. Liu, H.-S. Peng, C.-M. Tang, X.-D. Wang, X.-L. Wen, T.-S. Wen, Y.-C. Wu, B.-H. Zhan, Q.-H. Zhu, X.-J. Huang, W.-M. An, W.-H. Hung, C.-X. Tang, Y.-Z. Lin, X.-D. Wang, L.-M. Chen, H. Kotaki, M. Kando and K. Nakajima, *Chin. Phys. Lett.* **25**, 2938 (2008).
[88] N. A. M. Hafz, T. M. Jeong, I. W. Choi, S. K. Lee, K. H. Pae, V. V. Kulagin, J. H. Sung, T. J. Yu, K.-H. Hong, T. Hosokai, J. R. Cary, D.-K. Ko and J. Lee, *Nat. Photon.* **2**, 571 (2008).
[89] S. Kneip, S. R. Nagel, S. F. Martins, S. P. D. Mangles, C. Bellei, O. Chekhlov, R. J. Clarke, N. Delerue, E. J. Divall, G. Doucas, K. Ertel, F. Fiuza, R. Fonseca, P. Foster, S. J. Hawkes, C. J. Hooker, K. Krushelnick, W. B. Mori, C. A. J. Palmer, K. Ta Phuoc, P. P. Rajeev, J. Schreiber, M. J. V. Streeter, D. Urner, J. Vieira, L. O. Silva and Z. Najmudin, *Phys. Rev. Lett.* **103**, 035002 (2009).
[90] C. G. Durfee III and H. M. Milchberg, *Phys. Rev. Lett.* **71**, 2409 (1993).
[91] P. Volfbeyn, E. Esarey and W. P. Leemans, *Phys. Plas.* **6**, 2269 (1999).
[92] Y. F. Xiao, H. H. Chu, H. E. Tsai, C. H. Lee, J. Y. Lin, J. Wang and S. Y. Chen, *Phys. Plas.* **11**, L21 (2004).
[93] M. Levin, A. Pukhov, A. Zigler, K. Sugiyama, K. Nakajima, R. F. Hubbard, A. Ting, D. F. Gordon, P. Sprangle and D. Kaganovich, *Phys. Plas.* **13**, 083108 (2006).
[94] T. Kameshima, H. Kotaki, M. Kando, I. Daito, K. Kawase, Y. Fukuda, L. M. Chen, T. Homma, S. Kondo, T. Zh. Esirkepov, N. A. Bobrova, P. V. Sasorov and S. V. Bulanov, *Phys. Plas.* **16**, 093101 (2009).
[95] M. W. Liu, A. H. Deng, J. S. Liu, R. X. Li, J. C. Xu, C. Q. Xia, C. Wang, B. F. Shen, Z. Z. Xu and K. Nakajima, *Rev. Sci. Instrum.* **81**, 036107 (2010).
[96] A. H. Deng, M. W. Liu, J. S. Liu, X. M. Lu, C. Q. Xia, J. C. Xu, C. Wang, B. F. Shen, R. X. Li, Z. Z. Xu and K. Nakajima, *Plas. Sci. Technol.* **13**, 362 (2011).
[97] T. Hosokai, M. Kando, H. Dewa, H. Kotaki, S. Kondo, N. Hasegawa, K. Nakajima and K. Horioka, *Opt. Lett.* **25**, 10 (2000).
[98] D. J. Spence, A. Butler and S. M. Hooker, *J. Phys. B* **34**, 4103 (2001).
[99] A. Butler, D. J. Spence and S. M. Hooker, *Phys. Rev. Lett.* **89**, 185003 (2002).
[100] A. J. Gonsalves, T. P. Rowlands-Rees, B. H. P. Broks, J. J. A. M. van der Mullen and S. M. Hooker, *Phys. Rev. Lett.* **98**, 025002 (2007).
[101] D. G. Jang, M. S. Kim, I. H. Nam, H. S. Uhm and H. Suk, *Appl. Phys. Lett.* **99**, 141502 (2011).
[102] S. M. Wiggins, M. P. Reijnders, S. Abuazoum, K. Hart, G. Vieux, G. H. Welsh, R. C. Issac, X. Yang, D. R. Jones and D. A. Jaroszynski, *J. Plas. Phys.* **78**, 355 (2012).
[103] N. A. Bobrova, A. A. Esaulov, J.-I. Sakai, P. V. Sasorov, D. J. Spence, A. Butler, S. M. Hooker and S. V. Bulanov, *Phys. Rev. E* **65**, 016407 (2001).
[104] B. H. P. Broks, K. Garloff and J. J. A. M. van der Mullen, *Phys. Rev. E* **71**, 016401 (2005).
[105] T. P. Rowlands-Rees, C. Kamperidis, S. Kneip, A. J. Gonsalves, S. P. D. Mangles, J. G. Gallacher, E. Brunetti, T. Ibbotson, C. D. Murphy, P. S. Foster, M. J. V. Streeter, F. Budde, P. A. Norreys, D. A. Jaroszynski, K. Krushelnick, Z. Najmudin and S. M. Hooker, *Phys. Rev. Lett.* **100**, 105005 (2008).
[106] T. P. A. Ibbotson, N. Bourgeois, T. P. Rowlands-Rees, L. S. Caballero, S. I. Bajlekov, P. A. Walker, S. Kneip, S. P. D. Mangles, S. R. Nagel, C. A. J. Palmer, N. Delerue, G. Doucas, D. Urner, O. Chekhlov, R. J. Clarke, E. Divall, K. Ertel,

P. S. Foster, S. J. Hawkes, C. J. Hooker, B. Parry, P. P. Rajeev, M. J. V. Streeter and S. M. Hooker, *Phys. Rev. ST Accel. Beams* **13**, 031301 (2010).
[107] C. McGuffey, M. Levin, T. Matsuoka, V. Chvykov, G. Kalintchenko, P. Rousseau, V. Yanovsky, A. Zigler, A. Maksimchuk and K. Krushelnick, *Phys. Plas.* **16**, 113105 (2009).
[108] W. T. Wang, W. T. Li, J. S. Liu, C. Wang, Q. Chen, Z. J. Zhang, R. Qi, Y. X. Leng, X. Y. Liang, Y. Q. Liu, X. M. Lu, C. Wang, R. X. Li and Z. Z. Xu, *Appl. Phys. Lett.* **103**, 243501 (2013).
[109] W. T. Li, J. S. Liu, W. T. Wang, Z. J. Zhang, Q. Chen, Y. Tian, R. Qi, C. H. Yu, C. Wang, T. Tajima, R. X. Li and Z. Z. Xu, *Appl. Phys. Lett.* **104**, 093510 (2014).
[110] J. D. Jackson, *Classical Electrodynamics*, 3rd ed. (John Wiley & Sons, New York, 1999).
[111] P. Michel, C. B. Schroeder, B. A. Shadwick, E. Esarey and W. P. Leemans, *Phys. Rev. E* **74**, 026501 (2006).
[112] A. Deng, K. Nakajima, X. Zhang, H. Lu, B. Shen, J. Liu, R. Li and Z. Xu, *Laser Part. Beams* **30**, 281 (2012).
[113] A. Deng, K. Nakajima, J. Liu, B. Shen, X. Zhang, Y. Yu, W. Li, R. Li and Z. Xu, *Phys. Rev. ST Accel. Beams* **15**, 081303 (2012).
[114] S. Corde, K. Ta Phuoc, G. Lambert, R. Fitour, V. Malka and A. Rousse, *Rev. Mod. Phys.* **85**, 1 (2013).
[115] S. Kneip, C. McGuffey, F. Dollar, M. S. Bloom, V. Chvykov, G. Kalintchenko, K. Krushelnick, A. Maksimchuk, S. P. D. Mangles, T. Matsuoka, Z. Najmudin, C. A. J. Palmer, J. Schreiber, W. Schumaker, A. G. R. Thomas and V. Yanovsky, *Appl. Phys. Lett.* **99**, 093701 (2011).
[116] J. Wenz, S. Schleede, K. Khrennikov, M. Bech, P. Thibault, M. Heigoldt, F. Pfeiffer and S. Karsch, *Nat. Comms.* **6**, 7568 (2015).
[117] J. H. Jeon, K. Nakajima, H. T. Kim, Y. J. Rhee, V. B. Pathak, M. H. Cho, J. H. Shin, B. J. Yoo, S. H. Jo, K. W. Shin, C. Hojbota, L. J. Bae, J. Jung, M. S. Cho, J. H. Sung, S. K. Lee, B. I. Cho, I. W. Choi and C. H. Nam, *Phys. Plas.* **23**, 073105 (2016).
[118] Y. Ma, L. M. Chen, D. Z. Li, W. C. Yan, K. Huang, M. Chen, Z. M. Sheng, K. Nakajima, T. Tajima and J. Zhang, *Sci. Rep.* **6**, 30491 (2016).
[119] J. Pruet, D. P. McNabb, C. A. Hagmann, F. V. Hartemann and C. P. J. Barty, *J. Appl. Phys.* **99**, 123102 (2006).
[120] E. Esarey, B. A. Shadwick, P. Catravas and W. P. Leemans, *Phys. Rev. E* **65**, 056505 (2002).
[121] I. Kostyukov, S. Kiselev and A. Pukhov, *Phys. Plas.* **10**, 4818 (2003).
[122] J. H. Sung, S. K. Lee, T. J. Yu, T. M. Jeong and J. Lee, *Opt. Lett.* **35**, 3021 (2010).
[123] S. Agostinelli *et al.*, *Nucl. Instrum. Methods A* **506**, 250 (2003).
[124] A. L. Meadowcroft, C. D. Bentley and E. N. Stott, *Rev. Sci. Instrum.* **79**, 113102 (2008).
[125] O. Lundh, J. Lim, C. Rechatin, L. Ammoura, A. Ben-Ismail, X. Davoine, G. Gallot, J. P. Goddet, E. Lefebvre, V. Malka and J. Faure, *Nat. Phys.* **7**, 219 (2011).
[126] J. Osterhoff, A. Popp, Z. Major, B. Marx, T. P. Rowlands-Rees, M. Fuchs, M. Geissler, R. Horlein, B. Hidding, S. Becker, E. A. Peralta, U. Schramm, F. Gruner, D. Habs, F. Krausz, S. M. Hooker and S. Karsch, *Phys. Rev. Lett.* **101**, 085002 (2008).
[127] S. Steinke, J. van Tilborg, C. Benedetti, C. G. R. Geddes, C. B. Schroeder, J. Daniels, K. K. Swanson, A. J. Gonsalves, K. Nakamura, N. H. Matlis, B. H. Shaw, E. Esarey and W. P. Leemans, *Nature* **530**, 190 (2016).
[128] C. Pagania, E. L. Saldin, E. A. Schneidmiller and M. V. Yurkov, *Nucl. Instrum. Methods A* **463**, 9 (2001).
[129] P. Elleaume, J. Chavanne and B. Faatz, *Nucl. Instrum. Methods A* **455**, 503 (2000).
[130] G. Mourou, B. Brocklesby, T. Tajima and J. Limpert, *Nat. Photon* **7**, 258 (2013).
[131] A. Buck, M. Nicolai, K. Schmid, C. M. S. Sears, A. Sävert, J. M. Mikhailova, F. Krausz, M. C. Kaluza and L. Veisz, *Nat. Phys.* **7**, 543 (2011).
[132] J. K. Lim, P. Frigola, G. Travish, J. B. Rosenzweig, S. G. Anderson, W. J. Brown, J. S. Jacob, C. L. Robbins and A. M. Tremaine, *Phys. Rev. ST Accel. Beams* **8**, 0072401 (2005).
[133] Z. J. Zhang, W. T. Li, J. S. Liu, W. T. Wang, C. H. Yu, Y. Tian, K. Nakajima, A. H. Deng, R. Qi, C. Wang, Z. Y. Qin, M. Fang, J. Q. Liu, C. Q. Xia, R. X. Li and Z. Z. Xu, *Phys. Plas.* **23**, 053106 (2016).
[134] W. Ackermann, G. Asova, Y. Ayvazyan, *et al.*, *Nat. Photon.* **1**, 336 (2007).
[135] http://loasis.lbl.gov.
[136] https://portal.slac.stanford.edu/sites/ard_public/facet/Pages/default.aspx.
[137] http://awake.web.cern.ch/awake/#.
[138] K. Nakajima, *Eur. Phys. J. Spec. Top.* **223**, 999 (2014).
[139] S. F. Martins, R. A. Fonseca, W. Lu, W. B. Mori and L. O. Silva, *Nat. Phys.* **6**, 311 (2010).
[140] T. Tajima, *Proc. Jpn. Acad. Ser. B* **86**, 147 (2010).
[141] D. Strickl and G. Mourou, *Opt. Commun.* **56**, 219 (1985).

Kazuhisa Nakajima was involved in the superconducting RF accelerator R&D under Prof. Maury Tigner at Cornell University in 1981. For about a decade after 1983, he devoted himself to many accelerator physics studies on electron–positron colliders at KEK, including various beam instability problems and measurements of beam polarization in electron storage rings, as well as high-energy physics in monopole search experiments. In 1996, he initiated laser acceleration research for advanced accelerator R&D at JAERI. Thereafter his experimental research over 20 years was mainly performed by the collaboration with high-power laser facilities in China, South Korea and India, as well as Japan.

Reviews of Accelerator Science and Technology
Vol. 9 (2016) 63–83

DOI: 10.1142/S1793626816300036

Electron and Positron Beam–Driven Plasma Acceleration

Mark J. Hogan
SLAC National Accelerator Laboratory,
2575 Sand Hill Road,
Menlo Park, CA 94303, USA
hogan@slac.stanford.edu

Particle accelerators are the ultimate microscopes. They produce high energy beams of particles — or, in some cases, generate X-ray laser pulses — to probe the fundamental particles and forces that make up the universe and to explore the building blocks of life. But it takes huge accelerators, like the Large Hadron Collider or the two-mile-long SLAC linac, to generate beams with enough energy and resolving power. If we could achieve the same thing with accelerators just a few meters long, accelerators and particle colliders could be much smaller and cheaper. Since the first theoretical work in the early 1980s, an exciting series of experiments have aimed at accelerating electrons and positrons to high energies in a much shorter distance by having them "surf" on waves of hot, ionized gas like that found in fluorescent light tubes. Electron-beam-driven experiments have measured the integrated and dynamic aspects of plasma focusing, the bright flux of high energy betatron radiation photons, particle beam refraction at the plasma–neutral-gas interface, and the structure and amplitude of the accelerating wakefield. Gradients spanning kT/m to MT/m for focusing and 100 MeV/m to 50 GeV/m for acceleration have been excited in meter-long plasmas with densities of 10^{14}–$10^{17}\,\mathrm{cm}^{-3}$, respectively. Positron-beam-driven experiments have evidenced the more complex dynamic and integrated plasma focusing, 100 MeV/m to 5 GeV/m acceleration in linear and nonlinear plasma waves, and explored the dynamics of hollow channel plasma structures. Strongly beam-loaded plasma waves have accelerated beams of electrons and positrons with hundreds of pC of charge to over 5 GeV in meter scale plasmas with high efficiency and narrow energy spread. These "plasma wakefield acceleration" experiments have been mounted by a diverse group of accelerator, laser and plasma researchers from national laboratories and universities around the world. This article reviews the basic principles of plasma wakefield acceleration with electron and positron beams, the current state of understanding, the push for first applications and the long range R&D roadmap toward a high energy collider.

Keywords: Advanced accelerator; electron; positron; plasma; wakefield.

1. Introduction

The field of high energy physics (HEP) explores and discovers the laws of nature as they apply to the basic constituents of matter and the forces between them. To enable these discoveries, the development of particle accelerators at increasingly high energies is being pursued. Accelerator test facilities with high quality beams are essential for supporting the necessary research into new acceleration techniques with the potential to dramatically reduce the size and cost of future colliders. Recent research on the acceleration of particles by strong wakefields in plasma has demonstrated accelerating gradients of tens of billions of electron volts (GeV) per meter, several hundred times higher than in the metal structures currently used for accelerators. This technology holds great promise for high energy electron–positron colliders, but also for other science applications that can be realized at lower beam energies, such as compact free electron lasers or medical accelerators.

The Large Hadron Collider (LHC) at CERN is currently the highest energy accelerator in the world. It provides proton–proton collisions at an energy of about 13 TeV. Because protons are composite particles, the LHC energy will allow the production of particle states up to about 5 TeV.

The next high energy accelerator proposed after the LHC is an electron–positron collider: the International Linear Collider (ILC). The initial phase of the ILC is envisioned to have a total energy of 500 GeV, with the possibility of a subsequent increase in the energy to 1 TeV. The ILC would use superconducting niobium structures with an average accelerating gradient of 31.5 million electron volts per meter, and the overall length would be 31 km.

Both the LHC and the ILC are multibillion-dollar investments, due to the extensive civil construction needed to house the accelerators, and industrial scale production of advanced technology components. The time scale for such projects is long. The LHC took more than 30 years from concept to realization. The ILC builds on more than 30 years of R&D on linear colliders and 10 years of active detailed design. It will take another decade for construction and another decade or two of operation after that before an upgrade is considered. This sets the time scale for practical application of higher gradient accelerating technologies to high energy colliders. Techniques such as beam-driven plasma wakefield acceleration are aimed at achieving a gradient significantly higher than that of the ILC to keep the size and cost of even higher energy colliders in an affordable range.

For almost three decades there has been dedicated research on the acceleration of electrons by strong wakefields in plasma. Recent achievements of this R&D program have been notable, particularly the high gradient, high efficiency acceleration of an electron beam. This combination of accelerating fields hundreds of times greater than for the ILC design and high electrical efficiency makes it possible to envision more compact and affordable collider designs with the energy and beam power (luminosity) required for physics at the energy frontier.

2. Electron-Beam-Driven Plasmas

To understand why plasmas have such enormous potential for accelerating electrons, consider a one-dimensional relativistic plasma wave (electrostatic). The electric field can be obtained from Gauss law:

$$\nabla \cdot \boldsymbol{E} = 4\pi \frac{\rho}{\varepsilon_0}. \tag{1}$$

Assuming a plasma of density n_e and characteristic frequency ω_p, Eq. (1) becomes

$$k_p E_Z = \frac{\omega_p}{c} E_Z = \frac{n_e e}{\varepsilon_0}, \tag{2}$$

with an electric field of amplitude

$$E_Z = \left(\frac{m_e c^2}{\varepsilon_0}\right)^{\frac{1}{2}} n_e^{\frac{1}{2}}, \tag{3}$$

or, when rewritten in engineering units, can be approximated as

$$E_Z[\mathrm{GV/m}] \cong 100\sqrt{n_e[\mathrm{cm}^{-3}]}, \tag{4}$$

i.e. for a plasma of density $10^{16}\mathrm{e}^-/\mathrm{cm}^3$ we get an electric field of 10 GV/m, two orders of magnitude larger than the accelerating field in traditional accelerating cavities. Thus, we can see that plasma waves can produce very large amplitude electric fields, which can then in principle be used to accelerate electrons to a high energy in a relatively short distance. Plasma waves can be driven by intense laser pulses or tightly focused particle beams. In this section we will discuss the case of electrons, and in the next section we will look at the similarities and differences for positron beams.

2.1. *The plasma wakefield accelerator*

The basic electron-beam-driven plasma accelerator, also known as a plasma wakefield accelerator (PWFA), is illustrated in Fig. 1.

In this scenario a relativistic bunch of electrons (the drive bunch) is propagating from left to right into an ionized but electrically neutral plasma. The space charge fields of the electron bunch expel the plasma electrons, and if the beam density is greater than the plasma density, all of the plasma electrons will be expelled. This regime is known as the blowout regime. In most scenarios the more massive ions do not move significantly. After the passage of the drive bunch, the ions exert a restoring force that pulls the electrons back toward the axis of propagation, creating a space charge oscillation or wake. The ion cavity formed by this wake is commonly referred to as the bubble. As we will see in the following sections, the ion column within the bubble contains very strong focusing fields and the motion of the plasma electrons around the outside creates very strong accelerating fields. In a plasma wakefield accelerator, these fields are used to rapidly and efficiently transfer energy from the drive bunch to a trailing witness bunch copropagating in the back of the bubble.

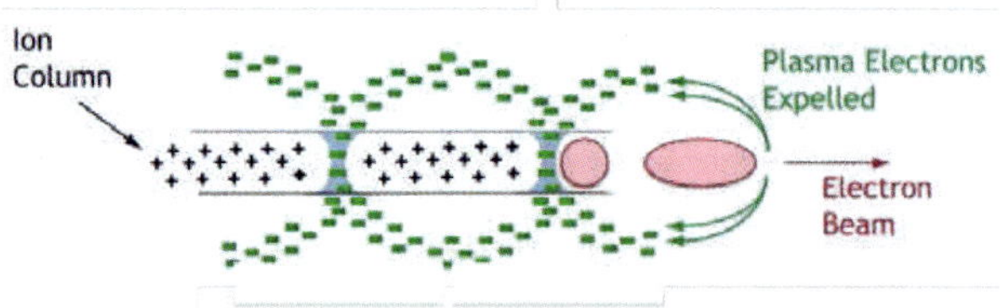

Fig. 1. Illustration of the basic principles of a plasma wakefield accelerator.

2.2. *The electron plasma frequency*

Plasma density perturbations created by a tightly focused electron beam have a time scale and physical dimension proportional to the electron plasma frequency. This single quantity will be useful for determining many characteristics of plasma accelerators and can be simply derived as follows.

Imagine an electron layer displaced in one dimension by a length delta. This creates something analogous to two capacitor plates with surface charge density

$$\sigma = en_e\delta \tag{5}$$

and an electric field between the two given by

$$E = \frac{\sigma}{\varepsilon_0} = \frac{en_e\delta}{\varepsilon_0}. \tag{6}$$

This creates a restoring force,

$$m_e\frac{dv}{dt} = -m_e\frac{d^2\delta}{dt^2} = -eE = \frac{e^2n_e\delta}{\varepsilon_0}, \tag{7}$$

that may be rewritten as a harmonic oscillator equation,

$$\frac{d^2\delta}{dt^2} + \omega_p^2\delta = 0, \tag{8}$$

with a characteristic electron plasma frequency.

$$\omega_p[\mathrm{s}^{-1}] \equiv \left(\frac{e^2n_e}{\varepsilon_0 m_e}\right)^{\frac{1}{2}} \cong 6\cdot 10^4\sqrt{n_e[\mathrm{cm}^{-3}]}. \tag{9}$$

The quantity c/ω_p is the plasma skin depth and the plasma wavelength is $2\pi c/\omega_p$, which can be approximated as

$$\lambda_p \cong 100\ \mu\mathrm{m}\cdot(n_p[\mathrm{cm}^{-3}]/10^{17})^{-\frac{1}{2}}. \tag{10}$$

With the scale of the plasma bubble in hand, we will now examine the origin and magnitude of the transverse and accelerating fields within it.

2.3. *Focusing in the ion column*

Consider a uniform ion density n_i equal to the initial plasma density n_0. The focusing forces will be transverse and a balance between the radial electric field and the $v\times B$ force which in cylindrical coordinates will be $\sim E_r - cB_{\mathrm{phi}}$. When the beam density is much greater than the plasma density it will produce a fully blown-out ion column. In the beam frame, there are then no currents to drive B_{phi} and the focusing is then simply obtained from Gauss' law for an infinite cylinder:

$$\boldsymbol{\nabla}\cdot\boldsymbol{E} = 4\pi\frac{\rho}{\varepsilon_0}, \tag{11}$$

which implies that

$$2\pi r dz E_r = \frac{\pi r^2 en_i}{\varepsilon_0}, \tag{12}$$

which then reduces to

$$E_r = \frac{1}{2}\frac{en_{e0}}{\varepsilon_0}r. \tag{13}$$

Note that the focusing is linear in radius and constitutes an ideal lens free of geometric aberrations.

2.4. *Propagation of a single electron*

The motion of a single relativistic electron in the radial electric field given by Eq. (13) will be

$$\gamma m\frac{dv_r}{dt} = F_r, \tag{14}$$

which implies that

$$\gamma mc^2\frac{d^2r}{dz^2} = e\frac{1}{2}\frac{en_{e0}}{\varepsilon_0}r, \tag{15}$$

which expands to

$$\frac{d^2r}{dz^2} = \frac{1}{2\gamma c^2}\frac{e^2n_{e0}}{m\varepsilon_0}r \tag{16}$$

and can then be rewritten using the definition of plasma frequency given by Eq. (9):

$$\frac{d^2r}{dz^2} = \frac{\omega_{pe}^2}{2\gamma c^2}r = \frac{k_{pe}^2}{2\gamma}r = k_\beta^2 r. \tag{17}$$

This is harmonic motion as long as there is no energy gain or loss and will have the form

$$\frac{d^2r}{dz^2} = k_\beta^2 r, \tag{18}$$

which implies that

$$r(z) = r_0 e^{ik_\beta z}. \tag{19}$$

It is also worth noting that a relativistic electron will oscillate with a wave number and frequency given by

$$k_\beta^2 = \frac{k_p^2}{2\gamma} \quad \text{or} \quad \omega_\beta = \omega_{pe}/\sqrt{2\gamma} \ll \omega_{pe}. \tag{20}$$

For large values of gamma, this large difference in oscillation frequency can be exploited to speed up numerical simulations through the so-called quasistatic approximation as used in the code QuickPIC [1, 2].

2.5. *Propagation of a beam of electrons*

The propagation of a beam of electrons can be described by the envelope equation [3]

$$\frac{d^2\sigma}{dz^2} + \left(K - \frac{\varepsilon^2}{\sigma^4}\right)\sigma = 0, \tag{21}$$

with

$$K \equiv k_\beta^2 = \frac{k_p^2}{2\gamma}. \tag{22}$$

Note that there will be no evolution of the beam spot size σ in the matched condition

$$\frac{d^2\sigma}{dz^2} = 0, \tag{23}$$

which implies that

$$\frac{\varepsilon^2}{\sigma^4} = \frac{1}{\beta^2}, \tag{24}$$

which can be rewritten to define a matched beta function:

$$\beta_{\text{matched}} = \frac{\sqrt{2\gamma}}{k_p} = \sqrt{2\gamma}\frac{c}{\omega_p}. \tag{25}$$

Note that there is a matched beta function for the ion column, not a matched beam size. The matched beam size is related to the matched beta function by the beam emittance. For example, for a plasma density of $10^{17}\text{e}^-/\text{cm}^3$, $c/\omega_p = 17\ \mu\text{m}$ and the matched beta will be 1 mm. For a normalized emittance of 1 μm and a beam energy of 1 GeV, the matched beam size σ will be 700 nm. Matching the electron beam to the strong focusing of the ion column enables the beam to propagate for extended distances in a tightly focused condition to sustain the conditions necessary for maintaining the blowout regime.

This model of plasma focusing has been verified in experiments over several orders of magnitude in plasma density [4–6], and even considered for application as a plasma lens [7, 8]. Consider the evolution of a beam's transverse size while propagating in a vacuum:

$$\sigma_r(z) = \sigma_{r0}\left(1 + \frac{\varepsilon^2 z^2}{\sigma_0^4}\right)^{\frac{1}{2}} = \sigma_{r0}\left(1 + \frac{\varepsilon^2}{\beta_0^2}\right)^{\frac{1}{2}}. \tag{26}$$

The addition of a plasma ion column will add a focusing element with a strength proportional to the length of the plasma and the square root of the plasma density. While it is not always practical to measure the beam size within the plasma, an equivalent measurement can be made by measuring the beam size downstream of the plasma and then varying the focusing by changing the density or the length. An example of such measurements, was reported in Refs. 4 and 5. From Eq. (26), a beam focused to a waist will then diverge with a hyperbolic profile in vacuum. If a plasma is then introduced and the beam beta function is larger than the matched value, and the term in the parentheses in (21) is greater than zero, the beam will begin to focus and the measured beam size downstream will decrease. As the focusing is increased, the beam will eventually focus to a waist within the plasma and again begin to diverge, resulting in a larger size downstream. As the focusing increases further, the beam will pinch multiple times within the plasma and the amplitude of the envelope oscillation will grow — see Fig. 2.

When a beam is injected into the plasma with a beta function equal to the matched condition described in Eq. (25), the beam will propagate at a constant size within the plasma. Figure 3 shows the measured beam size downstream of the plasma for the case where the beam is injected with a beta function less than or equal to the matched value. In this case the term in parentheses will be less than or equal to zero and the beam divergence at the plasma exit is a weak function of the plasma density.

The figures below confirm several salient features of the ion focusing: the focusing can be orders of magnitude stronger than for conventional quadrupole

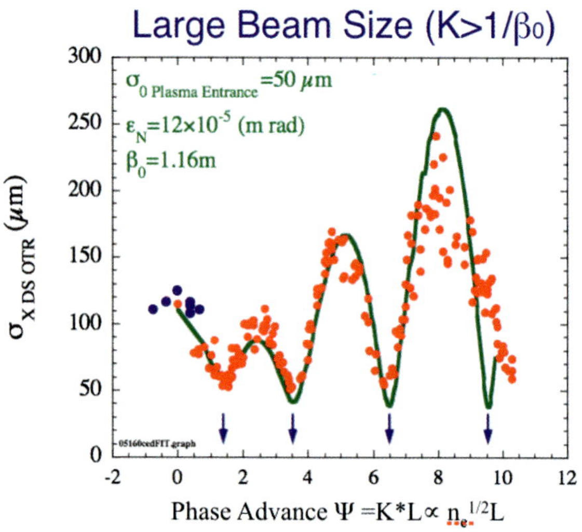

Fig. 2. Electron beam transverse size measured downstream of a plasma as a function of plasma density for the case of a beam not matched to the ion column focusing.

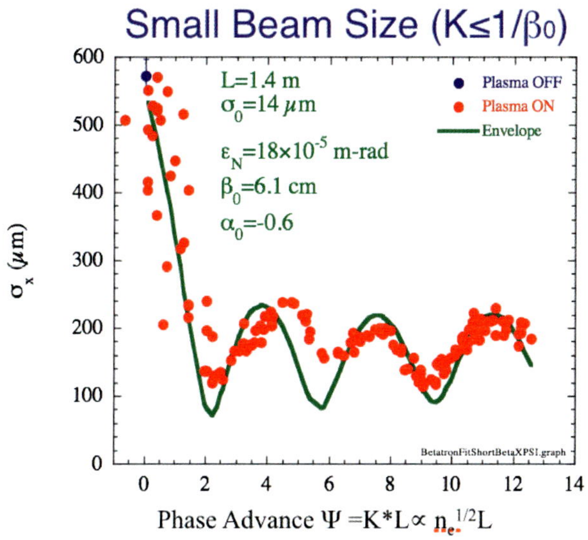

Fig. 3. Electron beam transverse size measured downstream of a plasma as a function of plasma density for the case of a beam matched to the ion column focusing at the highest density.

magnets (such as focusing gradient MT/m for 10^{17}e$^-$/cm^3), the focusing is well described by the simple model presented in Sec. 2.5, and for mismatched beams there will typically be multiple foci within the plasma even for highly relativistic beams.

When the electron bunch enters the initially neutral but fully ionized plasma, it takes a finite amount of time for the plasma electrons to blowout and for the pure ion column to form. Consequently, the head of the bunch does not experience the full ion column focusing force, and exhibits a lower number of betatron oscillations than the portion traveling in the pure ion column. The dynamic nature of the ion column formation and focusing for the head of the bunch has been measured in the experiments reported in Ref. 9.

Additional dynamics occur when the plasma column has a finite width. The deflection of the electron beam at the plasma boundary has also been observed [10]. The bending angle of the beam core trajectory follows a refraction law similar to Snell's law for a light beam at a dielectric boundary [11]. When the beam is initially misaligned and moving parallel to a neutral plasma column, at an initial distance of many plasma column radii, it is attracted into the column. Once the beam enters the plasma it drives a plasma wake similar to that of an electron beam entering the plasma column head-on [12].

2.6. *Accelerating fields — unloaded wakes*

The accelerating fields were described several decades ago [13, 14] and can be obtained from a combination of the momentum/force equation

$$\frac{\partial v}{\partial t} = -\frac{eE}{m}, \tag{27}$$

the continuity equation

$$\frac{\partial}{\partial t}\left(\frac{\partial n}{\partial t} + \nabla \cdot nv\right) = 0 \tag{28}$$

and the Poisson equation

$$\nabla \cdot E = -4\pi e(\partial n + n_b). \tag{29}$$

By changing variables with zeta $= z{-}ct$ and substituting k_p^2 for w_p^2/c^2 an expression for the perturbed plasma density can be obtained:

$$(\partial_\zeta^2 + k_p^2)\partial n = -k_p^2 n_b, \tag{30}$$

which acts as a driving term for the electric field:

$$(\nabla_\perp^2 + k_p^2)E_z = -4\pi e \nabla \partial n. \tag{31}$$

In the narrow beam limit ($k_p\sigma_r \ll 1$), an equation for the E_z, field behind the beam can be obtained:

$$E_Z = \frac{8\pi eN}{\sigma_z^2} u e^{-u}, \quad \text{with } u = k_p^2\sigma_z^2/2. \tag{32}$$

This expression is maximized when the bunch length is matched to the plasma density as

$$k_p\sigma_z = \sqrt{2} \tag{33}$$

and has the notable scaling

$$E_Z \propto n_p^{\frac{1}{2}} \propto \frac{N}{\sigma_z^2}. \tag{34}$$

In practical terms, this can be rewritten as

$$eE_z[\text{GeV/m}] \simeq 50\left(\frac{N}{2\cdot 10^{10}}\right)\left(\frac{30}{\sigma_z[\mu\text{m}]}\right)^2. \tag{35}$$

So, for example, a beam with 2×10^{10} electrons and 30 μm bunch length, when matched to a plasma density that satisfies Eq. (33), will have a peak accelerating gradient of 50 GeV/m.

Computer simulations have verified these scalings and can be used to accurately predict the shape of the accelerating wakefield within the plasma. An example of such a simulation is shown in Fig. 4. While the peak amplitude of the field predicted by Eq. (35) is correct, it can also be seen that in the

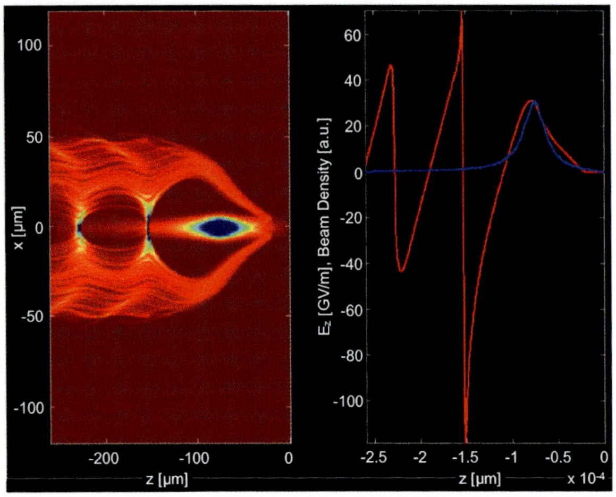

Fig. 4. QuickPIC simulation for the SLAC–UCLA E-167 experiment, which took place at the SLAC FFTB facility. The panel on the left shows the electron beam propagating into the plasma from left to right. The plasma electrons are expelled, forming an ion bubble around the injected electron beam. The panel on the right shows the on-axis electron beam's current profile (blue) and the longitudinal wakefield amplitude (red).

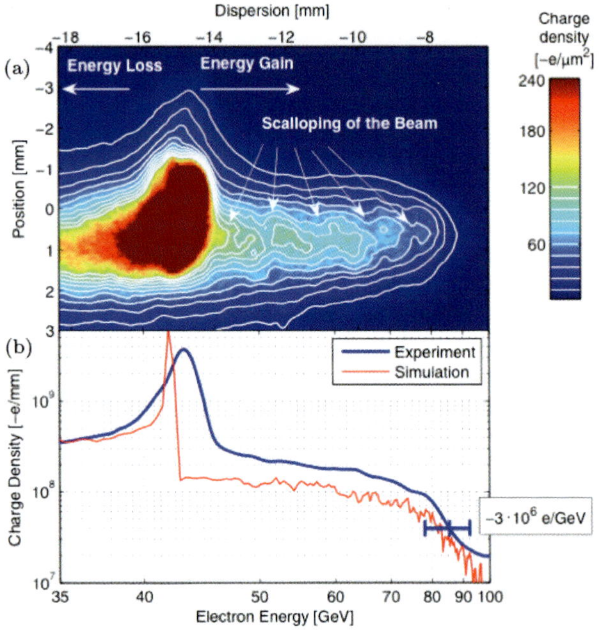

Fig. 5. The energy spectrum of the electrons in the 35–100 GeV range as first reported in Ref. 16. The dispersion (shown on the top axis) is inversely proportional to the particle energy (shown on the bottom axis). The head of the pulse, which is unaffected by the plasma, is at 43 GeV. The core of the pulse, which has lost energy driving the plasma wake, is dispersed partly out of the field of view of the camera. Particles in the back of the bunch, which have reached energies up to 85 GeV, are visible on the right. (b) Projection of the image in (a), shown in blue. The simulated energy spectrum is shown in red. The differences between the measured and the simulated spectrum near 42 GeV are due to an initial correlated energy spread of 1.5 GeV not included in the simulations. The horizontal error bar is due to the uncertainty in estimating the deflection angle and the spot size of the beam.

blowout regime the shape of the wakefield is not sinusoidal, but instead has more of a sawtooth shape in the rear of the bubble [15].

Accelerating fields of 50 GeV/m have been created and sustained over meter scale distances in the laboratory and used to double the energy of electrons from the 3 km SLAC linac in just 85 cm [16].

Figure 5 shows that while the amplitude of the accelerating field is impressive, it is also changing rapidly along the bunch, leading to a large spread in output energies. This can be remedied through beam loading.

2.7. *Accelerating fields — beam loading*

Beam-loading the plasma wakefield is important, for two reasons. Firstly, the presence of an appropriately placed witness bunch with sufficient peak current will modify the trajectories of the plasma electrons rushing back in toward the axis. It can be shown [17–20] that in the quasistatic approximation and in a frame comoving with the drive beam ($v = c$), for a cylindrically symmetric system $E_{\text{phi}} = B_z = B_r = 0$ and

$$E_z = -\frac{1}{c\varepsilon_0}\int_r^\infty dr j_r. \tag{36}$$

When the returning plasma electrons follow a linear trajectory, $E_z(r, z)$ will be constant and all of the electrons within the witness beam will gain energy at a constant rate, leading to a beam with a narrow spread in energies.

Beam loading is important for potential high energy physics applications. Particle colliders need to produce luminosity sufficient for studying the physics problems in question. The luminosity can be written as a function of beam power:

$$\mathcal{L} = \frac{P_b}{E_b}\left(\frac{N}{4\pi\sigma_x\sigma_y}\right). \tag{37}$$

At TeV energies, with practical numbers of particles per bunch ($\sim 10^{10}$) and submicron beam sizes at the IP, future linear colliders will need tens of megawatts of beam power to achieve the required luminosity of $\sim 10^{34}$. Such a large beam power requires that every stage of the accelerating process be highly efficient so as to keep the overall site power to reasonable values of a few hundred megawatts.

The theoretical work of Refs. 17–20 provides a prescription for tailoring the current profile of the drive and witness beams to flatten the wakefields within the plasma bubble. While the exotic profiles needed for perfect beam loading may be difficult to achieve in practice, these same papers also showed through numerical simulations that Gaussian bunches can still load the wake sufficiently to produce energy spreads of less than 1% and high efficiency. In the narrow beam limit ($k_p\sigma_r \ll 1; k_p\sigma_z \approx 1$), the amount of charge (Q) required for the witness beam to flatten the accelerating field of the wake to a value E_s is proportional to the radius of the bubble (R_b) to the fourth power,

$$Q[nC] = 0.047\frac{mc\omega_p}{eE_s}\sqrt{\frac{10^{16}}{n_p[\mathrm{cm}^{-3}]}}(k_pR_b)^4, \quad (38)$$

and, for example, to flatten the plasma wake to half the wave-breaking field $mc\omega_p/e$ in a plasma of density $10^{16}\mathrm{e}^-/\mathrm{cm}^3$ where the blowout radius is c/ω_p would require about 0.1 nC.

High gradient, high efficiency plasma wakefield acceleration has recently been demonstrated. In these experiments two appropriately spaced bunches were sent into the plasma. The drive beam lost energy driving the plasma wake and more than 30% of the energy extracted from the drive beam was transferred back to the witness beam, resulting in an energy gain of more than a GeV in a 30-cm-long plasma [21]. Subsequent experiments extended the plasma length and improved the interaction to boost the witness beam energy by up to 9 GeV in a 1.3-m-long plasma [22] with a 5% energy spread — see Fig. 6.

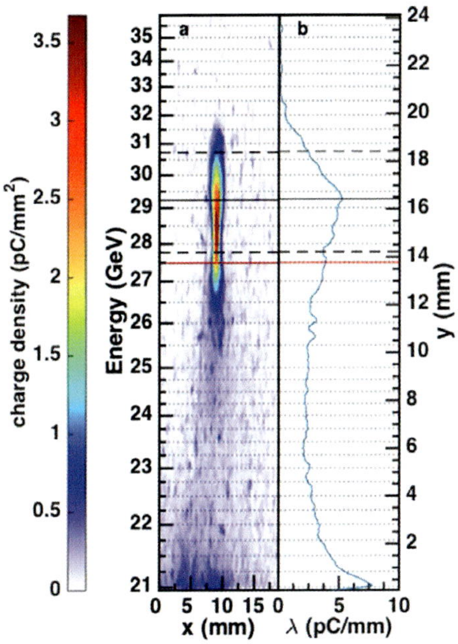

Fig. 6. An electron beam after gaining 9 GeV per particle in a 1.3-m-long electron beam-driven plasma wakefield accelerator as first reported in Ref. 22. The amount of charge accelerated in the spectral peak was 28.3 pC, and the root-mean-square energy spread was 5.0%. The left axis displays the energy calibration of the screen, and the right and bottom axes display the physical size of the beam on the screen. The color axis corresponds? to the charge density in units of pC mm^{-2}, represented on a linear scale. The horizontal lines represent centroid energy (red), the peak energy (solid black), and the values corresponding to the rms energy spread about the peak energy (dashed black).

For symmetric current profiles, the ratio of the peak accelerating field of the wake to the peak decelerating field of the wake cannot exceed 2 [23, 24]. Thus, for Gaussian beams (symmetric) the maximum energy boost to the witness beam will be limited to twice the drive beam energy. By shaping the current profile of the drive beam such that it ramps up slowly, over several plasma periods, and then terminates abruptly (much less than one plasma period), the transformer ratio can exceed 2 up to factors of 5 or more [25, 26]. For very high energy applications, higher transformer ratios could reduce the required number of plasma stages to reach a given energy, or reduce the required drive beam energy, both of which would offer greater flexibility and potentially simplify designs. For applications such as free electron lasers which require energy spreads of one part in 10^4 or better, tailoring the current profile of the witness beam for optimal wake loading will minimize the final energy spread.

2.8. *Higher transformer ratios*

Looking ahead, in the near future there will be attempts to take advantage of beam loading to optimize the acceleration process for higher efficiency, higher transformer ratios and reduced energy spread.

By flattening the wakefield across the drive beam and extracting energy at a constant rate, the PWFA process can be optimized for the highest efficiency and may continue until the drive beam has depleted nearly all of its stored energy.

3. Positron-Beam-Driven Plasmas

The blowout regime has many attractive features for electron acceleration: strong ideal focusing from the ion column, as well as large amplitude accelerating fields that are constant in radius and that can be loaded to be constant along the witness bunch. For positron beams, the situation is more complex.

3.1. *Positrons — transverse focusing forces*

Rather than expel the plasma electrons, a positron beam pulls the plasma electrons inward. Because the positron beam can pull plasma electrons in from an entire cubic skin depth of plasma, the plasma electron density can exceed the incoming positron beam density. As the electrons stream through the positron bunch and cross the axis, there can be large concentrations of charge with large focusing near the axis of propagation. The net effect is a focusing force that varies nonlinearly in both radius and position along the bunch. There is no simple model to describe the evolution of the bunch, so computer simulations are used to interpret experimental results.

For relatively low density meter scale plasmas, the variation in focusing along the bunch has been measured [27]. At larger plasma densities ($\sim 10^{14} e^-/cm^3$), there are a sufficient number of betatron oscillations such that the variation in focusing leads to phase mixing and filamentation of the positron beam phase space. The net effect is halo formation and emittance growth as measured in Ref. 28.

3.2. *Positrons — accelerating fields*

When the beam density is small compared to the plasma density, the perturbation to the plasma electrons will be small, the oscillations of the plasma electron density will be sinusoidal and the plasma response will be similar for an electron beam or a positron beam. An accelerating wakefield with peak amplitude ~80 MeV/m has been measured along a positron bunch for low plasma densities ($\sim 10^{12} e^-/cm^3$) in this so-called linear regime [29]. The wake amplitude can be increased somewhat in the so-called quasi-nonlinear regime [30], where the beam density is still greater than the plasma density, but the density of particles (electrons or positrons) per plasma skin depth is less than the plasma density.

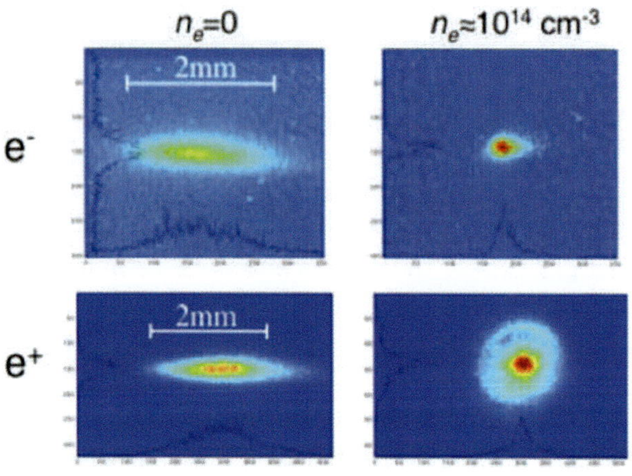

Fig. 7. The transverse beam profile for electron and positron beams with and without plasma focusing. The images on the left show typical profiles for a beam with asymmetric emittance focused to a round spot at the plasma entrance. When observed on a profile monitor several meters downstream, the beam has an elliptical profile with a beam extent proportional to the emittance in the horizontal and vertical transverse plane. When the 1.4-m-long plasma is turned on, the ion column tightly focuses the electron beam, as shown in the upper right panel. For a positron beam, the focusing is aberrated, leading to halo formation as seen in the lower right panel.

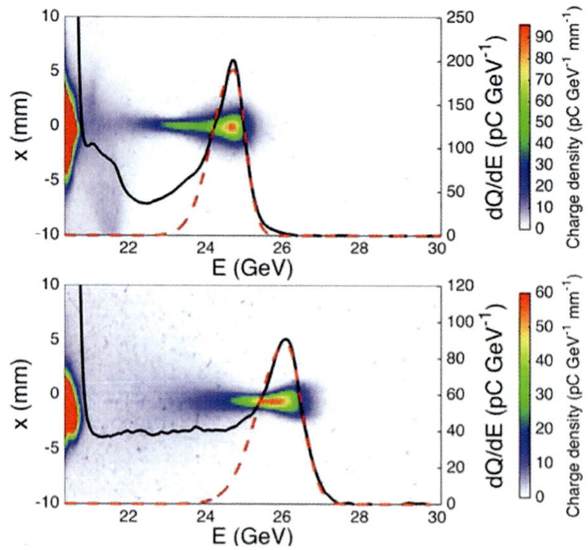

Fig. 8. Two examples of the accelerated portion of the spectrum of the positrons after the interaction with the plasma. In both panels, the density of charge per unit energy and length of the dispersed positron beam profile is shown in color, and the spectral charge density dQ/dE is represented by the black solid line (right scale). Asymmetric Gaussian fits to the peaks at the top and bottom are shown as red dashed lines. Particles that do not participate in the interaction appear to be saturated at the initial beam energy, 20.35 GeV.

A new regime for positron PWFA was recently discovered when the density-length product was increased by two more orders of magnitude. It was found that for certain positron beam parameters, a single positron bunch will interact with the plasma in such a way that it bifurcates into two distinct populations of positrons and then settles into a near-equilibrium condition. In this state, one set of positrons experience loaded transverse and longitudinal wakes such that they gain energy with a uniform accelerating gradient and exit the plasma with low divergence and no appreciable beam halo [31] — see Fig. 8.

The results were investigated and interpreted with the aid of computer simulations. When the single high density positron bunch enters the plasma, some of the plasma electrons that are drawn into the positron beam remain near the axis and copropagate with some of the positron beam. This results in a neutralization of the focusing forces and a loading of the longitudinal wakefield.

3.3. *Positrons — hollow channel plasmas*

An alternate technique for accelerating positrons in plasmas involves engineering the plasma to control the fields and accentuate the accelerating fields while eliminating or greatly reducing the transverse fields. A hollow channel plasma is a cylindrically symmetric plasma tube of plasma with a given plasma density in the walls and no plasma on the axis of propagation [32–35]. In this case a symmetric electron or positron beam propagating down the axis will feel no transverse forces, but the motion of the plasma electrons in the channel wall will create longitudinal fields that can accelerate a trailing bunch of particles. An example of the fields produced by such a hollow channel plasma is shown in Fig. 10.

Hollow channel plasmas have recently been created in the laboratory and the transverse and longitudinal fields have been characterized using high energy positron beams [36].

Hollow channel, or nearly hollow channel, plasmas have been proposed as solutions for accelerating positrons to very high energies for high energy physics applications [37]. Although the transverse fields are zero for symmetric beams propagating on axis, the fields grow rapidly when these conditions are not met and the requirements on beam symmetry

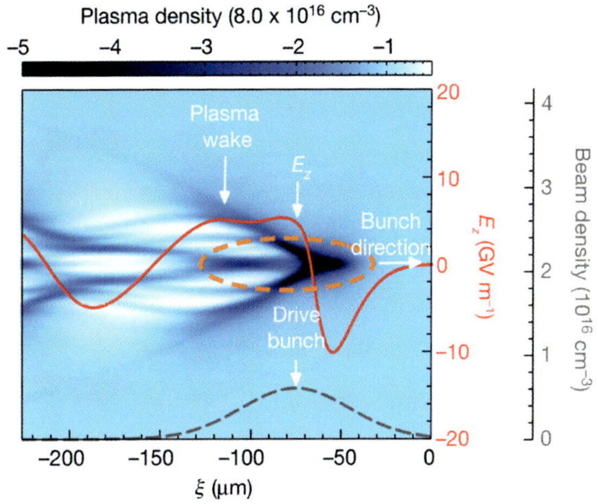

Fig. 9. Simulated plasma wakes driven by short and intense positron bunches. The electron plasma density in the y plane is shown after a propagation distance of 135 cm into the plasma, where y is the dimension transverse to the direction of motion of the bunch, and $z-ct$ is the dimension parallel to the motion. The one-third-of-the-maximum contour of the initial positron bunch density is represented by the orange dashed line, the on-axis density profile of the initial positron bunch by the gray dashed line, and the on-axis longitudinal electric field E_z by the red solid line. The beam and plasma parameters are the same as those in the experiment described in Ref. 31. The self-loaded plasma wake, where the trailing particles of a single bunch extract energy from the wake excited by the particles in the front. The color scale represents the perturbed plasma electron density (the plasma density prior to the passage of the beam is constant at 8×10^{16} cm^{-3}).

and alignment will be severe for collider applications and need to be studied. To mitigate these effects, as well as the emittance growth possible from multiple Coulomb scattering in the plasma [38], some focusing may be desirable and the plasma could also be a so-called quasihollow channel. Moving from proof-of-principle demonstration experiments to collider level beam parameters will also require the development of hollow channel plasma sources that are truly hollow, lacking both plasma and neutral gas on axis.

Finding a suitable regime to accelerate positrons in a plasma with high gradient (>1 GeV/m), high efficiency and minimal emittance growth remains a challenge for both beam- and laser-driven plasma accelerators.

4. Injection

Since the earliest theoretical works imagined the possibility of using plasma waves to create very high energy electrons in a very compact distance, high energy physics applications have been considered.

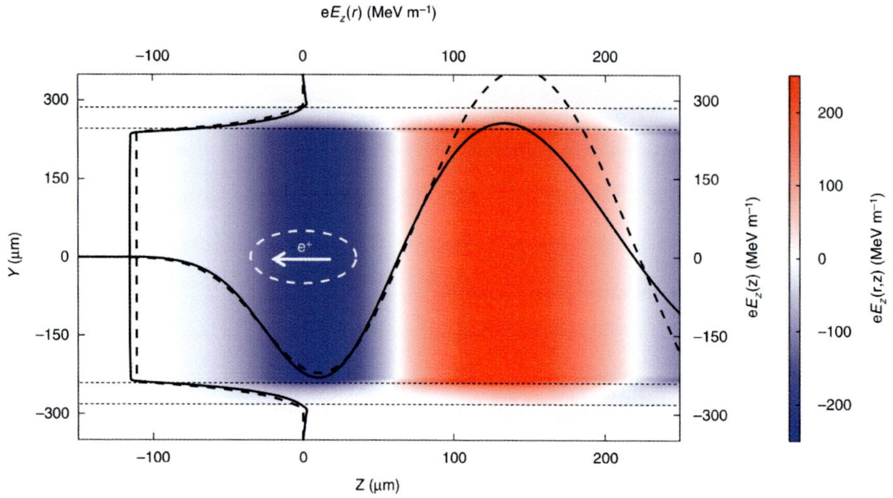

Fig. 10. Simulation of the longitudinal field in a hollow channel plasma accelerator. The E_z field is due to an ultra-relativistic positron beam driving a wake in a hollow channel plasma, as simulated by QuickPIC [1, 2]. The beam propagates to the left and the one-sigma contour of the beam is shown by the white dashed line. The black dotted lines at ±240 and ±290 mm are the inner and outer radii of the plasma channel, respectively. Lineouts of the simulated and calculated on-axis E_z field at $r = 0\mu$m and the radial variation in E_z at the peak decelerating field at $z = 11\mu$m are shown with solid and dashed black lines, respectively. The simulated and calculated fields show excellent agreement up to $z = 80\mu$m, where the charge separation of the plasma electrons on the surface of the channel becomes significant.

More recently, high quality multi-GeV electron beams have given birth to fourth generation light sources where the self-amplified spontaneous emission (SASE) instability [39, 40] is used to generate flashes of coherent X-rays with peak brilliance many orders of magnitude higher than for previous techniques [41].

FEL oscillators have operated at longer wavelengths for many decades [42]. In the oscillator configuration, high reflectivity mirrors on the upstream and downstream ends of a magnetic undulator(s) form a cavity that confines the radiation mode. The electron beam is passed through the cavity many times at a high repetition rate and the optical power builds up to the desired level. High reflectivity normal incidence mirrors do not exist for hard X-rays, so the SASE technique is used to generate the X-rays in a single pass through the undulators. Relativistic electrons in a magnetic undulator oscillate back and forth due to the $v \times B$ force and produce synchrotron radiation in a narrow cone around the forward direction. Certain wavelengths of radiation will slip past the oscillating electrons by exactly one radiation wavelength per undulator period. The transverse motion of the electrons can be aligned with the electric field of the copropagating radiation or opposed to the electric field. As a result, some electrons will gain energy and some will lose energy. The amplitude of the transverse motion of the lower energy electrons will be larger than that of the high energy electrons. The energy-dependent path length through the undulator then results in a bunching of the electrons in the direction of propagation. As the electrons bunch together, they radiate more strongly and the bunching process accelerates. For the SASE process to work effectively, the beam must have a sufficiently high 6D brightness — low normalized emittance and low uncorrelated energy spread.

The photocathode RF gun is the state of the art electron source that produces beams with sufficiently high brightness for SASE FELs like the SLAC LCLS to work [43]. In the photoinjector, a copper cathode is placed in a region of large electric field (typically ~100 MeV/m) inside an RF cavity. An ultraviolet laser pulse strikes the cathode and releases electrons into the electric field, where they are rapidly accelerated to relativistic energies. Typical photoinjector parameters are charge per bunch of a few hundred pC, MeV energy, ps duration, mm transverse size and μm normalized emittance. Before reaching the undulators, the photoinjector beam is repeatedly accelerated, focused and compressed to reach the necessary energy and peak current, all the while taking great care to avoid emittance growth from incoherent synchrotron radiation (ISR), coherent synchrotron radiation (CSR) and microbunching instabilities (MBI).

4.1. *Plasma injectors*

In plasma injectors, the RF cavity of the photoinjector is replaced by the bubble formed in the nonlinear blowout regime. In contrast to the RF photoinjectors, the plasma naturally favors beams with dimensions measured in microns and peak currents in the kiloamp range. Depending on the plasma density, the bubble can have accelerating fields of 10–100 GeV/m. This combination of parameters raises the possibility of making an electron beam with all the necessary parameters from a single short plasma accelerator — provided that an appropriate mechanism can be identified to control the release of the electrons into the plasma bubble.

4.2. *Ionization injection*

One of the first mechanisms observed to inject electrons into the wake of a plasma wakefield accelerator is ionization injection. When the plasma is derived from a gas species other than hydrogen or if it is made up of multiple gases, electrons can be injected into the wake as a result of secondary ionization [44, 45]. The secondary ionization can be triggered by any mechanism that creates a large-enough field to exceed the next available ionization potential, such as the wakefield itself or the radial field of the electron beam focused by the ion column [46]. Characteristics of the injected electrons have been measured [45], as well as the beam loading effect of injection on the plasma wake itself [46]. Because the beam driver in a plasma wakefield accelerator is a relativistic beam, the electrons released within the plasma bubble must acquire enough longitudinal momentum to become relativistic before they slip out of the wake of the drive beam. For a constant plasma density, the fulfillment this requirement will be determined by the location within the bubble where the electrons are released — see Fig. 11.

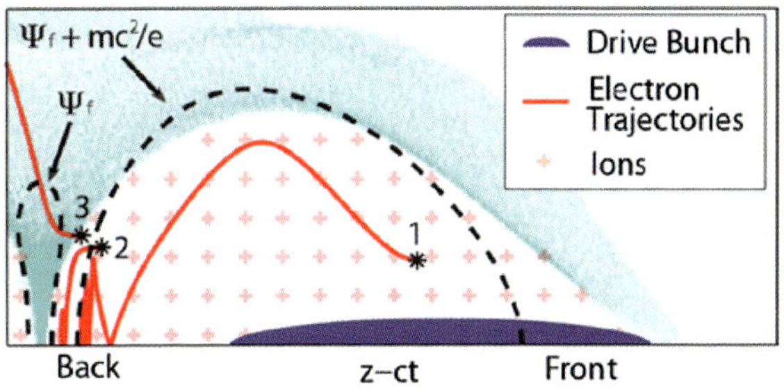

Fig. 11. Illustration of electron trapping, calculated with OSIRIS [47]. A few example helium electron trajectories overlie an image of lithium electron density (teal), where the asterisks display the ionization location, and the dotted lines represent psi contours. The electrons that escape their atoms within the $\Psi_f + mc2/e$ contour (1 and 2) become trapped, while electron 3 slips out of the wake.

4.3. *Trojan horse injection*

Trojan horse injection, also known as the underdense plasma photocathode, is a variant of ionization injection where electrons are released into the plasma wave by laser-triggered secondary ionization [48]. This technique requires two neutral species of gas or vapor — one with a relatively low ionization threshold (LIT) and the other with a relatively high ionization potential (HIT). A volume of the LIT medium is then ionized to create a neutral plasma prior to the arrival of the drive electron beam. Note that this process requires enough energy to ionize a volume many c/ω_p in diameter and long enough for the injected electrons to reach the desired final energy. A laser with an intensity greater than the ionization threshold for the HIT medium ($\sim 10^{14}$ W/cm^2) is then focused within the plasma bubble created by the drive beam and releases electrons into the plasma wake. This technique offers the advantage of precise control of the location, duration and spatial extent of the released electrons. Numerical simulations indicate that the quality of the resulting electron beam will depend on the details of the injection laser pulse [49]. Simultaneously focusing the ionization laser pulse to a few microns such that it fits within the plasma bubble, spatially overlapping this pulse with the electron drive beam and its wakefield to a few microns and synchronizing the laser to fire at a time within the plasma bubble to a few femtoseconds is quite demanding and pushes the state of the art in experimental techniques, and yet this technique has recently been demonstrated to inject up to 50 pC of electrons into a plasma wakefield and accelerate them to over 1 GeV in energy within a few-cm-long plasma [50].

To gain further control over alignment and stability of the injection process, it has recently been proposed to use two colliding laser pulses to reach sufficient intensity to trigger injection [51–53].

4.4. *Density downramp injection*

In density downramp injection, a sudden decrease in plasma density momentarily slows down the forward velocity of wake, providing enough time for plasma electrons from the sheath to be injected. The bubble size is related to the plasma wavelength, which is in turn proportional to the plasma density [Eq. (10)]. When the drive beam propagates from a region of high plasma density into a region of lower plasma density, the wavelength will increase and the length of the bubble will stretch, giving background plasma electrons an effective forward velocity relative to the wake. When the combination of the forward momentum and the forward velocity of the wave matches the beam driver, plasma electrons can be trapped. The momentum of the plasma electrons is a function of the strength of the blowout and favors high peak current (multikiloamp) drive beams. The effective forward velocity of the wake is a function of the gradient of the density transition — faster transitions are better, on the order of tens of percent within a plasma skin depth. Density transitions in the plasma have been achieved using porous membranes [54], gas expansion through apertures [55–57], or optically triggered [58].

5. Betatron Radiation

Synchrotron light sources are used for basic and applied research in physical, chemical and biological sciences, and in engineering. These employ magnetic undulators and wigglers to generate high brightness photon beams in the X-ray region using relativistic electron beams. An ion channel induced by an electron beam as it propagates through a plasma can also wiggle the beam electrons to generate X-ray radiation. Equation (19) showed that the betatron wavelength for a beam of electrons is proportional to the plasma wavelength, and the normalized ion column focusing strength can be written as

$$a_\beta = \gamma k_\beta r_\beta, \tag{39}$$

which can be approximated as

$$a_\beta \approx 0.13\sqrt{\gamma n[10^{18}\mathrm{cm}^{-3}]}\, r_\beta[\mu\mathrm{m}], \tag{40}$$

which then gives the critical frequency of the radiation approximated as

$$\hbar\omega_c[\mathrm{keV}] \approx 10^{-5}\gamma^2 n[10^{18}\mathrm{cm}^{-3}]r_\beta[\mu\mathrm{m}]. \tag{41}$$

For example, a 5 GeV electron beam with a 10 μm beam size in a $10^{17}\mathrm{e}^-/\mathrm{cm}^3$ plasma will have MeV critical energy photons.

An ion channel induced by the beam can wiggle beam electrons to produce x-radiation and betatron emission from a single 28.5 GeV beam propagating in a $10^{14}\mathrm{e}^-/\mathrm{cm}^3$ plasma has been measured centered around a wavelength of 14 keV [59]. The radiated power is expected to scale with the square of the plasma density [59] and has been measured as in Fig. 12.

$$\frac{dW_{\mathrm{loss}}}{dz} = \frac{1}{3} r_e m_e \gamma_b^2 \omega_\beta^2 K^2 \propto (\gamma_b^2, n_p^2, r_0^2). \tag{42}$$

Such plasma wigglers comprising only a plasma cell may offer simplicity of construction, flexibility in undulator wavelength, and potential savings in cost over the magnetic wiggler counterpart in future generation of light sources. The electron beam hosing

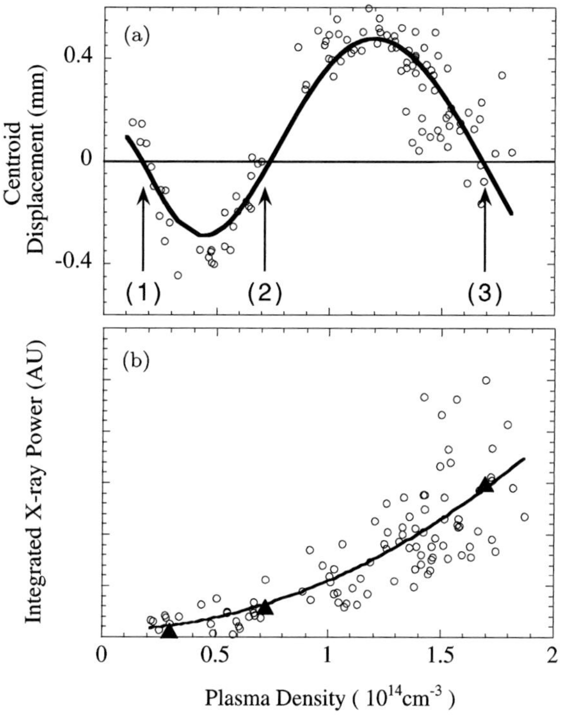

Fig. 12. (a) The predicted (solid line) and measured displacement of the beam centroid as a function of plasma density. (b) The estimated (triangles) and the measured (circles) X-ray energy in the 5–30 keV range as a function of plasma density. The solid line is the quadratic fit to the experimental data.

instability could ultimately limit the number of periods of oscillation by causing transverse breakup of the beam in such a device [60, 61].

High energy physicists use electron–positron collisions to validate the predictions of various field theories. The positrons needed for the collisions are currently produced by bombarding a high Z, solid target that is several radiation lengths thick with a high energy electron beam [62]. The resulting interaction creates bremsstrahlung X-ray photons which can interact with the atomic nuclei of the target, producing electron–positron pairs. These currently used "thick target," bremsstrahlung positron sources may well fail when scaled to meet the requirements of future linear colliders, due to thermal stress fatigue from large volumetric electron beam energy deposition into the target from ionization and multiple scattering. One way to reduce this problem is to produce the X-rays separately from a thin (less than one radiation length), high Z target where the X-ray conversion to electron–positron pairs takes place.

X-rays emitted by betatron motion of high energy electrons in a high strength plasma wiggler have been proposed as a scheme for generating positrons from a thin target. The basic principles and scaling laws for this novel scheme have been demonstrated in Ref. 63. Positrons in the energy range of 3–30 MeV were produced by X-rays emitted by betatron motion in a plasma wiggler of 28.5 GeV electrons. The extremely high strength plasma wiggler was an ion column induced by the electron beam as it propagated through and ionized dense lithium vapor. X-rays in the range of 1–50 MeV, directed as a forward cone angle of 0.1 mrad, collided with a 1.7 mm thick tungsten target to produce electron–positron pairs.

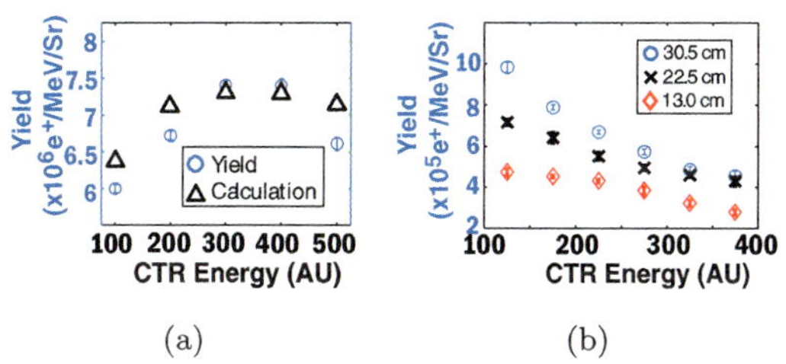

(a) (b)

Fig. 13. The integrated measured and calculated e$^+$ yield in the 4–20 MeV energy range versus upstream coherent transition radiation (UCTR) for n_p 10^{17} cm^{-3} and L_p 11 cm. (b) The integrated e$^+$ yield in the 27–30 MeV energy range versus UCTR for three different plasma lengths with $n_p = 2.7 \times 10^{17}$cm^{-3}.

The positron spectra were found to be strongly influenced by the plasma density and length as well as the electron bunch length. By characterizing the beam propagation in the ion column the influence of the plasma density, and length, and the electron beam bunch length, the positron yield was quantified with good agreement between the measured and calculated positron spectra.

Betatron radiation from electron beams oscillating in an ion column produced in laser wakefield acceleration [64] has been used for phase contrast imaging [65–67]. The inherent synchronization in a plasma accelerator driven by a laser beam also makes laser wakefield accelerators (LWFAs) natural candidates for inverse Compton sources, where a portion of the laser energy is back scattered off of the hundreds-of-MeV LWFA electrons [68].

Although the ion column focusing in a plasma wakefield accelerator can be used as a radiation source to produce bright beams of X- and γ-rays, the beam and plasma parameters must be optimized either for radiation production or for acceleration. Betatron radiation is unavoidable, but can be reduced by choosing a small beam radius — see Eq. (42). In future plasma-based accelerators operating with ultralow emittance beams, the matched beam radius will be small and the energy loss to betatron radiation will remain negligible compared to the energy gain.

6. Roadmap to the Future

Since the original idea for a new class of particle accelerators based on plasma wakefield acceleration surfaced over 30 years ago, there has been tremendous progress in understanding how to bring these concepts from idea to reality. Wakefields up to 50 GeV/m have been created and sustained in meter-long plasmas and used to stably propagate and accelerate electrons. A beam-driven plasma wakefield accelerator has been shown to accelerate a trailing electron bunch with >GeV/m fields and very high efficiency. Novel techniques have also been created and used to accelerate positron beams in plasma wakefield accelerators. In addition, plasmas are showing great potential as a source for a new generation of ultrahigh brightness electron beams. The rapid experimental progress of the last 15 years has the community looking ahead to potential applications of plasma wakefield acceleration. Near term

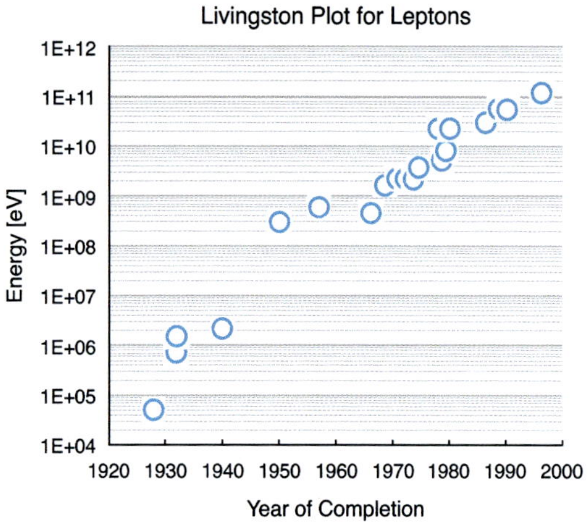

Fig. 14. The so-called Livingston plot of the energy of lepton accelerators plotted versus year of completion shows the exponential growth of particle energy that has continued for over seven decades.

applications, longer term applications and the facilities needed to continue developing the technology are the subject of this section, as well as Ref. 92.

6.1. *A plasma wakefield accelerator linear collider (PWFA-LC)*

The Livingston plot illustrates the remarkable progress in the energy reach of particle accelerators and colliders, see Fig. 14.

Since the 1930s accelerators have increased in their maximum energy by more than an order of magnitude every decade and profound discoveries about the nature of the universe have followed. The Livingston plot also shows how this progress was enabled by great strides in accelerator science and engineering. Accelerators were able to push the energy frontier by continually adopting new technologies.

The requirements for an electron–positron linear collider in the TeV energy range are well understood from decades of conceptual design work based on conventional RF cavity acceleration systems. Plasma wakefield acceleration provides much higher accelerating gradients and may offer a lower cost approach to achieving these energies. Reaching high energies is not enough for a collider — a collider must also meet the luminosity requirement imposed by the physics goals, which is the order of $10^{34}\mathrm{cm}^{-2}\mathrm{s}^{-1}$ at 1 TeV. As discussed in Sec. 2.7 and described by Eq. (37), high luminosity requires low emittance beams focused to nanometer size at high-enough repetition to achieve about 10 MW average power. A practical collider technology must have high power transfer efficiency into the beam to keep the overall site power down to an acceptable and affordable level.

One of the earliest examples of what a plasma-based collider might look like was published in Ref. 69. This concept proposed using a heavily loaded (high efficiency) L-band or S-band linac to create a stream of bunches that would then be distributed to power a series of plasma stages in two separate arms of a collider. Each plasma stage would rapidly and successively boost the energy of a main beam. The details of the collisions at the interaction point were not discussed and were assumed to be electron–electron, electron–gamma or gamma–gamma. Still, this concept illustrated many of the advantages and possibilities of producing a compact, electrically efficient collider for high energy physics research.

In 2009 a similar but more detailed approach was taken for the concept in Ref. 70. In this concept a highly efficient pulsed normal conducting linac is used to create a series of trains. Each train is split to power an electron arm and a positron arm of a collider. The pulse trains power a series of meter-long plasma cells that boost the energy successively to 1 TeV center of mass energy at the interaction point. This concept is illustrated in Fig. 15

An example set of beam parameters for this concept is given in Table 1.

In preparation for the 2013 Community Summer Study (CSS2013), also known as Snowmass, an alternate concept for a plasma wakefield accelerator–based linear collider was developed based on highly efficient superconducting linacs [71] and subsequently updated in Refs. 72 and 73. Parameters similar to those of ILC were chosen for the main

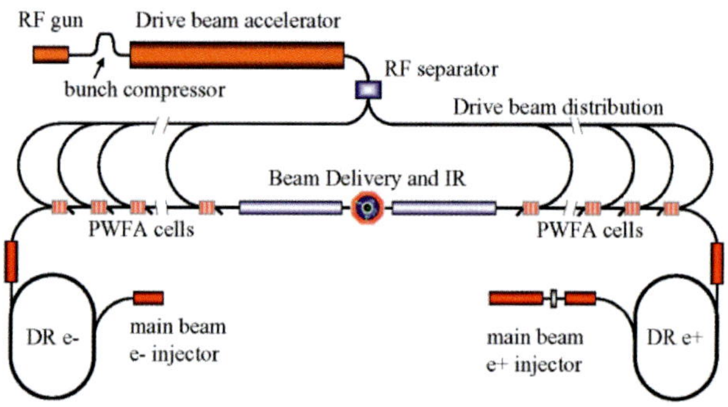

Fig. 15. The concept for a multistage PWFA-based linear collider.

Table 1. Key parameters of the conceptual multistage PWFA-based linear collider illustrated in Fig. 15.

Main beam bunch population	1×10^{10}
Main beam bunches per train	125
Main beam rep rate	100 Hz
Total power two main beams	20 MW
Drive beam energy	25 GeV
Drive beam peak current	2.3 A
Drive beam pulse length	10 μs
Average power drive beam	58 MW
Plasma density	1×10^{17} e^-/cm^3
Plasma accelerating gradient	25 GeV/m
Plasma cell length	1 m
Efficiency: drive-main beam	35%
Efficiency: wall-RF-drive	50% $\times$ 90% = 45%
Overall efficiency	15.7%
Wall plug power for accel.	127 MW
Site power estimate	170 MW
Main beam emittances x, y	2, 0.05 mm-mrad
Main beam size at IP x, y, z	0.14, 0.0032, 10 μm
Luminosity	3.5×10^{34} $cm^{-2}s^{-1}$
Luminosity in 1% energy	1.3×10^{34} $cm^{-2}s^{-1}$

beams, and the plasma cells assumed a 25 GeV main bunch energy gain per meter-long plasma stage for an overall effective gradient of 1 GV/m. This concept for a PWFA-based linear collider is shown schematically in Fig. 16, with key parameters provided in Table 2. As with previous concepts, this design adopted established concepts for the beam generation and focusing systems, taking advantage of the extensive R&D for conventional RF linear colliders such as the ILC [74] and CLIC [75] but with the advantage of potentially lower power consumption and cost.

The acceleration in the plasma cell is a single bunch process. This flexibility in the interval between bunches allows a pulse format with the main bunches colliding in CW mode at several kHz repetition frequency. Each main bunch is rapidly accelerated, with high energy transfer efficiency, in a series of meter-long plasma cells powered by the drive beams. The drive bunches are accelerated by a CW superconducting RF recirculating linac taking advantage of the impressive progress of the RF technology developed by the ILC. This choice provides excellent power efficiency and great flexibility in the number of bunches. The CW mode operation is advantageous from the view point of both the plasma source and the accelerator complex. It should be noted that the time structure of the PWFA-LC could be adapted for pulsed mode operation, without significant reductions in the efficiency of the accelerator complex. This would open up the possibility of an ILC energy upgrade to the multi-TeV energy range without significant modifications to the ILC facility.

In addition to using CW superconducting linacs, this concept traded the many turnaround beam lines used in the previous designs for a series of magnetic chicanes to create the necessary bunch timing. A variant of this concept is illustrated in Fig. 16.

Each of these collider concepts illustrates the enormous potential for high gradient plasma acceleration to drastically shrink the size of a future linear collider. These concepts also highlight many of the key research-and-development areas that must be explored before such concepts can be realized. Many of the topics specific to the plasma accelerating sections were enumerated in Ref. 71 and are summarized below:

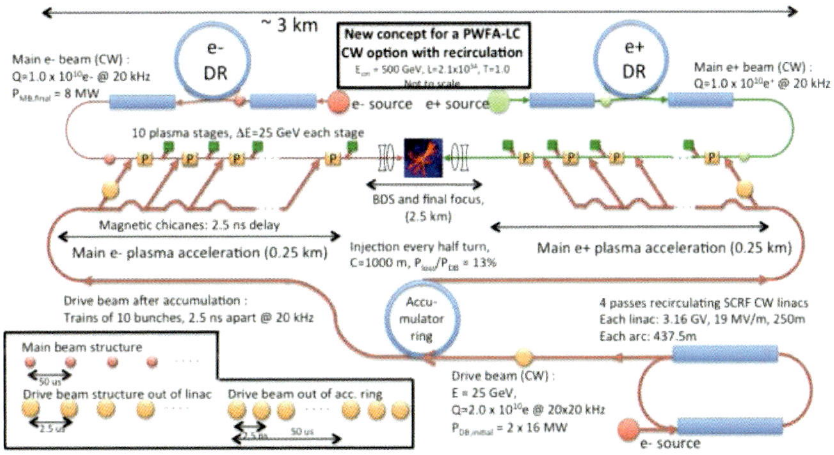

Fig. 16. Layout of a 500 GeV PWFA linear collider. Each main bunch is accelerated by 25 GeV in each of ten plasma stages. The plasma is driven by e^- bunches, generated by a SCRF CW recirculating linac, and distributed collinearly with the main beams.

Table 2. The key parameters for the plasma-wakefield based linear collider concept illustrated in Fig. 16 and described in detail in Ref. 71.

Energy at IP, CM	GeV	500
Mode		e+, e−
Polarization		No, yes
N, experimental bunch		1×10^{10}
Main beam bunches/train		1
Main beam bunch spacing	ns	5×10^4
Repetition rate	Hz	20,000
n exp. bunch/s	Hz	20,000
Avg. current in main beam	uA	32.04
Power in each main beam	W	8×10^6
Effective accelerating gradient	MV/m	1000
Overall length of each linac	m	250
BDS (both sides)	km	2.50
Overall facility length	km	3.00
IP parameters		
Exp. bunch gamepsX,	m	1×10^{-5}
Exp. bunch gamepsY,	m	3.5×10^{-8}
beta-x,	m	1.1×10^{-2}
beta-y,	m	1×10^{-4}
sigx,	m	4.74×10^{-7}
sigy,	m	2.67×10^{-9}
sigxp,	rad	4.31×10^{-5}
sigyp,	rad	2.67×10^{-5}
sigz,	m	2×10^{-5}
Y		2.39×10^{-1}
Dx		1.03×10^{-2}
Dy		1.83
Uave		0.48
delta_B	%	6.66
P_Beamstrahlung [W]		5.3×10^7
ngamma		0.73
Hdx		1.00
Hdy		4.62
Hd		1.7
Geo. lum	(cm^{-2} s^{-1})	1.25×10^{34}
Total lum.	(cm^{-2} s^{-1})	2.09×10^{34}
Int. lum.	(fb^{-1} per 2E7s)	418
Lum.1%		1.25×10^{34}

- Development of a concept for positron acceleration with high beam brightness
- High beam loading with both electrons and positrons (required for high efficiency)
- Beam acceleration with small energy spreads (required to achieve luminosity and luminosity spectrum)
- Preservation of small electron beam emittances (required to achieve luminosity) and mitigation of effects resulting from ion motion
- Preservation of small positron beam emittances (required to achieve luminosity) and mitigation of effects resulting from plasma electron collapse
- Average bunch repetition rates in the tens of kHz (required to achieve luminosity)
- Synchronization of multiple plasma stages to achieve the desired energy
- Optical beam matching between plasma acceleration stages and from plasma to beam delivery systems.

The issues delineated above naturally focus on the heart of the plasma collider concepts — the plasma cell. There are many additional challenges with making and distributing the appropriate drive beams as well as preserving the main beam quality between stages. These issues are discussed in further detail in Ref. 76.

6.2. *Plasma afterburners*

A plasma afterburner is a concept for using short plasma sections several meters in length to double the energy of a linear collider just before the collision point [37]. In this scenario the beams from each side of a linear collider are transformed into pairs of microbunches, with the first driving a plasma wake that accelerates the second. In addition, the luminosity of the energy-doubled collider could be maintained by employing plasma lenses to reduce the beam spot size before collision. The concept for a plasma afterburner is envisioned as a path to exploit the enormous investment made in an existing linear collider and use the benefits of plasma acceleration to provide an upgrade path to higher energy. With the exception of the issues associated with staging multiple plasma cells together, plasma afterburners have similar R&D goals to the stand alone collider designs discussed in Sec 6.1.

6.3. *Near term applications*

The need for novel techniques to produce high energy beams of electrons is most severe in high energy physics applications. Indeed, in terms of almost every parameter (final energy, beam power, emittance, positrons), collider applications are the most demanding. Still, there are other applications that would benefit from a reduction in the size and cost of the accelerators required to make GeV beams, e.g. inverse Compton sources and X-ray free electron lasers. The realization of any such near term application would address many technological issues, such as stability and tolerances that could inform larger scale future initiatives such as a plasma-based linear collider.

One area where free electron lasers have more stringent requirements than colliders is energy spread. A collider final focus system typically has an energy bandwidth of around 1%. Free electron lasers typically require the energy spread to be less than the FEL Pierce parameter, which is typically 10^{-4}. The state of the art for plasma accelerators is currently around 1%. There have been attempts to use and characterize applicability of beams produced by laser wakefield accelerators to drive soft X-ray undulator sources [77], but so far these sources have not progressed from spontaneous radiation to the high gain FEL regime. Alternate approaches to FEL undulator designs such as the transverse gradient undulator [78] may mitigate the tight requirement on energy spread. Other concepts seek to combine the attractive elements of plasma acceleration and injection with the availability of high power, high rep rate electron drivers. Figure 17 shows a concept for a series of X-ray FELs driven by an array of compact plasma accelerators.

In this concept, a single high efficiency high repetition rate electron linac provides individual bunches to a series of beam lines. Each beam line then manipulates the longitudinal phase space of the beam to produce a triangular current profile and drive a plasma wakefield with a transformer ratio of 5, as discussed in Sec. 2.8. Because this beam will not be used in the FEL, and only be used to power the plasma, the emittance and energy spread can

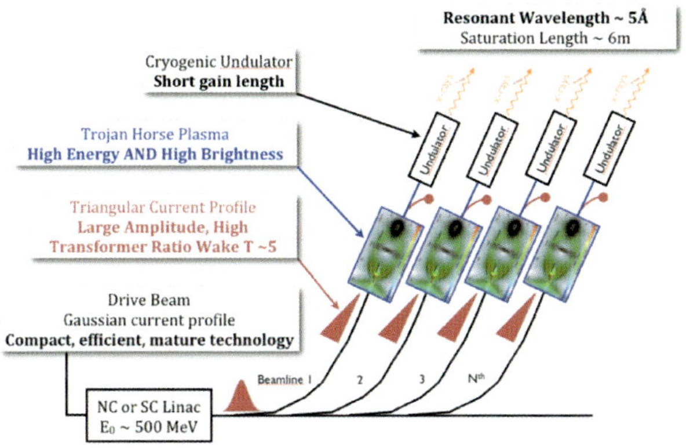

Fig. 17. A concept for a series of compact electron-beam-driven X-ray free electron lasers fed by independent high brightness plasma injectors.

be allowed to grow in the compression process to the limit of fitting inside the plasma bubble. Once this beam sets up the plasma wakefield, a laser pulse is fired, triggering the release of electrons in the plasma wake, as in Sec. 4.3. As a result, an ultra-high brightness beam is produced and accelerated to several times the drive beam energy in a few centimeters of plasma. This high brightness beam can then be sent to a compact undulator to lase at 5 Å and reach FEL saturation in less about 6 m — see Table 3 [73, 79].

As the demand for high brightness electron and photon beams continues to grow, the next decade will likely see some form of plasma wakefield accelerator deployed for a near term demonstration application.

6.4. *Connection to accelerator test facilities*

The progress in plasma wakefield acceleration is directly linked to the availability of accelerator test

Table 3. A set of conceptual parameters for the PWFA-injection-powered FELs illustrated in Fig. 17.

Drive beam	
Charge	3 nC
Energy	500 MeV
Rep rate	1 MHz
Bunch length	210μ m, ramped
Peak current	8.5 kA
Normalized emittance	2.25 mm-mrad
Trojan horse (plasma)	
Plasma density	10^{17} e$^-$/cc
Plasma length	20 cm
Transformer ratio	5
Trojan horse (beam)	
Charge	3 pC
Energy	2.5 GeV
Energy spread	2×10^{-4}
Normalized emittance	3×10^{-8} m-rad
Peak current	300 A
Bunch length	12 fs
Brightness	7×10^{17} A/m^{2}rad^2
Undulator parameters	
Period	9 mm
K	2
Number of periods (N)	660
Radiation parameters	
Wavelength	5.4 Å
Single pulse energy	50 μJ
Number of photons	$>10^{11}$
Peak power	1.6 GW

facilities. Indeed, understanding developed in conjunction with experimental efforts that first observed plasma wakefield acceleration at the Argonne National Lab Advanced Accelerator Test Facility in the late 1980s [80] and later mapped the plasma wakefield at the Argonne Advanced Wakefield Accelerator in the 1990s [81].

The pace of progress accelerated dramatically in the early 2000s, when a collaboration from UCLA, USC and SLAC began to use the electron and positron beams from the SLAC linac to study PWFA at the Final Focus Test Beam (FFTB) facility. The FFTB experiments produced many important first demonstrations, including breaking the 1 GeV energy gain barrier [15], energy doubling of 42 GeV electrons [16], transport and focusing of electrons in meter scale plasmas [5], and the first acceleration of positrons in a plasma [29].

The success of these experiments led to the construction and five-year operation of the Facility for Advanced Accelerator Experimental Tests (FACET) at SLAC [82, 83]. FACET operated as a National User Facility from 2012 to 2016 with a peer-reviewed proposal-driven science program. As with the FFTB experiments, strong university–national-lab collaborations produced many important first demonstrations related to plasma wakefield acceleration, such as high efficiency acceleration of electron and positron beams with narrow energy spread [21, 22, 31] and controlled injection into beam-driven plasma wakes using ionization injection [46] and the Trojan horse technique [50].

In addition to the FACET facility at SLAC studying electron and positron plasma wakefield acceleration, there are many laboratories around the world using high power lasers to study laser-driven plasma acceleration [84], and even a new beam line at CERN to study–proton-driven plasma wakefield acceleration [85–87]. The combined success of all of these programs is leading to increased international competition. At DESY in Germany a new beam line called FLASHForward (Future oriented wakefield accelerator research and development at FLASH) is being constructed off of the FLASH-II FEL beam line to study beam and laser driven plasma acceleration [88, 89]. At the SLAC National Accelerator Laboratory in the United States, a new National User Facility called FACET-II is being constructed using the middle kilometer of the SLAC linac [90]. The focus of these facilities will be basic physics research as well as more directed programs trying to demonstrate capabilities that will enable near term and longer term utilization of plasma wakefield acceleration for high energy physics and photon science applications (see Secs. 6.1–6.3).

6.5. *Simulation development*

Collider modeling, tolerance studies and optimization, and the development of first applications will continue to drive the development of, and benefit from, advances in simulation capabilities. New algorithms for more efficient codes and the continued speedup of available hardware will reduce the time necessary for modeling a given problem and allow the simulation of ever-more extreme parameters. For example, simulating a single meter stage with the resolution required to include ion motion requires 10^7+ core hours. Positron PWFA, beam breakup instabilities, ion motion and high brightness injection schemes will all benefit from higher precision simulations that can be run for long distances and include full physics, such as betatron radiation loss, ion motion, scattering, all ionization models, arbitrary beam and plasma profiles and polarization. Integration with accelerator and FEL codes will be necessary for complete system designs.

In the near term, simulations are needed to help design experiments and also to help interpret and guide experiments, including real time steering. Simulations are needed to investigate new ideas and compare concepts. High fidelity, full scale simulations are needed to explore parameters and concepts not accessible to current experiments. Simulations are also essential for understanding longer term ideas such as a PWFA-LC, where self-consistent beam loading scenarios for both e^- and e^+ with PWFA-LC emittances with/without tailored bunch shaping, transporting beams between stages, and multistages including tolerances on alignment and hosing all need to be modeled.

Some well-tested simulation software exists and needs to be enhanced. Examples are QuickPIC [1] and OSIRIS [47]. The roadmap for simulations will require codes running on next generation hardware, continued innovation and sharing of algorithms, improved fidelity and improved physics.

6.6. *Beam driver technology*

Plasma wakefield acceleration benefits from developments in the broader accelerator community. Research for the ILC helped develop superconducting accelerator structures that are now being deployed in many large scale accelerator projects, such as the European XFEL and the LCLS-II. Industrialization and commissioning experience with these systems makes it possible to envision highly efficient, high repetition rate beam drivers for PWFA using technology already in hand. In another example, the needs of the XFEL community have helped drive timing synchronization and precision from picoseconds in the 1990s to femtoseconds today and are driving quickly to attoseconds in the near future.

6.7. *Diagnostic developments*

Colliders and light sources have very demanding requirements. Diagnostics can help researchers understand the relevant physics within the plasma cell without the need to design all subsystems to collider level tolerances. Experiments vary parameters in a controlled manner, while other parameters jitter. Ideally, diagnostics measure every beam parameter every shot including orbit, charge, bunch length, transverse size, emittance, energy spectrum and phase space. In addition to the beam parameters, diagnostics need to measure plasma parameters such as density, uniformity, length, column width and temporal evolution. With proper diagnostics correlations can be plotted to ascertain a range of acceptable inputs. The small spatial extent and short timescales of the plasma wakefield accelerator processes make these very challenging measurements that require femtosecond time resolution and micron spatial resolution. The advanced accelerator community has a history of innovation in this area and this is a good opportunity to work together to develop common techniques.

6.8. *Plasma source development*

The experiments described elsewhere in this article demonstrate that there are already techniques available for creating plasmas that have the required density and length to provide the accelerating gradient and energy gain envisioned for future applications. The two most important areas for progress in plasma source development concern emittance preservation and repetition rate.

As shown in Secs. 2.3–2.5, the focusing in the plasma ion column is far stronger than for conventional magnetic lenses. To preserve the beam emittance, there needs to be a mechanism to match a beam with finite energy spread from the strong focusing in the ion column to a conventional optical lattice on either end. A theoretical framework has recently been developed that accomplishes this task by precisely tailoring the density profile on the upstream and downstream ends of the constant density plasma acceleration section [91]. Techniques for achieving this level of control in the laboratory are under active investigation by several groups.

The second area of needed development concerns the plasma recoverability from one pulse to the next. Even when it is operated with high beam loading, there will still be significant energy remaining in the plasma wake. For example, a 25 GeV drive beam containing 2.5×10^{10} e^- has 100 J of stored energy. For a collider operating at 15 kHz repetition rate, this represents 1.5 MW of beam power. If 10% of the drive beam energy remains in the plasma wake, this will effectively be 150 kW of energy or heat that must be dissipated before the next bunch arrives. Most experiments to date operate with repetition rates between 1 and 10 Hz. Plasma sources that can dissipate energy quickly while still providing the same plasma density, length and uniformity to the next bunch need to be developed for applications requiring much higher repetition rates.

7. Conclusion

The field of beam-driven plasma acceleration continues to mature, and near and longer term applications are being contemplated with ever greater specificity. Progress has been rapid as strong collaborations between national laboratories and university groups have combined a breadth of expertise in areas such as accelerators, lasers, plasmas, and computation to conduct ambitious experiments at state-of-the-art accelerator test facilities.

To chart the course for the next decade, the advanced accelerator community has recently come together at a series of workshops and developed a high level roadmap for beam-driven plasma acceleration [92]. Two areas of beam-plasma physics rose to the top of the list of research priorities for the next decade: emittance preservation and positron acceleration. Additional priorities include beam loading, higher transformer ratios, beam dynamics and tolerances, plasma source development, staging, and off-ramps and first applications. Continued progress will need powerful test facilities such as FACET-II and FLASHForward to continue to test new theories and designs. New diagnostics and more powerful simulation capabilities need to be developed and tested. Many common themes and research areas have been identified as opportunities for collaboration nationally and internationally. It is an exciting time for the field of plasma wakefield acceleration, and the years to come will no doubt see continued progress and discovery, the development of first applications and refined concepts for light sources and energy frontier colliders.

References

[1] C. Huang *et al.*, *J. Comp. Phys.* **217**, 658 (2006).
[2] W. An *et al.*, *J. Comput. Phys.* **250**, 165–177 (2013).
[3] K. G. Steffen, in *High Energy Beam Optics* (John Wiley, New York, 1965), p. 173.
[4] C. Clayton *et al.*, *Phys. Rev. Lett.* **88**, 154801 (2002).
[5] P. Muggli *et al.*, *Phys. Rev. Lett.* **93**, 014802 (2004).
[6] C. Thaury *et al.*, *Nat. Commun.* **6**, 6860 (2015).
[7] J. J. Su *et al.*, *Phys. Rev. A* **41**(6), 3321 (1990).
[8] J. S. T. Ng *et al.*, *Phys. Rev. Lett.* **87**, 244801 (2001).
[9] C. L. O'Connell *et al.*, *Phys. Rev. ST Accel. Beams* **5**, 121301 (2002).
[10] P. Muggli *et al.*, *Nature* **411**, 43 (2001).
[11] P. Muggli *et al.*, *Phys. Rev. ST Accel. Beams* **4**, 091301 (2001).
[12] E. Adli *et al.*, submitted for publication. Private communication (2016).
[13] P. Chen *et al.*, *Phys. Rev. Lett.* **54**, 693 (1985).
[14] J. B. Rosenzweig *et al.*, *Phys. Rev. A* **44**, R6189 (1991).
[15] M. J. Hogan *et al.*, *Phys. Rev. Lett.* **95**, 054802 (2005).
[16] I. Blumenfeld *et al.*, *Nature* **445**, 741 (2007).
[17] K. V. Lotov, *Phys. Rev. E* **69**, 046405 (2004).
[18] W. Lu *et al.*, *Phys. Rev. Lett.* **96**, 165002 (2006).
[19] M. Tzoufras *et al.*, *Phys. Rev. Lett.* **101**, 145002 (2008).
[20] M. Tzoufras *et al.*, *Phys. Plas.* **16**, 056705 (2009).
[21] M. Litos *et al.*, *Nature* **515**, 92–95 (2014).
[22] M. Litos *et al.*, *Plas. Phys. Control. Fus.* **58**, 034017 (2016).
[23] K. L. F. Bane *et al.*, *IEEE Trans. Nucl. Sci.* **32**, 3524 (1985).
[24] R. D. Ruth *et al.*, *Part. Accel.* **17**, 171 (1985).
[25] W. Lu *et al.*, *Proc. Particle Accelerator Conference* (Vancouver, B.C.; May 4–8, 2009).
[26] B. Jiang *et al.*, *Phys. Rev. Spec. Top. Accel. Beams* **15**, 1 (2012).
[27] M. J. Hogan *et al.*, *Phys. Rev. Lett.* **90**, 205002:1–4 (2003).
[28] P. Muggli *et al.*, *Phys. Rev. Lett.* **101**, 055001 (2008).
[29] B. Blue *et al.*, *Phys. Rev. Lett.* **90**, 214801:1–4 (2003).
[30] J. B. Rosenzweig *et al.*, *AIP Conf. Proc.* **1299**, 500–504 (2010).
[31] S. Corde *et al.*, *Nature* **524**, 442 (2015).
[32] T. C. Chiou *et al.*, *Phys. Plasmas* **2**, 310–318 (1995).
[33] T. C. Chiou *et al.*, *Phys. Rev. Lett.* **81**, 3411–3414 (1998).
[34] C. B. Schroeder *et al.*, *Phys. Rev. Lett.* **82**, 1177–1180 (1999).
[35] W. D. Kimura *et al.*, *Phys. Rev. Spec. Top. Accel. Beams* **14**, 041301 (2011).
[36] S. Gessner *et al.*, *Nat. Commun.* **7**, 11785 (2016).
[37] S. Lee *et al.*, *Phys. Rev. Spec. Top. Accel. Beams* **5**, 011001–1–4 (2002).
[38] N. Kirby *et al.*, *Proc. Particle Accelerator Conference* (Albuquerque, NM 2007), pp. 3097–3099.
[39] K. Kondratenko *et al.*, *Part. Accel.* **10**, 207–216 (1980).
[40] R. Bonifacio *et al.*, *Opt. Commun.* **50**, 373–378 (1984).
[41] P. Emma *et al.*, *Nat. Photon.* **4**, 641–647 (2010).
[42] J. Madey *et al.*, *J. Appl. Phys.* **42**, 1906–1913 (1971).
[43] D. H. Dowell *et al.*, presented at 9th European Particle Accelerator Conference (EPAC2004) (Lucerne, Switzerland; July 5–9, 2004).
[44] E. Oz *et al.*, *Phys. Rev. Lett.* **98**, 084801 (2007).
[45] N. A. Kirby, *Phys. Rev. Spec. Top.-Accel. Beams* **12**(5), 053101 (2009).
[46] N. Vafaei *et al.*, *Phys. Rev. Lett.* **112**, 025001 (2014).
[47] R. A. Fonseca *et al.*, in *Proc. Int. Conf. Computational Science — Part III* (Springer-Verlag, London, 2002), p. 342.
[48] B. Hidding *et al.*, *Phys. Rev. Lett.* **108**, 035001 (2012).
[49] Y. Xi *et al.*, *Phys. Rev. Spec. Top. Accel. Beams* **16**, 031303 (2013).
[50] A. Deng *et al.*, private communication (and to be published) (2016).
[51] F. Li *et al.*, *Phys. Rev. Lett.* **111**, 015003 (2013).
[52] X. L. Xu *et al.*, *Phys. Rev. Lett.* **112**, 035003 (2014).
[53] Y. Wan *et al.*, *Plasma Phys. Control. Fusion* **58**, 034015 (2016).
[54] M. C. Thompson *et al.*, *Phys. Rev. Spec. Top. Accel. Beams* **7**, 011301 (2004).
[55] H. Suk, N. Barov, J. B. Rosenzweig and E. Esarey, *Phys. Rev. Lett.* **86**, 1011 (2001).
[56] A. Martinez de la Ossa *et al.*, *Phys. Rev. Lett.* **111**, 245003 (2013).

[57] J. Grebenyuk, *Nucl. Instrum. Methods Phys. Res. Sec. A: Accelerators, Spectrometers, Detectors and Associated Equipment*, **740**, 11 (2014).
[58] G. Wittig, *Phys. Rev. Spec. Top. Accel. Beams* **18**, 081304 (2015).
[59] S. Wang *et al.*, *Phys. Rev. Lett.* **88**, 135004 (2002).
[60] D. Whittum *et al.*, *Phys. Rev. Lett.* **67**, 991 (1991).
[61] A. A. Geraci and D. H. Whittum, *Phys. Plasmas* **7**, 3431 (2000).
[62] *SLC Design Handbook* R. Erickson (ed.), (1984), Chap. 5, pp. 15–79 (unpublished).
[63] D. K. Johnson *et al.*, *Phys. Rev. Lett.* **97**, 175003: 1–4 (2006).
[64] S. Corde *et al.*, *Rev. Mod. Phys.* **85**, 1 (2013), arXiv:1301.5066v1 and references therein.
[65] S. Kneip *et al.*, *Nat. Phys.* **6**, 980–983 (2010).
[66] A. Rousse *et al.*, *Phys. Rev. Lett.* **93**, 135005 (2004).
[67] S. Cipiccia *et al.*, *Nat. Phys.* **7**, 867–871 (2011).
[68] K. Ta Phuoc *et al.*, *Nat. Photon.* **6**, 308–311 (2012).
[69] J. B. Rosenzweig *et al.*, *Nucl. Instrum. Methods A* **410**, 532 (1998).
[70] A. Seryi *et al.*, *Proc. Particle Accelerator Conference* (Vancouver, B.C.; May 4–8, 2009), pp. 1–3.
[71] E. Adli *et al.*, arXiv:1308.1145.
[72] E. Adli *et al.*, in *Proc. IPAC2013* (Shanghai, China, 2013), pp. 1613–1615.
[73] J. P. Delahaye *et al.*, in *Proc. IPAC2014* (Dresden, Germany, 2014), pp. 3791–3793.
[74] http://www.linearcollider.org/ILC/GDE/technical-design-documentation (2013).
[75] http://clic-study.org/accelerator/CLIC-ConceptDesignRep.php (2012).
[76] D. Schulte *et al.*, *Reviews of Accelerator Science and Technology*, current volume (2016).
[77] M. Fuchs *et al.*, *Nat. Phys.* **5**, 826 (2009).
[78] Z. Huang *et al.*, *Phys. Rev. Lett.* **109**, 204801 (2012).
[79] B. Hidding, J. B. Rosenzweig and Z. Huang, private communication (2013).
[80] J. B. Rosenzweig *et al.*, *Phys. Rev. Lett.* **61**, 1 (1988).
[81] N. Barov *et al.*, *Phys. Rev. Lett.* **80**, 1 (1998).
[82] M. J. Hogan *et al.*, *CERN Courier* (Mar. 2011), pp. 23–26.
[83] M. J. Hogan *et al.*, *New J. Phys.* **12**, 055030 (2010).
[84] Laser-driven plasma accelerators, in *Reviews of Accelerator Science and Technology*, current volume (2016).
[85] A. Caldwell *et al.*, *Nat. Phys.* **5**, 363–367 (2009).
[86] A. Caldwell *et al.*, CERN-SPSC-2013-013, SPSC-TDR-003.
[87] E. Gschwendtner *et al.*, in *Proc. IPAC2014* (Dresden, Germany), pp. 582–585.
[88] A. Aschikhin *et al.*, arXiv:1508.03192v2 (2015).
[89] A. Aschikhin *et al.*, *Nucl. Instrum. Methods A* **806**, 175–183 (2016).
[90] SLAC-R-1067.
[91] X. L. Xu *et al.*, *Phys. Rev. Lett.* **116**, 124801 (2016).
[92] E. Colby and L. K. Len, in *Reviews of Accelerator Science and Technology*, current volume (2016).

Mark J. Hogan is the Plasma Group Leader and the Deputy Director of the FACET Division at SLAC National Accelerator Laboratory. He earned his Ph.D. in Physics from UCLA in 1998, working on self-amplified spontaneous emission free electron laser experiments at UCLA and Los Alamos National Laboratory. He went to SLAC in 1998 to work on the first beam-driven plasma wakefield accelerator experiments and has served as the spokesperson for all follow-on experiments from the year 2000 to the present. He is currently in charge of the science program for the FACET National User Facility and is developing ideas for the follow-on facility FACET-II. His research interests involve the study of high energy beam, lasers, plasmas and their interaction with matter, as well as the physics of beam-driven plasma wakefield acceleration and the development of compact TeV linear colliders and X-ray free electron lasers based on plasma wakefield acceleration.

Reviews of Accelerator Science and Technology
Vol. 9 (2016) 85–104

DOI: 10.1142/S1793626816300048

World Scientific
www.worldscientific.com

Proton-Beam-Driven Plasma Acceleration

Erik Adli
Department of Physics, University of Oslo, N-0316 Oslo, Norway
erik.adli@fys.uio.no

Patric Muggli
Max Planck Institute for Physics, Munich, Germany
muggli@mpp.mpg.de

We describe the main ideas, promises and challenges related to proton-driven plasma wakefield acceleration. Existing high-energy proton beams have the potential to accelerate electron beams to the TeV scale in a single plasma stage. In order to drive a wake effectively the available beams must be either highly compressed or microbunched. The self-modulation instability has been suggested as a way to microbunch the proton beams. The AWAKE project at CERN is currently the only planned proton-driven plasma acceleration experiment. A self-modulated CERN SPS beam will be used to drive a plasma wake. We describe the design choices and experimental setup for AWAKE, and discuss briefly the short-term objectives as well as longer-term ideas for the project.

Keywords: Plasma wakefield acceleration; proton beams; CERN; AWAKE.

1. Introduction

A plasma wakefield accelerator is an energy transformer. A drive beam propagates through a plasma, creating a strong plasma wake and in the process losing energy to the plasma. Part of this energy can be used to accelerate a trailing beam, here called the "witness beam." The driver can be a high-intensity laser beam, where the ponderomotive force of the laser pulse drives the plasma wake [1, 2], or a charged particle beam, where the space charge force drives the plasma wake [3, 4]. For laser drivers, the technique is called "laser wakefield acceleration" (LWFA), while for particle beam drivers, the term "plasma wakefield acceleration" (PWFA) has been adopted. For both techniques, the amount of energy available for acceleration is given by the energy of the driver and the energy transfer efficiency from the drive beam to the witness beam. In order to efficiently drive the plasma with a particle bunch, the driver length σ_z should be such that $k_{\mathrm{pe}}\sigma_z \simeq \sqrt{2}$ [5], where $k_{\mathrm{pe}} = 2\pi/\lambda_{\mathrm{pe}}$, with λ_{pe} being the cold electron plasma wavelength. The transverse spot size σ_r should be $k_{\mathrm{pe}}\sigma_r \lesssim 1$ in order to avoid development of filamentation instability [6]. Figure 1(a) illustrates a drive–witness bunch configuration.

Assuming dephasing or emittance-driven head erosion [7] are not limiting factors, the drive beam can drive the wake until its energy has been depleted. At this point a new drive beam must be injected into a new plasma stage in order to accelerate the witness beam to higher energies, as illustrated in Fig. 1(b). Injection and ejection of the drive beam requires a potentially long and complex interstage beam line between the plasma stages [8], resulting in a reduction of the average accelerating gradient. The energy gain of a witness particle in a single plasma stage may also be limited by the transformer ratio, defined as the ratio of the accelerating field within the witness bunch to the decelerating field within the drive bunch. For a symmetric bunch and for linear wakes, the transformer ratio is limited to 2, and the witness particle energy gain limited to twice the drive particle energy. Simulations studies indicate that bunch shaping [4, 9] or bunch trains [4, 10] may provide larger transformer ratios. However, transformer ratios of 2 or more have so far not been demonstrated experimentally in PWFA two-bunch experiments [11], and high-transformer-ratio PWFA remains a field of active research. The energy gain of a witness bunch in a single stage may therefore be limited by either the drive particle energy or the drive beam total energy.

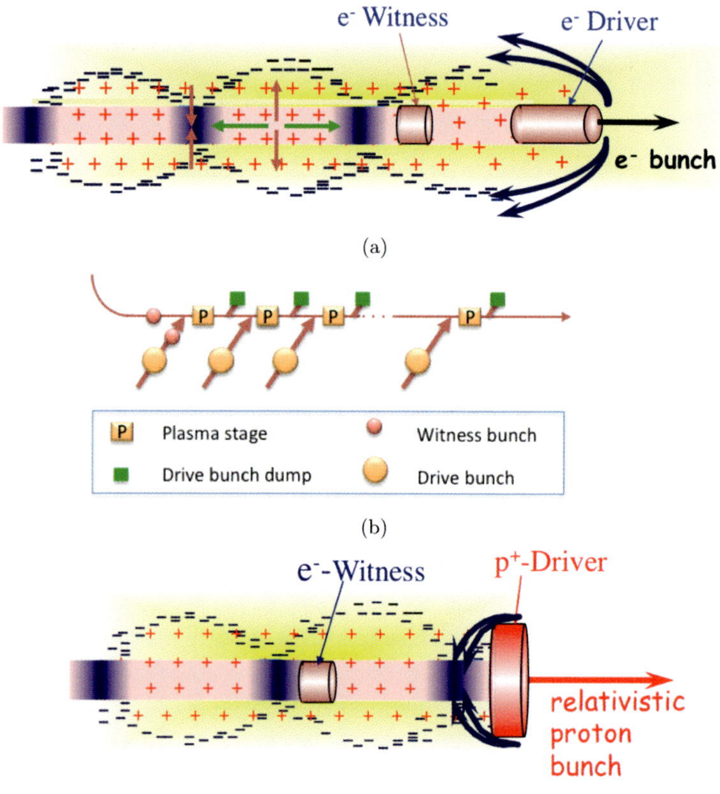

Fig. 1. (a) The PWFA principle: a drive beam drives a wake in a plasma. Part of the energy lost to the plasma can be used to accelerate a trailing beam. (b) When a drive beam is depleted and cannot drive the plasma anymore, it is ejected from the beam line and a new drive bunch can be injected into a new plasma stage. PD-PWFA has the potential for providing very-high-energy gain in a single stage. (c) A short proton bunch can generate a plasma wake structure similar to the electron-driven blowout regime, a structure ideal for accelerating electrons.

Following the successful demonstration of high-gradient acceleration in electron-beam-driven PWFA [4], ideas to use energetic proton beams as drivers came about — first written down in Refs. 12 and 13, illustrated in Fig. 2 and discussed in more detail below. Due to the heavy mass of protons and consequently low synchrotron radiation losses, it is possible to accelerate protons to very high energies in synchrotrons. In a single proton-driven PWFA stage, electron or positron bunches can potentially be accelerated to energies of the same order as the proton energy, without the need for high transformer ratios. A positively charged drive bunch pulls plasma electrons toward the axis as it propagates [14], as opposed to an electron drive bunch, which blows the plasma electrons away from the axis. If the electron drive current is large enough, plasma electrons may be completely blown away during the passage of the driver, forming an accelerating cavity as the blown-out plasma electrons are attracted back toward the ions; see Fig. 1(a). This operating regime is commonly known as the blowout regime and has been well studied for electron drivers, both theoretically and experimentally [4], and is ideal for

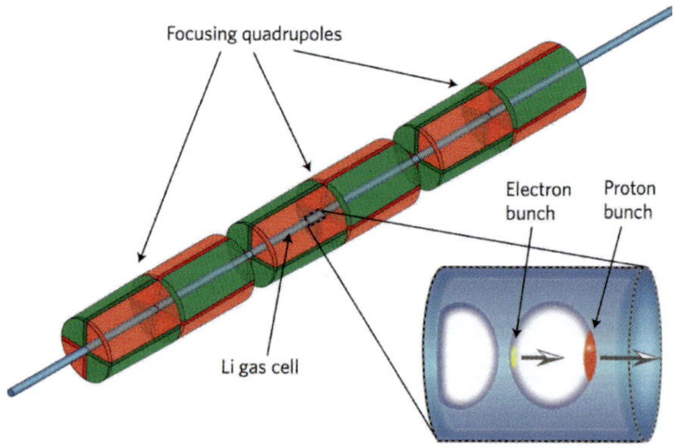

Fig. 2. Illustration of a PD-PWFA concept; a compressed version of the LHC bunch accelerates an electron bunch to 620 GeV in a single, 450-m-long plasma stage, with external focusing quadrupoles. (From Ref. 13.)

accelerating electrons. If the positively charged driver is sufficiently short, the plasma electrons will mostly overshoot behind the drive bunch and form a plasma cavity structure similar to that of an electron driver, as discussed in Refs. 13 and 15 and illustrated in Fig. 1(c). This is true for both positron and proton bunches. High-energy proton bunches are more easily and more efficiently produced than positron ones. An ideal drive beam for accelerating electron bunches to high energy in a single stage is therefore a short proton bunch with both high particle energy and sufficient total bunch energy.

2. High-Energy Proton Drivers

High-energy proton bunches produced by synchrotrons are available at CERN in Switzerland [16], Fermilab in the USA [17], J-PARC in Japan [18] and IHEP in Russia [19]. Table 1 summarizes the main beam parameters of proton synchrotrons potentially of interest for proton-driven PWFA (PD-PWFA), here defined as beams with a particle energy of at least 10 GeV. Figure 3 shows the CERN accelerator complex, which produces three of the highest-energy proton beams currently available; the Proton Synchrotron (PS), Super Proton Synchrotron (SPS) and Large Hadron Collider (LHC) beams. The LHC provides the highest particle energy available worldwide (7 TeV), while J-PARC provides the highest total bunch energy (130 kJ).

2.1. *Generating efficient drivers*

The scale of accelerating fields in plasmas is given by the wave-breaking field, of strength $\sim\sqrt{n_0/10^{16}\text{cm}^{-3}}10\,\text{GV/m}$ [20]. In order to obtain accelerating fields of order GV/m in PWFA, the plasma density needs to be $\gtrsim 10^{14}\,\text{cm}^{-3}$, with the corresponding plasma wavelength $\lambda_{\text{pe}} \lesssim$ few mm. This requires drive beam lengths, $k_{\text{pe}}\sigma_z \sim \sqrt{2}$, to be of the order of a few hundred μm (duration of ps). Since available high-energy proton beams have bunch lengths of the order of a few hundred mm (duration of ns), these bunches are several orders of magnitude too long to efficiently drive a plasma wake. Available proton bunches are long because until now there has been little need for short bunches, and because instabilities in synchrotrons increasingly become a problem with increasing beam current. To generate drivers for PD-PWFA, it has been suggested to either compress the entire beam or microbunch the beam in order to resonantly drive the wake. Both options are discussed in the following.

2.1.1. *Bunch compression*

In Ref. 16 a number of options for bunch length manipulation in synchrotrons are studied, including bunch rotation prior to extraction, for example by lattice manipulation and RF voltage changes, and 4D or 6D emittance exchange transformations, possibly involving fast-deflecting crab cavities. Reference 16 concludes that a single method is probably not sufficient to reduce the bunch length from the existing ns lengths to ps length, and would probably need to be combined with bunch compression after extraction from the synchrotron. After extraction, the bunch can be shortened in a magnetic bunch compressor, as is common for electron bunches. First, RF voltage

Table 1. Proton beams of 10 GeV or higher energy worldwide.

Synchrotron	Maximum particle energy (GeV)	Maximum bunch energy (kJ)	Minimum bunch duration (ns) rms	Relative energy spread (%) rms	Transverse emittance (mm mrad)
CERN PS, Geneva, Switzerland	26	0.54	1	0.1	3.0
CERN SPS, Geneva, Switzerland	450	8.3	0.4	0.03	3.5
CERN LHC, Geneva, Switzerland	7,000	124	0.25	0.01	3.75
Fermilab Main Injector, Batavia, Illinois, USA	120 (150)	2.1	0.6	0.04	5.4
J-PARC, Ibaraki prefecture, Japan	30	130	15	0.05	1
IHEP U70, Protvino, Russia	50 (70)	4.0 (5.6)	12	0.1	2

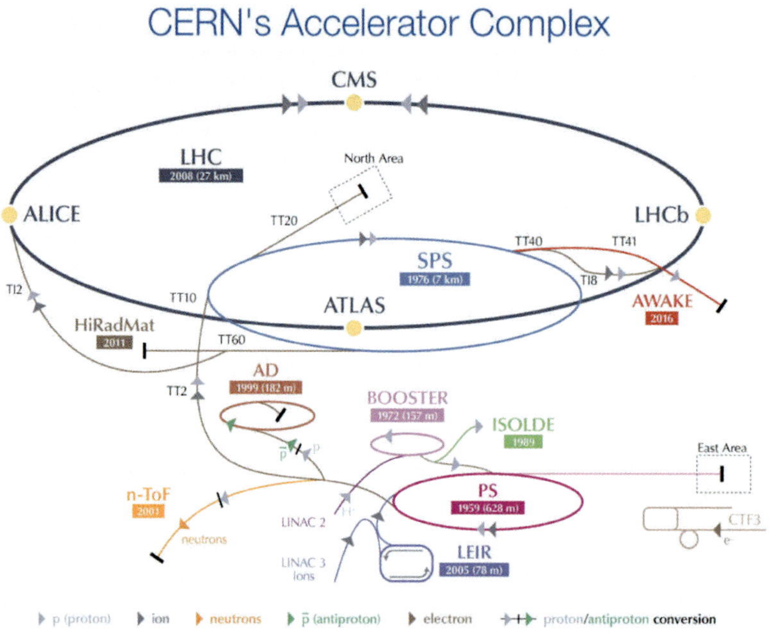

Fig. 3. The CERN accelerator complex produces three of the highest-energy proton beams currently available: the PS, SPS and LHC beams. The LHC bunches provide the highest particle energy available worldwide (7 TeV design energy). (Figure courtesy of CERN.)

is used to introduce an energy modulation varying along the bunch. Then, particles follow dispersive paths in a magnetic chicane or dogleg, compressing the bunch longitudinally, while increasing the uncorrelated energy spread. In Ref. 21 it is estimated that to compress a 1 TeV beam from 1 cm to 165 μm would require 65 GV of RF voltage, and in Ref. 22, 25 GV of RF voltage is estimated to compress the CERN SPS bunch from 1 cm to 350 μm. Figure 4 illustrates different compressions methods. While bunch compression of high-energy proton bunches may be interesting for future PD-PWFA applications, this would not necessarily be the preferred choice for the first PD-PWFA experiments, due to the size and cost of the RF system and magnetic compressors.

2.1.2. *Microbunching and self-modulation*

An effective drive beam can also be produced by microbunching of an initially long bunch, so that the wake of each microbunch adds to those set up by leading bunches [3, 10, 23]. The microbunching

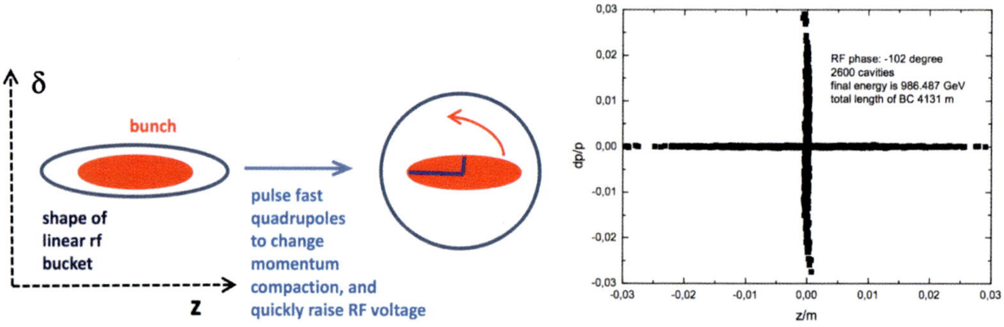

(a) (b)

Fig. 4. Proton bunches can be compressed by a number of different methods, such as (a) fast bunch rotation prior to extraction from the synchrotron, or (b) magnetic bunch compression after extraction; the plot shows the longitudinal phase space before (flat) and after (upright) compression. Even when one is combining different methods, the RF voltage required to generate proton drivers of ps duration would imply large and costly compression systems. [Figure (a) from Ref. 16; Fig. (b) from Ref. 21.]

spacing must be $\sim\lambda_{pe}$, in order to drive the wake resonantly. We first discuss the concept of self-modulation, as this has a particular interest for PD-PWFA. It was understood early that a transverse two-stream instability [24] could radially microbunch a long beam propagating in a plasma [25], with bunching period $\sim\lambda_{pe}$ — a process called the self-modulation instability (SMI). The equations for the SMI of a particle beam in a plasma are analogous to those of a laser beam in plasma, extensively studied in for example Refs. 26 and 27. The promise of PD-PWFA [13], combined with the technical challenges of compressing existing proton bunches to lengths of $\sim k_{pe}^{-1}$, leads to a renewed interest for the SMI in a particle beam. In Refs. 28 and 29 an analytical model for the SMI is developed, and mathematical expressions for the instability growth rates are provided. It is shown by numerical simulations that the microbunches generated by the SMI have the correct bunch-to-bunch distance to resonantly excite a large amplitude wake ($\sim$GV/m). It is furthermore shown that the SMI can be seeded, either by a precursor bunch [25] or by using a charge profile with a hard leading edge [28]. In order to synchronize the wakefields with an injected witness bunch, seeding is important. Also, if the SMI is not seeded the competing transverse hosing instability [24] may destroy the coherence of the plasma wakefields, strongly limiting the amplitude of the fields. References 28 and 30 show that by seeding the SMI the growth rate of the hosing instability may be greatly suppressed, with the wakefields mostly unperturbed. Figure 5, from Ref. 28, illustrates the effect of seeding by using a hard-edge half Gaussian as charge profile. For a regular Gaussian charge profile, the on-axis longitudinal field is suppressed by a factor of 5 with respect to the half Gaussian, as the former is subject to the competing hosing instability. Due to the simplicity of the SMI bunching scheme, where no extra bunching hardware or compression hardware is required, this concept was adopted for the first proposals for PD-PWFA experiments [31, 32].

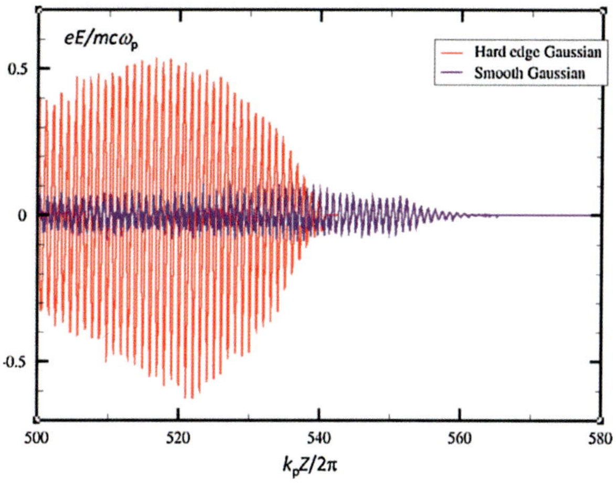

Fig. 5. The axial longitudinal wakefield in a plasma (the beam travels toward the right): in purple, for a regular Gaussian charge profile: in red, for a hard-edge half Gaussian where the cut seeds the self-modulation instability. Because of the competing hosing instability, the field generated by the regular Gaussian is suppressed by a factor of 5 with respect to hard-edge half Gaussian. (From Ref. 28.)

In numerical simulations with self-modulated drivers, it was observed that although a large amplitude wakefield is indeed produced once the SMI has developed, the proton beam continues to evolve as it propagates, eventually leading to decay of the longitudinal field [33]. It was furthermore observed that by having a step in the plasma density of a few percent, in the region where the SMI is developing, the proton beam evolution can be partly stabilized and the field decay mitigated [33]. Later work [34, 35] discuss the optimal location and amplitude of such a step as a function of train length, beam emittance and other parameters. However, whether a stepped plasma density profile is optimal to best stabilize the proton beam evolution is not fully understood. The field amplitudes arising from self-modulated beams, as well as the efficiency of the step, are parameter-dependent. In Ref. 36 the accelerating fields generated by the CERN PS, SPS and LHC beams are compared, both without and with a density step, and the results are summarized in Fig. 6. With a stepped plasma density profile, the wake from the LHC beam stabilizes at a field amplitude of about 1 GV/m, and can efficiently drive a wake for many kilometers, according to Ref. 36. These simulations were performed using the hybrid code LCODE [37, 38]. It was found that the peak fields are roughly proportional to the number of particles in the drive bunch and inversely proportional to the square of the transverse bunch size, which is in good correspondence with theory [5, 34]. However, in some cases, as discussed in Ref. 39, the wakefield amplitude is limited by nonlinear elongation of the wave period.

Alternatively, a train of microbunches with the desired frequency can be generated by longitudinal

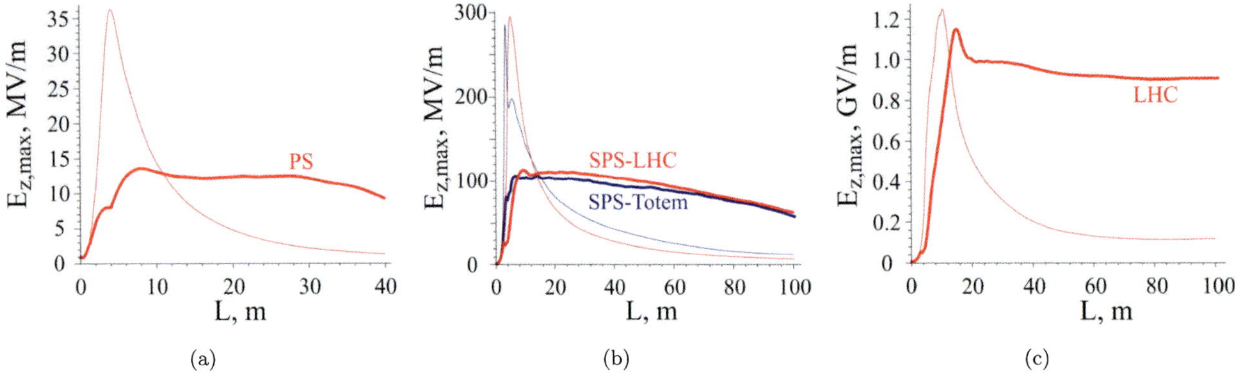

Fig. 6. The peak axial accelerating field from self-modulated proton bunches: (a) CERN PS, (b) CERN SPS and (c) CERN LHC. The thin lines show the fields for a uniform plasma profile, while the bold lines show the field when the plasma profile has a density increase in a step of a few percent after 1–3 m of propagation. This density step aids in stabilizing the proton beam evolution and the accelerating fields. All simulations are seeded by using a hard-cut half Gaussian charge profile. (From Ref. 36.)

bunching of the proton beam before it enters the plasma. A number of methods have been studied for generating microbunched drive trains, also because this is an important topic for free electron lasers. They include energy modulation followed by velocity bunching or magnetic compression, driving a photocathode with a pulsed laser train ("comb-laser") [41], *or* using metal masks to carve out a bunch pattern in one of the transverse dimensions, typically at low energy. After the mask, the transverse modulation is converted into a longitudinal modulation by either introducing correlated energy spread and high dispersion where the mask is located [42], or using emittance exchange lattices [43]. Longitudinal microbunching has the following advantages over self-modulation: it can provide more control over the initial bunch phase, there is no need to rely on an instability to grow in the plasma, and a larger percentage of the initial charge can be used to drive the wake. On the other hand, longitudinal bunching would require hardware upgrades of the existing facilities, for example large amounts of new RF power. Most of the above methods are not mature enough to be readily used without further R&D. For PD-PWFA applications, the plasma wavelength must be enforced to match the longitudinal bunching wavelength, putting constraints on the plasma density tolerances. Microbunching by self-modulation has the advantage that the bunch spacing is automatically adjusted to drive the plasma resonantly. The evolution over long distances of both self-modulated and longitudinally bunched trains is being investigated. Which option will work better in practice to generate the most efficient PD-PWFA drivers remains to be studied experimentally.

3. PD-PWFA Experiments

The first written concept of a PD-PWFA accelerator, "Design for an Advanced Accelerator Driven by a Proton (or Ion) Beam," is from 2003 [12]. In this work it is proposed to use a 1 TeV proton beam to accelerate electron or positron beams up to an energy of 500 GeV or more. 2D numerical simulations were performed assuming a short proton beam. The challenge of compressing available Fermilab proton beams to the required lengths was acknowledged. This work was not pursued further. A few years later, in 2009, Ref. 13 discussed various aspects of PD-PWFA, also assuming a short bunch of 100 μm length. Acceleration of 600 GeV electrons using a 1 TeV proton bunch was demonstrated by 2D and 3D numerical simulations. Strong external focusing quadrupoles were used to mitigate emittance expansion of the drive beam head. In Ref. 36 numerical studies showed that the nominal-length LHC beam, after undergoing self-modulation, could also accelerate a test electrons up to 6 TeV, without the use of external focusing quadrupoles. Following the promise of PD-PWFA in Refs. 13 and 36, a protocollaboration studied various options for drive beam and site locations, and a Letter of Intent for a "Demonstration Experiment in Proton-Driven Plasma Wakefield Acceleration" [31] was submitted to the CERN SPS and PS experiments committee (SPSC) in May 2011, with the aim of demonstrating PD-PWFA using the CERN SPS beam, relying on the self-modulation instability to produce an efficient driver. The experiment, named "The Advanced Proton-Driven Plasma Wakefield Acceleration Experiment," or AWAKE, was approved by the CERN SPSC and is at the time of writing undergoing commissioning. AWAKE is the only approved PD-PWFA experiment worldwide and will be discussed in more detail below.

PROTOPLASMA [32] at Fermilab was a proposed R&D project with many similarities to the AWAKE proposal [31], suggesting to use the 120 GeV Fermilab Main Injector beam as a self-modulated proton driver. This proposal was not approved and there are currently no proposals to perform PD-PWFA experiments at Fermilab. In Asia and Russia there are also currently no plans or proposals to perform PD-PWFA experiments.

3.1. *Proof-of-principle studies*

Proposed and ongoing PD-PWFA experiments depend on the self-modulation instability to generate a microbunch drive train from a long proton bunch. Also, electron or positron beams, long with respect to the plasma wavelength, may self-modulate in a plasma [44, 45]. Several experiments have been proposed and performed in order to study the SMI in electron and positron beams, since such beams have been more readily accessible at accelerator test facilities.

At the Brookhaven National Laboratory Accelerator Test Facility (ATF), an experiment to study the SMI as well as the seeding of the instability was performed [44]. In the experiment a 58 MeV electron bunch, about 1 mm long with a sharp rising edge, was sent into plasmas of varying densities, corresponding to 1–7 plasma wavelengths along the bunch. Evidence of energy modulation at the plasma wavelengths was demonstrated [46], as shown in Fig. 7(a). It was also shown that the sharp rising edge of the bunch seeds the phase of the wakefield, a necessary condition for the deterministic external injection of a witness bunch in an accelerator. The results show that the seeding of the instability is possible since the transverse wakefields necessary for the SMI seeding are always associated with the longitudinal ones observed in the experiment.

In Ref. 45 the SMI in the 20 GeV electron and positron beams available at the FACET test facility [47] is studied numerically, showing that the beams may self-modulate if $\sigma_z \gg \lambda_{\rm pe}$, and if the SMI is seeded in order to suppress the competing hosing instability. Experiments with the target of demonstrating the SMI in electron and positron

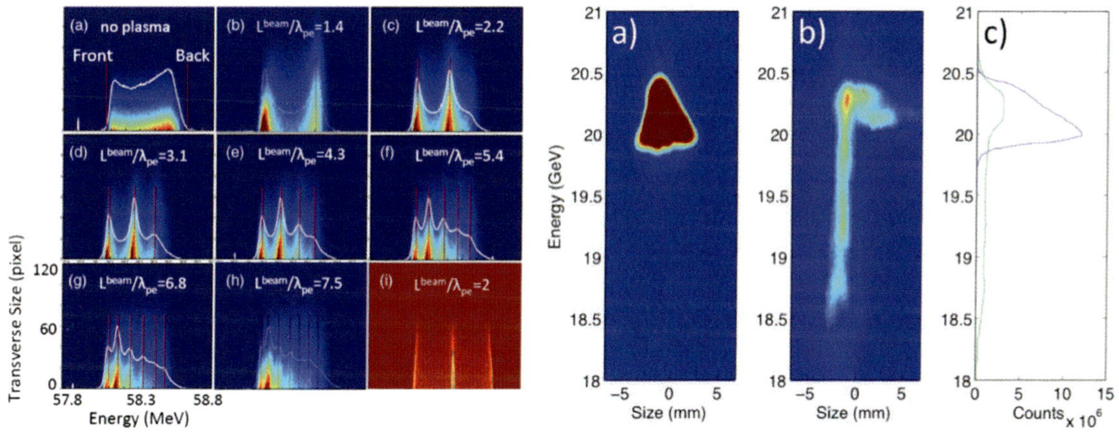

(a) (b)

Fig. 7. Proof-of-principle studies which aim at demonstrating the self-modulation instability in lepton beams, performed with long electron and positron bunches where $k_{\rm pe}\sigma_z \gg 1$. (a) At ATF/BNL, evidence of energy modulation along a long bunch, at the plasma wavelength, was observed. The experiment also demonstrated that a steeply rising charge profile seeded the fields, ensuring a deterministic phase reference [46]. (b) At FACET/SLAC, energy loss one order of magnitude stronger than that expected from linear theory was observed — about 2 GeV and about 100 MeV, respectively [48]. The measured 2 GeV deceleration is in reasonable agreement with numerical simulations. (c) The plasma source at PITZ/DESY, where experiments are underway with the aim of producing direct evidence of the SMI of electron beams [49]. (Figures from Refs. 46, 48 and 49.)

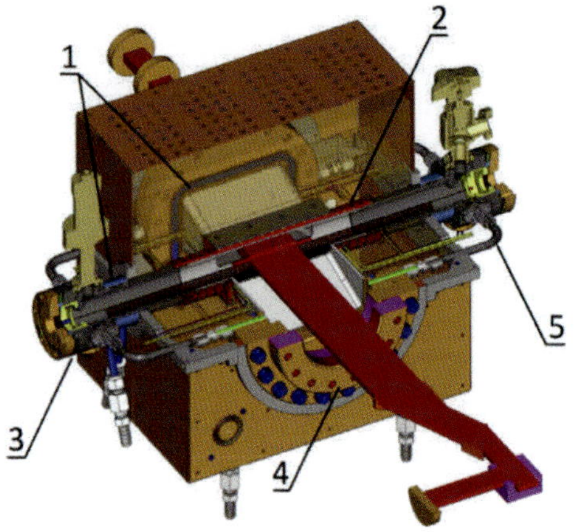

(c)

Fig. 7. (*Continued*)

beams at FACET were performed [48]. An about 1.5-mm-long bunch was used. The results [48] show that the energy spectra and the halo formation of a long electron bunch, having interacted with lithium plasma, are consistent with a bunch that has self-modulated. In particular, energy loss one order of magnitude stronger than that expected from linear theory was observed, about 2 GeV versus about 100 MeV, as illustrated in Fig. 7(b). The 2 GeV deceleration is in reasonable agreement with the predictions of the simulated self-modulation in Ref. 45. However, no direct observation of the SMI was reported in Ref. 48.

At the Photo Injector Test facility at DESY Zeuthen (PITZ), complementary experiments are underway to demonstrate self-modulation of an electron beam [49]. In 2015 a first set of experiments was conducted utilizing a novel cross-shaped lithium heat pipe oven and an ArF excimer laser for plasma generation. No self-modulation was observed in the first experimental run [49], for reasons thought to be related to plasma density and low ionization laser energy. The experiment is planned to continue in 2016, with an upgraded plasma cell and ionization laser beam to address shortcomings of the first experiment. Figure 7(c) shows a schematic of the PITZ plasma source, from Ref. 49.

Additional simulation studies related to PD-PWFA have been performed by various groups. Due to the long drive beams proposed for PD-PWFA, where wakefields are used after many plasma oscillations, ion motion is potentially a bigger concern for PD-PWFA than for electron-driven PWFA.

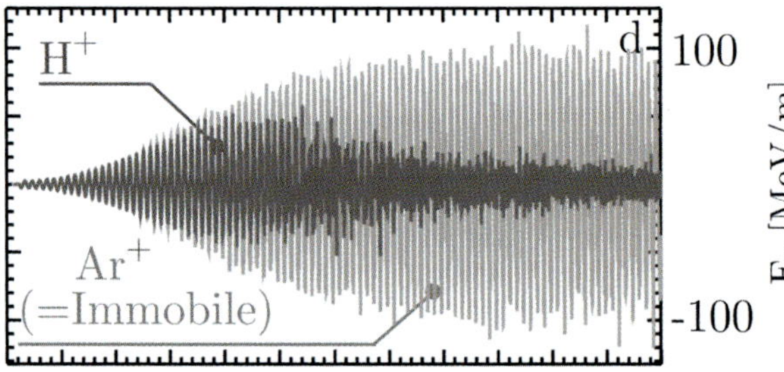

Fig. 8. The effect of ion motion in PD-PWFA, illustrated for the AWAKE parameter. For a hydrogen plasma the ion motion suppresses the resonant wake buildup (the beam travels toward the left). For an argon plasma the effect of the ions on the field is negligible. (From Ref. 50.)

References 50 and 51 show that the growth rate of the SMI, for AWAKE parameters, is reduced when using a hydrogen plasma, and unaffected when using an argon plasma. (See Fig. 8). Reference 52 indicates that the SMI may develop even if the plasma channel is narrow, $k_{pe}\sigma_r > 1$, and also in transversally inhomogeneous plasma channels. References 54 and 55 investigate proton-driven hollow channel acceleration. Recently, PD-PWFA study activities have also been initiated in South Korea [56]. In addition extensive simulation studies have been performed as preparation for the AWAKE experiment. These will be discussed in the next part of this review.

4. AWAKE

We now turn our attention to AWAKE, an experiment at CERN aimed at studying the driving of plasma wakefields with high-energy proton bunches [57–59].

4.1. *Overview*

As mentioned above, proton bunches are interesting plasma wakefield drivers because existing bunches carry a large amount of energy per particle (many GeV) and per bunch (many kilojoules). Staging of many plasma cells may therefore not be required for the witness bunch (electrons or positrons) to reach the TeV-energy scale [13]. However, high-energy proton bunches also have a large rigidity, which means that it is difficult to make them either very short or very tightly focused. This limits the plasma density that can be used in a PWFA since the bunch and plasma density must satisfy $k_{\mathrm{pe}}\sigma_z \simeq \sqrt{2}$ and $k_{\mathrm{pe}}\sigma_r \lesssim 1$ for effective wakefield driving. AWAKE therefore aims at reaching accelerating fields on the order of 1 GV/m in long plasmas (>10 m) with the electron density in the $10^{14}\,\mathrm{cm}^{-3}$ to $10^{15}\,\mathrm{cm}^{-3}$ range. Since available proton bunches are long, the $k_{\mathrm{pe}}\sigma_z \simeq \sqrt{2}$ condition is fulfilled by using the SMI, as explained above.

The first experiments will therefore focus on the study of the SMI of the proton bunch in the plasma (2016–17). Later experiments will use externally injected electrons to sample the wakefields and to be accelerated to the GeV-energy scale.

The AWAKE schematic is shown in Fig. 9, and numbers in the following text refer to numbers on the figure.

The AWAKE experiment uses the long proton bunch ($\sigma_z \simeq 12$ cm) produced by the CERN SPS, with a proton energy of 400 GeV. The proton bunch extracted from the SPS (1') is merged with the ionizing laser pulse. The laser pulse and the proton bunch travel together through the metal vapor cell (3–4). The proton bunch transverse size at the plasma entrance is $\sigma_r \simeq 220\,\mu$m. The laser pulse transverse size is around 1 mm. The copropagating ionization front shown in (2) provides the seeding of the SMI. Only the fraction of the bunch participating in SMI (behind the laser pulse, in the plasma) and wakefield excitation is shown in (1') and (4'). Growth of the SMI and the resulting self-modulation of the proton bunch occurs over the first 3–5 m of plasma (3). The modulated bunch (4') resonantly drives wakefields over the remaining length of plasma (4). The laser pulse is dumped (5) and the proton bunch radial modulation is measured using electro-optic sampling (EOS) diagnostics (6) and optical transition radiation diagnostics (7). An RF gun driven by a laser pulse derived from the ionizing laser system produces a witness electron bunch (8). The electron bunch can be injected into the wakefields, at an adjustable position in the plasma (9). After the plasma, the electron bunch energy spectrum is measured using a broad acceptance magnetic spectrometer (10).

Simulations show that, with the long proton bunch and the SMI, wakefields with amplitudes close to 1 GV/m can be excited in a plasma with a typical density of $7 \times 10^{14}\,\mathrm{cm}^{-3}$ [36, 59, 60]. The same simulations show that the electrons reach an energy of around 1 GeV (from their 16 MeV injection energy) with a finite (<10%) energy spread.

4.1.1. *Beam and site choice*

The experimental as well as the physics parameters are dependent on the beams and infrastructure

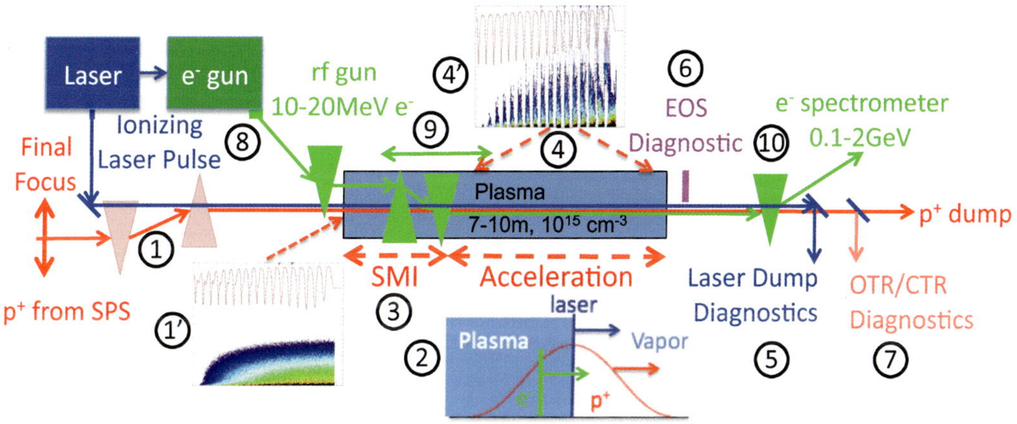

Fig. 9. Schematic of the AWAKE experiment. Numbers are used in the text.

available at CERN. Numerical simulations [36, 61] indicate that higher-energy bunches generally produce larger accelerating gradients over longer distances. Beams produced by the CERN PS at 24 GeV were therefore excluded. The beams produced by the CERN SPS at 400 GeV were chosen because of the lack of access to the LHC beams.

A number of sites around the SPS were considered and the facility previously known as CERN Neutrinos to Gran Sasso (CNGS) was chosen. It required a minimum of beam line and facility modifications to become the AWAKE facility and was also the financially most advantageous option still fulfilling the AWAKE requirements [58]. A schematic of the location of AWAKE is shown in Fig. 10.

4.1.2. *Proton bunch*

The AWAKE proton energy is 400 GeV, limited by the bending magnets in the transfer line to the experimental area. A single bunch has a population of 1–3 $\times\ 10^{11}$ particles and an rms length of $\sigma_z = 12\,\text{cm}$. The nominal beam normalized emittance is 3.6 mm-mrad and the relative energy spread 0.03%. With available quadrupole magnets the beam can be focused to $\sigma_r = 220\,\mu\text{m}$, equivalent to a beam beta function at its waist, β^*, of 4.5 m. The proton will initially be delivered to AWAKE every 30 s, with the possibility of being delivered every 7.2 s.

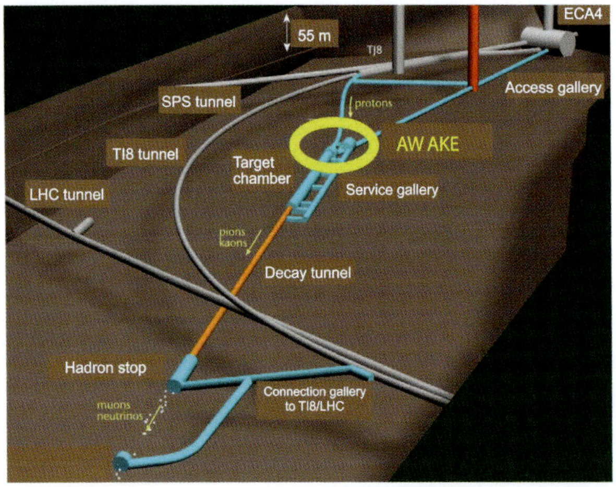

Fig. 10. Schematic of the AWAKE location at CERN. A single proton bunch is extracted from the SPS near one of the LHC extraction lines (TJ8). The plasma is located just upstream of the ex-CNGS target chamber location. The service gallery is used for AWAKE electronics and diagnostics because of the lower radiation level.

4.1.3. *Plasma*

The plasma parameters follow from the bunch parameters. To satisfy $k_{\text{pe}}\sigma_r < 1$, the plasma electron density must not exceed $7 \times 10^{14}\,\text{cm}^{-3}$. At the nominal electron plasma density of $7\times10^{14}\,\text{cm}^{-3}$, the relativistic plasma wave period is $\lambda_{\text{pe}} \simeq 1.3\,\text{mm}$ and the plasma frequency $f_{\text{pe}} = 237\,\text{GHz}$. Since $\lambda_{\text{pe}} \ll \sigma_z$, the relativistic proton bunch is subject to the SMI, which will modulate its density with a $\simeq \lambda_{\text{pe}}$ longitudinal period. The initial beam-to-plasma-density ratio is $n_{b0}/n_e \simeq 6 \times 10^{-3}$. The typical peak accelerating field that can be expected (with a single short bunch) is on the order of $\frac{n_{b0}}{n_e}$ times the wave-breaking field, $\sim\sqrt{n_0/10^{16}\,\text{cm}^{-3}}\,10\,\text{GV/m}$. This corresponds to $\simeq$14.5 MV/m. However, since $N_\lambda \simeq \sigma_z/\lambda_{\text{pe}}$ bunches drive the wakefields, accelerating fields on the order of 1.4 GV/m can be expected.

The plasma length is chosen to be 10 m or about $2\beta^*$. The plasma radius must be larger than the extent of the wakefields, although small-radius plasma may not strongly affect the SMI development [52]. The minimum plasma radius is 1 mm, on the order of λ_{pe}.

The SMI bunch train resonantly drives the plasma wakefields. This imposes stringent conditions on the plasma density uniformity and profile. While the SMI is relatively insensitive to plasma density variations [62, 63], the acceleration of witness particles over a long plasma is not. To be effectively accelerated the witness particles must remain within the accelerating *and* focusing phase of the wakefields, i.e. within about one quarter of a plasma period. A variation in plasma density δn_e translates into a variation of the plasma wavelength of $\delta\lambda_{\text{pe}} = -\frac{1}{2}\lambda_{\text{pe}}\frac{\delta n_e}{n_e}$. The variation in the relative position of the witness particles $N_\lambda\delta\lambda_{\text{pe}}$ must remain small compared to the extent of the accelerating and focusing region of the wakefields, i.e. $N_\lambda\delta\lambda_{\text{pe}} \ll \lambda_{\text{pe}}/4$. With the numbers above, this means that the plasma density uniformity must remain within 0.2% [62].

On-axis external injection of witness particles also imposes a condition on the plasma density ramp at the plasma entrance that will exist in actual plasma sources. Since the plasma electron density varies in the ramp, so does the period of the wakefields. The injected particles therefore find themselves in defocusing regions of the wakefields and exit the wakefields. Simulation results show that, with AWAKE parameters, when the injected particles cover the entire length of a wakefield period, most injected electrons are lost if the plasma density ramp length exceeds about 10 cm [45, 60]. The case of injected e^+ is slightly different, but the capture efficiency remains small (~5%). Note that a sharp density ramp is especially important at the plasma entrance, where the injected electrons have a low energy (10–20 MeV).

4.1.4. *SMI development and seeding*

As mentioned before, the transverse hosing instability is of concern because it has a growth rate similar to that of the SMI [28, 30, 45]. Due to the simulation challenges related to the AWAKE parameters, in particular the very long length of the drive bunch in terms of plasma skin depth $c/\omega_{\rm pe}$, no systematic study of the relative importance of the hosing instability with respect to SMI has been performed. However, simulation results indicate that without seeding the SMI does not develop to significant wakefield values over the AWAKE plasma length. SMI seeding is therefore mandatory for AWAKE and is included in the plasma source that is described next.

4.2. *Plasma source*

The plasma source that was chosen for AWAKE is a laser-ionized, rubidium (Rb) vapor source [65]. Figure 11 shows a schematic of the Rb source. Rb has a relatively low ionization potential (4.177 eV) for its first electrons and can therefore be field-ionized at laser pulse intensities larger than only 1.7×10^{12} W/cm^2. Field ionization frees the first electron of every Rb atom, thereby making the plasma electron density (and thus its uniformity) equal to that of the neutral vapor.

4.2.1. *Density uniformity*

Rb melts at 39.30° C. A vapor density $n_{\rm Rb}$ in the 1–10 × 10^{14} cm^{-3} range is obtained by heating Rb to 150–200° C, corresponding to a vapor pressure of 0.002–0.02 mbar. The vapor density uniformity is obtained by imposing a very uniform temperature along the vapor column (without flow). This is obtained by using a heat exchanger with a large heat transfer fluid flow. Figure 12 shows a picture of the heat exchanger as installed in the AWAKE facility. Measurements have shown that the temperature along the 10 m heat exchanger remains within 0.3° C around 220° C or ~ 500 K, satisfying the requirement $\delta T/T = \delta n_{\rm Rb}/n_{\rm Rb} < 0.2\%$ (or 1 K). Rb ions also have a relatively large mass (Rb atomic mass 85 or 87),

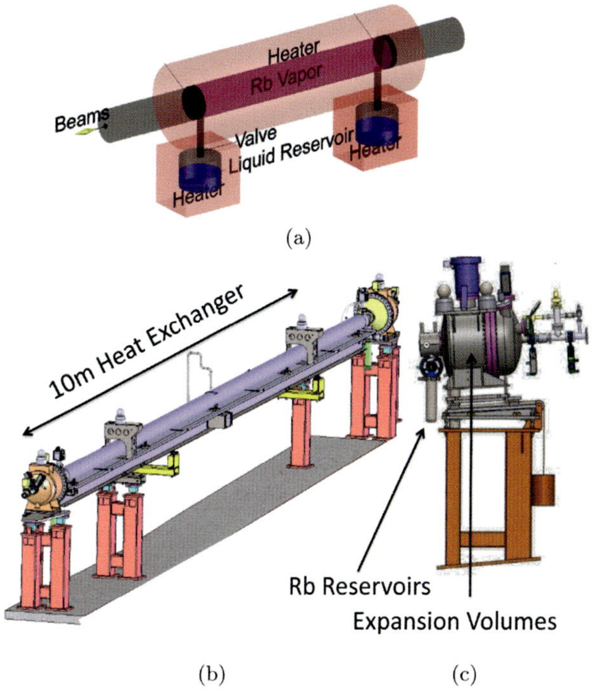

Fig. 11. (a) The principle of the vapor source: the Rb vapor flows from the reservoirs, and fills the inner tube, which is heated by a heat exchanger. (b) Technical drawing of the source, showing the Rb reservoirs (one at each end), the 10 m heat exchanger (outer tube), and the expansion volumes for the Rb vapor. (c) Detailed technical drawing of the source ends. Drawings courtesy of WDL Limited, UK and CERN.

Fig. 12. Picture of the heat exchanger part of the vapor source as installed in the AWAKE facility. The view is looking up towards the SPS. Picture courtesy of CERN.

which means that their large inertia can help mitigate the effects of ion motion [50].

Note that since the vapor and thus the plasma density are controlled through imposing temperatures on the system, one may envisage a vapor source with various temperatures corresponding to various densities. One could in particular realize a density step to stabilize the proton beam evolution mentioned in the previous section.

4.2.2. *Density ramp*

At the end of the 10 m heat exchanger, apertures allow the Rb vapor to expand into a large volume. Reservoirs at each end evaporate the Rb and (mostly) confine the vapor flow from the reservoir to the expansion volume, through the aperture. A theoretical model shows that the scale length for density variations along the system axis is on the order of the aperture diameter, which is 10 mm, and direct simulation Monte Carlo (DSMC) results confirm the length of the density ramp [59, 66].

4.2.3. *Density gradient*

Symmetry between the two ends of the vapor source cannot be guaranteed by design, which leads to a density gradient along the source. This density gradient can be controlled or eliminated by independently controlling the Rb reservoir temperatures. Plasma density gradients can be used to compensate for dephasing of the accelerated particles and the wakefields and reach larger final energies [67]. The vapor density at each end of the source is measured using white light interferometry near the D_2 optical transition of Rb [68].

4.2.4. *Vapor ionization and SMI seeding*

In AWAKE the vapor ionization to create the plasma and the seeding of the SMI are achieved using a short ($\simeq$100 fs) and intense laser pulse. The laser system is a commercial chirped pulse amplification (CPA) system with an erbium: doped oscillator at 780 nm with a bandwidth of $\pm$5 nm. The laser can produce up to 450 mJ/pulse at a repetition rate of 10 Hz. It is focused to a radius of 1 mm near the vapor source entrance by an adjustable telescope. The Rayleigh length of the beam is $\sim$4 m. Calculations [65] show that an energy on the order of 70 mJ is sufficient for ionization over the 10 m length and 1 mm minimum radius.

In order to seed the SMI, the laser pulse copropagates with the proton bunch and is placed within the proton bunch. [See (2), Fig. 9.] The abrupt plasma density creation (over a time much shorter than the plasma wave period, $\sim 1/\omega_{pe} > 3$ ps) effectively seeds the SMI [69]. The laser pulse intensity is too low to drive significant wakefields and does not seed the instability.

The laser oscillator will also produce the seed for the laser that will drive the RF photoinjector gun, discussed below.

4.3. *SMI detection*

The first experiments will be dedicated to the study of the SMI of the proton bunch along the plasma. The occurrence of the SMI results in defocusing of some of the proton bunch particles, in charge modulation of the bunch along its length as well as in the generation of wakefields sustained by an electron plasma density modulation. All three effects can in principle be measured.

The first evidence of SMI will come from comparing the transverse images of the proton bunch after the vapor source, with and without plasma [70]. The occurrence of the SMI leads to the formation of a halo of the defocused proton.

Two diagnostics for the proton bunch modulation are based on transition radiation (TR). TR is emitted when the proton traverses the interface between two dielectric media (vacuum and metal in AWAKE).

The radiation is prompt and incoherent in the optical regime, where the radiation wavelength is much larger than the bunch's typical size or modulation. It is known as optical transition radiation (OTR). The OTR can be imaged onto the entrance slit of a streak camera with ps resolution. Therefore, the light modulation expected from the charge density modulation along the bunch axis can be directly observed and its frequency measured. For the electron plasma density range of AWAKE, the modulation frequency (approximately equal to the plasma frequency) lies in the 100–300 GHz frequency range. It was shown experimentally that modulation frequencies in this range can be detected by the ps resolution streak camera [71]. The OTR light was mocked using two laser beams with beating frequencies in the above range and gated at the ns timescale (similar to the proton bunch duration). The modulation frequency is measured for the power spectrum of the Fourier transform of the streak camera image along the time axis. The start of the SMI development along the bunch can also be correlated with the location of the ionizing laser pulse to demonstrate the seeding effect. This is obtained by operating the streak camera in a slow mode with a ns timescale.

The TR is coherent for wavelengths longer than the bunch's typical size or modulation. It is then known as coherent transition radiation (CTR), and is often used to measure longitudinal bunch length. While the SMI leads to a radial charge modulation with longitudinal periodic modulation, calculations show that the CTR collected in a finite forward angle contains information about the SMI [72] and that it is emitted with significant power. This is because in AWAKE the peak longitudinal wakefield amplitude reaches a significant fraction of the wave-breaking value ($E_{z,\max}/E_{\mathrm{WB}} \sim 0.5$), and thus the p^+ bunch reaches full radial modulation, i.e. consists of truly longitudinally well-separated microbunches within the accelerating structure. In addition, the typical diverging angle of the defocused p^+ is $\sim$1 mrad and they are therefore at radii larger than λ_{pe} at the diagnostic and do not radiate coherently at the modulation wavelength. The modulation frequency will be measured using microwave heterodyne measurement techniques. The CTR signal of unknown frequency is mixed in a nonlinear microwave diode with a signal of known frequency, chosen close to the plasma frequency expected from the plasma density. The frequency difference is then directly measured with a large bandwidth oscilloscope ($\sim$40 GHz).

4.4. *Electron injection*

Although the wakefield phase velocity is relatively small ($\gamma_{\mathrm{phase}} \simeq \gamma_{\mathrm{proton}}$), the wakefield amplitude is also relatively small ($\sim$1 GV/m) and trapping of plasma electrons is unlikely. External injection is therefore *a priori* necessary. The first injection experiments will use an electron bunch longer than the wakefield periods. This eliminates the need for timing synchronization between the electron bunch and the wakefields at the wakefield period scale. In this case, the plasma will trap and self-select a subset of the injected bunch particles. Simulation results [59] indicate that GeV-energy beams with a narrow relative energy spread (a few %) can be obtained. The fraction of electrons trapped is small.

Trapping of plasma electrons might be possible using a shorter, lower-emittance proton bunch, which may be available after the LHC LS2, and a corresponding higher plasma density.

During the growth phase of the SMI, along the first $\sim$4 m of plasma, the phase velocity of the wakefields varies from smaller than that of the proton bunch to superluminous [60]. This effect, similarly to the plasma density variation mentioned above, can lead to the loss of the externally injected particles. Various injection geometries have therefore been considered for AWAKE, side injection, "oblique" injection and on-axis injection.

4.4.1. *Side electron injection*

Injection at a location along the plasma after the SMI has saturated has been proposed [75]. However, this scheme is not practical, since the plasma is contained in a closed vacuum pipe 40 mm in diameter and no screen or measurement is located inside the Rb vapor. It also requires a dipole magnet around the plasma source to bend the electron beam toward the plasma, where it must cross the wakefields to have a chance to be trapped.

4.4.2. *Oblique electron injection*

Oblique electron injection consists in having the electron beam cross the axis of the proton beam a short distance ($\sim$1 m) into the plasma [59]. The two beam vectors, and thus the crossing and injection point, can be determined using the two upstream screens. In this injection scheme, the electrons do not completely avoid the plasma region where the SMI grows. Figure 13 shows the electron energy spectrum expected with baseline experimental parameters and oblique injection.

4.4.3. *On-axis electron injection*

On-axis injection is the only geometry that can lead to low final emittance (assuming that emittance is otherwise preserved along the plasma). However, it is also the most sensitive to the wakefield evolution during the SMI growth, since electrons travel within the wakefields from the beginning of the plasma.

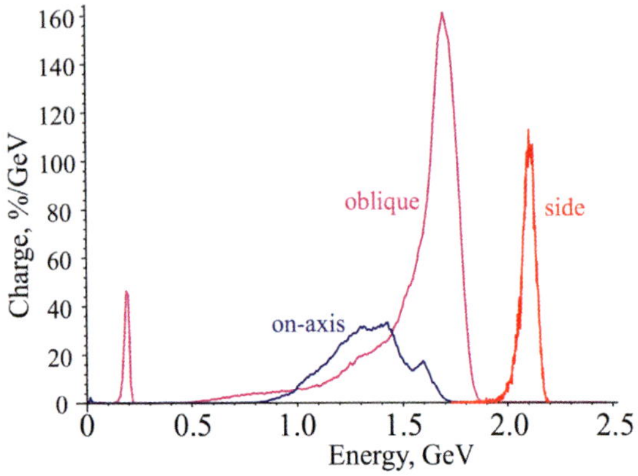

Fig. 13. The final simulated energy spectra of electrons in cases of the side, on-axis and oblique injection methods. (From Ref. 59.)

Simulation results predict small trapping efficiency in this geometry; however, it will be tested experimentally, due to its simplicity and potential for lower final emittance. Operation with two plasma sources may avoid this issue.

4.4.4. *Electron injector*

The electron injector is an S-band, 2.5-cell RF photoinjector gun developed by the CLIC project [73]. The unloaded accelerating gradient is ~100 MV/m for a klystron power of 20 MW. The photocathode is driven by a laser system using the same oscillator as the ionizing laser, thereby insuring synchronization of the gun's RF oscillator and laser oscillator train at the sub-100 fs level. Synchronization between the ionizing laser pulse and the electron bunch is of sub-ps level. The gun is followed by a $2\pi/3$ S-band traveling accelerating structure powered by the same klystron as the gun. It delivers a 9.6 MV/m gradient for a 15 MW input power. The final electron bunch energy is adjustable up to 20 MeV. The bunch charge is adjustable up to about 200 pC. The minimum normalized emittance is 2 mm-mrad and the bunch length is typically 4 ps. The electron bunch is transported to the plasma, where it can be focused to a transverse size of ~250 μm. The transport line can accommodate two of the injection schemes described above: oblique and on-axis injection. It has enough flexibility to place the beam waist at various locations over the first few meters of the plasma. Two screens placed in the common electron and proton beam line upstream of the plasma and equipped with various screens allow for overlap in space and time or for relative position and angle measurements between the beams. The beam line is well diagnosed with RF beam position monitors, pepper-pot emittance measurement, etc. A schematic of the integration of the RF gun in the AWAKE facility is shown in Fig. 14.

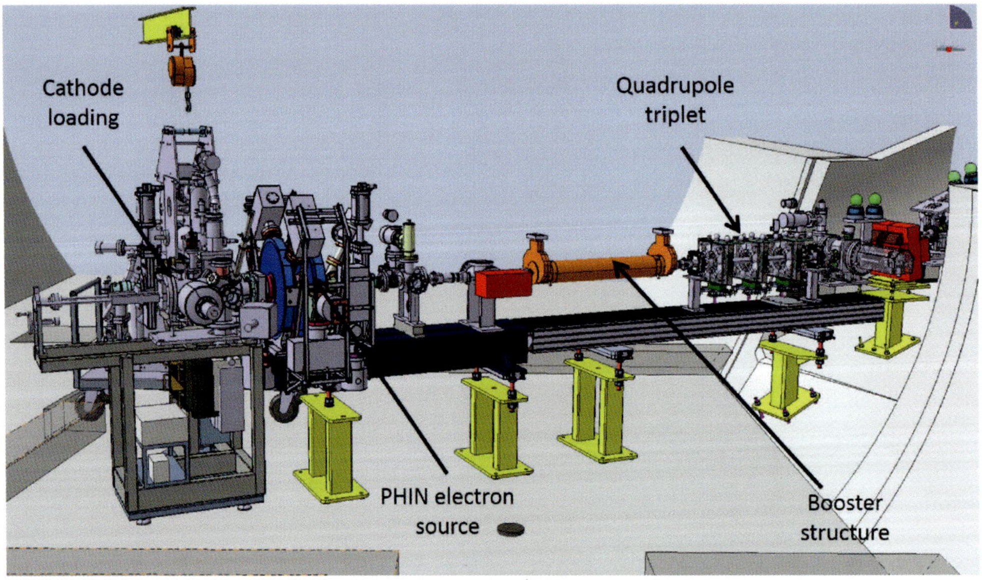

Fig. 14. Schematic of the RF gun (PHIN) integrated into the AWAKE facility (from Ref. 58). The electrons travel from the gun toward the proton beam line on the right-hand side through a dedicated tunnel.

4.4.5. *Energy gain*

For AWAKE the initial beam-to-plasma-density ratio is around 6×10^{-3}, much less than unity. This means that the initial beam–plasma interaction is in the linear PWFA regime. After development of the SMI and where the electron injection should occur, approximately one proton bunch rms length behind the seeding laser pulse, the wakefields reach a fraction of the wave-breaking field [39]. The accelerating fields are in this case described by nonlinear theory [4, 5] and are strongly dependent on the longitudinal position, leading to a significant final energy spread for a finite length injected bunch.

The various electron injection geometries are expected to yield final energies in the GeV range with a few-% relative energy spread. The AWAKE electron spectrometer is based on a dipole magnet available at CERN [78]. The magnet provides a bending field with an integrated strength of 1.3–1.6 Tm. It is equipped with a wide, ~1.5 m screen, so as to be able to observe a wide energy range (0.5–2.5 MeV) in a single event. It is also equipped with a quadrupole doublet, in order to image the beam at the plasma entrance onto the screen. With the doublet powered, the energy resolution reaches the sub-% level over a narrow energy range. Focusing in the vertical (nondispersive) plane strongly increases the signal-to-background ratio. The background signal originates from the secondary particles produced by protons in upstream screens.

4.4.6. *Beam loading*

The energy spread of an accelerated electron bunch could potentially be improved by appropriate beam loading of the wakefield, a well-known method to decrease the energy spread in PWFA [74] as well as conventional accelerators. Beam loading may only be considered for the on-axis injection scheme, since the wakefields of the injected bunch are on the order of those of the drive bunch and thus significantly perturb them, potentially preventing trapping in the off-axis schemes. Initial numerical simulations consider the case of on-axis injection into a separate plasma cell located after the initial plasma cell in which the SMI develops [76]. Simulation results obtained with a premodulated, shorter test proton bunch indicated that beam loading indeed reduces the final energy spread [77]. The possibility of producing a short witness bunch that can properly load the wakefield, for an already-modulated driver, is one of the motivations for running with two plasma sources for Run 2 of AWAKE [76].

4.5. *Integration*

Integrating into an existing facility all the various systems necessary for AWAKE is a challenge. Figure 15 shows a schematic of the AWAKE facility with its main components. The proton bunch extracted from the SPS is focused at the entrance of the vapor cell. It travels through a horizontal chicane that allows for the injection of the ionizing laser pulse along its final trajectory, as shown in Fig. 9. The laser is hosted in a parallel tunnel transformed into a clean and temperature-controlled room. The laser beam is transported to the proton beam line, where light bleeding through a mirror is used for a "virtual plasma" analysis. Screens placed upstream

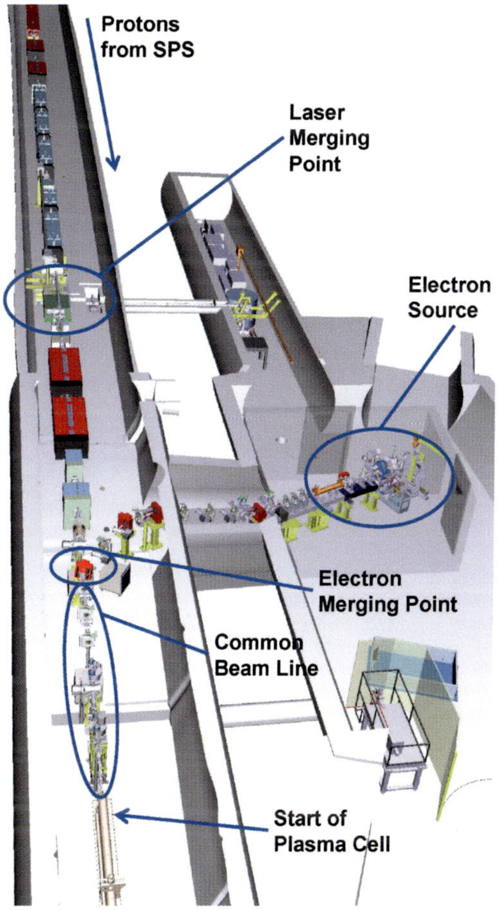

Fig. 15. Integration of the experimental apparatus in the AWAKE facility. (From Ref. 58.)

and downstream of the plasma source allow spatial alignment of the proton and laser beams. One of the upstream screens is used to time the proton bunch and the laser beam using OTR and a streak camera. The proton beam diagnostics are located downstream of the plasma and consist of four screens. They are protected from the intense laser pulse by a light-tight laser beam dump. Two screens, separated by ~8 m, will be used to measure the details of the bunch transverse profile. Two will be used for OTR and CTR diagnostics.

The electron beam is transported toward the proton beam line, where its trajectory can be made collinear with, parallel to, or with an angle with respect to that of the proton beam. The light from the electron spectrometer screen is transported to another parallel tunnel where the CCD camera is located, shielded for the proton tunnel radiation.

4.5.1. *Synchronization*

On one hand and as discussed above, the laser pulse used for ionization that sets the initial phase of the wakefield and that will drive the electron gun photocathode are derived from a single laser oscillator and can be synchronized in a relatively standard way since these both use CW oscillator sources that can be phase-locked. On the other hand, the SPS is a "single-shot" machine that requires around 4 s to produce a single proton bunch and whose RF phase is constantly changed.

Therefore, once the SPS produces a 400 GeV bunch for AWAKE, the bunch is stored for up to 500 ms to allow the phase-locking of the "transient" SPS RF at ~200 MHz to the laser mode-locker frequency of ~88 MHz. The long storage time is necessary to ensure that the SPS proton bunch can be extracted synchronously with the 10 Hz high-power laser pulse. Specific frequency multipliers and dividers had to be developed to lock the no-harmonic frequencies of the various systems. A phase-compensated fiber system was also developed to transmit and maintain the relative phase between AWAKE and the SPS RF over more than 3 km.

4.6. ***AWAKE: future steps***

The AWAKE experiment as described above is scheduled to run until the end of 2018, the time of the start of the LHC Long Shutdown 2. Experience has already shown that the development and integration of an experiment of the scale of AWAKE takes years. Work has therefore started toward exploring and defining experiments and studies that will take place after this shutdown.

The most relevant topics are:

- The production of a high-quality (low emittance and energy spread), high-energy (>1 GeV) electron bunch;
- The development of plasma sources scalable to very long lengths;
- Short and long-term applications of the acceleration scheme.

These are briefly discussed below.

4.6.1. *Witness bunch acceleration*

Production of a high-quality particle bunch is one of the holy grails of plasma-based particle accelerators. It seems clear that when one is using the SMI of the drive bunch to reach large wakefield amplitudes and for reasons described above, it is necessary to separate the self-modulation plasma from the acceleration plasma.

The separation and operation of two independent plasma sources allows, in principle, on-axis injection of external particles after the SMI has evolved. However, this also brings a number of challenges that were avoided in the first phase of AWAKE.

The self-modulated proton bunch evolves once it exits the plasma and enters another. The beam and plasma therefore have to find a new equilibrium between their modulated densities. This leads to some evolution of the wakefields in phase and amplitude at the entrance of the second source, which also has its own density ramp.

Since the beam and plasma form a resonant system, the plasma density between the two sources must be adjusted (and reproducible) at the (sub-)% level.

The optimum witness bunch parameters must be determined. The advantage of operation at relatively low plasma density (compared to other plasma-based accelerators) is that the size of the accelerating structure is relatively large (mm). The size of the structure determines the size of the witness bunch as well as the tolerances in terms of spatial and temporal alignment. For example, the larger structure size means that reasonably long bunches (tens of microns) can be used, possibly benefiting from beam loading.

Alternatively, the very-short, very-high-peak-current bunches typically produced by LWFA could be considered for injection into the accelerating structure and have advantageous properties [79]: high-current and short compared to the accelerating structure, intrinsic synchronization (derived from the same laser), high energy of injection, etc. The source itself is also more compact than an RF-based one.

4.6.2. *Development of scalable plasma sources*

The current AWAKE plasma source meets many of the requirements, such as density uniformity, a short entrance ramp, stability and reproducibility, seeding of the SMI, or the possibility of producing a density step. However, it uses one TW-class laser beam for every 10–20 m, which for long plasmas would be costly and would require staging of the plasma in order to inject fresh laser beams. Alternative plasma source technologies, with potentially better scalability, are therefore being investigated.

4.6.3. *Helicon source*

Plasma can be created by nonresonant coupling of an RF wave (e.g. industry 13.6 MHz) to a weakly magnetized plasma in a helicon source. A helicon unit cell plasma source may thus consist of two coils to produce a uniform longitudinal magnetic field and of an RF antenna wrapped around the vacuum chamber. Helicon sources are thus interesting because they are modular, with only one entrance and one exit for the proton and electron bunches. These sources usually operate at densities two orders of magnitude lower than densities relevant to AWAKE. However, it was recently demonstrated at the Institute for Plasma Physics–Greifswald that a density of $7 \times 10^{14}\,\mathrm{cm}^{-3}$ can be reached. The issues of density uniformity and reproducibility have to be addressed and will be the focus of a new effort between the Institute for Plasma Physics–Greifswald, the Swiss Plasma Center–Lausanne, CERN and AWAKE.

4.6.4. *Discharge source*

Electrical discharges between two electrodes separated by long distances ($\sim$10 m) have already produced densities relevant to AWAKE. They are much simpler and more affordable devices than a TW-class laser. It is, however, not clear whether the density uniformity and reproducibility are commensurate with the SMI requirements. The issue of space between discharges of finite length must also be resolved.

4.7. *Toward applications*

While AWAKE is in its final construction phase at the time of writing, for experiments *on* SMI studies in late 2016 and electron external injection studies in 2017–18, AWAKE is preparing Run 2 [76], scheduled for after the LHC LS2 (2021). This preparation includes a first scientific program that could be realized within the current facility. Possible themes include: operation with shorter, lower-emittance proton bunches in higher-density plasmas; increasing the accelerating gradient; operation with two plasma sources (for SMI and acceleration); external injection of a short electron bunch in the second plasma source; and acceleration of an electron bunch with narrow energy spread and low emittance, as discussed above. Acceleration of positron bunches is also envisaged.

At the same time, mid- and long-term plans are also considered. An appealing path toward PD-PWFA applications is to use the existing CERN proton beams not only as plasma drivers, but also for HEP applications in the form of a high-energy electron–proton (eP) collider fitting on the CERN site. A basic concept for a PD-PWFA–based eP collider has been studied in Ref. 80. The principle is shown in Fig. 16. An LHC bunch could potentially drive electrons up to TeV energies [36], while an SPS bunch could potentially drive electron energies similar to those of the proposed LHeC collider (electron energies of 60–100 GeV) [81]. A PD-PWFA as proposed in Ref. 80 would have a several-orders-of-magnitude lower luminosity compared to the LHeC proposal [81], due to the relatively low repetition rate of the available proton beams used to drive the wakefields. However, a significant advantage of targeting eP applications is that the requirements for electron beam emittance are orders of magnitude more relaxed than for electron–positron linear colliders [81–83], imposing less demanding needs for emittance preservation during the plasma acceleration. Since the AWAKE experiment uses the SPS beam for the proof-of-principle experiments, the path

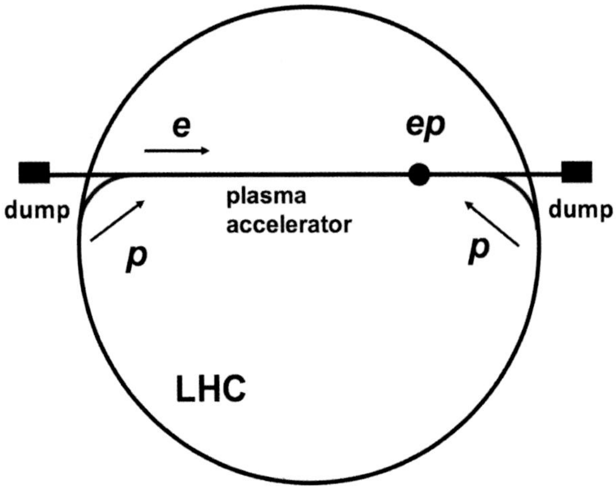

Fig. 16. A basic concept for a PD-PWFA–based electron–proton collider. One LHC bunch is used to accelerate an electron bunch to TeV energies before the electron bunch collides with a second LHC bunch. This type of application maximizes the use of existing proton accelerator infrastructure at CERN. (From Ref. 80.)

from AWAKE experiments to eP collisions with LHeC-energy bunches could potentially be reasonably short, without a need for large intermediate test facilities.

Acknowledgments

We gratefully acknowledge useful discussions with Konstantin Lotov concerning the self-modulation instability as well as AWAKE simulations in general, with Allen Caldwell about the early phases of AWAKE, and with Vladimir Shiltsev for information about PROTOPLASMA. Furthermore, we thank Ioanis Kourbanis, Sergey Ivanov and Susumu Igarashi for providing the main parameters for the Fermilab Main Injector, IHEP U70 and J-PARC synchrotrons, respectively.

References

[1] T. Tajima and J. M. Dawson, Laser electron accelerator, *Phys. Rev. Lett.* **43**, 267 (1979).
[2] Laser-driven plasma acceleration, in *RAST*, this volume (2016).
[3] P. Chen, J. Dawson, R. Huff and T. Katsouleas, Acceleration of electrons by the interaction of a bunched electron beam with a plasma, *Phys. Rev. Lett.* **54**, 693 (1985).
[4] Electron-beam-driven plasma acceleration, in *RAST*, this volume (2016).
[5] W. Lu, C. Huang, M. M. Zhou, W. B. Mori and T. Katsouleas, *Phys. Plas.* **12**, 063101 (2005).
[6] B. Allen, V. Yakimenko, M. Babzien, M. Fedurin, K. Kusche and P. Muggli, Experimental study of current filamentation instability, *Phys. Rev. Lett.* **109**, 185007 (2012).
[7] M. Zhou *et al.*, Beam head erosion in self-ionized plasma wakefield accelerators, in *Proc. PAC07* (2007), p. 3064.
[8] C. A. Lindstrøm *et al.*, Staging optics considerations for a plasma wakefield acceleration linear collider, *Nucl. Instrum. Methods in Phys. Res. A* **829**, 224 (2016).
[9] W. Lu *et al.*, High transformer ratio PWFA for application on XFELs, in *Proc. PAC09* (2009), p. 3028.
[10] E. Kallos *et al.*, Plasma wakefield acceleration utilizing multiple electron bunches, in *Proc. PAC07* (1997), p. 3070.
[11] I. Blumenfeld *et al.*, Scaling of the longitudinal electric field and transformer ratio in a nonlinear plasma wakefield accelerator, *Phys. Rev. ST Accel. Beams* **10**, 111301 (2010).
[12] B. E. Blue, Plasma wakefield acceleration of an intense positron beam, Ph.D. thesis (UCLA, 2003).
[13] A. Caldwell, K. Lotov, A. Pukhov and F. Simon, Proton-driven plasma-wakefield acceleration, *Nat. Phys.* **5**, 363 (2009).
[14] S. Lee, T. Katsouleas, R. G. Hemker, E. S. Dodd and W. B. Mori, Plasma-wakefield acceleration of a positron beam, *Phys. Rev. E* **64**, 045501 *R* (2011).
[15] K. V. Lotov, Simulation of proton-driven plasma wakefield acceleration, *Phys. Rev. ST Accel. Beams* **13**, 041301 (2010).
[16] R. Assmann *et al.*, Generation of short proton bunches in the Cern Accelerator Complex, in *Proc. PAC09* (2009), p. 4542.
[17] S. Nagaitsev, Fermilab Antiproton Source, Recycler Ring, and Main Injector. FERMILAB-FN-0957-AD (2013).
[18] T. Koseki *et al.*, Beam commissioning and operation of the J-PARC main ring synchrotron, *Prog. Theor. Exp. Phys.* **02**, B004 (2012).
[19] S. Ivanov, Accelerator complex U70 of IHEP: status and upgrades, in *Proc. RuPAC2014* (2014), TUX02.
[20] J. M. Dawson, Nonlinear electron oscillations in a cold plasma, *Phys. Rev.* **113**, 383 (1959).
[21] A. Caldwell *et al.*, Preliminary study of proton driven plasma wakefield acceleration, in *Proc. PAC09* (2009), p. 4551.
[22] G. Xia and A. Caldwell, Producing short proton bunch for driving plasma wakefield acceleration, in *Proc. IPAC10* (2010), p. 4395.
[23] P. Schutt, T. Weiland and V. M. Tsakanov, On the wakefield acceleration using the sequence of driving bunches. DESY M-88-13 (1988), p. 12.
[24] D. H. Whittum, Transverse two-stream instability of a beam with a Bennett profile, *Phys. Plas.* **4**, 1154 (1997).

[25] K. Lotov, Instability of long driving beams in plasma wakefield accelerators, in *Proc. EPAC08* (1998), p. 806.

[26] E. Esarey, J. Krall and P. Sprangle, Envelope analysis of intense laser pulse self-modulation in plasmas, *Phys. Rev. Lett.* **72**, 2887 (1994).

[27] W. B. Mori, The physics of the nonlinear optics of plasmas at relativistic intensities for short-pulse lasers, *IEEE Trans. Plas. Sci.* **33**, 1942 (1997).

[28] N. Kumar, A. Pukhov and K. Lotov, Self-modulation instability of a long proton bunch in plasmas, *Phys. Rev. Lett.* **104**, 255003 (2010).

[29] A. Pukhov *et al.*, Phase velocity and particle injection in a self-modulated proton-driven plasma wakefield accelerator, *Phys. Rev. Lett.* **107**, 145003 (2011).

[30] J. Vieira, P. Muggli and W. B. Mori, Hosing instability in self-modulated plasma wakefields, *Phys. Rev. Lett.* **112**, 205001 (2014).

[31] E. Adli *et al.*, Letter of intent for a demonstration experiment in proton-driven plasma wakefield acceleration. CERN-SPSC-2011-020 (2011).

[32] J. C. T. Thangaraj *et al.*, PROTOPLASMA — proton-driven plasma-wakefield experiment at Fermilab: stages and approach. FERMILAB-CONF-12-409-APC (2012).

[33] K. V. Lotov, Controlled self-modulation of high energy beams in a plasma, *Phys. Plas.* **18**, 024501 (2011).

[34] K. V. Lotov, Effect of beam emittance on self-modulation of long beams in plasma wakefield accelerators, *Phys. Plas.* **22**, 123107 (2015).

[35] K. V. Lotov, Physics of beam self-modulation in plasma wakefield accelerators, *Phys. Plas.* **22**, 103110 (2015).

[36] A. Caldwell and K. Lotov, Plasma wakefield acceleration with a modulated proton bunch, *Phys. Plas.* **5**, 103101 (2011).

[37] K. V. Lotov, Simulation of ultrarelativistic beam dynamics in plasma wakefield accelerator, *Phys. Plas.* **5**, 785 (1998).

[38] K. V. Lotov, Fine wakefield structure in the blowout regime of plasma wakefield accelerators, *Phys. Rev. ST Accel. Beams* **6**, 061301 (2003).

[39] K. V. Lotov, Excitation of two-dimensional plasma wakefields by trains of equidistant particle bunches, *Phys. Plas.* **20**, 083119 (2013).

[40] P. Piot *et al.*, Generation and characterization of electron bunches with ramped current profiles in a dual-frequency superconducting linear accelerator, *Phys. Rev. Lett.* **108**, 034801 (2012).

[41] M. Boscoloa, M. Ferrario, I. Boscolob, F. Castellib and S. Cialdi, Generation of short THz bunch trains in a RF photoinjector, *Nucl. Instrum. Methods in Phys. Res. A* **577**, 409 (2007).

[42] P. Muggli, V. Yakimenko, M. Babzien, E. Kallos and K. P. Kusche, Generation of trains of electron microbunches with adjustable subpicosecond spacing, *Phys. Rev. Lett.* **101**, 054801 (2008).

[43] Y.-E. Sun *et al.*, Tunable subpicosecond electron-bunch-train generation using a transverse-to-longitudinal phase-space exchange technique, *Phys. Rev. Lett.* **105**, 234801 (2010).

[44] Y. Fang, W. Mori and P. Muggli, Numerical study of self-modulation instability of 1 nC electron bunch at ATF, *AIP Conf. Proc.* **1507**, 559 (2012).

[45] J. Vieira, Y. Fang, W. B. Mori, L. O. Silva and P. Muggli, Transverse self-modulation of ultra-relativistic lepton beams in the plasma wakefield accelerator, *Phys. Plas.* **19**, 063105 (2012).

[46] Y. Fang *et al.*, Seeding of self-modulation instability of a long electron bunch in a plasma, *Phys. Rev. Lett.* **112**, 045001 (2014).

[47] Facility for Advanced Accelerator Experimental Tests at SLAC. http://facet.slac.stanford.edu.

[48] E. Adli *et al.*, Progress of plasma wakefield self-modulation experiments at FACET, *Nucl. Instrum. Methods in Phys. Res. A* **829**, 334 (2016).

[49] O. Lishilin *et al.*, First results of the plasma wakefield acceleration experiment at PITZ, *Nucl. Instrum. Methods in Phys. Res. A* **829**, 37 (2016).

[50] J. Vieira, R. A. Fonseca, W. B. Mori and L. O. Silva, Ion motion in self-modulated plasma wakefield accelerators, *Phys. Rev. Lett.* **109**, 145005 (2012).

[51] J. Vieira, R. A. Fonseca, W. B. Mori and L. O. Silva, Ion motion in the wake driven by long particle bunches in plasmas, *Phys. Plas.* **21**, 056705 (2014).

[52] Y. Fang, J. Vieira, L. D. Amorim, W. Mori and P. Muggli, The effect of plasma radius and profile on the development of self-modulation instability of electron bunches, *Phys. Plas.* **21**, 056703 (2014).

[53] L. Yi *et al.*, Scheme for proton-driven plasma-wakefield acceleration of positively charged particles in a hollow plasma channel, *Phys. Rev. ST Accel. Beams* **16**, 071301 (2013).

[54] L. Yi *et al.*, Scheme for proton-driven plasma-wakefield acceleration of positively charged particles in a hollow plasma channel, *Phys. Rev. ST Accel. Beams* **16**, 071301 (2013).

[55] L. Q. Yi *et al.*, Positron acceleration in a hollow plasma channel up to TeV regime, *Sci. Rep.* **4**, 4171 (2014).

[56] M. Chung *et al.*, Studies of the self-modulation and other instabilities in proton beam–driven plasma wakefield accelerators. URSI Asia-Pacific Radio Science Conference, Seoul, South Korea (2016).

[57] R. Assmann *et al.*, Proton-driven plasma wakefield acceleration: a path to the future of high-energy particle physics, *Plas. Phys. Control. Fus.* **56**, 084013 (2014).

[58] E. Gschwendtner *et al.*, AWAKE, the advanced proton driven plasma wakefield acceleration experiment at CERN, *Nucl. Instrum. Methods in Phys. Res. A* **829**, 76 (2016).

[59] A. Caldwell *et al.*, Path to AWAKE: evolution of the concept, *Nucl. Instrum. Methods in Phys. Res. A* **829**, 3 (2016).
[60] K. V. Lotov *et al.*, Electron trapping and acceleration by the plasma wakefield of a self-modulating proton beam, *Phys. Plas.* **21**, 123116 (2014).
[61] K. V. Lotov *et al.*, Parameter sensitivity of plasma wakefields driven by self-modulating proton beams, *Phys. Plas.* **21**, 083107 (2014).
[62] K. Lotov *et al.*, Effect of plasma inhomogeneity on plasma wakefield acceleration driven by long bunches, *Phys. Plas.* **20**, 013102 (2013).
[63] C. B. Schroeder *et al.*, Growth and phase velocity of self-modulated beam-driven plasma waves, *Phys. Rev. Lett.* **107**, 145002 (2011).
[64] K. V. Lotov, Controlled self-modulation of high energy beams in a plasma, *Phys. Plas.* **18**, 024511 (2011).
[65] E. Oz and P. Muggli, A novel Rb vapor plasma source for plasma wakefield accelerators, *Nucl. Instrum. Methods in Phys. Res. A* **740**(11), 197 (2014).
[66] G. Plyushchev and R. Kersevan, to be published.
[67] A. Petrenko, K. Lotov and A. Sosedkin, Numerical studies of electron acceleration behind self-modulating proton beam in plasma with a density gradient, *Nucl. Instrum. Methods in Phys. Res. A* **829**, 63 (2016).
[68] E. Oz *et al.*, An acccurate Rb density measurement method for a plasma wakefield accelerator experiment using a novel Rb reservoir, *Nucl. Instrum. Methods in Phys. Res. A* **829**, 321 (2016).
[69] D. F. Gordon *et al.*, Seeding of the forward Raman instability by ionization fronts and Raman backscatter, *Phys. Rev. E* **64**, 046404 (2001).
[70] M. Turner *et al.*, Indirect self-modulation instability measurement concept for the AWAKE proton beam, *Nucl. Instrum. Methods in Phys. Res. A* **829**, 314 (2016).
[71] K. Rieger, Master's thesis (Technical University of Munich, 2015).
[72] M. Martyanov, to be published.
[73] G. Bienvenu, CTF3 3 GHz RF gun test at CERN. CTF3-Note-093 (2009).
[74] M. Tzoufras *et al.*, Beam loading in the nonlinear regime of plasma-based acceleration, *Phys. Rev. Lett.* **101**, 145002 (2008).
[75] K. V. Lotov, Optimum angle for side injection of electrons into linear plasma wakefields, *J. Plas. Phys.* **78**(4), 455 (2012).
[76] E. Adli, Towards Awake applications: electron beam acceleration in a proton driven plasma wake, in *Proc. IPAC'16* (Busan, South Korea, 2016), p. 2557.
[77] V. K. Berglyd Olsen *et al.*, Loading of a plasma-wakefield accelerator section driven by a self-modulated proton bunch, in *Proc. IPAC'15* (Richmond, VA, USA, 2015), p. 2551.
[78] L. Deacon *et al.*, A spectrometer for proton driven plasma accelerated electrons at Awake — recent developments, in *Proc. IPAC'16* (Busan, South Korea, 2016), p. 2605.
[79] P. Muggli *et al.*, Injection of a LWFA electron bunch in a PWFA driven by a self-modulated proton bunch, in *Proc. IPAC'14* (Dresden, Germany, 2014), p. 1470.
[80] A. Caldwell and M. Wing, VHEeP: a very high energy electron–proton collider. arXiv: 1606.00783 [hep-ex].
[81] LHeC Study Group, CERN-OPEN-2012-01 (2012).
[82] CLIC Conceptual Design Report (2012).
[83] International Linear Collider Technical Design Report (2013).

Erik Adli did his Ph.D. research on linear colliders at CERN and his postdoctoral work on PWFA experiments at FACET/SLAC. He is currently an associate professor at the University of Oslo, Norway, where he is leading the Norwegian accelerator activities toward the AWAKE and CLIC projects at CERN, and toward ESS.

Patric Muggli is leading the Future Accelerators group at the Max Planck Institute for Physics in Munich, Germany. The main focus of his current research is the AWAKE project at CERN. He is a fellow of the American Physical Society and of IEEE. He received the IEEE Particle Accelerator Science and Technology Award in 2011.

Reviews of Accelerator Science and Technology
Vol. 9 (2016) 105–126

DOI: 10.1142/S179362681630005X

Dielectric Laser Accelerators: Designs, Experiments, and Applications

K. P. Wootton

SLAC National Accelerator Laboratory,
2575 Sand Hill Rd, Menlo Park, California 94025, USA
wootton@slac.stanford.edu

J. McNeur

Department of Physics,
Friedrich-Alexander-Universität Erlangen-Nürnberg,
Staudtstrasse 1, 91058 Erlangen, Germany
joshua.mcneur@fau.de

K. J. Leedle

Department of Electrical Engineering,
Stanford University,
450 Serra Mall, Stanford, California 94305, USA
kleedle@stanford.edu

Novel laser-powered accelerating structures at the miniaturized scale of an optical wavelength ($\sim 1\mu$m) open a pathway to high repetition rate, attosecond scale electron bunches that can be accelerated with gradients exceeding 1 GeV/m. Although the theoretical and computational study of dielectric laser accelerators dates back many decades, recently the first experimental realizations of this novel class of accelerators have been demonstrated. We review recent developments in fabrication, testing, and demonstration of these micron scale devices. In particular, prospects for applications of this accelerator technology are evaluated.

Keywords: Dielectric laser accelerators; electron linear accelerators; free electron lasers; lithography; attosecond science.

1. Introduction

Since the first conceptualization of a laser-powered electron linear accelerator (linac) [1], various laser electron linac schemes have been proposed and demonstrated. In this review, we summarize the crucial technological developments and underlying physical principles of recently demonstrated microstructure dielectric laser accelerators (DLAs). Building on prior reviews [2–7], in the present work we emphasize recent results and potential near-term and future technological opportunities for these miniaturized electron linacs.

This article is organized into several main sections. The motivation of the DLA is first outlined through the historical context of relevant scientific and technological developments. We then present an overview of various proposed microstructure geometries. The operating principles of demonstrated microstructures and recent experimental results are summarized. Finally, we outline prospects for near-term applications of DLA components fabricated and integrated "on a chip."

2. Historical Context

Criteria for the acceleration of charged particles by an electromagnetic field are given by the Lawson–Woodward theorem [8–11]. Succinctly, charged particles will not be accelerated by an electromagnetic wave if all of the following criteria are met [12]:

(i) The particle is in vacuum, far from any material surface;
(ii) The field is a far field (plane wave), integrated over an infinite interaction length;
(iii) The particle is relativistic;
(iv) No static electric or magnetic fields are present, yielding a particle trajectory which is approximately straight;

(v) The interaction is between the electromagnetic field and a single particle.

Thus, acceleration of charged particles requires that one or more of these conditions be broken. Several strategies can be adopted, in particular the introduction of a near-field structure, with feature sizes governed by the drive wavelength, which modifies the phase of incident wave fronts [13, 14]. This is the approach used in conventional radio frequency linacs, employing a conducting near-field structure [8]. A comprehensive summary of acceleration theorems and their relevance to various acceleration schemes can be found in Ref. 15.

One approach to accelerating charged particles using a laser is the abrupt termination of an incident field at a boundary [16–19]. Several experimental demonstrations have been made of electron acceleration by laser pulses at single-boundary structures, evading the Lawson–Woodward theorem [20–24]. Such structures may prove useful as a diagnostic of femtosecond (and shorter) electron–photon interactions [25–27]; however, they present practical difficulties in staging as a high gradient laser-driven linac.

Several early proposals employing open structures were based on reflective (all-metallic) gratings, using a wavelength scaling of existing linac designs [15, 28, 29]. The acceleration mechanism employed in those demonstrations was the inverse Smith–Purcell effect [30].

The earliest proposal for a staged dielectric laser accelerator and focusing device was based on a periodic structure [31, 32]. It employed absorptive thin metal gratings deposited onto flat dielectric surfaces. This structure functioned through the use of an amplitude mask with a period matching the wavelength of the incident electric field. Conceptually similar radially symmetric geometries were subsequently proposed [33, 34], though the assumed geometry of structures significantly larger than λ prevents their acceleration of relativistic particles [35]. Ultimately, it was not clear that metal deposited on dielectric surfaces was going to have the required laser damage threshold for high gradient accelerators.

Laser-driven structure-based accelerators were considered to be composed of either planar dielectrics [8, 36, 37], or metallic gratings as a wavelength-scaled miniature linac [14, 38, 39]. Evanescent fields near a single grating or dielectric slab could be used to accelerate subrelativistic charged particles; however, the accelerating modes of the inverse Smith–Purcell effect were synchronous only with subrelativistic particle velocities [8]. What was required was local confinement of the evanescent fields using a dual-sided structure [8]. The enabling breakthrough was combining the grating design with dielectric materials. An early proposal suggested to modulate the phase of a laser field by a stepped grating using the refractive index of a dielectric [40]; however, that study significantly overestimated the material damage threshold.

Initial work on DLAs was stunted by a misunderstanding of the damage mechanism: proposals were either pessimistic about the damage threshold of dielectrics (200−400 MeV/m [8, 35]) or overly optimistic [40]. The material damage mechanism of ultrafast laser pulses (ps and fs) is surface ablation, rather than being due to impurities in the bulk material (as with ns laser pulses). As a result, the material damage threshold (in terms of the maximum field gradient sustainable in the structure material) is significantly higher for ultrashort pulses [41–43]. With an understanding of the damage mechanism of the structure [44, 45], engineered DLAs have emerged.

Thus, the time is ripe for the development of accelerators powered by ultrafast laser technology. Recent technological developments in lasers have made infrared femtosecond lasers readily available [46]. When based on a fiber architecture, such lasers will be uniquely scalable for particle accelerator applications [47]. Lasers are also highly desirable as an accelerator power source owing to their high energy efficiency [48, 49], an important consideration in planned future accelerators [50, 51]. Due to the linear interaction of the accelerating fields in DLAs, their utilization is potentially both efficient and scalable.

Furthermore, microfabrication techniques driven by the semiconductor industry have advanced to the point of feasible fabrication of DLA microstructures [52–54]. The μm scale wavelength results in feature sizes of order 400 nm: such structures are readily fabricated using conventional ultraviolet lithography. In addition, electron beam lithography is a convenient fabrication process for expedient prototyping of new structure geometries [55]. DLAs have emerged at the interface of these blossoming technologies.

3. Microstructure Geometries

In the case of the planar grating structure and even in more general terms, DLAs operate via laser-induced excitation of an electromagnetic field mode with a nonzero electric component parallel to the velocity of electrons traversing the mode. For long distance, synchronous interaction to occur, the phase velocity of the excited mode must match the electron velocity. A wide variety of nanostructured dielectric geometries have been considered to efficiently excite such near-field modes.

3.1. *Bragg mirrors*

One such geometry consists of two planar Bragg mirrors separated by a vacuum channel through which electrons travel [56]. Each Bragg mirror is designed in such a way that the light reflecting off each interface within the mirror constructively interferes, leading to field confinement in the vacuum channel. A defect is introduced by changing the thickness of the dielectric layers adjacent to the channel. This allows control of the phase velocity of the accelerating mode confined therein in order to satisfy the synchronicity condition. Light can be coupled into the cavity via dielectric coupling slots [57].

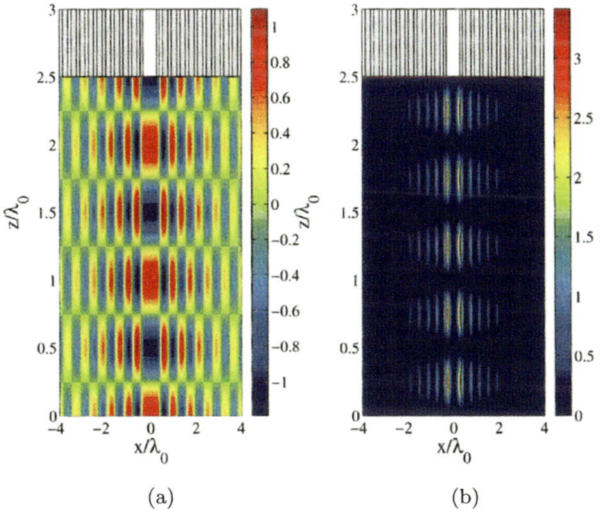

Fig. 1. Profile of fields within a planar Bragg mirror structure, with the distributed Bragg reflector profile illustrated at the top. The electron and laser beam propagate in z, with the vacuum channel centered at $x = 0$. (a) Longitudinal electric field, E_z, normalized by the peak field. (b) Poynting vector, normalized by $|E_0|^2/\eta_0$, where η_0 is the intrinsic impedance. [Reprinted figure (with permission) from A. Mizrahi and L. Schächter, *Phys. Rev. E* **70**, 016505 (2004). © 2004 by the American Physical Society.]

This planar Bragg geometry has been extensively simulated (yielding the fields shown in Fig. 1) and fabricated [58–60]. Due to the resonant nature of this design, accelerating gradients as much as ten times the incident field are possible [58].

3.2. *Slab geometry*

The refraction of an inclined incident laser beam on a slab structure leads to a speed-of-light accelerating mode [8, 36, 61]. A structure employing such refraction at the interface of a silicon–silicon carbide sample has been proposed [62]. Because silicon carbide is a polar semiconductor, its dielectric permittivity is negative for the mid-infrared wavelengths of CO_2 lasers. As a result, at the interface between a bulk silicon substrate and a silicon carbide thin film, evanescent waves are supported. Schemes in which laser light is coupled into this structure via gratings and prisms have additionally been proposed [63, 64]. They have been tested for compatibility with CO_2 lasers.

3.3. *Photonic band gap structures*

An alternative means of confining an accelerating field mode spatially is found through the use of photonic band gap (PBG) crystals, such as the hexagonal rod lattice shown in Fig. 2 [65]. A defect is introduced into the lattice in the form of a large hole in its center, in which a transverse magnetic (TM) accelerating mode along which electrons travel is excited [66–68]. In such a structure, the radius of the lattice holes, the lattice periodicity, the radius of

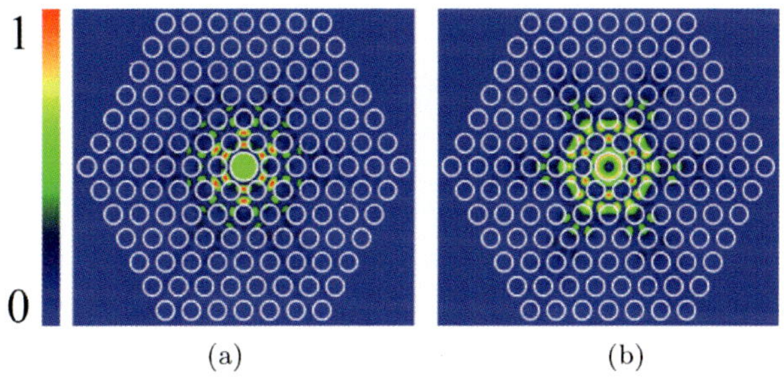

Fig. 2. A two-dimensional photonic crystal lattice which supports a TM speed-of-light accelerating mode. Longitudinal (a) and radial (b) electric field magnitude of the speed-of-light mode. The color scale indicates field relative magnitude. [Adapted figure (under CC-BY 3.0 license) from R. J. Noble, J. E. Spencer and B. T. Kuhlmey, *Phys. Rev. ST Accel. Beams* **14**, 121303 (2011). Published 2011 by the American Physical Society.]

the defect hole, and the permittivity of the material affect the band gap properties. These parameters are carefully controlled to ensure that the TM mode confined in the defect has a phase velocity matching the velocity of accelerated electrons.

Various schemes for coupling laser light into modes propagating in the fiber have been studied using simulations [68, 69] and experiments [65, 70]. The accelerating mode can also be excited through an electron beam wake, which is another technique for direct characterization of speed-of-light fiber modes [71, 72].

Two-dimensional PBG structures have been studied as DLAs [73]. These structures are infinite in one direction transverse to the electron velocity. A separate dual grating structure was also proposed employing ponderomotive focusing of electrons in addition to acceleration [74–76].

Three-dimensional PBG structures — such as the so-called woodpile geometry shown in Fig. 3 — have also been proposed [77, 78]. The design is tailored in such a way that a TM mode is well confined to the defect channel in the center of the photonic crystal. Indeed, a 300 MeV/m unloaded field gradient is expected along the electron axis of propagation. Further, the design has been fabricated via the sequential assembly of etched silicon layers to form a woodpile structure with a defect (electron beam channel) [79–81]. The structure has also been fabricated using 3D direct laser write [82]. Although this structure is amenable to the addition of ancillary devices such as couplers [81, 83, 84], coupling of light into the structure requires fabrication that has not yet been achieved due to restrictive error tolerances.

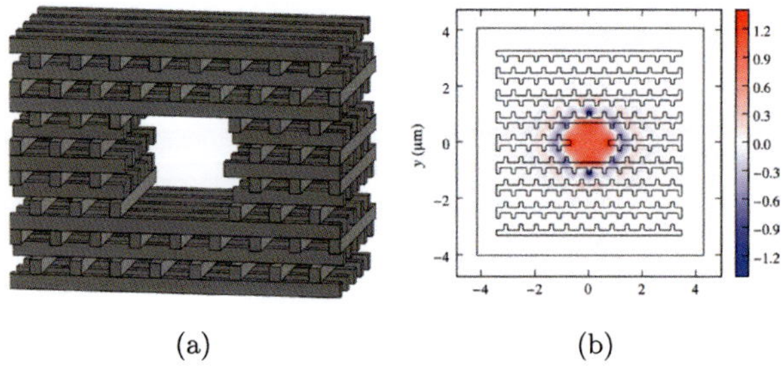

Fig. 3. A three-dimensional photonic crystal lattice. (a) Illustration of the woodpile structure. (b) Spatial profile of the confined accelerating field E_z, normalized to the electric field amplitude at the center of the vacuum defect. [Adapted figure (under CC-BY 3.0 license) from B. M. Cowan, *Phys. Rev. ST Accel. Beams* **11**, 011301 (2008). Published 2008 by the American Physical Society.]

3.4. *Plasmonic structures*

An alternative to all-dielectric structures is the use of metal surfaces to create resonant structures [85]. A structured metal surface with antenna slots at periodic spacings offers a means to locally enhance the electric field 10–15 times the incident laser field [86, 87].

Recently, plasmonic metasurfaces have been proposed as an engineered laser accelerator surface [88–90]. The proposed structure is composed of a staggered array of slots and antennas in a thin metallic surface in order to enhance the accelerating gradient. A critical parameter to be measured will be the metasurface damage threshold at the operating laser wavelength.

3.5. *Grating structures*

Grating DLA structures of various geometries are transversely driven by pulse-front-tilted or multiple drive laser beams to excite a grating mode. Several grating geometries have been proposed [91], with basic variations illustrated in Fig. 4.

Elliptical and rectangular dual pillar designs [Figs. 4(a) and 4(b)] are fabricated with 2D lithography from a monolithic substrate [92, 97]. They have high refractive contrast and can produce a variety of photonic crystal modes, and elliptical dual pillar gratings have been experimentally demonstrated [92]. Elliptical dual pillar gratings have relaxed

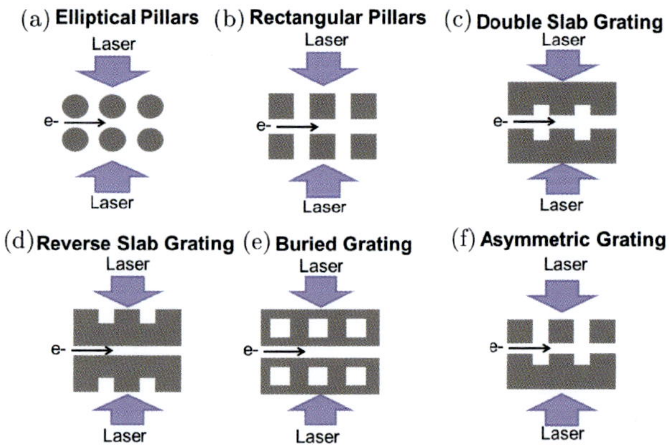

Fig. 4. Basic designs of two-dimensional gratings suitable for building dielectric laser accelerators. (a) Elliptical dual pillar grating [92], (b) rectangular dual pillar grating [93, 94], (c) dual slab grating [95, 96], (d) reverse slab grating [97], (e) buried grating structure [98, 99], (f) asymmetric grating (any combination of the above, including elliptical and rectangular features).

fabrication requirements compared to rectangular pillars, although monolithic 1D rectangular dual pillar gratings can be fabricated using a wafer heterostructure and selective etches. Alternatively, a rectangular dual grating could be fabricated by overlapping two orthogonal 1D gratings to obtain very sharp corners in a 2D monolithic grating design.

The dual slab grating [Fig. 4(c)] uses 1D gratings and has low aspect ratio etches [95, 96, 100]. It also has a solid substrate attached to the gratings for efficient heat removal and optical alignment, but requires wafer bonding. Both the dual slab grating and the reverse slab grating [Fig. 4(d); Ref. 97] could be fabricated monolithically with 2D top-down lithography. The 2D reverse slab grating and the buried grating [Fig. 4(e); Refs. 98 and 99] both have the advantage of having solid and smooth vacuum channel sidewalls for the electrons, which could provide advantages for reduced electron scattering, better charge control and integration with on-chip waveguides. Finally, any of the previously described gratings can be mixed and matched to make an asymmetric grating [e.g. Fig. 4(f)], which could have advantages for specific applications. For example, a rectangular pillar grating combined with a slab grating provides a monolithic, 1D dual grating structure.

Resonance effects in the thickness of single gratings have also been explored in simulation [101]. In that study, the substrate thickness was $\sim\lambda$, which could be particularly relevant for optimization of pillar structures. Further work remains on exploring the effect of so-called guided mode resonances in the case of finite laser pulse lengths, transverse laser spot sizes, and finite geometries.

3.6. *Deflecting and focusing structures*

Deflecting structures are integral components of particle accelerators [102–104]. Indeed, undulators for free electron lasers have been designed using deflecting microstructures illuminated by alternating pulse-front-tilted laser beams to define each pole [102, 105]. Alternatively, the undulator could utilize lithographically defined undulator poles, requiring only a single pulse-front-tilted laser beam as the driver [106].

In addition, periodic focusing of the electron beam must be provided. Several focusing arrangements using laser-driven grating structures have been proposed. In such arrangements, focusing is provided by symmetric pumping of the structure using two lasers [107, 108]. Additionally, a single grating focusing structure based on a parabolic grating tooth geometry has recently been proposed and demonstrated [109].

4. Fundamentals of Grating and Pillar Structures

Since 2013, three particular DLA geometries have been used to demonstrate dielectric laser acceleration experimentally. They are the single grating [110, 111], dual grating [95, 96, 112, 113], and dual pillar structures [92, 94, 114]. We describe each in the following sections.

4.1. *Single grating*

Due to ease of fabrication and experimental testing, the single grating dielectric geometry shown in Fig. 5 and its variants (discussed below) have recently been tested. When illuminated by a perpendicular laser pulse (see Fig. 5), a near-field diffracted field pattern is excited above the surface of the grating. This field pattern can be decomposed into a Fourier sum of traveling waves in the z direction in Fig. 5. In the geometry where $\alpha = \psi = \theta = 0$, the field profile of each of these traveling waves can be analytically described as [115]

$$E_z = \frac{icB_0}{\kappa\gamma} e^{i(k_{nz}z + k_{ny}y - \omega t + \phi)}, \tag{1}$$

$$E_y = \frac{-cB_0}{\beta} e^{i(k_{nz}z + k_{ny}y - \omega t + \phi)}, \tag{2}$$

$$B_x = B_0 e^{i(k_{nz}z + k_{ny}y - \omega t + \phi)}. \tag{3}$$

Here, c is the speed of light, B_0 is the amplitude of the transverse magnetic field, κ is the ratio of the grating periodicity to the incident wavelength, $\gamma = (1 - \kappa^2)^{-1/2}$, ω is the circular frequency of the incident laser, and $k_{nz} = n\omega/\kappa c$, $k_{ny} = k_{nz}/\gamma$, where n is the index of the mode. In order for the fields to synchronously accelerate electrons traveling in the z direction above the grating (and injected at the appropriate phase), the phase velocity of the $n = 1$ traveling wave, $\omega/k_{1z} = \kappa c$, equals the injection velocity of electrons, βc. A limiting factor of the single grating DLA arises from the transverse decay of the accelerating fields in the y direction, dictated by $k_y = k_z/\gamma = \omega/\gamma\kappa c$. Thus, the accelerating field has a decay length of $\beta\lambda\gamma/2\pi$.

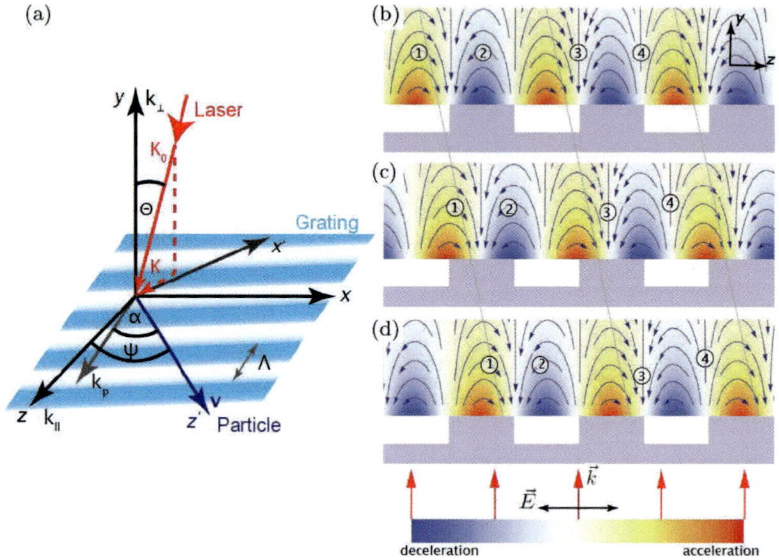

Fig. 5. (a) The geometrical setup of a laser incident upon the single grating. (b)–(d) Three snapshots of four charged particles traversing a single grating, illuminated by a laser. The simulated field pattern is shown. Depending on the injection phase of the positron in the excited accelerating wave, the positron can be accelerated, decelerated or deflected. [Adapted figure (with permission) from J. Breuer, J. McNeur and P. Hommelhoff, *J. Phys. B* **47**, 234004 (2014). © IOP Publishing.]

4.2. *Dual grating*

The evanescent fields of a single grating structure are not suitable for acceleration of relativistic particles ($\gamma \gg 1$) [8]. This is because the position of the evanescent mode with speed-of-light velocity occurs at a distance of ∞ from the grating surface, where the amplitude of the field is zero [110, 116, 117]. However, the fields between two dielectric surfaces (or gratings) are suitable because the superposition of evanescent waves traveling in opposite directions (along the y axis — either due to reflection off the second surface or a second incident source) results in a hyperbolic cosine transverse profile, rather than an exponential decay [110, 116]. The first proposal of such a geometry was as a miniature linac [8]. A design employing dielectric dual gratings was subsequently proposed [40]. Recently, dual grating structures designed for fs lasers were developed, as illustrated in Fig. 6.

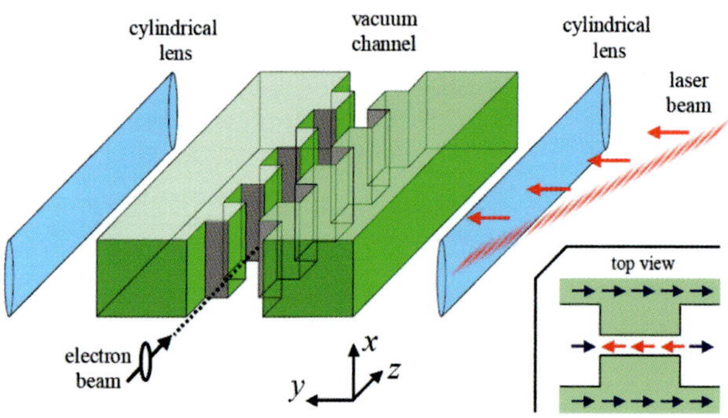

Fig. 6. Dual grating dielectric laser accelerator microstructure, illuminated from one side by a pulse-front-tilted laser beam. [Adapted (under CC-BY 3.0 license) from T. Plettner, P. P. Lu and R. L. Byer, *Phys. Rev. ST Accel. Beams* **9**, 111301 (2006). Published 2006 by the American Physical Society.]

The dual grating structure operates as a phase reset structure [40]. In the simplest geometry, the accelerator structure is a dielectric slab with a binary grating of structure period $\Lambda = \beta\lambda$. The "phase reset" arises from the phase difference between refraction through a dielectric pillar of refractive index $n > 1$, compared to a vacuum region with $n = 1$. The grating structure is designed to support evanescent grating modes propagating upward and downward through the structure. In the case of a pure TM mode accelerating structure ($\alpha = \psi = 0$), within a structure period these diffraction modes U_n, V_n, and W_n in the z, y, and x directions, respectively, can be denoted by [103]

$$U_n(y) = u_{n,+}e^{+\Gamma_n y} + u_{n,-}e^{-\Gamma_n y}, \tag{4}$$

$$V_n(y) = v_{n,+}e^{+\Gamma_n y} + v_{n,-}e^{-\Gamma_n y}, \tag{5}$$

$$W_n(y) = w_{n,+}e^{+\Gamma_n y} + w_{n,-}e^{-\Gamma_n y}, \tag{6}$$

where the modes u, v, and w are the amplitudes of the upward and downward (reflected and transmitted) propagating diffraction modes in the z, y, and x directions, respectively. The coefficient of the nth grating diffraction order Γ_n is given by

$$\Gamma_n = \sqrt{(nk_p - k\sin\theta)^2 - k^2}, \tag{7}$$

where θ is the angle of incidence of the laser to the structure (typically $\theta = 0$), $k = 2\pi/\lambda$, and $k_p = 2\pi/\Lambda$, where Λ is the structure period. Summing over all modes, the fields can be described by the Fourier series [103]

$$E_z(y,z,t) = \sum_{n=-\infty}^{+\infty} U_n(y)e^{iz(nk_p - k\sin\theta)}e^{ikct-i\phi}, \tag{8}$$

$$E_y(y,z,t) = \sum_{n=-\infty}^{+\infty} V_n(y)e^{iz(nk_p - k\sin\theta)}e^{ikct-i\phi}, \tag{9}$$

$$B_x(y,z,t) = \sum_{n=-\infty}^{+\infty} W_n(y)e^{iz(nk_p - k\sin\theta)}e^{ikct-i\phi}, \tag{10}$$

where ϕ corresponds to the phase between the electron and the incident laser. Assuming that the electron travels in the $\vec{z}$ direction, then the average force on a particle is [103]

$$\langle F_z\rangle = q\mathbb{R}\left\{\frac{1}{s(T)}\int_0^{s(T)} \sum_{n=-\infty}^{+\infty} U_n(y)e^{iks/\beta - i\phi} \times e^{is(nk_p - k\sin\theta)}\mathrm{d}s\right\}, \tag{11}$$

$$\langle F_y\rangle = q\mathbb{R}\left\{\frac{1}{s(T)}\int_0^{s(T)} \sum_{n=-\infty}^{+\infty} V_n(y)e^{iks/\beta - i\phi} \times e^{is(nk_p - k\sin\theta)}\mathrm{d}s\right\}, \tag{12}$$

$$\langle F_x\rangle = 0, \tag{13}$$

where $s(T) = \beta cT$ is the distance traveled by a particle in time T. In practice, relativistic TM mode structures are designed such that $\langle F_z\rangle \gg \langle F_y\rangle$ for relativistic electrons. The appearance of a small but nonzero F_y occurs because the structure is pumped asymmetrically from one side only. With two laser beams (incident simultaneously from above and below), there is cancelation in this transverse force arising from symmetry in the upward and downward propagating laser modes [103, 116]. Experimental demonstrations of relativistic beams in DLAs have not observed measurable deflection from an asymmetrically pumped accelerating structure [96, 118]. However, significant transverse deflecting forces can be imposed on the particle if the grating is rotated such that $\alpha \neq 0$ [102, 103].

Structures are designed to operate with peak electric fields provided by ultrafast femtosecond laser pulses. Pulses of order 100 fs are routinely available, but limit the interaction length to $\sim$30 μm (of the order of 50 DLA periods). The interaction region can be extended by employing pulse front tilt [91, 95]. Pulse front tilt of an incident laser pulse is illustrated in Fig. 6, occurring when an optical pulse encounters a dispersive element (diffractive or refractive).

Many numerical optimization studies of the dual grating structure have been undertaken, for different laser wavelengths, dielectric materials, and electron beam energies [107, 112, 113, 119–121]. These optimizations principally use finite difference time domain (FDTD) or finite difference frequency domain (FDFD) electromagnetic simulations. They are usefully employed to identify appropriate microfabrication techniques.

From the perspective of electron beam quality, the impedance that microfabricated corrugated dielectric accelerator structures present to the beam is significant [122]. Related accelerator structures have been proposed, whereby the phase gratings are recessed with a smooth, continuous vacuum channel surface for the electron beam [62–64, 98, 123]. Additionally, the proposed structure could be fabricated by etching the structure in one dimension into the wafer. This eliminates the need to bond wafers, which can introduce phase misalignment between the top and bottom gratings, reducing the accelerating gradient [99].

4.3. *Dual pillars*

Two rows of dielectric rods represent the simplest implementation of a dual grating dielectric laser accelerator with the minimum material used. Additionally, due to the high refractive index contrast of a dual pillar structure and the lack of any substrate material affecting the optical mode, a wide variety of optical modes can be engineered with this simple geometry. Particle accelerators based on dual pillar gratings can be traced back to Palmer [39] for radio frequency applications with metallic rods. Rows of dielectric rods can be used to create high reflectivity mirrors for transverse magnetically polarized light, creating a strongly modulated optical near field for electron acceleration. Moreover, dual pillar gratings can create a variety of coupled accelerator modes with hyperbolic cosine or hyperbolic sine mode profiles in the same way as the dual slab grating.

The constituent block of a dual pillar structure is a single row of pillars that will produce evanescent inverse Smith–Purcell grating modes on either side [92]. The thickness of the pillars controls the optical delay across each pillar and hence interference between the incident and exciting optical fields interacting with the pillar grating. With sufficient refractive index contrast (for example, silicon:vacuum has 3.5:1 refractive index contrast), a pillar can produce a $\pi/2$ phase modulation between the vacuum gaps between pillars and the pillars themselves, canceling out the transmitted optical field and creating a high reflectivity grating with a strongly modulated evanescent periodic mode. Using partially transmissive gratings and/or dual-sided illumination allows excitation of a hyperbolic coupled mode between the two rows of pillars with ultrashort laser pulses.

5. Recent Experimental Results

Several recent experiments have demonstrated electron acceleration using DLAs of various geometries. These experimental results are outlined briefly.

5.1. *Single grating demonstrations*

Many variations of these dielectric single gratings have been tested with subrelativistic electron beams. An experimental setup used to test such structures with a 20–30 keV electron beam is detailed elsewhere [110]. At the time of those first experiments, the apparatus consisted of a scanning electron microscope serving as a DC electron source, a Ti:sapphire long cavity oscillator (generating pulses with a pulse length of 110 fs, a central wavelength of 787 nm, a pulse energy of 150 nJ, and a repetition rate of 2.7 MHz), a four-axis actuator stage to align the electron beam and laser pulses to the DLA, and a detection system consisting of a high pass energy filter (a retarding field spectrometer) and a microchannel plate detector.

In practice, the voltage of the retarding field spectrometer is set higher than the injected electron energy divided by the electron charge. For instance, if the injected energy is 25 keV, the spectrometer voltage is set to 26 kV so that only electrons which have gained an energy of 1 keV or more as they pass the DLA traverse the energy filter and reach the MCP detector.

This setup was used for the first proof-of-principle dielectric laser acceleration of subrelativistic electrons [111]. In that experiment, a fused silica single grating with a periodicity of 750 nm was illuminated by the aforementioned Ti:sapphire laser. This periodicity, limited by fabrication capabilities, meant that the third spatial harmonic was used for accelerating electrons with an injection energy of approximately 28 keV. The dependence of the resulting acceleration signal on the spectrometer voltage, the laser polarization, the distance of the electron beam from the surface of the grating, and the injection energy of the electrons is shown in Fig. 7.

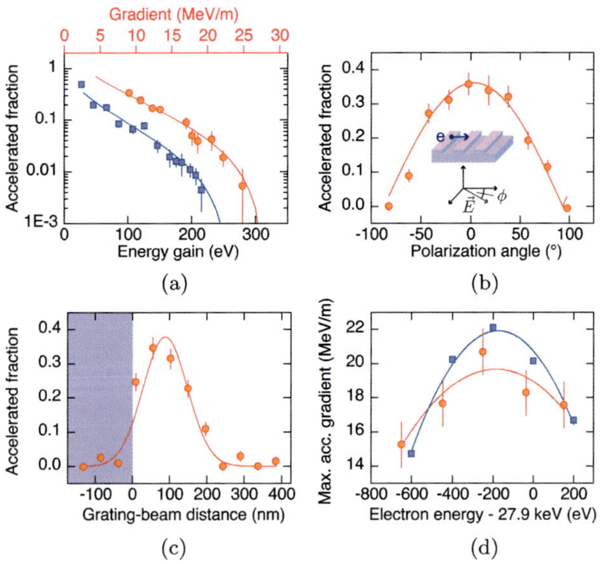

Fig. 7. The detected electron count rate for the experimental setup described in Sec. 5.1 as a function of the spectrometer voltage (a), the polarization of the incident laser (b), the distance of the electron beam centroid from the surface of the grating (c), and the injection energy of the electrons (d). [Reprinted figure (with permission) from J. Breuer and P. Hommelhoff, *Phys. Rev. Lett.* **111**, 134803 (2013). © 2013 by the American Physical Society.]

A significant step enabling new fabrication techniques for DLAs was the first demonstration of acceleration and deflection using a DLA fabricated from silicon [124], in which a 96.3 keV electron beam was accelerated with a gradient of 218 MeV/m and deflected with a gradient of 160 MeV/m. The structure used in that experiment was a single grating with 490 nm pitch designed to phase-match with the fundamental spatial harmonic of a 130 fs, 5 nJ Ti:sapphire oscillator at 907 nm wavelength. This corresponded to longitudinal energy gains of up to 1.2 keV for acceleration and transverse energy gains of 0.9 keV for deflection. Acceleration and deflection were measured simultaneously using an imaging spectrometer. Silicon is a desirable DLA material principally because of the ubiquity of fabrication techniques that can be employed to integrate the accelerator with other optical components, such as on-chip waveguides. Moreover, its relatively high index of refraction at optical and near-infrared wavelengths leads to efficient coupling of incident laser light into the desired accelerating mode. Since that experiment, several other DLA structures have been fabricated in silicon, some of which are described below.

The electron beam experimental setup in Ref. 111 was recently upgraded via the use of Si single gratings (with a period of 620 nm) with a Tm fiber laser operating at a central wavelength of 2 μm, a pulse energy of 1 μJ, a pulse length of 500 fs, and a repetition rate of 1 MHz. The periodicity of the grating and the incident wavelength ensured that 28 keV electrons would be synchronously accelerated by the first harmonic above the grating. Moreover, the transverse decay of the accelerating mode scales with the incident wavelength, so that the size of the region of strong accelerating fields is more than twice as large when compared with the previous proof-of-principle experiment, lending motivation for working with 2 μm lasers in DLA experiments. Electrons with energy gains approaching 1.5 keV were detected — a fivefold increase when compared to the proof-of-principle experiment [111].

The smallest feature size that can be fabricated using conventional lithography is of the order of 300 nm, which constrains the range of electron beam energies that can be accelerated by the first spatial harmonic of the structures. However, lower electron energies (velocities) can be accelerated at DLAs by using harmonics of the fundamental spatial mode. Experiments at single gratings with the third, fourth, and fifth harmonics have demonstrated acceleration of electrons with kinetic energy as low as 9.6 keV with an acceleration gradient of 7.5 MeV/m [125, 126].

For a grating geometry with a fixed periodicity, the velocity of accelerated subrelativistic electrons will eventually no longer match the phase velocity of the accelerating traveling wave. Thus, electrons slip in phase with respect to the traveling wave. Electrons injected at accelerating phases encounter decelerating phases after a length determined by the electron energy, laser wavelength, and field gradient. This must be compensated for by either frequency-chirping the incident laser pulse or modifying the acceleration structure design, where the phase velocity of the accelerating mode has to match the particle velocity at all longitudinal coordinates. A design, based on lithography-based structural chirping, that aims to achieve this is shown in Fig. 8.

In the chirped structural design [109], the distance between grating teeth depends on the longitudinal coordinate z as $\lambda_p = \lambda_{p0} + az$, where $\lambda_{p0} = 620\,\text{nm}$ is the starting grating period and a

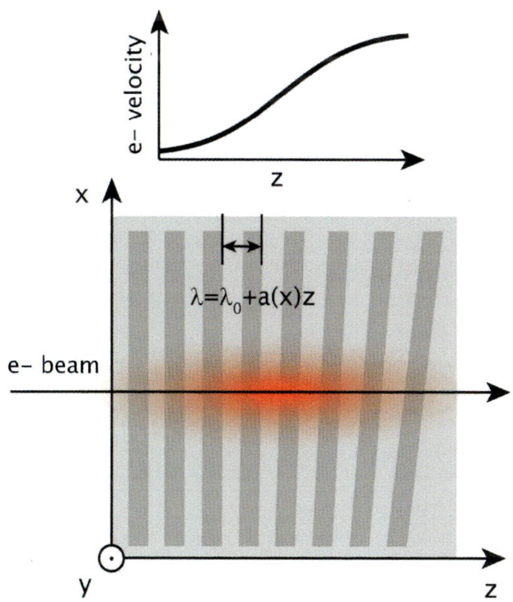

Fig. 8. The chirped structural design that overcomes the so-called dephasing of accelerated electrons. The grating period increases as electrons travel in the positive z direction. Further, the slope of increasing grating period (a) is a function of the position over which the electrons pass the structure in the x direction, allowing this parameter to be tuned *in situ* [109].

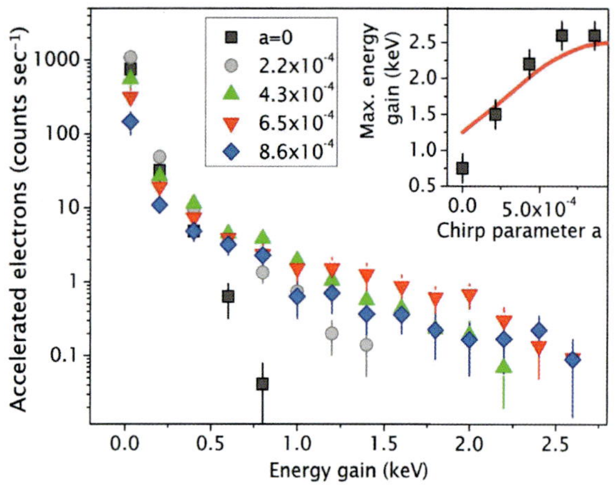

Fig. 9. The accelerated count rate as a function of the minimum electron energy gain for various chirp parameters a. The dephasing limit is overcome for chirp parameters $a > 0$ [109].

the chirp parameter. In Fig. 9 the accelerated electron current as a function of $eU_s - E_{k0}$ is shown for different values of a but the same laser parameters, with e the electron charge and U_s the DC voltage applied to the spectrometer. The maximum energy gain is now limited by the incident field amplitude of $E_{\max} = 1.3\,\text{GV/m}$ and the interaction distance, which in turn is given by the laser spot radius $w = 29 \pm 3\,\mu\text{m}$. Without any structural chirp ($a = 0$), dephasing limits the maximum achievable energy gain to $\Delta E_k = 0.80\,\text{keV}$, in agreement with theory [115]. For $a = 6.5 \times 10^{-4}$, the maximum energy gain is more than tripled to $\Delta E_k = 2.6\,\text{keV}$, corresponding to an acceleration gradient of 69 MeV/m.

5.2. *Dual grating demonstrations*

Dual grating experiments have been conducted using relativistic electron beam energies in the range of 8–60 MeV. The incident electron beams were accelerated by conventional radio frequency linacs upstream of the DLA [127–131]. Several facilities for dual grating DLA experiments with nonrelativistic [132–135] and relativistic electron beams have subsequently been proposed [136–140].

The first demonstration of dielectric laser acceleration at relativistic energy was performed using a dual grating structure [96, 141, 142]. The structures were fabricated from fused silica [53, 54, 143], and driven by a $\lambda = 800\,\text{nm}$ Ti:sapphire laser with 1 ps pulse duration. In that experiment, a maximum electron accelerating gradient of 350 MeV/m was observed.

A factor of DLAs that was not accounted for at the time of early theoretical research is that the peak breakdown field of the material is strongly dependent upon the pulse duration [41]. In effect, for a given incident fluence, reducing the incident laser pulse duration increases the electric field of the accelerator. This was experimentally demonstrated recently through acceleration of relativistic electrons using fs scale laser pulses at a dual grating structure [118]. Employing laser pulses of 90 fs duration, a peak accelerating gradient of $690 \pm 100\,\text{MeV/m}$ was observed. Results are illustrated in Fig. 10.

In experiments accelerating electron beams using DLAs, the energy spread of the incident electron beam is modulated. This arises because the incident electron bunch of ~100 fs duration is much longer than the DLA structure with a period of ~3 fs. An appropriately bright source with microbunching elements (Sec. 6.2) is necessary for matching the restrictive longitudinal phase acceptance of DLA buckets. At relativistic energy, a phase-stable train of electron microbunches has been achieved using an inverse free electron laser, resulting in net acceleration after interaction with an optical scale accelerator [23, 24]. Sources designed for the shortest electron bunches have been demonstrated using velocity bunching [144], which may also be combined with nanotip photocathode sources with demonstrated low transverse emittance [145–147].

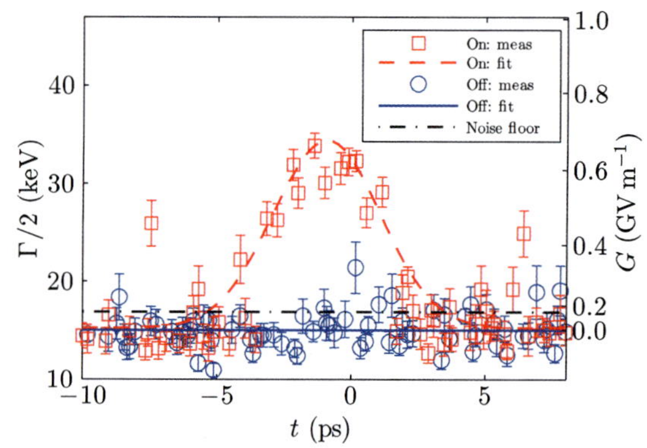

Fig. 10. Scanning the time of arrival of the accelerating fs laser pulse relative to the time of arrival of the electron bunch. With the laser on, the half width at half maximum of the electron beam energy distribution ($\Gamma/2$) was increased. A peak accelerating gradient of 690 ± 100 MeV/m was observed. [Reprinted figure (with permission) from K. P. Wootton *et al.*, *Opt. Lett.* **41**, 2696 (2016). © 2016 by the Optical Society of America.]

5.3. *Resonant structures*

A Bragg mirror structure was developed as a slab-symmetric, resonant DLA. The bandpass of the resonant structure was demonstrated from optical transmission through the structure [58]. Electron beam tests were performed at the E-163 experimental end station at SLAC with a 60 MeV beam and ~1 ps Ti:sapphire laser pulses [52, 58, 59, 148].

In experiments searching for acceleration with the structure, broadening of the transmitted energy spread was observed corresponding to an accelerating gradient of 28 MeV/m [149]. Further advances in the precision of the fabricated structures may be necessary in order to efficiently couple into the speed-of-light resonant accelerating modes.

5.4. *Dual pillar demonstrations*

The first monolithic dual grating DLA device demonstrated was a dual pillar structure made from a single piece of silicon [92]. Monolithic fabrication on a single wafer allows precise lithographic alignment of multiple structures and has strong advantages for future integrated accelerator devices. An example of a fabricated dual pillar structure is illustrated in Fig. 11.

Silicon dual pillar grating structures were demonstrated at subrelativistic electron energies of 86.5 keV and 96.3 keV with a few-nanojoule Ti:sapphire oscillator drive laser [92]. By tailoring the pillar dimensions, both an inverse Smith–Purcell accelerating mode (exponential transverse profile of the mode, analogous to a single grating) and an evanescent coupled mode (cosh transverse profile, analogous to a dual grating) were demonstrated. The inverse Smith–Purcell mode structure used a highly reflective pillar grating and a single drive laser, which only produced measurable excitation of one row of pillars. This structure produced maximum accelerating gradients of 370 ± 40 MeV/m and 255 ± 25 MeV/m deflecting gradients for longitudinal energy gains of 2.05 ± 0.02 keV and transverse energy gains of 1.43 ± 0.1 keV, respectively. The experimental maximum acceleration and deflection observed approached 35% and 24% of the incident laser field strength, respectively.

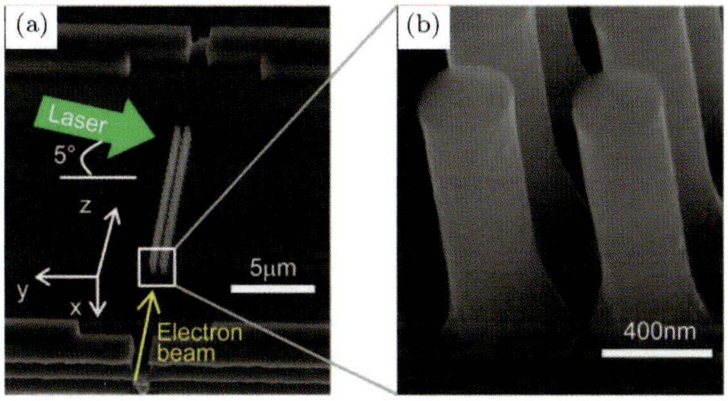

Fig. 11. Dual pillar structures fabricated in silicon for acceleration of subrelativistic electron beams. [Reprinted figure (with permission) from K. J. Leedle *et al.*, *Opt. Lett.* **40**, 4344 (2015). © 2015 by the Optical Society of America.]

The evanescent coupled mode dual pillar structure used a partially reflective pillar grating to allow a single drive laser to excite the evanescent fields on both sides of the accelerator channel to form a cosh accelerating mode in the channel. This coupled cosh mode produced uniform accelerating gradients of up to 196 ± 20 MeV/m with little beam deflection for electrons in the center of the coupled mode [92]. This was the first demonstration of a coupled mode with low deflection at subrelativistic electron energies. Many of the other subrelativistic experiments described are limited in laser–electron interaction length by the transverse deflection of the laser fields. Future experiments with optimized structures and two drive lasers could enable higher acceleration gradients with a more robust low deflection hyperbolic cosh mode over extended accelerator lengths.

5.5. *Two-stage acceleration demonstrations*

In order to verify the viability of phase-controlled staged acceleration, the authors of Ref. 109 employed an optical variation to the setup described in Sec. 5.1. An interferometer-like setup is introduced that splits the incoming laser pulse into two pulses, which are then focused to two separate spatial regions on the single grating structure (see Fig. 12). Electrons encounter two laser pulses separated in time by 200 fs.

By varying the relative arrival time of the two pulses with a delay line with 200-attosecond precision, the experiment shifts between two regimes: one in which electrons that are accelerated in the first pulse are decelerated in the second pulse and one in which electrons are accelerated in both pulses, enhancing the total acceleration signal. The oscillation between these regimes, evidence of phase-controlled staging, is shown in Fig. 13. This has been used to imprint sub-fs interactions on an electron

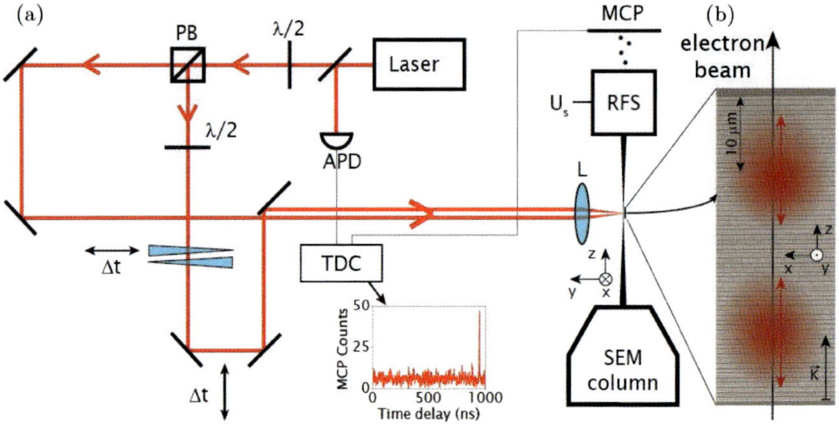

Fig. 12. Optical setup to create two distinct interaction regions for a single grating silicon DLA [125].

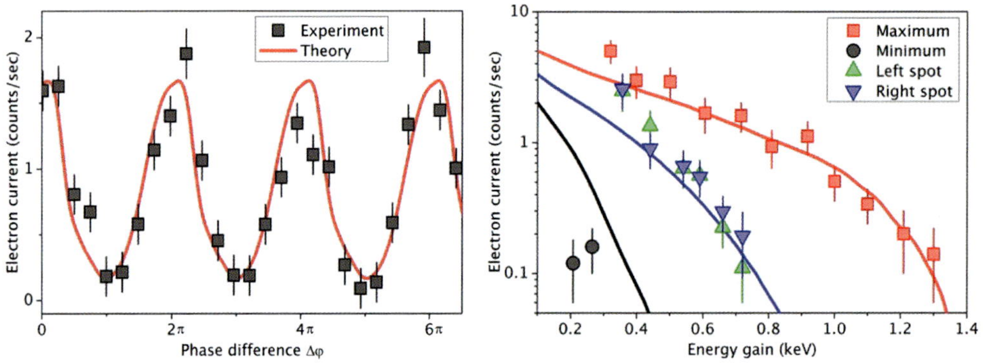

Fig. 13. The accelerated signal as a function of the relative phase of two laser pulses illuminating two distinct interaction regions. The signal oscillates between suppression and enhancement of acceleration. Moreover, the energy gain is roughly doubled in the enhanced regime (red) compared to interaction with each region (blue, green) [109].

beam [150]. Moreover, this staging demonstrates the linear scaling of the energy gain with the interaction distance and with the number of stages.

5.6. *Focusing structure*

Silicon gratings have also been used to demonstrate focusing of a subrelativistic electron beam, employing a design shown in Fig. 14 [109]. In a single grating structure, a deflecting mode can be excited by tilting the grating with an angle α with respect to the electron beam, as proposed in Ref. 102. As the deflection angle φ of the electron depends linearly on the grating tilt angle α, focusing can be achieved by a DLA microstructure with grating teeth with a parabolic transverse profile (so that the tilt angle of the grating tooth depends on the distance from the center of the design axis as $\alpha = \sigma x$, where σ is a constant and x is the distance from the design axis).

The focusing structure was demonstrated by translating an electron beam spatially in x across the structure, and observing the deflected position of the electron beam downstream. Hence the deflected angle φ could be measured, as illustrated in Fig. 14. The deflected angle φ was observed to be proportional to the offset x. That is to say, linear focusing was observed with $\varphi = bx$, with focal length $f = 1/b$. Transverse focusing and longitudinal acceleration are optical-phase-dependent in this DLA structure, and because electrons arrive randomly in the laser optical phase, the focusing strength and acceleration differ for electrons, depending on their optical phase. Effective focal lengths of the structure were determined for electrons with minimum energy gains of 200, 400, and 600 eV as 500 ± 20, 260 ± 10, and $190 \pm 10\,\mu$m, respectively, for a structure with a profile given by $\sigma = 0.39\,\mu\text{m}^{-1}$.

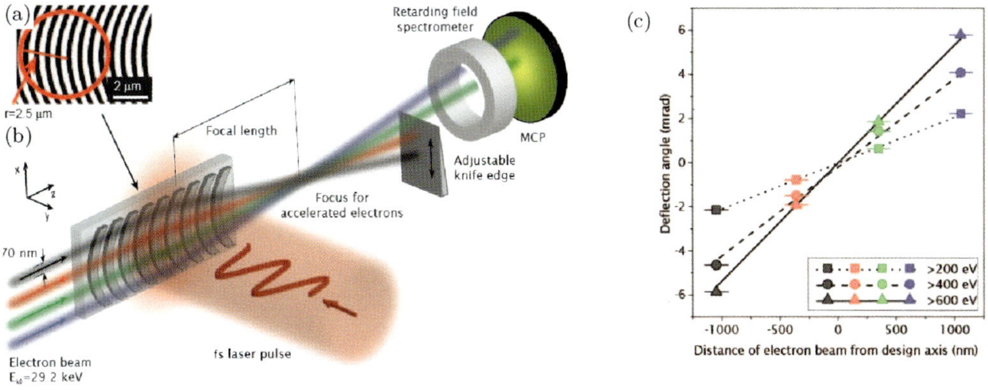

Fig. 14. A focusing DLA structure. (a) The radius of curvature of the grating teeth is illustrated. (b) The focal length of the structure was measured by translating the electron beam and a knife edge. (c) Different accelerated electron energies are focused with different focusing gradients, corresponding to the deflecting force at the appropriate optical phase [109].

6. Potential Applications

A schematic illustrating the basic components of a DLA-based accelerator (Fig. 15) consists of multiple stages of acceleration and focusing elements illuminated by a single laser source. Moreover, the laser source could trigger a photocathode (e.g. a tungsten or lanthanum hexaboride nanotip), which would then serve as an electron source for the multistage DLA. A future accelerator based on this architecture could enable many different applications [151]. Many interrelated questions remain, including the achievable pulse length and bunch charge from a laser-triggered nanotip, the handshake energy between the cathode and the first dielectric acceleration stage, and beam current stability of such a nanotip source [126]. However, a simple scaling argument can provide an estimate of the beam parameters for different applications.

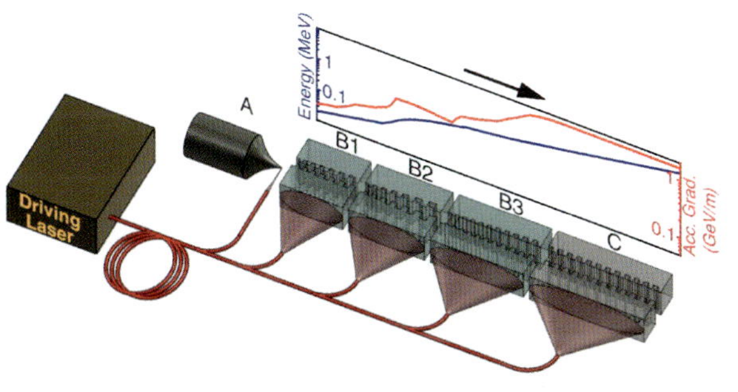

Fig. 15. Schematic illustration of an accelerator based on a nanotip emitter and successive DLA stages. [Reprinted figure (with permission) from J. Breuer and P. Hommelhoff, *Phys. Rev. Lett.* **111**, 134803 (2013). © 2013 by the American Physical Society.]

For a 30 keV electron beam, such a device can support 80 electrons per accelerating bucket [115], where we assume a 2 μm drive wavelength, a 400 as bunch length, and an emittance-limited beam in which the perveance is treated as a perturbation. For a laser with a pulse length of 600 fs and a flat top temporal profile, 7200 electrons per laser pulse can be supported. Further, for high repetition rate lasers (>1 MHz, like in commercially available solid state lasers), >10 nA of average current can be supported in a multistage DLA. Moreover, the supported charge per bunch scales favorably with higher electron energy. Indeed, an optimal charge of >1 fC per bucket and >10 fC per bunch train is feasible for relativistic ($\beta \sim 1$) electrons [48, 49].

6.1. *Medical therapy*

Even modest nA current in a mm scale MeV electron source would be of great benefit for medical patient treatment. While protons can be used to irradiate tissue deep inside the body, medical electron beams are limited to treatment of superficial cancers and diseases (within 6 cm of the patient's surface) due to the energy deposition along the path from the surface to the tumor [152]. These beams operate at electron energies of 6–20 MeV. The generation of such an electron beam presently requires radio frequency accelerators with a typical acceleration length of 1 m, and meter-long beam delivery systems. The acceleration length is determined by the acceleration gradient, which is about 10–20 MeV/m, while the delivery system must enable localization and positioning of the electron beam to the tumor.

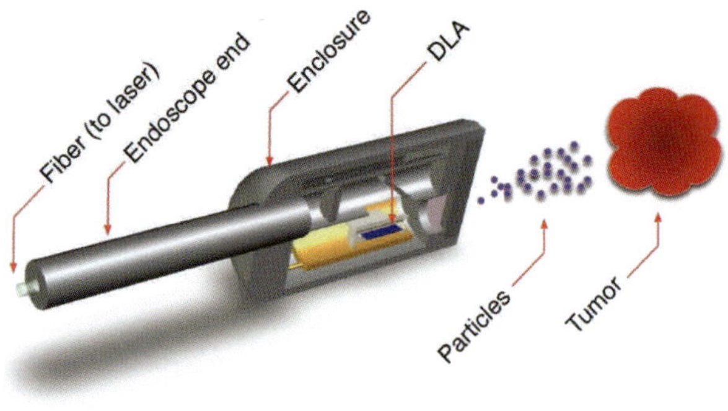

Fig. 16. Hypothetical schematic for the incorporation of a monolithic multistage DLA on an endoscope for the purpose of tumor irradiation. [Reprinted figure (with permission) from [R. J. England *et al.*, *Rev. Mod. Phys.* **86**, 1337 (2014). © 2014 by the American Physical Society.]

With a DLA and expected gradients exceeding 1 GeV/m, the generation of MeV electron beams over an acceleration length in the millimeter-to-centimeter range should be highly possible. With a compact source with micron scale beams the beam delivery complexities can be removed, and a significant improvement in the irradiation localization can be achieved. While a new form of tumor treatment might yield the largest benefit to society, other forms of medical applications may also be realized by this technology. Examples are end plate etching neurons, improving neuronal vesicle release and nerve function, as well as localized electron irradiation cleaning of sensitive devices (such as endoscopes with interior electronics).

With an active acceleration length on the millimeter-to-centimeter scale one may envision a multi-MeV electron beam generation device of the size of the tip of a pen, as illustrated schematically in Fig. 16. This would enable wholly new forms of electron beam treatments for clinicians, where, for example, the electron beam generation device is brought inside the body of a patient [153]. Because

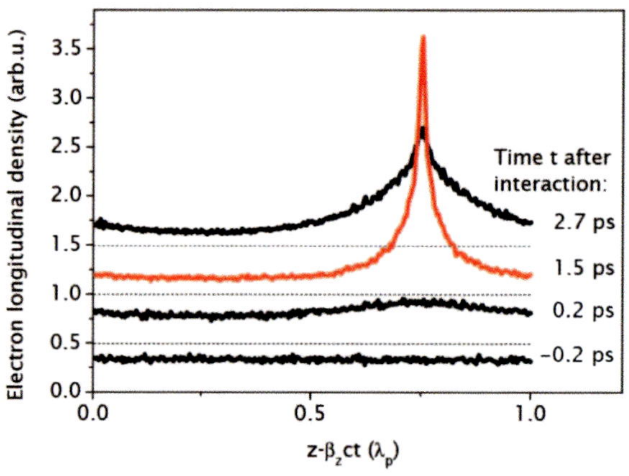

(a)

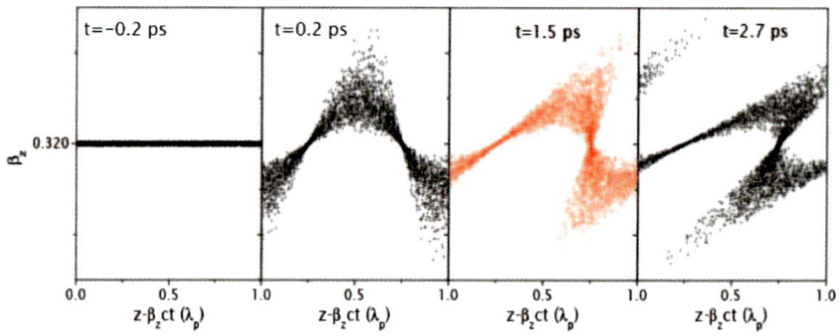

(b)

Fig. 17. Simulation of longitudinal velocity bunching of a subrelativistic electron beam downstream of interaction with a DLA. (a) Longitudinal electron density at various times t after interaction with a DLA. It can be seen that optimal bunching occurs approximately 1.5 ps downstream of the interaction, with an FWHM less than 200 as. For clarity, the zero baseline (dotted line) is offset vertically by 0.5 arb. units between successive times. (b) Electron beam longitudinal phase space distributions corresponding to (a).

of the potentially close proximity of the electron source device to the cancerous tissue, electron energies in the range of 1–5 MeV seem very promising, as the electrons are absorbed in the range of a few mm to about 1 cm [154]. Because of the small exposed volume, an electron current in the range of nA is sufficient to deliver a single massive dose over an exposure time on the order of 1 s, ideal for handheld or robotic surgery, with easily removed heat loads in the Watt regime. This could be transformative for intraoperative electron radiation therapy (IOERT), a highly successful electron-beam-based precision cancer treatment employed for decades already [155–158]). So far, the space requirements of conventional accelerators have prevented full exploitation of the therapeutic potentialities of IOERT in the clinic.

6.2. *Suboptical cycle interaction*

In addition to the generation of relativistic electrons, DLAs have the capability of imprinting sub-optical cycle energy modulation onto free electrons. Hence, DLAs could be used to create suboptical cycle electron bunches. The subcycle velocity modulation induced by the interaction of subrelativistic electrons and laser-induced accelerating fields leads to ballistic microbunching if the electron beam drifts following the acceleration interaction. The simulated electron beam phase space for a one-optical-cycle-long subset of the bunch is shown in Fig. 17 at different times after interaction with the silicon single grating described above [150]. At maximum compression, the bunch full width at half maximum drops below 200 attoseconds for a 30 keV starting electron energy.

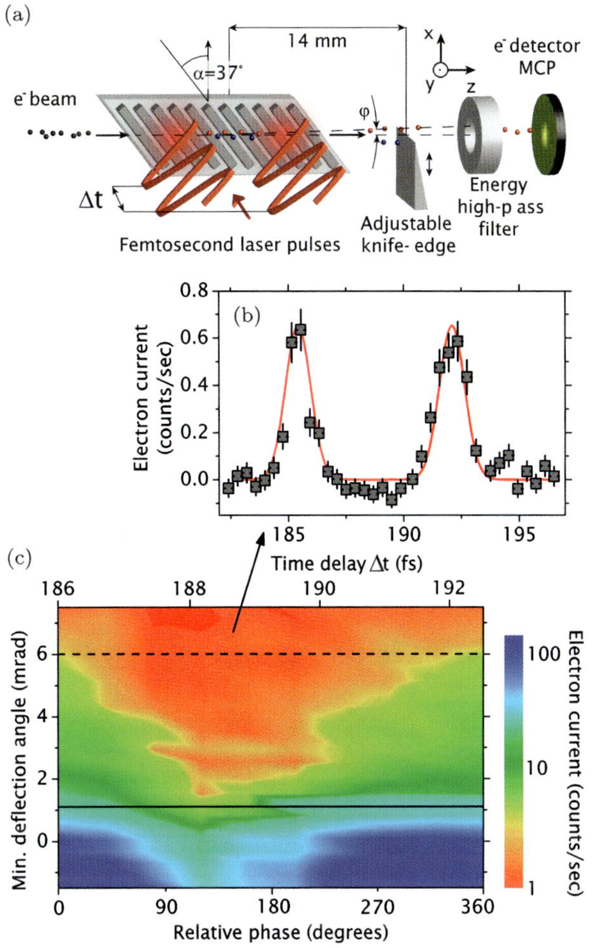

Fig. 18. The design and results of a two-stage deflection experiment. Two sequential regions of a tilted single grating DLA are illuminated by laser pulses with a controllable relative phase delay. By monitoring the accelerated signal that is also deflected past an adjustable knife edge as a function of the relative phase delay of the pulses illuminating the two regions, feature sizes on the order of 1 fs are detected in the energy-modulated electron beam [150].

This velocity bunching approach places stringent requirements on the interaction strength and electron energy in order to form a single microbunch for attosecond electron diffraction or other experiments. However, filtering the electron energy spectrum after interaction at two stages of deflecting structures enables the use of longer laser pulses to select out a single short temporal window. Consider the conceptual illustration in Fig. 18.

For this deflecting structure, electrons injected on crest are not only accelerated in z but also deflected in x. Moreover, the deflecting force imparted onto the electrons depends on their injection phase. By illuminating two regions of the deflecting structure and varying the relative arrival time of the two 2 μm laser pulses illuminating the regions, the demonstrated timing resolution of such a diagnostic is revealed to be 1.3 fs. Employing ultrashort incident laser pulses (such as a 20 fs laser pulse length with a central wavelength of 2 μm) and "gratings" consisting of only one grating tooth, a subcycle energy modulation can be imprinted on a passing electron beam. By then using a high pass energy filter such as the high pass spectrometer in Fig. 18, this subcycle (attosecond scale) feature could be isolated from the rest of the electron beam in a suboptical cycle pump–probe experiment [150].

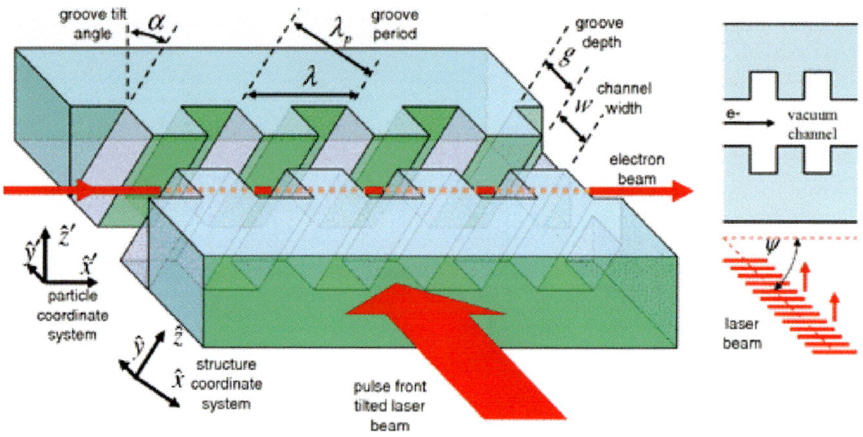

Fig. 19. A deflecting structure, driven by a pulse-front-tilted laser beam synchronous with the relativistic electron velocity. [Reprinted figure (under CC-BY 3.0 license) from T. Plettner and R. L. Byer, *Phys. Rev. ST Accel. Beams* **11**, 030704 (2008). Published 2008 by the American Physical Society.]

To realize a practical attosecond DLA streak camera, the evanescent decay of field strength at the single grating design must be overcome. With a single grating, the deflection angle of electrons is not only correlated with their arrival time at the interaction region but also with their arrival position with respect to the surface of the grating. To this end, a double grating design with a grating tooth tilt angle (shown in Fig. 19) is promising [102].

6.3. *Attosecond photon generation*

Extreme ultraviolet (EUV) sources are one option for lithographic fabrication of increasingly miniaturized microprocessor components [159, 160]. Over many decades, conventional accelerator light sources have matured to provide routine access to electromagnetic radiation from infrared through to hard X-rays [161, 162]. The development of light sources targeted to EUV wavelengths presents a unique opportunity for future accelerator-based light sources [163–165]. A coherent light source was an early proposed application of microstructure accelerators [166]. Other ideas have been developed taking advantage of the unique attosecond bunch format produced by DLAs [167, 168].

The critical hardware component is the deflecting structure [102, 104, 105, 116]. With an electron traveling at a transverse angle to the accelerating direction of a DLA, a nonzero deflection force results [91] as presented in Sec. 6.2. An undulator composed of dielectric deflecting structures is desirable because the structures accommodate undulator periods of $\sim$10–100 μm, with deflecting fields exceeding 1.3 GeV/m, equivalent to a magnetic field of $\sim$4 T [102]. Such a structure can be optimized for deflection transverse to the direction of electron propagation rather than longitudinal acceleration [169, 170]. An undulator can be formed by staging several such structures sequentially, as illustrated in Fig. 20.

Technological advances in lithography allow several simplifications to be made. Lithographic control of the microstructure phase means that single-sided illumination of the undulator poles could be employed instead of alternating at each pole [106].

6.4. *Beam position monitor*

To optimize accelerators with multiple stages and components, particle beam diagnostics are required.

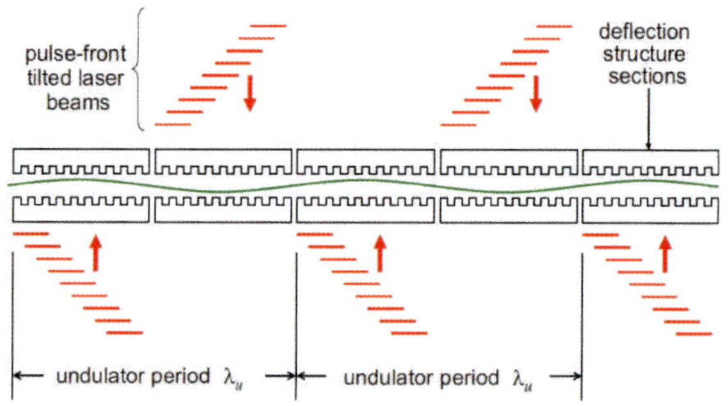

Fig. 20. An undulator composed of successive stages of laser-driven deflecting structures. [Reprinted from T. Plettner and R. L. Byer, *Nucl. Instrum. Methods Phys. Res., Sect. A* **593**, 63 (2008), with permission from Elsevier. © 2008 by Elsevier.]

In particular, diagnostics of the transverse beam position and distribution are necessary for reliable operation. Given that the transverse profile of electron beam channels in DLAs is on the order of ~1 μm, beam diagnostics with fundamental resolution on the scale of nanometers are required.

A candidate scheme for a nanometer resolution transverse beam position monitor involves the inverse acceleration process. Compared to acceleration of electrons by pumping a DLA structure with laser fields, the inverse process occurs when an electron beam passes through the structure and radiates energy. The evanescent wakefields generated by the electron beam couple to the DLA structure, resulting in the emission of real photons as an electron passes through. This has been developed as a passive structure beam position monitor [108, 171–173], and as a beam-driven photon source [174]. The dual grating structure used is illustrated in Fig. 21.

Electrons within the vacuum channel of the dual grating DLA emit radiation perpendicular to the surface of the grating with a central wavelength matching the periodicity of the grating. The dual grating geometry ensures that the periodicity of the grating depends on the transverse position of the beam in x. Thus, by monitoring the central wavelength and bandwidth of the emitted radiation, the centroid position and transverse spot size of the electron beam can be deduced. Using the 60 MeV, 20-μm-wide electron beam from the Next Linear Collider Test Accelerator at SLAC, a spatial resolution of 10 μm was achieved [173]. With smaller electron beams, a spatial resolution of 0.7 nm is expected, which would be useful for the control of beams through successive stages of DLAs.

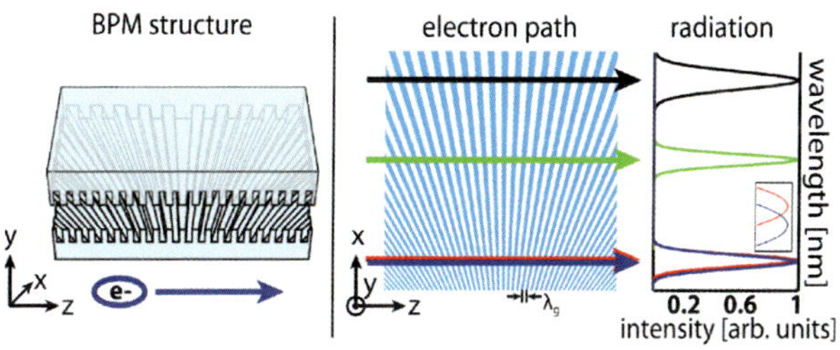

Fig. 21. Passive beam position monitor structure, based on dual grating DLAs with a period that varies in the transverse coordinate x. The beam position is determined from the wavelength of light emitted when electrons pass through the channel. [Reprinted figure (with permission) from K. Soong *et al.*, *Opt. Lett.* **39**, 4747 (2014). © 2014 by the Optical Society of America.]

In addition, an active BPM structure has recently been demonstrated [175]. By pumping the structure with laser photons, cross-correlation of the arrival of the laser pulse and electron bunch at the structure was observed with a spatial resolution of ~450 nm and a temporal resolution of ~480 fs. The experiment was conducted with a subrelativistic electron beam, but the technique can be extended to relativistic energies through the use of a double-sided structure.

6.5. *TeV scale linear collider*

We note that DLAs may be a viable technology candidate for a future 10 TeV electron–positron collider, as is argued in greater detail in Refs. 6 and 151. To reach such center-of-mass energies with microwave technology, a main linac over 100 km in length would be necessary; however, a DLA linac with a length of 10 km operating at a GeV/m acceleration gradient could achieve the same energy. The required luminosity of such a collider ($10^{36}\,\text{cm}^{-2}\,\text{s}^{-1}$ [6]) favors the DLA bunch format of low bunch charge operated at high repetition rates to avoid beamstrahlung energy spread at the interaction point, which scales with the bunch charge [151]. Lastly, we note that the existence of high repetition rate solid state lasers could lead to electrical wall plug efficiency exceeding 10%, which is very favorable for proposed collider schemes [48, 49].

7. Summary

Dielectric laser accelerators compose a subfield of novel laser-based accelerators that, due to decades of innovative design development and laser and nanofabrication progress, has recently blossomed with experimental demonstrations of advanced beam diagnostics and manipulation as well as accelerating gradients approaching 1 GeV/m. With further advances, applications related to photon generation, radiative oncology, and ultrafast metrology may soon be realized. We have summarized here the history of the field, the many designs conceived within, and their recent experimental results.

Acknowledgments

The authors are grateful to Prof. Robert L. Byer (Stanford University), Dr. R. Joel England (SLAC National Accelerator Laboratory), and Prof.

Peter Hommelhoff (Friedrich-Alexander-Universität Erlangen-Nürnberg) for comments that greatly improved the manuscript.

This work was supported by the US Department of Energy under Contract DE-AC02-76SF00515, the US Air Force Office of Scientific Research under grant FA9550-14-1-0190, the European Research Council under grant 616823 NearFieldAtto, and by the Gordon and Betty Moore Foundation under grant GBMF4744 (Accelerator on a Chip).

References

[1] K. Shimoda, *Appl. Opt.* **1**, 33 (1962).
[2] J. D. Lawson, in *Generation of High Fields for Particle Acceleration to Very-High Energies*, Tech. Rep. CERN-1985-007, CERN (Geneva, Switzerland, 1985), pp. 3–12.
[3] J. Arnesson and F. K. Kneubühl, *Infrared Phys.* **25**, 121 (1985).
[4] G. Travish and R. B. Yoder, *Proc. SPIE* **8079**, 80790K (2011).
[5] G. Travish, *AIP Conf. Proc.* **1507**, 85 (2012).
[6] R. J. England *et al.*, *Rev. Mod. Phys.* **86**, 1337 (2014).
[7] R. J. England, *IEEE J. Sel. Top. Quantum Electron.* **22**, 4401007 (2016).
[8] J. D. Lawson, *Laser Accelerators?*, Tech. Rep. RL-75-043, Rutherford Laboratory (Chilton, Oxon, UK, 1975).
[9] J. D. Lawson, *IEEE Trans. Nucl. Sci.* **26**, 4217 (1979).
[10] J. D. Lawson, *Part. Accel.* **10**, 73 (1980).
[11] P. M. Woodward and J. D. Lawson, *J. Inst. Electr. Eng. — Part III: Radio Commun. Eng.* **95**, 363 (1948).
[12] R. B. Palmer, An introduction to acceleration mechanisms, in *Frontiers of Particle Beams*, eds. M. Month and S. Turner, Lecture Notes in Physics, Vol. 296 (Springer, Berlin, Heidelberg, 1988), pp. 607–635.
[13] R. B. Palmer, *Part. Accel.* **11**, 81 (1980).
[14] R. B. Palmer, *IEEE Trans. Nucl. Sci.* **28**, 3370 (1981).
[15] R. B. Palmer, *AIP Conf. Proc.* **335**, 90 (1995).
[16] E. Esarey, P. Sprangle and J. Krall, *Phys. Rev. E* **52**, 5443 (1995).
[17] T. Plettner, *Analysis of Laser-Driven Particle Acceleration from Planar Infinite Conductive Boundaries*, Tech. Rep. SLAC-PUB-11637, Stanford Linear Accelerator Center (Menlo Park, CA, USA, 2006).
[18] T. Plettner, *Analysis of Laser-Driven Particle Acceleration from Planar Transparent Boundaries*, Tech. Rep. SLAC-PUB-11800, Stanford Linear Accelerator Center (Menlo Park, CA, USA, 2006).
[19] T. Plettner, *Field-Analysis of Vacuum Free-Space Laser Acceleration from Rough-Surface and Absorbing Thin Boundaries*, Tech. Rep. SLAC-PUB-13188, Stanford Linear Accelerator Center (Menlo Park, CA, USA, 2008).
[20] Y. C. Huang *et al.*, *Nucl. Instrum. Methods Phys. Res., Sect. A* **407**, 316 (1998).
[21] T. Plettner *et al.*, *Phys. Rev. ST Accel. Beams* **8**, 121301 (2005).
[22] T. Plettner *et al.*, *Phys. Rev. Lett.* **95**, 134801 (2005).
[23] C. M. S. Sears *et al.*, *Phys. Rev. ST Accel. Beams* **11**, 101301 (2008).
[24] C. M. S. Sears *et al.*, *Phys. Rev. ST Accel. Beams* **11**, 061301 (2008).
[25] F. O. Kirchner *et al.*, *Nat. Photon.* **8**, 52 (2014).
[26] A. Gliserin, M. Walbran and P. Baum, *Rev. Sci. Instrum.* **87**, 033302 (2016).
[27] C. Kealhofer *et al.*, *Science* **352**, 429 (2016).
[28] Y. Takeda and I. Matsui, *Nucl. Instrum. Methods* **62**, 306 (1968).
[29] K. Mizuno, S. Ono and O. Shimoe, *Nature* **253**, 184 (1975).
[30] K. Mizuno *et al.*, *Nature* **328**, 45 (1987).
[31] A. W. Lohmann, Particle Accelerator Utilizing Coherent Light (1966). US Patent 3267383. Filed 27th May 1963, issued 16th Aug. 1966.
[32] A. W. Lohmann, Particle Accelerator (1967). Canadian Patent 773768. Issued 12th Dec. 1967.
[33] P. S. Csonka, *Part. Accel.* **5**, 129 (1973).
[34] F. Scarlat and M. Argeşanu, *IEEE Trans. Nucl. Sci.* **24**, 1651 (1977).
[35] J. D. Lawson, *Physica Energiae Fortis et Physica Nuclearis* **5**, 339 (1981). [J. D. Lawson, *A Survey of Some Ideas for Accelerators Using Laser Light*, Tech. Rep. RL-81-030, Rutherford and Appleton Laboratories (Chilton, Oxon, UK, 1980).]
[36] B. R. Frandsen, S. A. Glasgow and J. B. Peatross, *Laser Phys.* **16**, 1311 (2006).
[37] F. Toufexis, T. Tang and S. G. Tantawi, in *Proc. 36th International Free Electron Laser Conference* (JACoW, Geneva, Switzerland, Aug. 2014), pp. 131–136.
[38] A. M. Sessler, *IEEE Trans. Nucl. Sci.* **30**, 3145 (1983).
[39] R. B. Palmer, in *New Developments in Particle Acceleration Techniques*, ed. S. Turner, Tech. Rep. CERN-1987-011-V-2, CERN, Vol. 2 (Geneva, Switzerland, 1987), pp. 633–641. [R. B. Palmer, *Open Accelerating Structures*, Tech. Rep. SLAC-PUB-4161, Stanford Linear Accelerator Center (Stanford, CA, USA, 1986).]
[40] H. Peng and J. Zhuang, in *Proc. 1979 Linear Accelerator Conference*, ed. R. L. Witkover, Tech. Rep. BNL-51134, Brookhaven National Laboratory (Upton, NY, USA, 1979), pp. 257–261.
[41] B. C. Stuart *et al.*, *Phys. Rev. Lett.* **74**, 2248 (1995).

[42] Y. M. Oh *et al.*, *Int. J. Heat Mass Transf.* **49**, 1493 (2006).
[43] K. Soong *et al.*, *AIP Conf. Proc.* **1507**, 511 (2012).
[44] R. L. Byer, *AIP Conf. Proc.* **193**, 4 (1989).
[45] D. Zheng and R. L. Byer, *AIP Conf. Proc.* **335**, 156 (1995).
[46] W. Sibbett, A. A. Lagatsky and C. T. A. Brown, *Opt. Express* **20**, 6989 (2012).
[47] G. Mourou *et al.*, *Nat. Photon.* **7**, 258 (2013).
[48] R. H. Siemann, *Phys. Rev. ST Accel. Beams* **7**, 061303 (2004).
[49] Y. C. Neil Na, R. H. Siemann and R. L. Byer, *Phys. Rev. ST Accel. Beams* **8**, 031301 (2005).
[50] M. Seidel *et al.*, in *Proc. 6th International Particle Accelerator Conference* (JACoW, Geneva, Switzerland, 2015), pp. 2428–2433.
[51] T. Parker and P. Peck, in *Proc. 55th ICFA Advanced Beam Dynamics Workshop on High Luminosity Circular* e^+e^- *Colliders* (Geneva, Switzerland, 2014), pp. 54–58.
[52] J. Zhou *et al.*, *Proc. SPIE* **8139**, 81390S (2011).
[53] E. A. Peralta *et al.*, in *Proc. 2011 Particle Accelerator Conference* (PAC'11 OC/IEEE, Piscataway, NJ, USA, 2011), pp. 280–282.
[54] E. A. Peralta *et al.*, *AIP Conf. Proc.* **1507**, 169 (2012).
[55] K. Suzuki, *Microlithography: Science and Technology*, 2nd edn. (CRC, Boca Raton, FL, USA, 2007), Ch. 6, pp. 329–360.
[56] A. Mizrahi and L. Schächter, *Phys. Rev. E* **70**, 016505 (2004).
[57] Z. Zhang, S. G. Tantawi and R. D. Ruth, *Phys. Rev. ST Accel. Beams* **8**, 071302 (2005).
[58] J. McNeur *et al.*, *AIP Conf. Proc.* **1507**, 470 (2012).
[59] J. McNeur *et al.*, in *Proc. 4th International Particle Accelerator Conference* (JACoW, Geneva, Switzerland, 2013), pp. 1307–1309.
[60] E. B. Sozer *et al.*, in *2013 19th IEEE Pulsed Power Conference (PPC)* (IEEE, Piscataway, NJ, USA, 2013), pp. 1–5.
[61] J. D. Lawson and M. Tigner, *Annu. Rev. Nucl. Part. Sci.* **34**, 99 (1984).
[62] G. Shvets and S. Kalmykov, *AIP Conf. Proc.* **737**, 983 (2004).
[63] S. Kalmykov *et al.*, *AIP Conf. Proc.* **1086**, 538 (2009).
[64] B. Neuner III *et al.*, *Phys. Rev. ST Accel. Beams* **15**, 031302 (2012).
[65] R. J. Noble, J. E. Spencer and B. T. Kuhlmey, *Phys. Rev. ST Accel. Beams* **14**, 121303 (2011).
[66] X. E. Lin, *Phys. Rev. ST Accel. Beams* **4**, 051301 (2001).
[67] R. J. Noble *et al.*, in *Proc. 2007 IEEE Particle Accelerator Conference* (IEEE, Piscataway, NJ, USA, 2007), pp. 3103–3105.
[68] C.-K. Ng *et al.*, *Phys. Rev. ST Accel. Beams* **13**, 121301 (2010).
[69] R. J. England *et al.*, *IEEE J. Sel. Top. Quantum Electron.* **22**, 4401109 (2016).
[70] R. J. England *et al.*, in *Proc. 2011 Particle Accelerator Conference* (PAC'11 OC/IEEE, Piscataway, NJ, USA, 2011), pp. 2067–2070.
[71] C. M. S. Sears *et al.*, in *Proc. 2007 IEEE Particle Accelerator Conference* (IEEE, Piscataway, NJ, USA, Jun. 2007), pp. 3106–3108.
[72] R. J. England *et al.*, *AIP Conf. Proc.* **1086**, 550 (2009).
[73] B. M. Cowan, *Phys. Rev. ST Accel. Beams* **6**, 101301 (2003).
[74] B. Naranjo *et al.*, *Phys. Rev. Lett.* **109**, 164803 (2012).
[75] B. Naranjo *et al.*, *AIP Conf. Proc.* **1507**, 488 (2012).
[76] J. B. Rosenzweig *et al.*, *AIP Conf. Proc.* **1507**, 493 (2012).
[77] B. Cowan, in *Proc. 2005 Particle Accelerator Conference* (IEEE, Piscataway, NJ, USA, 2005), pp. 731–733.
[78] B. M. Cowan, *Phys. Rev. ST Accel. Beams* **11**, 011301 (2008).
[79] C. McGuinness, E. Colby and R. L. Byer, *J. Mod. Opt.* **56**, 2142 (2009).
[80] C.-H. Lee, Novel fabrication of a 17-layer 3D silicon-based woodpile structure for dielectric laser-driven accelerator. Ph.D. thesis, Purdue University (West Lafayette, IN, USA, 2015).
[81] Z. Wu *et al.*, *IEEE J. Sel. Top. Quantum Electron.* **22**, 4400909 (2016).
[82] I. Staude *et al.*, *Opt. Express* **20**, 5607 (2012).
[83] Z. Wu *et al.*, *Phys. Rev. ST Accel. Beams* **17**, 081301 (2014).
[84] Z. Wu *et al.*, in *Proc. 6th International Particle Accelerator Conference* (JACoW, Geneva, Switzerland, 2015), pp. 2702–2704.
[85] J. Rosenzweig, A. Murokh and C. Pellegrini, *Phys. Rev. Lett.* **74**, 2467 (1995).
[86] R. B. Yoder and J. B. Rosenzweig, *Phys. Rev. ST Accel. Beams* **8**, 111301 (2005).
[87] R. B. Yoder and J. B. Rosenzweig, *Int. J. Mod. Phys. A* **22**, 4363 (2007).
[88] D. Bar-Lev and J. Scheuer, *Phys. Rev. ST Accel. Beams* **17**, 121302 (2014).
[89] D. Bar-Lev and J. Scheuer, in *Proc. CLEO: QELS–Fundamental Science* (OSA, Washington, DC, USA, 2015), p. FW4C.8.
[90] J. Scheuer, D. Bar-Lev and A. Gover, in *High-Brightness Sources and Light-Driven Interactions* (OSA, Washington, DC, USA, 2016), p. EM9A.4.
[91] T. Plettner, *Phase-Synchronicity Conditions from Pulse-Front Tilted Laser Beams on One-Dimensional Periodic Structures and Proposed Laser-Driven Deflection*, Tech. Rep. SLAC-PUB-12458, Stanford Linear Accelerator Center (Menlo Park, CA, USA, 2007).
[92] K. J. Leedle *et al.*, *Opt. Lett.* **40**, 4344 (2015).

[93] A. Aimidula *et al.*, in *Proc. 4th International Particle Accelerator Conference* (JACoW, Geneva, Switzerland, 2013), pp. 1283–1285.
[94] A. Aimidula *et al.*, *Phys. Plasmas* **21**, 023110 (2014).
[95] T. Plettner, P. P. Lu and R. L. Byer, *Phys. Rev. ST Accel. Beams* **9**, 111301 (2006).
[96] E. A. Peralta *et al.*, *Nature* **503**, 91 (2013).
[97] A. Aimidula *et al.*, *Nucl. Instrum. Methods Phys. Res., Sect. A* **740**, 108 (2014).
[98] C.-M. Chang and O. Solgaard, *Appl. Phys. Lett.* **104**, 184102 (2014).
[99] A. Ceballos *et al.*, in *Proc. 6th International Particle Accelerator Conference* (JACoW, Geneva, Switzerland, 2015), pp. 2717–2719.
[100] T. Plettner, R. L. Byer and P. P. Lu, in *Proc. 2007 IEEE Particle Accelerator Conference* (IEEE, Piscataway, NJ, USA, 2007), pp. 3175–3177.
[101] A. Szczepkowicz, *Appl. Opt.* **55**, 2634 (2016).
[102] T. Plettner and R. L. Byer, *Phys. Rev. ST Accel. Beams* **11**, 030704 (2008).
[103] T. Plettner *et al.*, *Phys. Rev. ST Accel. Beams* **12**, 101302 (2009).
[104] T. Plettner and R. L. Byer, *Laser-Driven Deflection and Methods Involving Charged Particle Beams* (2011). US Patent 7994472. Filed 16 Jun. 2009, issued 9 Aug. 2011.
[105] T. Plettner and R. L. Byer, *Nucl. Instrum. Methods Phys. Res., Sect. A* **593**, 63 (2008).
[106] K. P. Wootton *et al.*, in *Proc. 6th International Particle Accelerator Conference* (JACoW, Geneva, Switzerland, 2015), pp. 2698–2701.
[107] K. Soong *et al.*, in *Proc. 2011 Particle Accelerator Conference* (PAC'11 OC/IEEE, Piscataway, NJ, USA, 2011), pp. 292–294.
[108] K. Soong *et al.*, *AIP Conf. Proc.* **1507**, 516 (2012).
[109] J. McNeur *et al.*, Elements of a dielectric laser accelerator (2016). arXiv:1604.07684 [physics.acc-ph].
[110] J. Breuer *et al.*, *Phys. Rev. ST Accel. Beams* **17**, 021301 (2014).
[111] J. Breuer and P. Hommelhoff, *Phys. Rev. Lett.* **111**, 134803 (2013).
[112] K. Koyama *et al.*, *J. Phys. B* **47**, 234005 (2014).
[113] Y. Wei *et al.*, *Phys. Procedia* **77**, 50 (2015).
[114] Z. Wu *et al.*, in *Proc. North American Particle Accelerator Conference 2013* (JACoW, Geneva, Switzerland, 2013), pp. 141–143.
[115] J. Breuer, J. McNeur and P. Hommelhoff, *J. Phys. B* **47**, 234004 (2014).
[116] T. Plettner, R. L. Byer and B. Montazeri, *J. Mod. Opt.* **58**, 1518 (2011).
[117] J. D. Lawson, *Contemp. Phys.* **11**, 575 (1970).
[118] K. P. Wootton *et al.*, *Opt. Lett.* **41**, 2696 (2016).
[119] K. Koyama *et al.*, in *Proc. 3rd International Particle Accelerator Conference* (IEEE, Piscataway, NJ, 2012), pp. 2763–2765.
[120] K. Koyama *et al.*, *Proc. SPIE* **8779**, 877910 (2013).
[121] Y. Wei *et al.*, in *Proc. 6th International Particle Accelerator Conference* (JACoW, Geneva, Switzerland, 2015), pp. 2618–2620.
[122] B. Montazeri Najafabadi *et al.*, *AIP Conf. Proc.* **1507**, 476 (2012).
[123] S. Kalmykov *et al.*, *Philos. Trans. R. Soc. A* **364**, 725 (2006).
[124] K. J. Leedle *et al.*, *Optica* **2**, 158 (2015).
[125] J. McNeur *et al.*, *Nucl. Instrum. Methods Phys. Res., Sect. A* **829**, 50 (2016).
[126] J. McNeur *et al.*, *J. Phys. B* **49**, 034006 (2016).
[127] R. L. Byer *et al.*, in *Proc. 1999 Particle Accelerator Conference* (IEEE, Piscataway, NJ, 1999), pp. 321–324.
[128] R. Noble *et al.*, in *Proc. 2003 Particle Accelerator Conference* (IEEE, Piscataway, NJ, USA, 2003), pp. 1858–1860.
[129] E. Colby *et al.*, in *Proc. Particle Accelerator Conference 2005* (IEEE, Piscataway, NJ, USA, May 2005), pp. 2024–2026.
[130] E. Colby *et al.*, in *Proc. Particle Accelerator Conference 2007* (IEEE, Piscataway, NJ, USA, June 2007), pp. 3115–3117.
[131] P. Musumeci, J. Moody and G. Gatti, in *Proc. Particle Accelerator Conference 2007* (IEEE, Piscataway, NJ, 2007), pp. 2751–2753.
[132] S. Otsuki *et al.*, in *Proc. 5th International Particle Accelerator Conference* (JACoW, Geneva, Switzerland, 2014), pp. 1434–1436.
[133] M. Uesaka *et al.*, *IOP Conf. Ser.: Mater. Sci. Eng.* **79**, 012015 (2015).
[134] K. Koyama *et al.*, in *Proc. 7th International Particle Accelerator Conference* (JACoW, Geneva, Switzerland, May 2016), pp. 1585–1587.
[135] H. Okamoto *et al.*, in *Proc. 7th International Particle Accelerator Conference* (JACoW, Geneva, Switzerland, May 2016), pp. 4070–4072.
[136] P. A. McIntosh *et al.*, in *Proc. 4th International Particle Accelerator Conference* (JACoW, Geneva, Switzerland, 2013), pp. 3708–3710.
[137] Y. Wei *et al.*, in *Proc. 7th International Particle Accelerator Conference* (JACoW, Geneva, Switzerland, 2016), pp. 1970–1972.
[138] U. Dorda *et al.*, *Nucl. Instrum. Methods Phys. Res., Sect. A* **829**, 233 (2016).
[139] J. Zhu *et al.*, *Phys. Rev. Accel. Beams* **19**, 054401 (2016).
[140] J. Zhu *et al.*, in *Proc. 7th International Particle Accelerator Conference* (JACoW, Geneva, Switzerland, 2016), pp. 3244–3246.
[141] E. A. Peralta, Accelerator on a chip: Design fabrication, and demonstration of grating-based dielectric microstructures for laser-driven acceleration of electrons. Ph.D. thesis, Stanford University (2015).
[142] K. Soong, Particle accelerator on a wafer: demonstration of electron acceleration and diagnostics with microstructures. Ph.D. thesis, Stanford University (2014).

[143] R. J. England *et al.*, in *Proc. 3rd International Particle Accelerator Conference* (IEEE, Piscataway, NJ, USA, 2012), pp. 1050–1053.
[144] X. H. Lu *et al.*, *Phys. Rev. ST Accel. Beams* **18**, 032802 (2015).
[145] P. Hommelhoff *et al.*, *Ultramicrosc.* **109**, 423 (2009).
[146] T. Juffmann *et al.*, *Phys. Rev. Lett.* **115**, 264803 (2015).
[147] F. Süßmann, M. F. Kling and P. Hommelhoff, *From Attosecond Control of Electrons at Nano-Objects to Laser-Driven Electron Accelerators*, in *Attosecond Nanophysics*, eds. P. Hommelhoff and M. F. Kling (Wiley–VCH Verlag, Weinheim, Germany, 2015), Ch. 6, pp. 155–196.
[148] G. Travish *et al.*, in *Proc. 3rd International Particle Accelerator Conference* (IEEE, Piscataway, NJ, USA, May 2012), pp. 2738–2740.
[149] R. B. Yoder *et al.*, in *Proc. 6th International Particle Accelerator Conference* (JACoW, Geneva, Switzerland, 2015), pp. 2624–2626.
[150] M. Kozák *et al.*, Optical gating and streaking of free-electrons with attosecond precision (2015), arXiv:1512.04394 [physics.optics].
[151] P. Bermel *et al.*, *Nucl. Instrum. Methods Phys. Res., Sect. A* **734** (Part A), 51 (2014).
[152] K. R. Hogstrom and P. R. Almond, *Phys. Med. Biol.* **51**, R455 (2006).
[153] J. E. Clayton and P. A. Lovoi, Laser Accelerator Driven Particle Brachytherapy Devices, Systems and Methods (2015). US Patent 9022914 B2. Filed 30 Sep. 2011, issued 5 May 2015.
[154] H. O. Wyckoff *et al.*, *Radiation Dosimetry: Electron Beams with Energies Between 1 and 50 MeV*, Tech. Rep. 35, International Commission on Radiation Units and Measurements (Bethesda, MD, USA, 1984).
[155] M. Abe and M. Takahashi, *Int. J. Radiat. Oncol. Biol. Phys.* **7**, 863 (1981).
[156] J. R. Palta *et al.*, *Int. J. Radiat. Oncol. Biol. Phys.* **33**, 725 (1995).
[157] C. G. Willett, B. G. Czito and D. S. Tyler, *J. Clin. Oncol.* **25**, 971 (2007).
[158] J. S. Vaidya *et al.*, *Lancet* **376**, 91 (2010).
[159] M. Totzeck *et al.*, *Nat. Photon.* **1**, 629 (2007).
[160] G. E. Moore, *Electron.* **38**, 114 (1965).
[161] Z. T. Zhao, *Rev. Accel. Sci. Technol.* **3**, 57 (2010).
[162] M. E. Couprie, *J. Electron Spectrosc. Relat. Phenom.* **196**, 3 (2014).
[163] P. Naulleau, in *Proc. 6th International Particle Accelerator Conference* (JACoW, Geneva, Switzerland, 2015) p. MOYGB1.
[164] B. E. Carlsten *et al.*, *Nucl. Instrum. Methods Phys. Res., Sect. A* **622**, 657 (2010).
[165] W. A. Barletta *et al.*, *Nucl. Instrum. Methods Phys. Res., Sect. A* **618**, 69 (2010).
[166] P. S. Csonka, *Part. Accel.* **7**, 9 (1975).
[167] Y.-C. Huang *et al.*, *Phys. Rev. ST Accel. Beams* **18**, 080701 (2015).
[168] T. Plettner, *DC High-Voltage Super-Radiant Free-Electron Based EUV Source* (2015). US Patent 9053833 B2. Filed 27 Feb. 2013, issued 9 Jun. 2015.
[169] R. J. England *et al.*, in *Proc. 2013 North American Particle Accelerator Conference* (JACoW, Geneva, Switzerland, 2013), pp. 129–131.
[170] R. J. England, in *High-Brightness Sources and Light-Driven Interactions* (OSA, Washington, DC, USA, 2016), p. ET3A.2.
[171] K. Soong and R. L. Byer, *Opt. Lett.* **37**, 975 (2012).
[172] K. Soong *et al.*, in *Proc. 2013 North American Particle Accelerator Conference* (JACoW, Geneva, Switzerland, 2013), pp. 138–140.
[173] K. Soong *et al.*, *Opt. Lett.* **39**, 4747 (2014).
[174] J.-K. So *et al.*, *ACS Photon.* **2**, 1236 (2015).
[175] M. Kozák *et al.*, *Opt. Lett.* **41**, 3435 (2016).

K. P. Wootton earned his B.Sc. (Hons.) and B.E. (Hons.) degrees from Monash University in 2010, and his Ph.D. degree in Physics from the University of Melbourne in 2014. Presently, he is an Experimental Research Associate at SLAC National Accelerator Laboratory. His research interests include dielectric laser accelerators, ultralow vertical emittance electron beams, spin-polarized beams, and beam diagnostics based on synchrotron light and undulator radiation. Dr. Wootton is a member of the Australian Institute of Physics and the American Physical Society.

J. McNeur earned his B.Sc. in Physics and B.Sc. in Mathematics from the University of Chicago in 2007. He earned his Ph.D. from the University of California — Los Angeles in 2014. He is currently a Postdoctoral Scholar in the Laser Physics Department at the Friedrich-Alexander-Universität Erlangen-Nürnberg, with research interests including novel high gradient acceleration techniques, ultrafast electron sources, nanophotonics, and attosecond physics.

K. J. Leedle earned his B.S. in Electrical Engineering from the University of Wisconsin–Madison in 2010. He was awarded his Ph.D. degree in Electrical Engineering from Stanford University in 2016. He is currently a Postdoctoral Scholar in the Department of Electrical Engineering at Stanford University. His research interests include dielectric laser accelerators, ultrafast optics, electron optics, and nanophotonics. He is a member of the Optical Society.

Reviews of Accelerator Science and Technology
Vol. 9 (2016) 127–149

DOI: 10.1142/S1793626816300061

Dielectric Wakefield Accelerators

Chunguang Jing
Euclid Techlabs, LLC, 365 Remington Blvd., Bolingbrook, Illinois, 60440, USA
Accelerator R&D Group, Argonne National Laboratory, 9700 S. Cass Ave., Lemont, Illinois, 60439, USA
c.jing@euclidtechlabs.com

Dielectric-structure-based wakefield acceleration provides a viable approach to achieving the luminosity, efficiency, and cost requirements of a future linear collider as well as future x-ray light sources. This technology is capable of accelerating electrons and positrons at the substantially high gradients needed. Important progress in the development of dielectric wakefield accelerators has been made both experimentally and theoretically in the past few years. In this article we provide an overview of the basics of dielectric wakefield acceleration and major developments to date.

Keywords: Dielectric; wakefield; accelerator.

1. Introduction

1.1. *Motivation*

Charged particle acceleration at substantially higher gradients than in today's conventional accelerators is necessary to open up new possibilities in fundamental research. A TeV class lepton linear collider has been a desirable high-energy-physics machine for decades (due to the precision with which it can characterize particle physics events), yet it has been built under constraints of construction and operational cost. The enabling acceleration technology for a linear collider must feature a high effective gradient, high efficiency, and low fabrication cost. Over the past three decades much progress has been made for the metallic accelerators, including ~100 MV/m of X-band room temperature RF accelerators [1–3] and ~35 MV/m of superconducting RF accelerators [4]. Meanwhile, a large number of advanced accelerator concepts have been considered [5] but current R&D has converged in a few promising directions [6]. The dielectric wakefield accelerator (DWA), also known as the dielectric-loaded accelerator (DLA), is one of them. The DWA has the advantages of simplicity of fabrication and a potentially higher breakdown threshold than copper but with comparable shunt impedance (thanks to the advancement of low loss microwave ceramic materials).

1.2. *Historical use of dielectric materials in accelerators*

The study of RF-driven dielectric-loaded waveguides for particle acceleration can be traced back to the late 1940s [7–10]. Two types of structure were investigated during that period. As shown in Fig. 1, one concept used a dielectric-lined circular waveguide, and the other a dielectric-disk-loaded circular waveguide. The dielectric-lined waveguide is much simpler but was very lossy due to the unavailability at that time of low loss dielectric materials in the microwave regime. Quartz was used by Cohn's group [11] but the experimental results were not available. During the 1950s, two groups in the UK — the Atomic Energy Research Establishment (AERE) and Queen Mary College of the University of London — put significant efforts into developing dielectric-disk-loaded accelerators [12–14]. The major achievements presented in Ref. 5 include the first beam acceleration in dielectric-loaded accelerators (electrons were accelerated from 45 keV to 1.48 MeV using a 3 GHz linac at a gradient of 2.6 MeV/m, with a 1 MW magnetron operated at 2 μs pulse width and 50 pulses per second) and the first observation and study of multipactor-induced anomalous power loss. The researchers at the University of London attained a 10 MV/m gradient in a 3 GHz

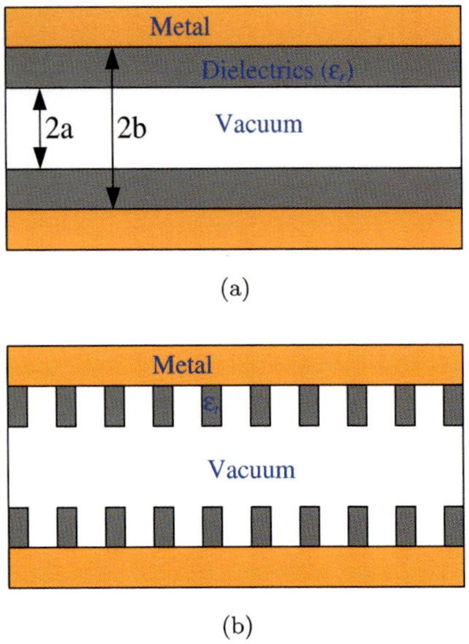

Fig. 1. Two major types of dielectric-loaded waveguides that have been considered as particle accelerators: (a) dielectric-lined circular waveguide; (b) dielectric-disk-loaded circular waveguide.

dielectric-disk-loaded cavity powered by a 2 MW magnetron before the occurrence of RF breakdown at the ceramic (TiO_2; $\varepsilon_r = 90$; $\tan\delta = 3 \times 10^{-4}$) and metal joint [13]. This record gradient was later increased to 30 MV/m (limited by the available RF power) through the use of ceramic glazing with green resistor enamel consisting mainly of lead borate, a few thousandths of an inch thick [14].

Since that time, metallic disk-loaded accelerators have prevailed in accelerator research and development, due to the high quality factor and reliability. Thanks to the availability of low loss microwave ceramics and the geometrical simplicity, in the last two decades the theoretical and experimental investigations into DLA structures for application to wakefield acceleration (predominantly the dielectric-lined waveguide) have been intensively revived [15–18]. In addition to conventional circular DLA structures, rectangular structures have also received much interest [19, 20]. Some related research programs, like the Argonne Wakefield Accelerator (AWA) [21], the Argonne National Laboratory / Naval Research Laboratory joint compact dielectric-loaded accelerator [22], the DULY ceramic RF power extractor [23], Yale's rectangular DLA structure–based wakefield accelerator [24], the SLAC W-band planar DLA structure [25], etc., have made significant experimental progress.

2. Theory of Dielectric Wakefield Accelerators

2.1. *Electromagnetics of DLA structures*

In this article, we focus on the dielectric-lined circular waveguide as the most commonly used dielectric particle accelerator concept. The general solution — using a decomposition of the fields in terms of $\exp(in\phi)$, where ϕ is the azimuthal coordinate — for the electric and magnetic field components in the dielectric-lined circular waveguide of Fig. 1(a) is

$$E_z = \begin{cases} B_1 J_n(k_1 r)\cos(n\phi)e^{j(\omega t-\beta z)}, & 0 \le r < a, \\ B_2\left[J_n(k_2 r) - \dfrac{J_n(k_2 b)}{Y_n(k_2 b)}Y_n(k_2 r)\right] & \\ \quad \cos(n\phi)e^{j(\omega t-\beta z)}, & a \le r \le b, \end{cases}$$

$$H_z = \begin{cases} A_1 J_n(k_1 r)\sin(n\phi)e^{j(\omega t-\beta z)}, & 0 \le r < a, \\ A_2\left[J_n(k_2 r) - \dfrac{J'_n(k_2 b)}{Y'_n(k_2 b)}Y_n(k_2 r)\right] & \\ \quad \sin(n\phi)e^{j(\omega t-\beta z)}, & a \le r \le b, \end{cases}$$

$$E_\phi = \begin{cases} \left[\dfrac{j\omega\mu_0}{k_1}A_1 J'_n(k_1 r) + \dfrac{j\beta n}{k_1^2 r}B_1 J_n(k_1 r)\right] \\ \quad \sin(n\phi)e^{j(\omega t-\beta z)}, \quad 0 \le r < a, \\ \left[\dfrac{j\omega\mu_0}{k_2}A_2 G'_{nn}(k_2 r) + \dfrac{j\beta n}{k_2^2 r}B_2 F_{nn}(k_2 r)\right] \\ \quad \sin(n\phi)e^{j(\omega t-\beta z)}, \quad a < r \le b, \end{cases}$$

$$E_r = \begin{cases} \left[-\dfrac{j\omega\mu_0 n}{k_1^2 r}A_1 J_n(k_1 r) - \dfrac{j\beta}{k_1}B_1 J'_n(k_1 r)\right] \\ \quad \cos(n\phi)e^{j(\omega t-\beta z)}, \quad 0 \le r < a, \\ \left[-\dfrac{j\omega\mu_0 n}{k_2^2 r}A_2 G_{nn}(k_2 r) - \dfrac{j\beta}{k_2}B_2 F'_{nn}(k_2 r)\right] \\ \quad \cos(n\phi)e^{j(\omega t-\beta z)}, \quad a < r \le b, \end{cases}$$

$$H_\phi = \begin{cases} \left[-\dfrac{j\omega\varepsilon_0}{k_1}B_1 J'_n(k_1 r) - \dfrac{j\beta n}{k_1^2 r}A_1 J_n(k_1 r)\right] \\ \quad \cos(n\phi)e^{j(\omega t-\beta z)}, \quad 0 \le r < a, \\ \left[-\dfrac{j\omega\varepsilon_r\varepsilon_0}{k_2}B_2 F'_{nn}(k_2 r) - \dfrac{j\beta n}{k_2^2 r}A_2 G_{nn}(k_2 r)\right] \\ \quad \cos(n\phi)e^{j(\omega t-\beta z)}, \quad a < r \le b, \end{cases}$$

$$H_r = \begin{cases} \left[-\frac{j\omega\varepsilon_0 n}{k_1^2 r} B_1 J_n(k_1 r) - \frac{j\beta}{k_1} A_1 J_n'(k_1 r)\right] \\ \quad \sin(n\phi) e^{j(\omega t - \beta z)}, \quad 0 \le r < a, \\ \left[-\frac{j\omega\varepsilon_r\varepsilon_0 n}{k_2^2 r} B_2 F_{nn}(k_2 r) - \frac{j\beta}{k_2} A_2 G_{nn}'(k_2 r)\right] \\ \quad \sin(n\phi) e^{j(\omega t - \beta z)}, \quad a < r \le b. \end{cases} \tag{1}$$

B_1, B_2, A_1, and A_2 are the field amplitudes in region I (vacuum) and region II (dielectric), respectively; v_p is the phase velocity of the wave propagating inside the tube; c is the speed of light in free space. k_1 and k_2 are the cutoff wave numbers in each region, and β is the wave propagation constant:

$$\begin{aligned} k_1 &= \omega\sqrt{\frac{1}{c^2} - \frac{1}{v_p^2}}, \\ k_2 &= \omega\sqrt{\frac{\varepsilon_r}{c^2} - \frac{1}{v_p^2}}, \\ \beta^2 &= k_0^2 - k_1^2 = \varepsilon_r k_0^2 - k_2^2, \\ k_0^2 &= \omega^2 \mu_0 \varepsilon_0. \end{aligned} \tag{2}$$

The special functions F_{nn}, G_{nn} and their derivatives are algebraic combinations of Bessel functions given in the following equation.

$$\begin{aligned} F_{nn}(k_2 r) &= J_n(k_2 r) - \frac{J_n(k_2 b)}{Y_n(k_2 b)} Y_n(k_2 r), \\ G_{nn}(k_2 r) &= J_n(k_2 r) - \frac{J_n'(k_2 b)}{Y_n'(k_2 b)} Y_n(k_2 r), \\ F_{nn}'(k_2 r) &= J_n'(k_2 r) - \frac{J_n(k_2 b)}{Y_n(k_2 b)} Y_n'(k_2 r), \\ G_{nn}'(k_2 r) &= J_n'(k_2 r) - \frac{J_n'(k_2 b)}{Y_n'(k_2 b)} Y_n'(k_2 r). \end{aligned} \tag{3}$$

Using the boundary conditions at $r = a(E_z, E_\phi, H_\phi$, and H_z continuous), the dispersion relation for an EM wave propagating in this waveguide can be obtained as

$$\begin{aligned} &\left[\frac{\mu_0}{k_1 a}\frac{J_n'(k_1 a)}{J_n(k_1 a)} - \frac{\mu_0}{k_2 a}\frac{G_{nn}'(k_2 a)}{G_{nn}(k_2 a)}\right] \\ &\quad\times\left[\frac{\varepsilon_0}{k_1 a}\frac{J_n'(k_1 a)}{J_n(k_1 a)} - \frac{\varepsilon_r\varepsilon_0}{k_2 a}\frac{F_{nn}'(k_2 a)}{F_{nn}(k_2 a)}\right] \\ &= \frac{(\beta n\omega)^2}{(k_1 k_2 a)^4}(\mu_0\varepsilon_r\varepsilon_0 - \mu_0\varepsilon_0)^2. \end{aligned} \tag{4}$$

The symmetry of each mode is denoted by the subscript nm, where n is the azimuthal mode number and m the number of radial nodes. For particle accelerator applications, the TM_{01} mode is usually chosen because of its field pattern. Equation (5) gives the expression for the TM_{01} mode fields, which have only three components

$$\begin{aligned} E_z &= \begin{cases} B_1 J_0(k_1 r) e^{j(\omega t - \beta z)}, & 0 \le r < a, \\ B_2 F_{00}(k_2 r) e^{j(\omega t - \beta z)}, & a \le r \le b, \end{cases} \\ E_r &= \begin{cases} -\frac{j\beta}{k_1} B_1 J_0'(k_1 r) e^{j(\omega t - \beta z)}, & 0 \le r < a, \\ -\frac{j\beta}{k_2} B_2 F_{00}'(k_2 r) e^{j(\omega t - \beta z)}, & a < r \le b, \end{cases} \\ H_\phi &= \begin{cases} -\frac{j\omega\varepsilon_0}{k_1} B_1 J_0'(k_1 r) e^{j(\omega t - \beta z)}, & 0 \le r < a, \\ -\frac{j\omega\varepsilon_r\varepsilon_0}{k_2} B_2 F_{00}'(k_2 r) e^{j(\omega t - \beta z)}, & a < r \le b. \end{cases} \end{aligned} \tag{5}$$

The electric fields of the TM_{01} mode given in Eq. (5) have very interesting characteristics. When the phase velocity of the TM_{01} wave is equal to c (synchronous condition), the longitudinal electric field E_z is independent of the radial coordinate in the vacuum region because the cutoff wave number k_1 becomes zero from Eq. (2). This implies (via the Panofsky–Wenzel theorem [26]) that there are no focusing and defocusing forces for a relativistic particle traveling inside the vacuum region, critical for emittance preservation in high energy linear accelerators. Figure 2 shows the variation of E_z along the radius at $z = 0$.

The dispersion relation of the TM_{01} mode reduces from Eq. (4) to

$$\left[\frac{1}{k_1}\frac{J_0'(k_1 a)}{J_0(k_1 a)} - \frac{\varepsilon_r}{k_2}\frac{F_{00}'(k_2 a)}{F_{00}(k_2 a)}\right] = 0. \tag{6}$$

From Eq. (2) we know that the phase velocity of the guided wave in this waveguide can be slowed down below the speed of light because of the high dielectric constant ceramic loading. Meanwhile, the inner radius a is chosen based on the requirements of beam dynamics. In general the beam aperture is bounded by the minimum value, which is defined by the tolerable deflecting force (smaller a, larger deflecting force to the beam), and the maximum value, which is defined by the tolerable gradient (larger a, smaller gradient). Thus, governed by the mode dispersion relation [Eq. (6)], the outer radius b can be adjusted accordingly to achieve the desired v_p.

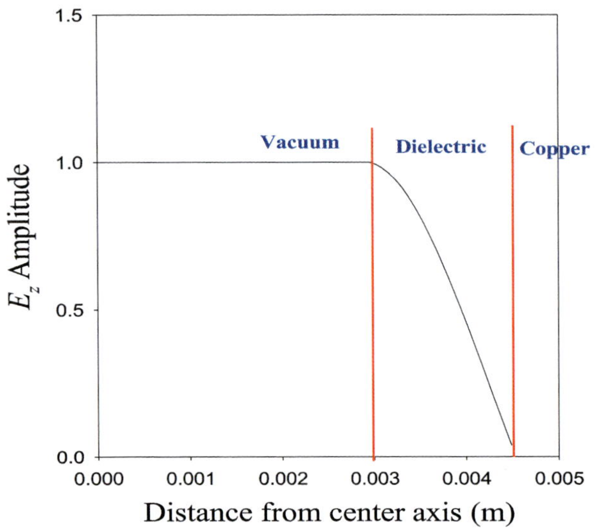

Fig. 2. The electric field amplitude E_z for the TM_{01} mode along the radius of the dielectric-lined circular waveguide shown in Fig. 1(a), with $a = 3\,\text{mm}$, $\varepsilon_r = 20$, and $b = 4.57\,\text{mm}$. The phase velocity of the TM_{01} wave is equal to c.

The beam-deflecting modes (hybrid modes) in dielectric-lined circular waveguides are the HEM_{nm} modes. The lowest frequency deflecting mode is the HEM_{11} mode, and it is the most harmful mode for an accelerating structure since it exerts a transverse force on the beam particles [27, 28]. If hybrid modes are present at a significant level, the deflecting forces can result in beam breakup (BBU) [29].

Figure 3 shows the dispersion curves of the propagation constant β versus frequency for the TM_{01} and HEM_{11} modes in a dielectric-lined circular waveguide with $a = 3\,\text{mm}$, $b = 4.57\,\text{mm}$, and $\varepsilon_r = 20$. When $\beta = 0$, the corresponding frequency is the cut-off frequency of each mode. The intersections of the two dispersion curves with the speed-of-light line ($2\pi f = c\beta$) indicate that the propagation constant of the HEM_{11} mode is smaller than that of the TM_{01} mode at the locations where these two modes have the same phase velocity, c. This implies that the amplitude of the transverse wakefields induced by particle beams passing through such a dielectric-lined waveguide is smaller than that of the longitudinal wakefield [27] because all the beam-driven modes have the same phase velocity ($= v_{\text{beam}}$).

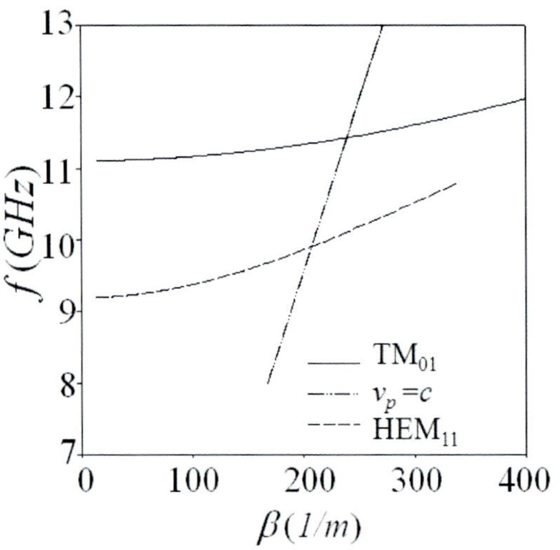

Fig. 3. Propagation constants of the TM_{01} and HEM_{11} modes versus frequency in a dielectric-lined circular waveguide with $a = 3\,\text{mm}$, $b = 4.57\,\text{mm}$, and $\varepsilon_r = 20$.

2.2. *Wakefields*

When a relativistic electron (or positron) bunch passes near a material boundary, EM fields generated by the head of the bunch affect the trailing bunch particles through the short range wakefields or those of the following bunches through the long range wakefields. The wakefields give the voltage change experienced by a test particle within or behind the bunch per unit charge of the driving bunch. The energy supplied to the wakefields comes at the cost of the kinetic energy lost by the bunch particles. In the case of a metallic boundary, both the electrical resistance of the metal and the geometric variation of the boundary contribute to the wakefields [29, 30]. In the case of a dielectric boundary, the Cherenkov radiation condition will be satisfied when the particle velocity $\beta = v/c > \varepsilon_r^{-0.5}$ (ε_r is the relative dielectric constant of the medium), in which case the beam will generate wakefields. In fact, a periodic metallic boundary, which has a large ratio of the beam aperture to the corrugation depth on the boundary, can be equivalently converted to an engineered material with specific permittivity and permeability [31]. In the case of (metallic) accelerating structures, the resistive wall contribution to the wakes is considerably weaker than that of the geometrical variation of the structure wall; similarly, the presence of a dielectric material in accelerating structures can, in general, be ignored.

Normally, the calculation of wakefields is tedious and cumbersome. The complete solution to the inhomogeneous wave propagation equations is required. For wakefields in circular dielectric-loaded accelerating structures, a clear derivation and some results can be found in Refs. 27 and 28. (It is worth pointing out that a new method to obtain the Green's function of a dielectric-lined waveguide of arbitrary shape has been developed recently [32, 33].) Instead of finding the Green's function directly, we will briefly summarize an approach to calculating the wakefield in DWA structures using intrinsic waveguide modes.

Rather than solving the bunch-excited fields directly through Maxwell's equations, the wakefields can be expressed as a sum over the waveguide parameter r/Q for each mode [34, 35]. r/Q is a figure of merit for particle accelerators and a function of the structure geometry only defined as the shunt impedance over the quality factor of an accelerator with a unit of ohm per meter. The mode selection for the wakefield calculation is based on the relation between the system response and the signal excitation. The guided wave modes in the waveguide system selected to compute wakefields are determined by the criterion of pattern matching the electromagnetic fields. Mode matching between the excited wakefield mode (classified by the leading order azimuthal harmonic) and the guided wave modes (waveguide intrinsic modes) allows them to have similar spatial distributions. For example, the TM_{0n} mode will match the longitudinal wakefield in a circular DWA structure. The transverse wakefields are also obtained from the Panofsky–Wenzel theorem [26]. Equation (7) shows the relation of the loss factor of a point charged particle for each mode (k_{l_i}) and the r over Q for each guided wave mode at the accelerating waveguide:

$$k_{l_i} = \frac{\omega_i}{4}\left(\frac{r}{Q}\right)_i. \tag{7}$$

The subscript i indicates the ith mode, ω is the resonant frequency, and r/Q is the geometry-dependent parameter for charged particles traveling *on axis* for each intrinsic mode (here we only consider longitudinal wakefields). r/Q is obtained by computing the stored energy U per unit length and the accelerating electrical field E_z:

$$\left[\frac{r}{Q}\right]_i = \left[\frac{|E_z|^2}{\omega U}\right]_i. \tag{8}$$

Then the longitudinal wake function (Green's function) can be simplified to

$$w_z = \sum_i 2k_{l_i}\cos(\omega_i t). \tag{9}$$

The summation of modes is an infinite series, but it can be truncated because of the finite length of the drive bunch, as in the case of most wakefield calculations. Finally, the wakefield (wake potential per unit length) of an electron bunch which has a current distribution of $I(t)$ can be calculated through a convolution integral:

$$W_z(t) = -\int_{-\infty}^{t} I(\tau)w_z(t-\tau)d\tau. \tag{10}$$

In addition, it is necessary to take the effect of the group velocity (v_g) into account in a wakefield calculation if the device is a high group velocity structure. Details can be found in Ref. 36.

From Eq. (10) the accelerating gradient in a DWA structure is determined by the peak current of the electron bunch and the loss factor (equivalent to the impedance of the structure). The impedance of the structure strongly depends on the inner radius of the DWA structure as well as the dielectric constant. A smaller beam channel yields higher impedance, but the effect of the dielectric constant is more complicated. In general, a lower dielectric constant leads to higher impedance. For an electron bunch with a given distribution (so its frequency bandwidth is fixed), there exist two operational design regimes for a wakefield accelerator: single mode and multimode operation. In single mode operation, the structure is designed in such a way that only the lowest mode is present inside the bandwidth of the electron bunch. Figure 4 shows an example where the longitudinal wakefield and its spectrum are calculated for a beam of 1 nC and 50 micron bunch length passing through a quartz structure with an inner radius of 0.2 mm and an outer radius of 0.232 mm. As in the case of an overmoded waveguide, multiple modes will be excited if we just increase the outer radius of the dielectric cylinder used to calculate the wakefield in Fig. 4; multimode results are shown in Fig. 5. Comparing Figs. 4 and 5, we find that the wakefield gradient is the same for the two different structures. In fact, it is generally true that the wakefield gradient is the same for single mode and multimode structures as long as their inner radius and dielectric constants are the same [32]. We note that a multimode structure may have an advantage regarding RF breakdown since it is exposed to a high gradient for a shorter time than a single mode structure. On the other hand, a single mode structure has an advantage regarding a higher transformer ratio for the same constraints. Lastly, the two structures have

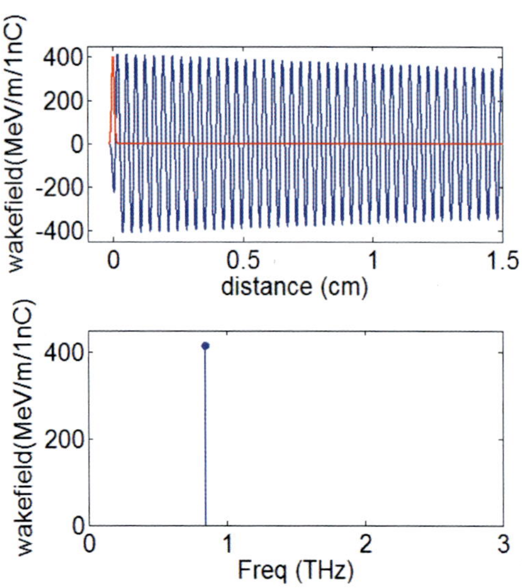

Fig. 4. Single mode wakefield excitation due to a 1 nC, 50 micron rms length electron bunch passing through a DLA structure. The bunch travels toward the left. The red line in the top figure represents the current distribution. The geometry of the waveguide is shown in Fig. 1 with $a = 0.2$ mm, $\varepsilon_r = 3.75$, and $b = 0.232$ mm. The phase velocity of the TM_{01} wave is equal to c at 850 GHz. Note that the maximum values of E_r and E_z satisfy $E_{\mathrm{rmax}}/E_{\mathrm{zmax}} = (\pi a)/\lambda$, where λ is the wavelength.

similar RF losses. This is surprising at first, since the high fields are present for a shorter time in the multimode structure, but this is compensated for by its greater number of higher order modes. Detailed

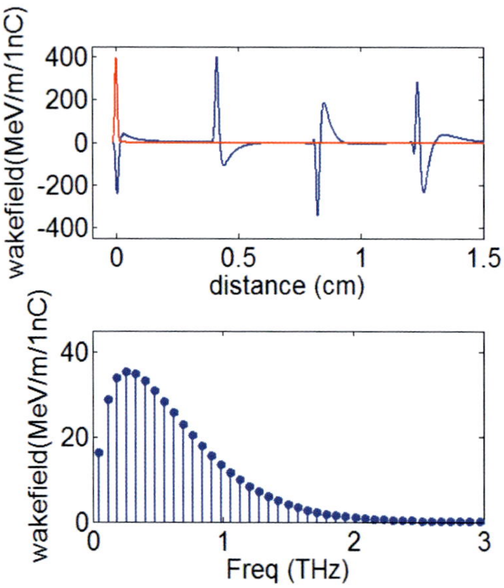

Fig. 5. Multimode wakefield excitation due to a 1 nC, 50 micron rms length electron bunch passing through a similar structure used in Fig. 4 but with a larger outer radius (1.42 mm).

discussions are beyond the scope of this article, but interested readers may see Ref. 37 for more details.

The longitudinal wakefield scales as $W_z \propto Q/a^2$, where Q is the drive bunch charge. This scaling implies that a high gradient can be achieved by either increasing Q or decreasing a. The transverse wakefield scales as $W_\perp \propto Q/a^3$ $(\perp = x, y)$, and thus decreasing a to increase the gradient will increase the BBU instability even more quickly. Figure 6 shows an example of calculated wakefields generated by a Gaussian bunch traversing a dielectric-lined waveguide with two different sets of parameters. Longitudinal wakes remain the same in both cases (by choosing different beam apertures and intensities),

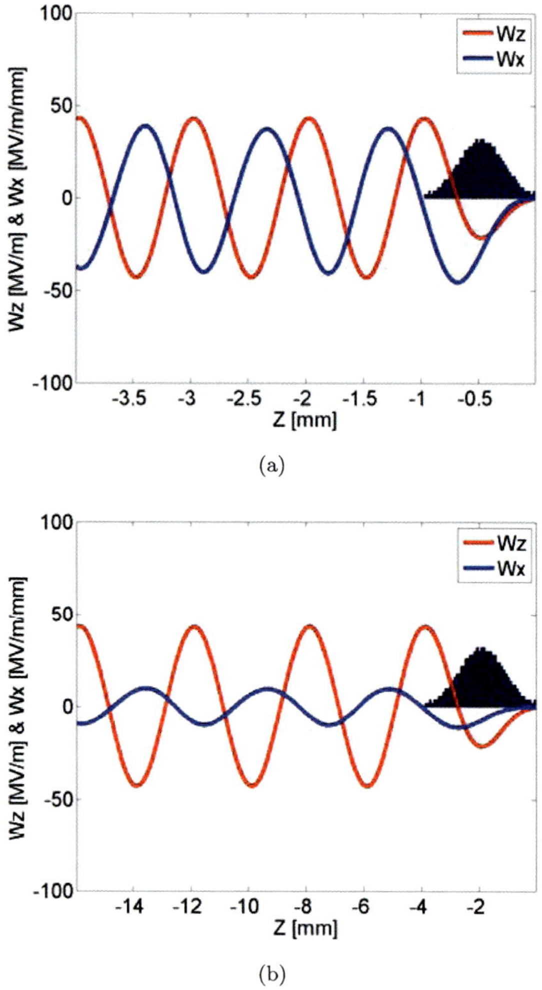

Fig. 6. Calculated wakefields in two cases. Only a single mode is considered. (a) case 1: $a = 0.5$ mm, $Q = 1$ nC, freq. = 300 GHz; (b) case 2: $a = 2$ mm, $Q = 16$ nC, freq. = 75 GHz. In both cases the ratio of bunch length over wavelength remains the same, i.e. 0.2.

but the transverse wake is much weaker in the case of a larger beam aperture.

3. The Two-Beam DWA in Principle

The two-beam accelerator (TBA), shown in Fig. 7, is a modified approach to wakefield acceleration. The drive beam energy is extracted into an RF pulse, which is then used to accelerate the main beam (or witness beam) in a separate accelerating structure. Ongoing efforts toward a TeV scale electron–positron linear collider from the Compact Linear Collider (CLIC) team is based on a room temperature TBA scheme, using an RF pulse length of 240 ns and a loaded gradient of ∼100 MV/m. Although the physics behind RF breakdown is not fully understood, observations in extensive experiments reveal that the RF breakdown threshold in accelerating structures increases as the RF pulse length decreases [38]. In the present room temperature, high gradient accelerator designs, a gradient of ∼100 MV/m with a pulse length in the range of 200–400 ns is usually the limit of normal operations. If the pulse length is reduced to ∼20 ns, a 200–300 MV/m operational accelerating gradient may be achievable.

One common concern related to the short RF pulse accelerator concept is the RF overhead, which is defined by the ratio of the RF transient time (i.e. the filling time in a traveling wave accelerator and the rise/fall time of the RF pulse) over the total RF pulse length in one RF pulse. The RF overhead is directly linked to the RF-to-beam efficiency, as shown in the following equation:

$$\eta_{\mathrm{rf-beam}} = \frac{I_{\mathrm{beam}} E_{\mathrm{load}} L_S}{P_{\mathrm{rf}}} \times \frac{T_{\mathrm{beam}}}{T_{\mathrm{rf}}}, \quad (11)$$

where I_{beam} is the beam current in the pulse, E_{load} the loaded gradient, L_s the structure length, P_{rf} the RF power in the pulse, T_{beam} the beam duration in the pulse, and T_{rf} the RF pulse width. A small RF overhead (i.e. $T_{\mathrm{rf}} - T_{\mathrm{beam}}$) contributes significantly to improving the RF-to-beam efficiency. In order to achieve a competitive RF-to-beam efficiency, a few strategies can be considered in a short pulse collider design: (1) using the two-beam acceleration approach in the main linacs to avoid the slow rise time of klystrons; (2) using broadband accelerating structures; (3) design of the main linac with a relatively large group velocity and a relatively short length to reduce the filling time; (4) design of the main linac with a relatively high frequency and optimal beam loading to improve the RF-to-beam efficiency.

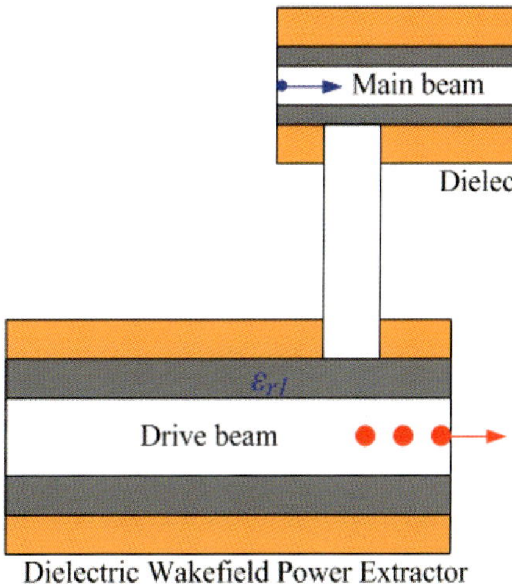

Fig. 7. Schematic of the dielectric TBA.

3.1. *The wakefield RF pulse in a finite length dielectric wakefield power extractor*

The wakefield calculation presented above is carried out assuming an infinitely long wakefield device. In this subsection, we briefly describe RF pulse generation using a coherently stacked wakefield radiated from a bunch train traversing a finite length DWA structure, such as the wakefield power extractor in a TBA unit. It will be shown in the next section that dielectric-based wakefield power extractors can be built at a rather low cost and with the convenience of transverse mode damping, which can alone be a powerful device for other applications.

A relativistic-beam-based wakefield power extractor usually consists of two parts: a decelerator section to transfer the drive beam's kinetic energy to the EM wave, and a coupling section to efficiently extract the EM wave. The decelerator is in general a constant impedance, high group velocity structure. When a high current bunch train passes through it, a high power RF packet moving in the same direction as the beam will be generated. Figure 7 shows the formation of an RF packet in this process. We assume that a drive bunch traveling in the $+z$ direction enters a decelerator of length L at the moment $t = 0$ and location $z = 0$ [Fig. 8(a)]. At the time t when the bunch is still inside the decelerator [Fig. 8(b)],

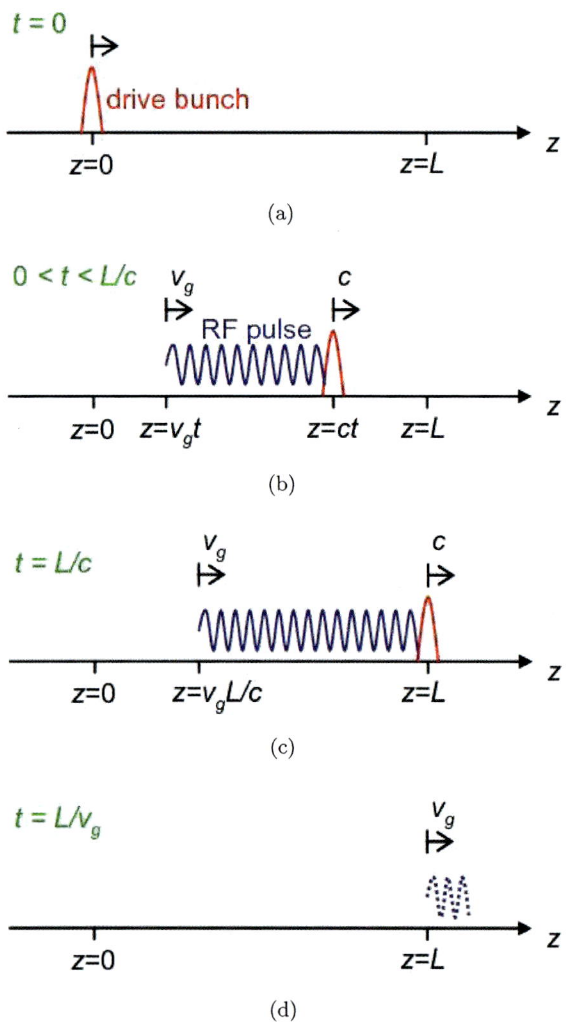

Fig. 8. The RF pulse generated in the decelerator by a single drive bunch: (a)–(d) snapshots of a single bunch and the excited RF pulse inside the decelerator region $0 \leq z \leq L$.

the head of the RF pulse (moving at c) is located at $z = ct$, while the tail of the RF pulse (moving at v_g) is located at $z = v_g t$. When the bunch (and the head of the RF pulse) reaches the downstream end ($z = L$) of the decelerator [Fig. 8(c)] at time is $t = L/c$, the tail of the RF pulse has reached the position $z = v_g L/c$. At this moment, the generated RF pulse would begin to be extracted through an RF coupler (not shown) at $z = L$, but in this heuristic description we imagine the RF packet propagating out of the end of the decelerator at $z = L$. The tail of the RF pulse exits the decelerator [Fig. 8(d)] at time $t = L/v_g$. The duration of the RF pulse is the difference between the moment the bunch reaches the exit ($t = L/c$) and the moment the tail of the RF pulse reaches the exit ($t = v_g L/c$). Therefore, the RF pulse duration for a single bunch excitation is given by $\tau_s = L(1 - \beta_g)/v_g$, where $\beta_g = v_g/c$.

Now let us consider the field buildup process due to bunch train excitation. If a second bunch reaches the exit after the first bunch but before the tail of the first RF pulse, then there will be a region where the RF pulses of the two bunches overlap. To create constructive interference (in-phase superposition) of the excited RF fields, the frequency of the excited RF mode (f_0) is chosen to be a harmonic of the bunch frequency, $1/T_b$, for a train of n bunches evenly spaced in time by T_b. The buildup in time of the amplitude of the longitudinal electrical field caused by a bunch train (E_{at}) at the exit of the decelerator ($z = L$) is shown in Fig. 9. (Note: This is the field that would be observed by an antenna located at $z = L$ and we have now redefined the time origin so that $t = 0$ corresponds to the arrival of the first bunch at $z = L$.) The RF pulses excited by individual bunches are labeled in color. The increasing amplitude of the net field (E_{at}) is due to their superposition. The time structure of the bunch train RF pulse (at $z = L$) consists of a rise time, a flat top, and a fall time, all of which are determined by the overlap of the individual single bunch RF pulses. The rise time is determined by the number of bunches whose RF pulses overlap with the first RF pulse and is equal to ceiling$(\tau_s/T_b) - 1$. For example, if the single bunch pulse length is $\tau_s = 2.4$ ns and the bunches are spaced at $T_b = 0.769$ ns, then there will be ceiling$(\tau_s/T_b) - 1 = 3$ trailing bunches whose RF pulses overlap with the first RF pulse. In general, the number of RF pulses that overlap at the exit of the decelerator is $N =$ ceiling(τ_s/T_b). From Fig. 9, it can be seen that for a long bunch train with $n \geq N$, the rise time t_r can be written as $t_r = (N - 1) \times T_b$. Similarly, the fall time is $t_f = t_r$. The steady state

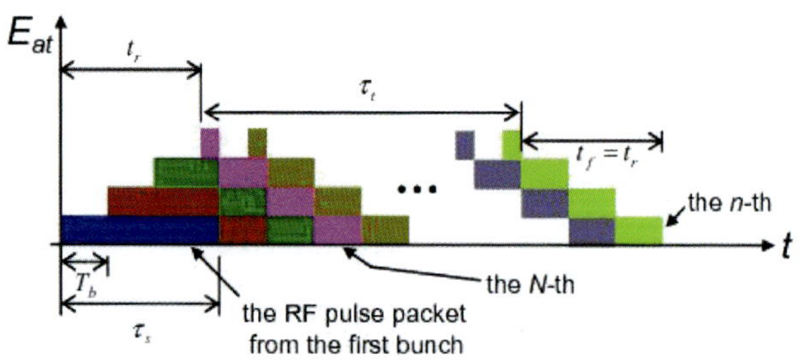

Fig. 9. The RF pulse developed by a bunch train in a DWA, taking into account finite group velocity effects. Pulses excited by different bunches are labeled with different colors.

(flat top) duration of the RF pulse is simply the total length duration minus the rise time and the fall time, i.e. $\tau_t = (n - 1) \times T_b + \tau_s - t_{tr}$. Note that for a very long train ($n \gg N$) the pulse length approaches $\tau_t \approx nT_b$.

The RF power generated by an ultrarelativistic bunch train in the decelerator is given by [39]

$$P = \frac{1}{4}\frac{\omega}{v_g}\frac{r}{Q}L^2 I^2 F^2 \left(\frac{1 - e^{-\alpha L}}{\alpha L}\right)^2, \quad (12)$$

where ω is the angular frequency of the RF mode, α is the attenuation per unit length, and F is the single bunch form factor — the Fourier transform of the bunch shape. For a Gaussian bunch with an rms bunch length σ_z, the bunch form factor can be calculated as

$$F = \exp\left[\frac{-\left(\frac{\omega}{c}\sigma_z\right)^2}{2}\right]. \quad (13)$$

3.2. *Dielectric accelerating structure*

As shown in Fig. 1(a), DLA structures can be made as simple as a dielectric tube surrounded by a conducting cylinder. Most of these devices use uniform, linear ceramic tubes so that they work as constant impedance accelerating structures. The TM_{01} mode (traveling wave accelerator) or TM_{01n} mode (standing wave) are the fundamental accelerating modes. Unlike conventional metallic accelerating structures, DLA structures do not require any structure periodicity to slow the phase velocity of the guided wave below the speed of light. For a given radius, the phase velocity of each guided wave mode in a DLA structure is governed by the dielectric constant of the material and its wall thickness. In general, for a given a/λ, the DLA structure can be made much smaller than a disk-loaded accelerating structure due to the high dielectric constant of the material used. This small size may be favored for applications with tight space requirements. It also facilitates the use of quadrupole lenses or a permanent solenoid around the structure to prevent the beam from breaking up in the case of high current beam acceleration. For example, given $a/\lambda = 0.156$, the diameter of a cell of a 120° traveling wave disk-loaded X-band accelerating structure is larger than 2 cm, but the outer diameter of the dielectric tube in the equivalent DLA structure is only 1 cm if the dielectric constant of the tube is 20.

However, the choice of dielectric constant should be considered as being far more complicated than the requirement of a reduced transverse dimension. It is strongly linked with other accelerating parameters as well. A very good estimate of the group velocity of a traveling wave DLA structure is given by $v_g/c \approx 1/\varepsilon_r$ (c is the speed of light and ε_r is the relative dielectric constant of the loaded material) when a/λ is less than 15%. The quality factor of a DLA structure can be roughly estimated as $Q \approx 5.22\times10^4/\{\sqrt{[f\langle\text{GHz}\rangle\times(\varepsilon_r-1)]}+5.22\times10^4\times\tan\delta\}$, where f is the frequency of the TM_{01} mode in units of GHz and $\tan\delta$ is the loss tangent of the dielectric material.

A DLA is generally best suited for use as a short structure powered by short, high frequency RF pulses, which reduces the probability of RF breakdown but requires only a short filling time to achieve a high RF-to-beam efficiency. Consider the rough design of an X-band traveling wave DLA structure. From the group velocity estimate, we choose $\varepsilon_r = 10$ for a structure with $v_g = 10\%\,c$. Then Q can be estimated to be about 3400. Since the size of the beam opening ($2a$) is independent of Q but closely coupled to the R/Q and shunt impedance R, we can choose a as small as the beam emittance allows to obtain both large R/Q and R values. Meanwhile, the filling time remains short since the group velocity is unrelated to the beam opening. It should be pointed out that a/λ of DLA structures cannot be chosen too small since it will increase the wakefield and thus the risk of beam breakup.

Transverse wakefield damping in DLA structures has been well studied [40]. Its implementation is rather simple: an axially slotted copper jacket is filled with RF absorber, or metallized strips on the outer surface of the dielectric tube are surrounded with RF absorber.

The axial slots interrupt the azimuthal current flow present in the higher order modes, but interfere only weakly with the TM_{01} accelerating mode. The low Q of the DLA structures also helps the damping of long range wakefields.

Figure 10 shows the a/λ dependence of a few parameters of X-band (11.7 GHz) traveling wave DLA structures. Three different dielectric constants are used in the plots, representing three commonly used low loss materials: alumina ($\varepsilon_r = 10; \tan\delta = 1 \times 10^{-4}$), MgCaTiO ($\varepsilon_r = 20; \tan\delta = 1 \times 10^{-4}$), and quartz ($\varepsilon_r = 3.8; \tan\delta = 6 \times 10^{-5}$). Figure 10(a)

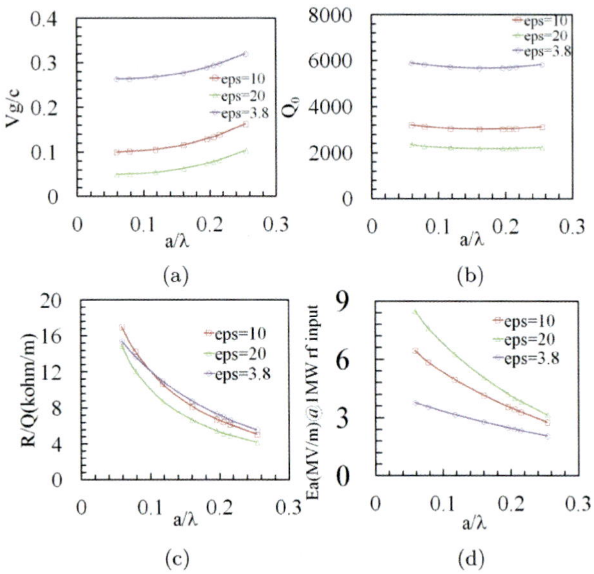

Fig. 10. Dependence of parameters in X-band DLA structures on the scaling parameter a/λ: (a) group velocity, (b) quality factor, (c) R/Q, (d) accelerating gradient.

clearly shows that the group velocity is bounded by the reciprocal of the dielectric constant as the beam aperture decreases. The quality factor of a DLA structure is the combined contribution of metal wall losses and dielectric losses, but it remains nearly constant across different beam channel sizes [Fig. 10(b)]. From Figs. 10(c) and 10(d) we can see that the R/Q values for different dielectrics can approach or cross each other for small beam channels but the gradient in a higher ε_r DLA structure is always higher. This is because of the strong dependence of the gradient E_a on the group velocity: $E_a^2 = [R/Q] \times 2\pi f \times P/v_g$, where P is the input RF power. Again, low values of v_g lead to a long RF filling time, which then requires a longer RF pulse to power the structure and accelerate the beam. The strong pulse length dependence of the RF breakdown rate will limit the accelerating gradient. However, in an application that does not need to operate with the gradient at a high risk level, a high dielectric constant material will be preferred since it will provide a higher gradient for a given RF power.

4. The Two-Beam DWA in Practice

The main facility carrying out experiments on the short pulse dielectric TBA is the Argonne Wakefield Accelerator (AWA) facility [41]. It can provide a high intensity drive beam that is generated by an 8 MeV photocathode (cesium telluride) RF gun, operating at 1.3 GHz, then boosted to 70 MeV. It produces a variety of bunch trains of up to 32 electron bunches and up to 60 nC per bunch (although not simultaneously), which leads to 10 GW beam power available to be extracted. The AWA also hosts a 10 MeV witness beam line. The drive and witness beams propagate in opposite directions. A beam line switchyard allows wakefield experiments to be performed using either the collinear configuration, in which the drive and witness bunches travel along the same structure, or the two-beam-accelerator configuration, in which RF power is transferred from the drive beam decelerating structure to the witness beam accelerating structure, by means of a waveguide.

4.1. *Tests of dielectric wakefield power extractors*

To date, dielectric-based wakefield power extractors (DWPEs) have been developed and tested at the AWA facility in three different frequency bands: C-band [42], X-band [43], and K-band [44]. Work on dielectric wakefield radiation in the THz regime will be discussed later in the article. In this subsection, we briefly discuss the development of K-band (26 GHz) and X-band (12 GHz) DWPEs.

Shown in Fig. 11, major design parameters of the 26 GHz DWPE are summarized in Table 1. The whole power extractor is designed to be machined in three parts: dielectric-loaded waveguide, coupler block, and downstream beam channel. After machining, the three parts are brazed together with flanges at the three open ends. The flange and corresponding gasket for the RF output port are specially designed to allow high power RF transmission without an RF discontinuity while a maintaining a high vacuum. Three 10-cm-long Forsterite tubes, with a

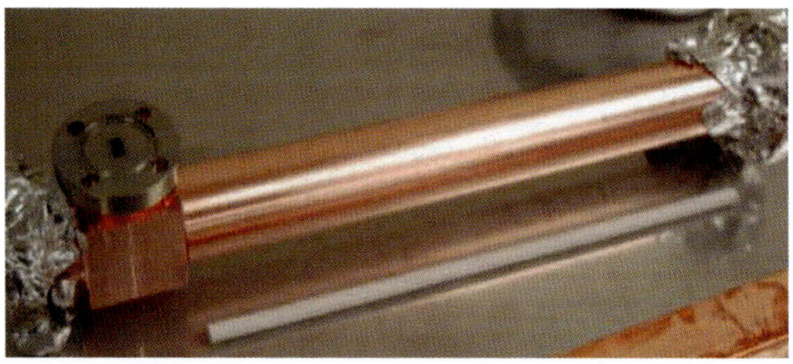

Fig. 11. The 26 GHz dielectric-based wakefield power extractor and dielectric tubes prior to loading.

Table 1. Design parameters of the 26 GHz dielectric-loaded decelerator.

Parameter	Symbol	Value
RF	f (GHz)	26
Inner radius of dielectric tube	a (mm)	3.5
Outer radius of dielectric tube	b (mm)	4.534
Dielectric tube length	L (cm)	30
Relative permittivity of dielectric	ε_r	6.64
Group velocity of RF mode	v_g (%c)	25
Loss tangent of ceramic	$\tan\delta$	1×10^{-4}
Quality factor (metallic wall and dielectric losses)	Q	2950
r/Q per unit length of decelerator	(Ω/m)	9788
Generated RF power (20 nC/bunch; $\sigma_z = 1.5$ mm; $T_b = 769$ ps)	(MW)	148

dielectric constant of 6.64 and $Q = 10{,}000$ at 10 GHz, are loaded in the circular waveguide to generate a 26 GHz high power RF wave with the ultrarelativistic drive beam. Note that there is no transverse mode damping feature in this prototype DWPE. A transverse mode damping scheme is discussed with the X-band DWPE below.

The structure was tested at the AWA facility. Figure 12 shows the RF pulse generated by a train of 16 bunches. We can see that a 10 ns flat top 26 GHz RF pulse has been achieved (the measured frequency shown is offset in the plot by 16 GHz). The highest power measured is 35 MW, driven by a train of 16 bunches, 12 nC per bunch [45].

Figure 13 shows the full-featured X-band (12 GHz) DWPE [43]. The whole structure is designed in three separate parts: dielectric-loaded waveguide, RF choke adaptor to match the impedance, and output coupler. In order to damp the

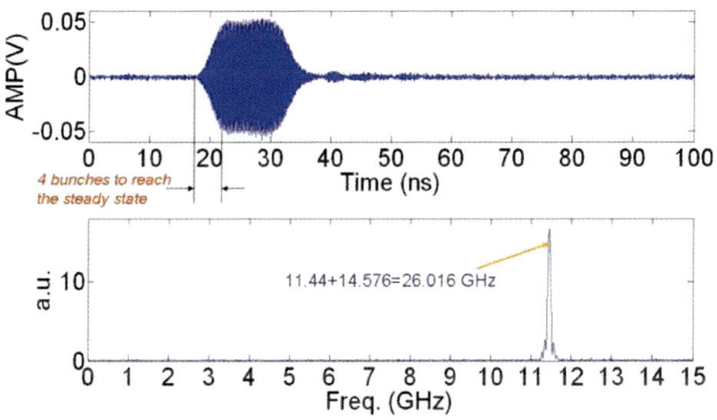

Fig. 12. The down-converted RF signal and its frequency spectrum obtained in the beam test of the 26 GHz wakefield power extractor. A 14.576 GHz local oscillator was used to down-convert the signal. A bunch train consisting of 16 bunches was used to drive the structure.

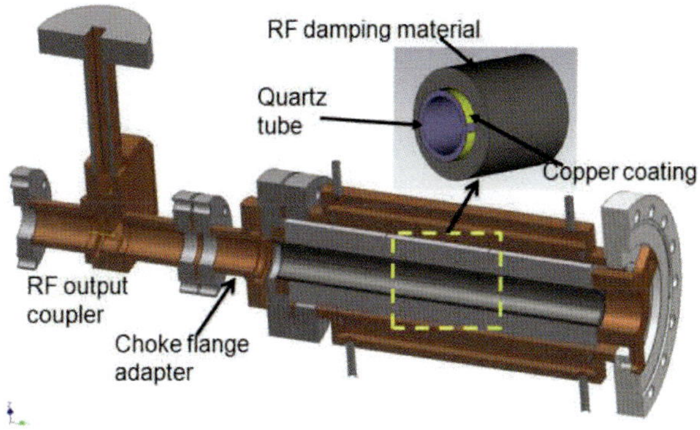

(a)

(b)

Fig. 13. The 12 GHz DWPE: (a) engineering design; (b) fabrication [46].

transverse wakefield, the dielectric (quartz) tube was metallized with 100 microns of sputtered copper on the outer surface of the tube. Then, a number of slots along the copper coating were etched away so that only axial surface currents were allowed to flow along the copper inner surface. The non-axisymmetric hybrid modes that comprise all six cylindrical field components will require azimuthal surface currents on the region of the conductor boundary and thus will leak out past the slots. A lossy SiC-AlN tube located outside the slots will heavily attenuate the hybrid modes. Meanwhile, the TM_{01} (or, generally, TM_{0n}) modes will maintain good propagation characteristics, with only a slight deterioration of Q due to the increased surface current intensity between the slots compared to a continuous conducting boundary. Since the ratio of the total slot width to the overall circumference is small, this perturbation of the TM_{0n} modes can be ignored. The deflecting modes (hybrid modes) in a dielectric-lined circular waveguide are the HEM_{nm} modes. The lowest deflecting mode is the HEM_{11} mode at 11.139 GHz for this dielectric PETS design. The calculated Q of this dipole mode drops from 9000 without damping down to 20, using the damping method described here.

The 12 GHz DWPE has been tested with high power RF (40 MW) and with beam. For detailed reports, see Ref. 46.

4.2. *Tests of dielectric accelerators*

Externally powered DLA structures have been tested extensively in the X-band by a joint effort of among AWA, NRL (Naval Research laboratory), and Euclid Techlabs [47, 48]. In these long RF pulse (~200 ns) experimental studies, the severe multipactor loading limits the achievable gradient (<20 MV/m). However, in the DLA case, there is a strong component of the RF electric field normal to the dielectric surface. This differs from the case of a single surface multipactor on an RF window, where the RF electric fields are tangential. This regime of the multipactor was described and analyzed in Ref. 49. At some threshold value of the RF field, seed electrons will impact the surface at sufficient energy to generate secondary electrons with a secondary electron emission coefficient that exceeds unity, resulting in a secondary electron avalanche. The saturation process for this avalanche relates to the effect of the space charge electric fields that are created by the buildup of the secondary electron cloud since, in the cylindrical geometry of a DLA structure, the secondary electron space charge cloud produces an electric field that accelerates the secondary electrons radially outward and back toward the dielectric surface. A large enough radial space charge electric field, compared to the RF electric field component that creates the multipactor, leads to saturation of the multipactor process, and thus to a steady state loading of the RF transmission through the DLA structure. In the steady state, the RF power is absorbed by the secondary electrons, and lost through collisions of the electrons with the walls of the structure.

The fraction of the RF power absorbed at saturation in the DLA has been found to be as high as 30% of the incident power per unit length, depending on the material. Up to a 20% reduction of multipactor loading was demonstrated in previous experiments through the use of TiN coatings applied using either physical vapor deposition (PVD) or atomic layer deposition (ALD) [47–49]. The use of geometric modifications, particularly the introduction of surface grooves to reduce the multipactor in DLA structures, has also been proposed [50]. An approach using an applied solenoid field to completely eliminate the multipactor in DLA structures has been suggested [51, 52]. This approach is potentially more attractive than other techniques, since it can maintain all the advantages of typical DLA structures, and is also independent of the accelerator parameters other than the operating frequency. In some cases, the applied field could be provided by permanent magnets placed outside the DLA structures. The demonstration of the multipactor being fully suppressed in an RF-driven DLA structure using an external solenoid field, shown in Fig. 14, is reported in Refs. 53 and 54. This eliminates one of the most important hurdles toward practical dielectric-based RF linear accelerators.

Since the secondary electron yield of dielectric material will decrease below 1 when the impact energy is higher than a certain value (material-dependent). In principle, when the gradient in DLA structures is higher than a certain level it may operate out of the multipactor regime. The short pulse TBA is favorable in regard to handling the ultra-high RF power and fast rise time. A 26 GHz dielectric accelerator (the accelerator part of the 26 GHz short pulse TBA prototype module) was developed [55]. Its major parameters are summarized in Table 2. For this proof-of-principle structure, we chose 10 cm as the effective length to reduce machining difficulties. As shown in Fig. 15(d), the single piece dielectric structure eliminates dielectric breakdown at the joints. The dielectric tube has two flared ends to

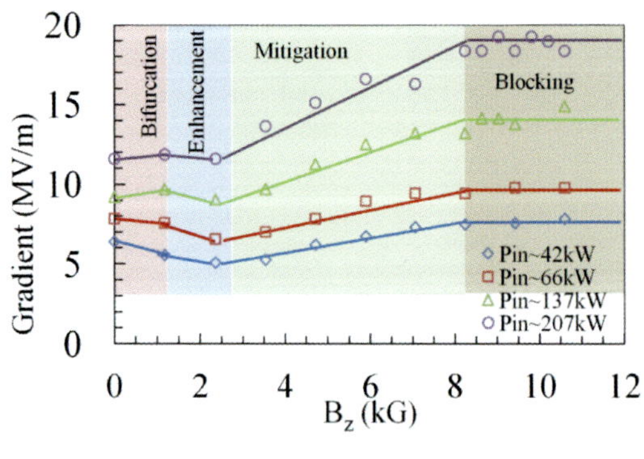

Fig. 14. The dependence of the peak gradient on axis upon the external axial magnetic field strengths and RF input power levels in an X-band short standing wave DLA [54].

Table 2. Parameters of the 26 GHz dielectric-loaded accelerating structure.

Parameter	Value
ID/OD of dielectric tube	3 mm/5.026 mm
Dielectric constant	9.7 (alumina)
Loss tangent	1×10^{-4}
Effective length of dielectric tube	100 mm
Shunt impedance of TM_{01} mode	50.44 (MΩ/m)
Group velocity	$0.1115c$
R/Q of TM_{01} mode	21983 (Ω/m)
Q of TM_{01} mode	2295

match the impedance, which is critical for achieving the broad passband required for short pulse, high gradient acceleration. The alumina tube was coated with 100-μm-thick sputtered copper on the outer surface [Fig. 15(b)]. We finished a straightforward engineering design of the 26 GHz short pulse DLA structure. As shown in Fig. 15(b), the same length of circular waveguide houses the metallized dielectric tube. The dielectric insert is held in position by specially designed gaskets, which have a slightly reduced inner diameter compared to the flange openings. A pumping port at the center of the copper housing provides sufficient vacuum pumping. Two RF couplers were also designed and fabricated [see Fig. 15(c)].

This structure was tested with high power RF generated by the 26 GHz wakefield power extractor. A maximum of 37 MW RF power was measured out of the power extractor, equivalent to a 54 MV/m gradient in the DLA structure. No breakdowns were ever observed. RF pulse lengths in the range, of 5–15 ns were explored during the experiment, varied using the number of electron bunches in the train [56].

5. The Concept of a Short Pulse TBA Collider

5.1. *Overview*

A 3 TeV linear collider scheme (i.e. the Argonne Flexible Linear Collider) has been proposed and is moving forward under a major joint effort by AWA and Euclid Techlabs (for details see Ref. 57). A simplified layout is shown in Fig. 16. It consists of ten 150 GeV stages on one side of the machine. Each 150 GeV stage is made up of 50 discrete 3 GeV modules sharing one drive beam source. This resembles the CLIC scheme, but with a few crucial differences. First, in each 150 GeV stage, 1000 ($= 50 \times 20$) short (∼24 ns) micro drive pulses pass through 50 modules with a local beam pulse repetition rate of 20 (also representing a 20 5-μs-long macro bunch train). Each module provides a 3 GeV gain, which adds up to 150 GeV after 50 modules. Overall, these 1000 micro drive beam pulses, organized by 20 repetitive

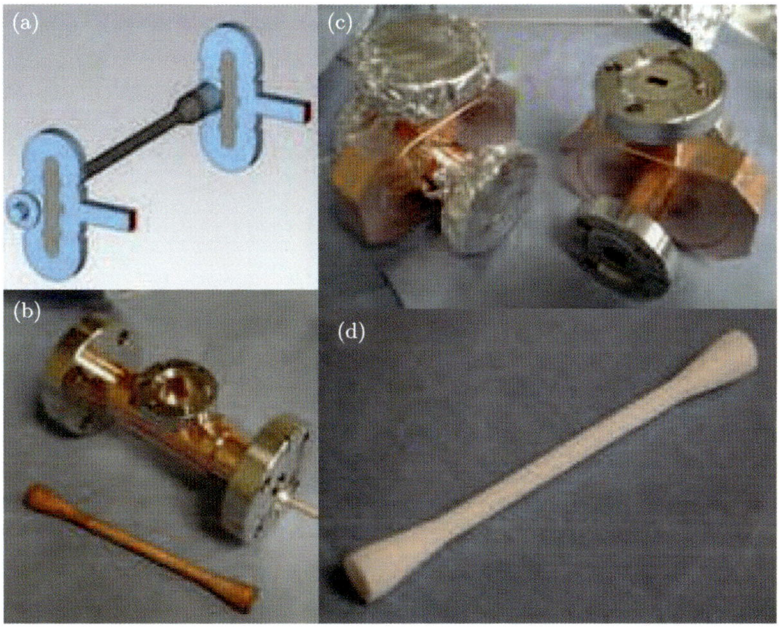

Fig. 15. (a) Design of the 26 GHz short pulse dielectric accelerator; (b) dielectric accelerator and metallized alumina tube; (c) two 26 GHz couplers; (d) the alumina tube prior to metallization [55].

5-μs-long macro bunches, form a 100 μs giant beam pulse. The machine repetition rate is 5 Hz.

To match the local beam pulse repetition rate of 20, the main beam consists of 20 short beam pulses in the same 100 μs period of time. The main beam current inside a pulse is 6.5 A, corresponding to 0.5 nC per bunch, one bunch for two RF cycles. The purpose of introducing the local drive beam repetition rate is to increase the average power. Another obvious difference from the CLIC approach is that the drive beam in this short pulse scheme is generated by a 1.3 GHz RF photoinjector in each 150 GeV stage with a high QE cathode, which can provide 50 nC/bunch with a bunch separation of 769 ps (32 sequential bunches form an ~24 ns micro drive pulse). Finally, to achieve a high RF-to-beam efficiency in the main linacs under short RF pulse conditions, we chose a high frequency (26 GHz), high group velocity (~11% c) dielectric-based structure (broadband RF coupling), which in turn provides an ~270 MV/m gradient (with the assumption of no breakdowns at this level in a 20 ns duration pulse), ~ 9 ns filling time, and ~3 ns rise time. Table 3 summarizes some of the preliminary design parameters.

As shown in Fig. 16, each 150 GeV stage can be broken into 50 3 GeV modules. Every 3 GeV module contains 38 0.3-m-long two-beam accelerator pairs (wakefield power extractor and main accelerator). The drive beam energy is 0.86 GeV at the entrance of the first wakefield power extractor. It loses 20.5 MeV for every wakefield power extractor and provides 1.33 GW, 22 ns RF output. After 38 extractors, the drive beam energy drops to 80 MeV and is then dumped at the end of the 3 GeV module. The total length of one module is 15 m, which matches the time periodicity of micro drive bunches (100 ns). The drive beam pulses are evenly distributed along a common transport line in parallel but travel in the direction opposite to the main beam. The distance between drive bunches has to be twice the length of one accelerating structure to ensure that the arrival time at the entrance of the wakefield power extractor matches the main beam. Within the 3 GeV module, different RF delays are used among the TBA pairs to synchronize the timing between the drive and the main beam.

Table 3. Some preliminary design parameters of the Argonne Flexible Linear Collider.

Parameters	Value
Main linac frequency	26 GHz
Drive linac frequency	1.3 GHz
Main linac loaded gradient	267 MV/m
Average drive beam current (one side)	80 mA
Average drive beam power (one side)	68.8 MW
Average main beam current (one side)	10.4 μA
Average main beam power (one side)	15.6 MW

In this design, the drive beam is delivered to each module by a fast kicker (high voltage stripline or traveling wave deflector). The intramodule synchronization condition is that the RF from the ith decelerator must be delayed by $t_{di} = 2 \times (i-1) \times L_s/c$, where $\{i = 1 \ldots m\}$, m is the number of decelerator–accelerator structure pairs in each module and L_s is the length of a single structure. In intramodule staging, the same drive beam passes through all decelerators within a single module. The RF generated by the drive beam in the decelerator is transferred through an RF waveguide to the accelerator to accelerate the main beam. The drive beam and the main beam pass through the decelerator–accelerator structure pairs in the opposite direction, i.e. the main beam passes through the first accelerator first and the mth accelerator last, while the drive beam passes through the mth accelerator first and the first accelerator last. At the moment when the RF from the first decelerator fills the first accelerating structure, the main beam is launched. In order for the RF to reach the ith accelerator before the $(i+1)$th accelerator, the RF from the later waveguide must be delayed by $2L_s/c$ relative to the former. At the exit of decelerator #1 the drive beam is dumped after exhausting most of its energy.

5.2. *Short pulse TBA research activities*

Intensive research activities on the short pulse TBA have been carried out lately, including high power RF generation using wakefield power extractors, demonstrations of high gradient TBA, and demonstrations of staged wakefield acceleration.

The experimental setup is shown in Fig. 17. It consists of two successive stages. An ~70 MeV high charge drive beam passes through two power extractors from right to left, as shown in the picture. The generated power is delivered to the short accelerator through a bow-shaped WR90 waveguide. The other side of the accelerator is terminated by a SiC-based RF load. RF power can be monitored by field probes

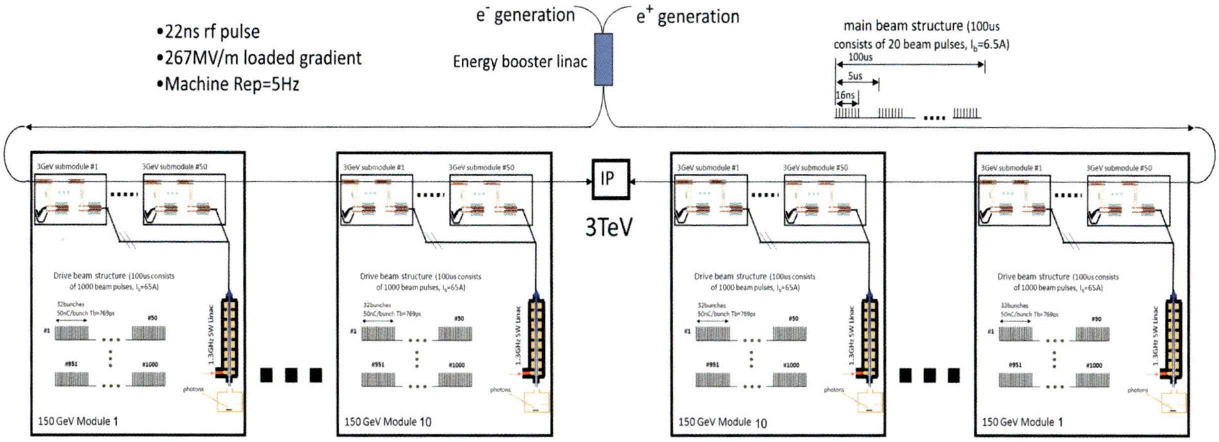

Fig. 16. The conceptual layout of the Argonne Flexible Collider.

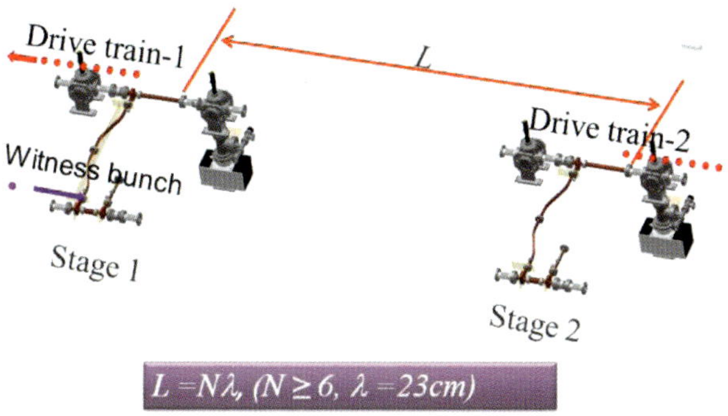

Fig. 17. Experimental configuration of the staged short pulse TBA at the AWA facility. Note that the space between two TBA units contains beam-focusing optics (not shown).

located at the output of the power extractor and the load. At an synchronized time, an ~10 MeV, 1 nC witness bunch is injected into the accelerator from left to right (shown in the picture) to gain the energy.

The latest results include ~200 MW, ~10 ns RF pulse generation; ~150 MeV/m with the single stage TBA and the staged TBA, in which a 0.5 nC bunch gained an equal amount of energy in two stages (~2.4 MeV per stage, corresponding to an average, acceleration gradient of ~70 MeV/m) [58]. It should be pointed out that, instead of dielectric structures, the short pulse TBA and staging experiment at AWA used metallic structures. Tests of dielectric TBAs will follow in the next step [59]. The technique is scalable to a staged acceleration at 200–300 MeV/m by using a GeV scale drive beam. Such a development will considerably reduce both the cost and the footprint of a future high energy physics collider as well as future x-ray light sources.

6. The Collinear DWA in Principle

The underlying principle of the collinear dielectric wakefield accelerator (CDWA) is that a short and relatively low energy but high current *drive* beam creates high accelerating fields in a dielectric wakefield structure, which can then be used to accelerate a trailing (*witness*) beam to high energy. The collinear configuration has the two beams traversing the accelerator along the same trajectory, as opposed to the two-beam acceleration (TBA) scheme, where the beams pass through different structures (decelerator and accelerator). In TBA, the wakefield energy is extracted from the decelerator to provide RF power for a conventional accelerator. On the other hand, the collinear wakefield accelerator directly transfers the energy from the drive bunch radiation to the trailing main bunch.

6.1. *The transformer ratio in the CDWA*

A central figure of merit in a collinear wakefield acceleration scheme is the accelerating gradient. The presence of a high accelerating gradient is always attractive because it will reduce the length of the main linac, which is the primary determining factor for the facility size and cost. Equation (10) in Sec. 2.2 indicates that the accelerating gradient can reach the GV/m level if a short, intense bunch traverses a CDWA structure operating in the THz regime. Another important parameter that influences the performance of a collinear wakefield accelerator is the transformer ratio, TR ≡ (maximum energy gain of the witness bunch)/(maximum energy loss of the drive bunch). This characterizes the energy transfer efficiency from the drive beam to the accelerated (witness) beam, and is ultimately related to the overall efficiency (under some constraints, such as fixed beam loading, or the same wakefield structure). Obviously, a high value of the transformer ratio is preferred.

The final energy of the witness bunch is the product of the gradient and the distance, L, that the drive bunch and the witness bunch have propagated in the structure. The overall wall plug efficiency of the wakefield accelerator is proportional to the fraction of the energy extracted from the drive bunch, $\eta_D = 1 - \mathrm{KE}_f/\mathrm{KE}_0$, where KE_0 is the initial kinetic energy, KE_f the final energy, and η_D is proportional to the propagation length of the drive bunch. For example, considering a $\mathrm{KE}_0 = 1\,\mathrm{GeV}$ drive bunch that generates a gradient of $1\,\mathrm{GV/m}$ and decelerates at $0.5\,\mathrm{GeV/m}$ (TR = 2), the maximum propagation length is $2\,\mathrm{m}$ when $\eta_D = 100\%$. Over the same distance, the witness bunch will be accelerated by $2\,\mathrm{GeV}$. (The beam loading by the witness beam is neglected since we assume its charge to be low.) For the same charge witness bunch, if the drive bunch maintains the same gradient behind the bunch but the deceleration is reduced to $0.1\,\mathrm{GeV/m}$ (TR = 10), then the witness bunch can gain $10\,\mathrm{GeV}$.

However, the gradient and transformer ratio appear to have a complicated relation. Numerical simulations of single bunch wakefield excitation reveal the dependence of the peak accelerating gradient (W^+) and TR on the bunch length σ_z (Fig. 18) [60]. As one would expect, the peak accelerating gradient increases as the bunch length decreases, but the behavior of TR as a function of σ_z/λ (λ represents the wavelength of the fundamental accelerating mode) is more complicated than the monotonically decreasing behavior of W^+. The TR is peaked near $\sigma_z = \lambda/4$, while dropping off toward 1 for bunch lengths on either side of $\lambda/4$.

While the TR is less than 2 under very general conditions — linear media; a relativistic, longitudinally symmetric drive bunch; and identical paths through the system of both drive and witness beams [61] — there are several techniques that can achieve TR > 2 by breaking the symmetry of the drive beam [62–68].

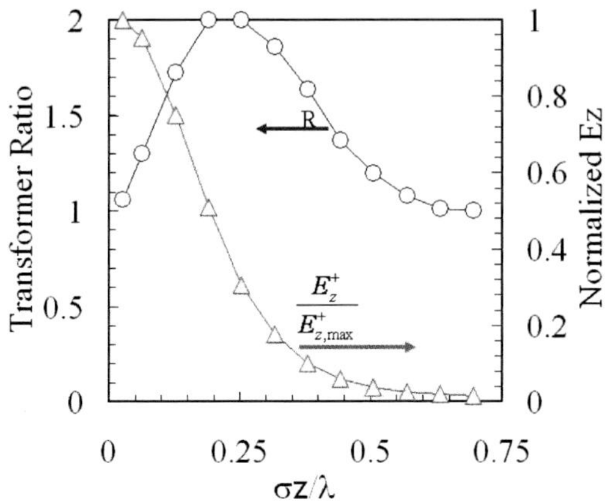

Fig. 18. The transformer ratio, TR, and peak accelerating field, E_z^+, as a function of the normalized bunch length (σ_z/λ). The plot is from Ref. 60. (Single mode operation was assumed in the calculation.)

The ramped bunch train technique (a train of electron bunches with a specific ramped-up charge ratio but equal time separation, i.e. an amplitude-modulated bunch train) has been experimentally demonstrated [65], and TR ~3.4 was obtained [66]. However, it appears to be a great challenge to achieve a higher TR using a longer bunch train, due to the tight requirements on the spatial separation and charge ratio among bunches in the train. These technical difficulties can be greatly reduced in an approach using a single shaped drive bunch [67, 68].

Figure 19 represents three typical bunch shapes — ST (single triangular), DT (double triangular), and quadratic — which have been found to significantly enhance the TR. However, the current density distribution of these shaped bunches is assumed to have a steep cutoff edge at the bunch tail and a perfect current ramp-up shape. In reality, space charge effects, coherent synchrotron radiation, and energy divergence of the bunch will diffuse the particles in phase space and cause difficulties in generating such ideal bunches in experiments, and thus the TR can differ significantly from the theoretical values.

Techniques to produce these shaped bunches have been significantly improved (although challenges still remain) since this approach to enhance the TR in collinear wakefield acceleration was first raised. Methods that have been explored to generate a shaped bunch include the compression of a time–energy chirped bunch with an anisochronous dogleg beam line [69]; temporal shaping of the drive-laser intensity in a photoinjector source [70]; manipulating

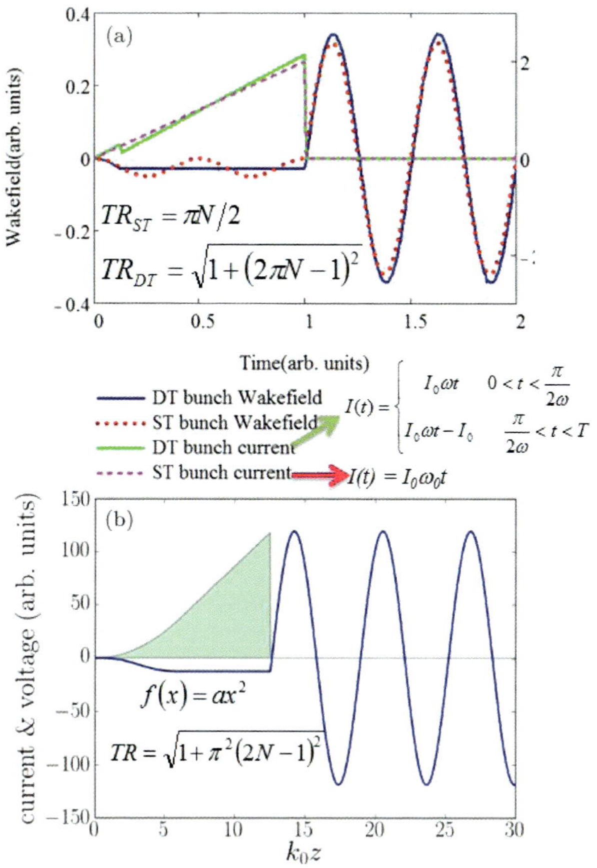

Fig. 19. (a) Wakefields for the ideal DT and ST bunches traversing a single mode wakefield device [67]; (b) an example of a "quadratic" current profile given by Ref. 68.

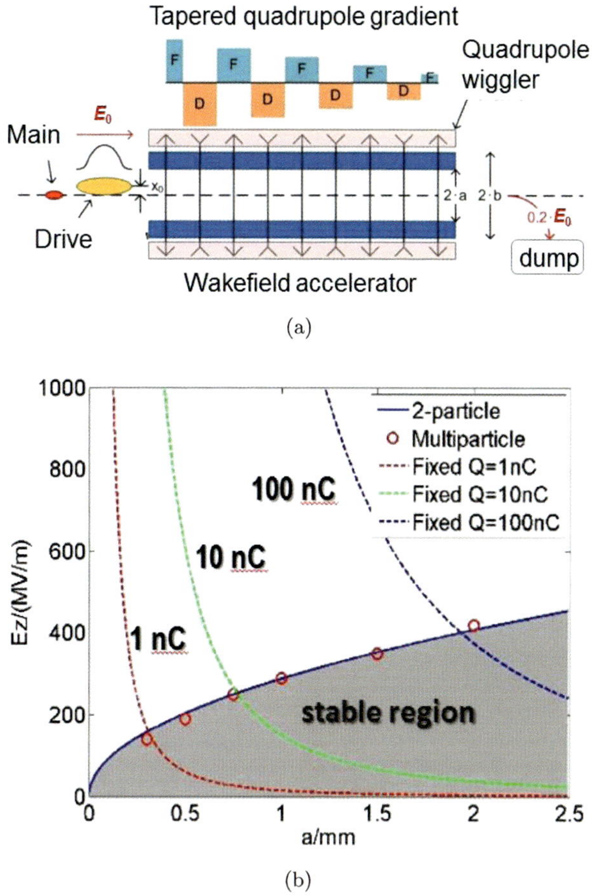

Fig. 20. (a) Schematic of a structure-based wakefield accelerator. The quadrupole fields are tapered to match the energy loss of the drive bunch along the accelerator. The final energy of the drive bunch at the end of the structure is 20% of its initial energy. (b) The maximum attainable energy gain in the structure-based wakefield accelerator limited by BBU as a function of the accelerating structure inner radius.

the longitudinal phase space nonlinearities induced from linacs operating at different frequencies [71]; a mask to select a transverse shape and then transform it to longitudinal space with an emittance exchange (EEX) beam line [72, 73]; laser pulse shaping combined with acceleration and compression [68]; and self-wake-induced bunch shaping [74].

6.2. *A sustained CDWA*

In order to increase the TR while maintaining the same gradient at the same wakefield structure, not only is specific bunch shaping required but the charge of the bunch also needs to be increased. This leads to a higher transverse wake, which may result in BBU at a much shorter propagation length. Hence η_D may drop to an unacceptable level for a large scale accelerator facility.

A quadrupole wiggler [composed of alternating focusing and defocusing quadrupoles, schematically shown in Fig. 20(a)] has been considered to confine the beam in the wakefield structure for long distances [75]. Lately, this idea has been revisited and improved by using an energy chirp in the drive bunch entering the wakefield structure [76]. In order to better understand the main parameter dependencies, a simple two-particle model was developed that was later found to correctly model BBU for a drive bunch with a Gaussian longitudinal distribution. Using this model we determine the maximum achievable acceleration gradient as a function of the accelerating structure inner radius a, as shown in Fig. 20(b). A gray area on this plot represents a stable region, and a white area an unstable region. The border curve scales as $a^{0.5}$, in good agreement with the results of multiparticle tracking, shown by the circles. In this simple model, the ultimate extent of the stability region is limited by a maximum attainable strength for the quadrupole field (which in this study was scaled as a^{-1} from the value of 1 T/mm at $a = 1$ mm). The dashed lines show scaling of the acceleration gradient when BBU is ignored.

This result shown in Fig. 20(b) can be used to guide the accelerator designer in the choice of a DWA structure. If a high charge is available (such as a drive bunch of $Q = 100$ nC), then a maximum gradient of 400 MV/m with structure radius $a = 2$ mm can be attained for sustainable wakefield acceleration. At the other end of the spectrum, if the drive bunch charge available is only 1.0 nC, as is typical for an SRF linac, then the maximum gradient is 150 MV/m for $a = 0.3$ mm.

The maximum acceleration gradient can be increased if one is willing to sacrifice the efficiency (by lowering the requirement for η_D) and the total energy gain of the accelerator. This is because the charge must be raised to increase the gradient, which increases BBU and therefore reduces the propagation length, L. For example, the gradient can be increased from ~290 MV/m to ~580 MV/m by doubling the charge in the reference case ($a = 1$ mm). However, L is then reduced from 85 cm to 30 cm, and thus η_D drops from 80% to 58% and the total energy gain decreases from 245 MeV to 175 MeV.

7. The CDWA in Practice

The pioneering experimental work on the CDWA study started in the late 1980s [17]. Significant progress has been made in the development of high gradient CDWAs since then: GV/m gradients at ~1 THz and 100 MV/m at 10 GHz have been achieved [77, 78]. As a consequence, CDWAs are emerging as a promising future supplemental alternative to plasma wakefield accelerators, offering the possibility of very high accelerating gradients for positron acceleration and high repetition rate particle acceleration, etc.

However, due to the lack of facilities which are capable of producing an intense drive beam and a synchronized witness beam simultaneously, high gradient wakefield acceleration has not been demonstrated until recently (~300 MeV/m of acceleration measured at FACET [79]). On the other hand, a number of CDWA-related technologies [65, 66, 80], structure variations [81–84], and other applications [85–88] have been experimentally explored.

Figure 21 shows one method to construct a CDWA using a quartz capillary (operating in the THz regime). It is metallized with a few μm of copper using a sputtering technique. The metallized quartz tube can then be mounted on a fixture in a vacuum chamber as a CDWA.

Other than the cylindrical CDWA, slab DWAs are widely studied because they have a less severe transverse wake while driving by a flat sheet beam [68]. Figure 22 shows an experimental study of a CVD (chemical vapor deposition) diamond-based CDWA [75]. CVD diamond has a very low microwave loss tangent (~1×10^{-3} in the THz regime), the highest available thermal diffusivity (~5 times better than copper at room temperature), and a high

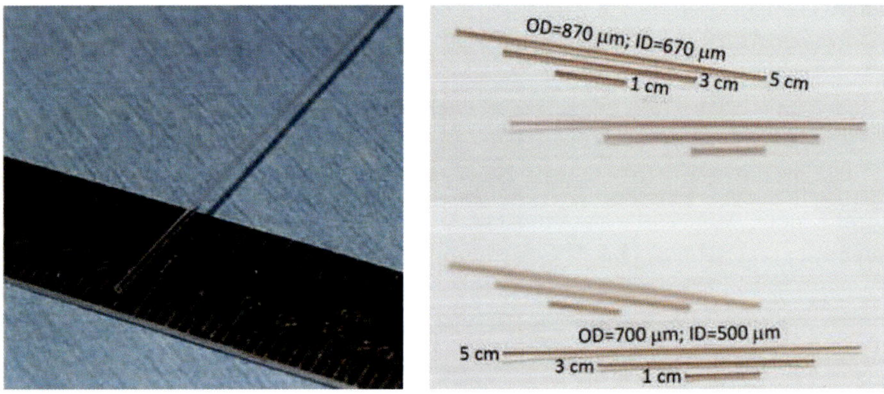

(a) (b)

Fig. 21. (a) A quartz capillary; (b) metallized quartz capillaries (different lengths and transverse sizes); (c) end-on picture of the tube; (d) SEM image of a 1.7 μm metallized layer.

(c) (d)

Fig. 21. (*Continued*)

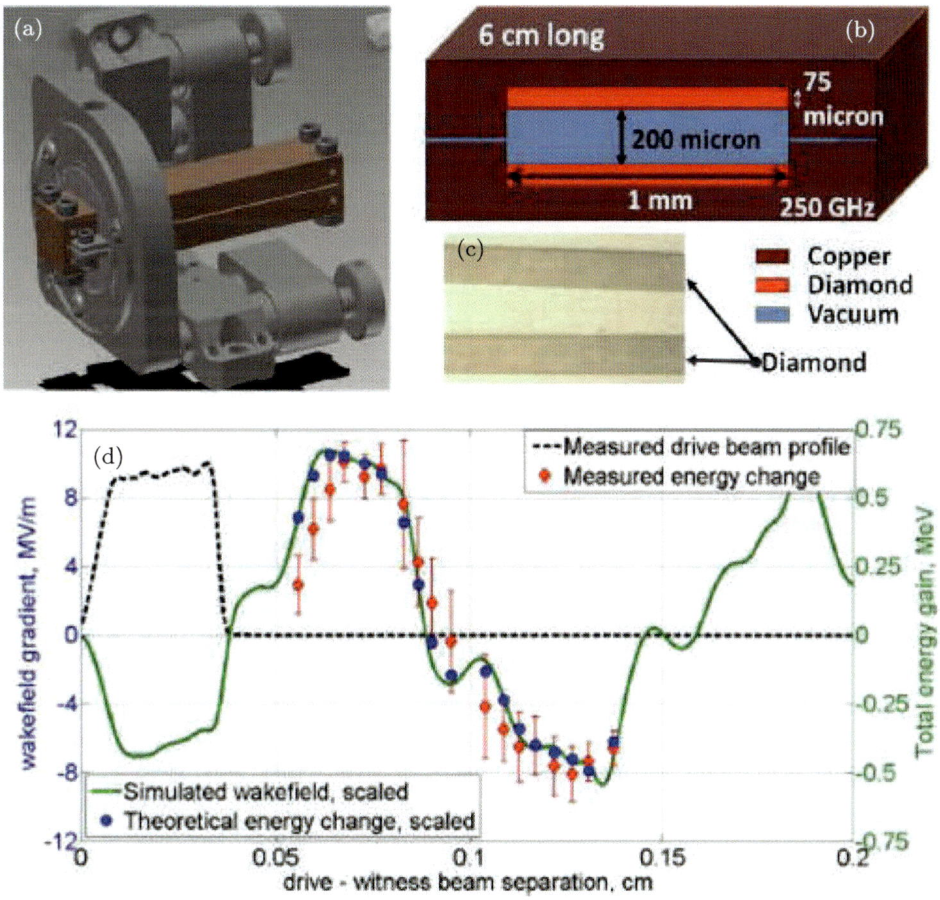

Fig. 22. (a) Structure mounted on a positioning holder; (b) structure dimensions; (c) photo of diamond after the experiment, exhibiting no visible damage; (d) energy gain of the witness beam as a function of separation from the drive beam. The shape of the experimentally mapped wake (diamonds) agrees well with the theoretical (scaled) prediction (circles) based on the wakefield (solid line) calculated for the drive beam current distribution (dashed line) [81].

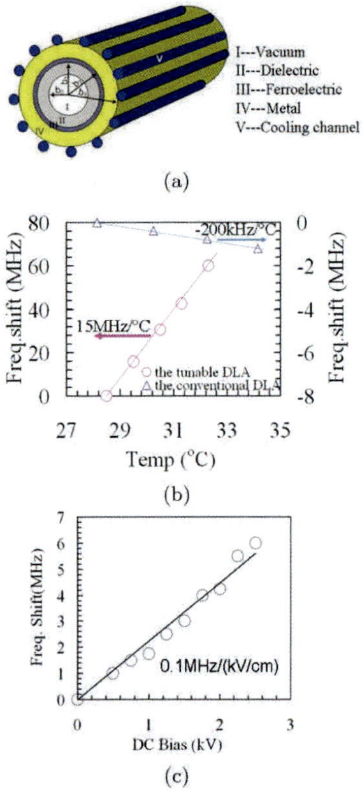

Fig. 23. (a) The tunable DWA structure; (b) comparison of the temperature dependence of the 14 GHz tunable and conventional DWA structures as measured using the network analyzer. (c) bench test configuration of a DC voltage tunable DLA structure. $b_0 = 4.79$ mm; $b_1 = 6.99$ mm ($\varepsilon_{r1} = 6.8$), $b_2 = 7.49$ mm ($\varepsilon_{r2} = 310$ at 20° C) [80].

RF breakdown field (potentially at the level of ~10 GV/m). Limited by the drive beam current, the work presented in Fig. 22 leads to a greater understanding of wakefield acceleration rather than a significant high gradient.

While the quest for a high gradient has been the primary focus of the field, synchronization between the electron beam velocity and the phase velocity of the accelerating field is also necessary since it is required in order for the bunch to continuously gain energy through differrent wakefield structures. Achieving precise synchronization requires a method of frequency-tuning the accelerating structure since the frequency of the assembled accelerating structure will, in general, differ from the design due to various sources of error.

One approach to tuning the frequency of the DLA structure, which was originally suggested in Ref. 89, is to add a thin layer of nonlinear ferroelectric material between the layer of conventional ceramic and the conducting outer sleeve [see Fig. 22(a)]. The dielectric constant of the ferroelectric layer can be varied with either temperature or DC voltage. Therefore, the effective dielectric constant inside the metal waveguide changes and hence so does the frequency of the CDWA structure. The first test of a tunable CDWA was reported in Ref. 80.

The results plotted in Figs. 22(b) and 22(c) show a significant positive temperature tunability coefficient of 15 MHz/°C and a DC-bias-induced fast response tuning coefficient of ~0.1 MHz/(kV/cm) for the structure incorporating the ferroelectric layer. For comparison, a slightly negative frequency response coefficient (−200 kHz/°C) was measured for a conventional DWA structure with a similar construction but without a ferroelectric layer. This frequency shift is caused by the thermal expansion of the copper tube. This technology can in principle be extended to the CDWA operating in the THz regime by adding a layer of ferroelectric material on top of regular dielectric tubes through various thin film coating techniques, like RF magnetron sputtering.

8. The Concept of a CDWA-Based FEL

A new concept for a soft x-ray free electron laser (FEL) facility based on a CDWA is proposed as a joint effort by a number of institutes [90, 91]. One of the goals discussed there is a rapid acceleration of a low charge witness electron bunch to a few GeV in one accelerating stage using a single 400 MeV high charge drive bunch.

The overall facility layout is shown in Fig. 24. The scheme consists of ten parallel FELs and ten parallel CDWAs that supply the FELs with ~2 GeV electrons. The CDWAs share a single 650 MHz cw SRF linac that accelerates 8 nC drive and 250 pC main bunches to ~400 MeV. Because of a need for tight synchronization between the drive and main bunches (of the order of 250 fs), both bunches are obtained from a single source bunch, as discussed below. The repetition rate of these duets of drive and main bunches in the CWA is limited to ~50 kHz. This keeps the average wakefield power dissipation in the accelerating structure at a manageable level, as shown below. Thus, the bunch repetition rate in the cw SRF linac is ~500 kHz. A spreader switchyard distributes electron bunches to the ten CDWAs on a rotating basis (or as needed). The drive bunch has

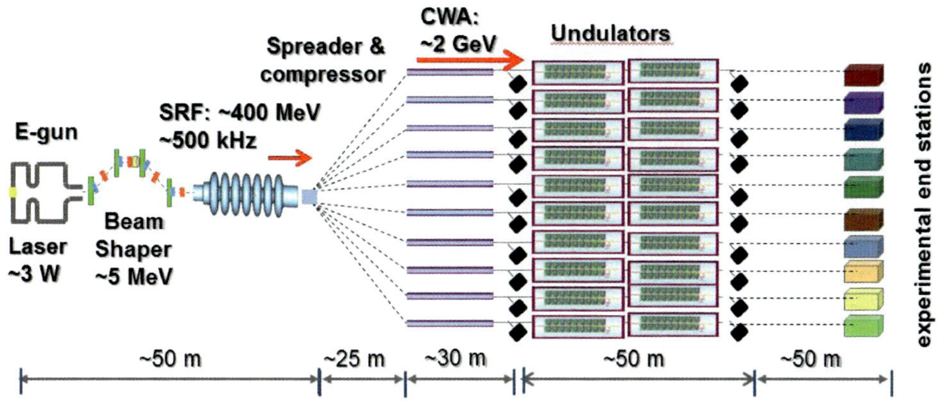

Fig. 24. Schematic of the FEL facility showing (not to scale) the electron gun, bunch shaping section, cw superconducting linac, spreader and transport lines, array of collinear wakefield accelerators, undulator arrays, and x-ray beam lines.

a special-shaped current profile to enhance the TR. Obtaining a short drive bunch ~3.3 ps requires off-crest acceleration of the already-preshaped bunch in the SRF linac producing an ~10% energy chirp at the end and an approximately tenfold bunch compression in each spreader beam line. The energy chirp that remains after the compression assists in suppression of the single bunch breakup instability.

The conceptual design has been continually making progress since its inception. The latest progress is presented in Ref. 92.

9. Conclusion

Dielectric materials have been widely used in the development of two-beam and collinear wakefield accelerators. The simple geometry and potentially high gradient of the DWA make it a very attractive approach for a compact high gradient linear accelerator. In this article we have summarized the operational principles and main experimental efforts in DWA research and development. Conceptual designs of two large scale facilities have been presented as possible future scientific applications. Limited by the available space and the author's knowledge, many other DWA applications could not be covered here. Readers may be interested in additional topics like dielectric-based low energy industrial or medical accelerators, DWA THz radiation sources, and DWA-based beam diagnostic tools [94–97].

Acknowledgments

The author thanks all colleagues and collaborators for their help in the DWA research which he has been involved in. He especially thanks Wei Gai for the fruitful discussions during the process of drafting this article. He also thanks US DOE funding agencies for supporting the DWA projects mentioned in this article and DWA research in general.

References

[1] CLIC Conceptual Design Report (CDR) (12 Oct. 2012), http://project-clic-cdr.web.cern.ch/project-CLIC-CDR/CDR_Volume1.pdf.
[2] T. Higo, in *Proc. Linear Accelerator Conference LINAC 2010* (Tsukuba, Japan, 2010), pp. 1038–1042.
[3] V. Dolgashev, S. Tantawi, Y. Higashi and B. Spataro, Geometric dependence of radio-frequency breakdown in normal conducting accelerating structures, *Appl. Phys. Lett.* **97**, 171501 (2010).
[4] F. Wang, C. Adolphsen and C. Nantista, Performance limiting effects in X-band accelerators, *Phys. Rev. ST Accel. Beams* **14**, 010401 (2011).
[5] ILC Technical Design Report (2013); https://www.linearcollider.org/ILC/Publications/Technical-Design-Report.
[6] C. Joshi, The Los Alamos Laser Acceleration of Particles Workshop and beginning of the advanced accelerator concepts field, *AIP Conf. Proc.* **1507**, 61 (2012).
[7] Accelerator R&D Subpanel Report, "Accelerating discovery: a strategic plan for accelerator R&D in the U.S.," Chap. 9, Apr. 2015. http://science.energy.gov/~/media/hep/hepap/pdf/Reports/Accelerator_RD_Subpanel_Report.pdf.
[8] S. Frankel, *J. Appl. Phys.* **18**, 650 (1947).
[9] G. G. Bruck and E. R. Wicher, *J. Appl. Phys.* **18**, 766 (1947).
[10] R. B. R.-Shersby-Harvie, *Nature* **162**, 890 (1948).
[11] G. Flesher and G. Cohn, *AIEE Trans.* **70**, 887 (1951).

[12] G. I. Cohn and G. T. Flesher, *IIT Rep.* **2** (1952).
[13] R. B. R.-Shersby-Harvie, L. B. Mullett, W. Walkinshaw, J. S. Bell and B. G. Loach, A theoretical and experimental investigation of anisotropic dielectric-loaded linear electron accelerators, *Proc. IEE B* **104**, 273–292 (1957).
[14] G. B. Walker and N. D. West, *Proc. IEE C* **104**, 381 (1957).
[15] G. B. Walker and E. L. Lewis, *Nature* **181**, 38 (1958).
[16] B. Zhang, Microwave inverse Cerenkov accelerator, in *Proc. 7th Advanced Accelerator Concepts Workshop*; *AIP Proc.* **398** (1996).
[17] W. Gai, R. Konecny and J. Simpson, Externally powered dielectric loaded waveguides as accelerating structures, in *Proc. 1997 Particle Accelerator Conference* (Vancouver, 1997), pp. 636–638.
[18] W. Gai, P. Schoessow, B. Cole, R. Konecny, J. Norem, J. Rosenzweig and J. Simpson, *Phys. Rev. Lett.* **61**, 2756 (1988).
[19] P. Schoessow, M. E. Conde, W. Gai, R. Konecny, J. Power and J. Simpson, *J. Appl. Phys.* **84**, 663 (1998).
[20] A. Tremaine, J. Rosenzweig and P. Schoessow, *Phys. Rev. E* **56**, 7204 (1997).
[21] T.-B. Zhang, J. Hirshfield, T. Marshall and B. Hafizi, *Phys. Rev. E* **56**, 4647 (1997).
[22] W. Gai, Performance of the Argonne Wakefield Accelerator Facility and initial experimental results, in *Proc. 7th Advanced Accelerator Concepts Workshop*; *AIP Proc.* **398** (1996).
[23] S. Gold, A. K. Kinkead, O. A. Nezhevenko, V. P. Yakovlev, J. L. Hirshfield, A. L. Vikharev, O. A. Ivanov, S. V. Kuzikov, A. M. Gorbachev, V. A. Isaev, W. Gai, R. Konecny and J. G. Power, High power accelerator R&D at the NRL 11.424 GHz Magnicon Facility, in *Proc. 10th Advanced Accelerator Concepts*; *AIP Conf. Proc.* **647** (Mandalay Beach, CA, June 2002), pp. 439–447.
[24] D. Yu, D. Newsham and A. Smirnov, 21 GHz ceramic RF power extractor, in *Proc. 10th Advanced Accelerator Concepts*; *AIP Conf. Proc.* **647** (Mandalay Beach, CA, June 2002), pp. 484–505.
[25] T.-B. Zhang, T. C. Marshall and J. L. Hirshfield, *IEEE Trans. Plas. Sci.* **26**, 787 (1998).
[26] M. E. Hill, C. Adolphsen, W. Baumgartner, R. S. Callin, X. E. Lin, M. Seidel, T. Slaton and D. H. Whittum, *Phys. Rev. Lett.* **87**, 094801 (2001).
[27] W. K. H. Panofsky and W. A. Wenzel, *Rev. Sci. Instrum.* **27**, 967 (1956).
[28] M. Rosing and W. Gai, *Phys. Rev. D* **42**, 1829 (1990).
[29] K.-Y. Ng, *Phys. Rev. D* **42**, 1819 (1990).
[30] T. P. Wangler, *RF Linear Accelerators* (John Wiley & Sons, 1998), pp. 314–353.
[31] A. W. Chao, *Physics of Collective Beam Instabilities in High Energy Accelerators* (Wiley, New York, 1993).
[32] G. Stupakov and K. L. F. Bane, *Phys. Rev. ST Accel. Beams* **15**, 124401 (2012).
[33] S. S. Baturin and A. D. Kanareykin, *Phys. Rev. Lett.* **113**, 214801 (2014).
[34] S. S. Baturin and A. D. Kanareykin, *Phys. Rev. Accel. Beams* **19**, 051001 (2016).
[35] H. Braun, CERN-CLIC Note 364 (1998), pp. 150–158.
[36] D. Whittum, SLAC-PUB-7802 (1998), pp. 52–54.
[37] C. Jing, WF Note WF-222.
[38] J. Power *et al.*, *Phys. Rev. E* **60**, 6061 (1999).
[39] F. Wang *et al.*, in *Proc. Advanced Accelerator Concepts: 14th Workshop*, eds. S. Gold and G. Nusinovich (2010), pp. 280–285.
[40] F. Gao *et al.*, *Phys. Rev. ST Accel. Beams* **11**, 041301 (2008).
[41] W. Gai and C.-H. Ho, *J. Appl. Phys.* **70**, 3955 (1991).
[42] www.hep.anl.gov/awa.
[43] F. Gao *et al.*, *IEEE Nucl. Sci.* **56**, 1492 (2009).
[44] C. Jing *et al.*, An X-band dielectric-based wakefield power extractor, in *Proc. 2013 Int. Particle Accelerator Conference* (Shanghai, 2013), pp. 2908–2910.
[45] C. Jing *et al.*, Dielectric-based wakefield power extractor, in *Proc. 14th Advanced Accelerator Concepts*; *AIP Conf. Proc.* **1299** (Annapolis, MD, June 2010), pp. 370–375.
[46] Final technical report of the SBIR project entitled "Development of a dielectric-based short RF pulse two beam accelerator prototype module." Awardee: Euclid Techlabs LLC, under Contract #DE-SC0004320.
[47] Final technical report of the SBIR project entitled "Development of a 12 GHz dielectric-based wakefield power extractor for CLIC applications." Awardee: Euclid Techlabs LLC, under Contract #DE-SC0004322.
[48] C. Jing *et al.*, *IEEE Trans. Plas. Sci.* **33**, 1155 (2005).
[49] C. Jing *et al.*, *IEEE Trans. Plas. Sci.* **38**(6), 1354 (2010).
[50] J. G. Power *et al.*, *Phys. Rev. Lett.* **92**, 164801 (2004).
[51] S. Antipov, C. Jing, A. Kanareykin, P. Schoessow, W. Gai, J. G. Power, S. Doran and B. Feng, in *Proc. 2011 Particle Accelerator Conference* (New York), pp. 310–312.
[52] C. Chang, J. Verboncoeur, S. Tantawi and C. Jing, *J. Appl. Phys.* **110**, 063304 (2011).
[53] O. V. Sinitsyn, G. S. Nusinovich and T. M. Antonsen, Jr., *AIP Conf. Proc.* **1507**, 505 (2012).
[54] C. Jing *et al.*, *Appl. Phys. Lett.* **103**, 213503 (2013).
[55] C. Jing, S. H. Gold, R. Fischer and W. Gai, *Appl. Phys. Lett.* **108**, 193501 (2016).
[56] C. Jing *et al.*, Short RF pulse linear collider, in *Proc. 2012 Int. Particle Accelerator Conference* (New Orleans, 2012), pp. 1924–1926.

[57] Final technical report of the SBIR project entitled "Development of a dielectric-based short RF pulse two beam accelerator prototype module." Awardee: Euclid Techlabs LLC, under Contract #DE-SC0004320.
[58] C. Jing and W. Gai, WF Note 239, http://www.hep.anl.gov/awa/links/wfnotes.htm.
[59] C. Jing *et al.*, submitted to *Phys. Rev. Lett.*
[60] Contact Dr. Wei Gai at wg@anl.gov for the AWA mission and experimental plan.
[61] J. Power *et al.*, WF Note WF-187, http://www.hep.anl.gov/awa/links/wfnotes.htm.
[62] P. B. Wilson, in *Proc. 13th SLAC Summer Inst. in Particle Physics*, SLAC Report No. 296, pp. 273–295.
[63] K. L. F. Bane *et al.*, SLAC-PUB-3662.
[64] P. Chen *et al.*, *Phys. Rev. Lett.* **56**, 1252 (1986).
[65] V. M. Tsakanov, *Nucl. Instrum. Methods. Phys. Res. A* **432**, 202 (1999).
[66] C. Jing *et al.*, *Phys. Rev. Lett.* **98**, 144801-1,-4 (2007).
[67] C. Jing *et al.*, *Phys. Rev. ST Accel. Beams* **14**, 021302 (2011).
[68] B. Jiang *et al.*, *Phys. Rev. ST Accel. Beams* **15**, 011301 (2012).
[69] F. Lemery and P. Piot, *Phys. Rev. ST Accel. Beams* **18**, 081301 (2015).
[70] R. England, J. Rosenzweig and G. Travish, *Phys. Rev. Lett.* **100**, 214802 (2008).
[71] M. Cornacchia *et al.*, *Phys. Rev. ST Accel. Beams* **9**, 120701 (2006).
[72] P. Piot *et al.*, *Phys. Rev. Lett.* **108**, 034801 (2012).
[73] P. Piot, Y.-E. Sun and M. Rihaouic, *AIP Conf. Proc.* **1086**, 677.
[74] G. H. Ha *et al.*, submitted to *Phys. Rev. Lett.*; also refer to Ha's talk at IPAC 2016.
[75] G. Andonian *et al.*, submitted to *Phys. Rev. Lett.*; also refer to Andonian's talk at the 2nd European Advanced Accelerator Concepts Workshop, Working Group #3 (13–19 Sep. 2015, La Biodola, Isola d'Elba).
[76] W. Gai *et al.*, *Phys. Rev. E* **55**, 3481 (1997).
[77] C. Li *et al.*, *Phys. Rev. ST Accel. Beams* **17**, 091302 (2014).
[78] M. C. Thompson *et al.*, *Phys. Rev. Lett.* **100**, 214801 (2008).
[79] M. E. Conde *et al.*, in *Proc. PAC07* (2007), pp. 1899–1903.
[80] B. O'Shea *et al.*, submitted to *Nat. Commun.* (2016).
[81] C. Jing *et al.*, *Phys. Rev. Lett.* **106**, 164802 (2011).
[82] S. Antipov *et al.*, *Appl. Phys. Lett.* **100**, 132910 (2012).
[83] S. Antipov *et al.*, *Diam. Relat. Mat.* **54**, 15 (2015).
[84] G. Andonian *et al.*, *Phys. Rev. Lett.* **113**, 264801 (2014).
[85] S. V. Shchelkunov *et al.*, *Phys. Rev. ST Accel. Beams* **15**, 031301 (2012).
[86] A. M. Cook *et al.*, *Phys. Rev. Lett.* **103**, 095003 (2009).
[87] G. Andonian *et al.*, *Appl. Phys. Lett.* **98**, 202901 (2011).
[88] S. Antipov *et al.*, *Phys. Rev. Lett.* **111**, 134802 (2013).
[89] S. Antipov *et al.*, *Phys. Rev. Lett.* **112**, 144801 (2014).
[90] A. Kanareykin *et al.*, in *Proc. Advanced Accelerator Concepts: 10th Workshop*, eds. C. Clayton and P. Muggli (2002), pp. 565–576.
[91] C. Jing *et al.*, A compact soft x-ray free-electron laser facility based on a dielectric wakefield accelerator. APS Light Source Note, ANL/APS/LS-332 (Aug. 2012).
[92] A. Zholents *et al.*, in *Proc. 2014 Free-Electron Laser Conference* (Basel, Switzerland, 2014), pp. 993–998.
[93] A. Zholents *et al.*, *Nucl. Instrum. Methods in Phys. Res. Sec. A*, in press (2016). DOI: 10.1016/j.nima.2016.02.003.
[94] Dielectric loaded particle accelerator (for subrelativistic beam). US Patent US 20150230326 A1.
[95] W. Li, Y. Lu, Z. He, Q. Jia and L. Wang, *Opt. Lett.* **41**(11), 2458 (2016).
[96] W. Li, Z. He and Q. Jia, *Phys. Rev. ST Accel. Beams* **17**, 042801 (2014).
[97] S. Bettoni, P. Craievich, A. A. Lutman and M. Pedrozzi, *Phys. Rev. ST Accel. Beams* **19**, 021304 (2016).

Chunguang Jing received his Ph.D. in 2005 from the Illinois Institute of Technology. Supported by the Laboratory Director Graduate Appointee, he performed his thesis work at Argonne Wakefield Accelerator Facility of Argonne National Laboratory. Since graduation, he has been a Senior Researcher at Euclid Techlabs LLC, and he was promoted to VP for Engineering in 2012. He has been a Guest Scientist at Argonne National Laboratory since 2005. Dr. Jing's research interests are focused on dielectric accelerators, wakefield acceleration, electron sources, etc.; he has been Principal Investigator of 12 DOE SBIR projects. He has authored or coauthored two book chapters, over 30 refereed journal papers, and 100 conference papers. He holds 2 US patents.

Reviews of Accelerator Science and Technology
Vol. 9 (2016) 151–163

DOI: 10.1142/S1793626816300073

Laser Technology for Advanced Acceleration: Accelerating Beyond TeV

Jonathan Wheeler and Gérard Mourou
International Center for Zetta-Exawatt Science and Technology,
École Polytechnique,
Route de Saclay, Palaiseau, 91128, France
**jonathan.wheeler@polytechnique.edu*
†gerard.mourou@polytechnique.edu

Toshiki Tajima
Department of Physics and Astronomy,
University of California,
Irvine, California, 92697, USA
ttajima@uci.edu

The implementation of the suggestion of thin film compression (TFC) allows the newest class of high power, ultrafast laser pulses (typically 20 fs at near-infrared wavelengths) to be compressed to the limit of a single-cycle laser pulse (2 fs). Its simplicity and high efficiency, as well as its accessibility to a single-cycle laser pulse, introduce a new regime of laser–plasma interaction that enhances laser acceleration. Single-cycle laser acceleration of ions is a far more efficient and coherent process than the known laser-ion acceleration mechanisms. The TFC-derived single-cycle optical pulse is capable of inducing a single-cycle X-ray laser pulse (with a far shorter pulse length and thus an extremely high intensity) through relativistic compression. The application of such an X-ray pulse leads to the novel regime of laser wakefield acceleration of electrons in the X-ray regime, yielding a prospect of "TeV on a chip." This possibility of single-cycle X-ray pulses heralds zeptosecond and EW lasers (and zeptoscience). The additional invention of the coherent amplification network (CAN) fiber laser pushes the frontier of high repetition, high efficiency lasers, which are the hallmark of needed applications such as laser-driven LWFA colliders and other, societal applications. CAN addresses the crucial aspect of intense lasers that have traditionally lacked the above properties.

Keywords: Thin film compression; laser wakefield acceleration; "TeV on a chip"; zeptoscience.

1. Introduction

Since the initial discussion by Tajima and Dawson on laser wakefield acceleration (LWFA) [1], plasma wakefields have been known to support the field gradients necessary for the next generation accelerators. The introduction of plasma as an accelerating medium rather than accelerating in a vacuum surrounded by conventional RF technology is in part forced by the current necessity to increase the accelerating gradient beyond the breakdown limit of materials maintaining the acceleration gradients. This places an upper limit on the acceleration gradient achievable and translates into accelerations to higher energies requiring ever-increasing facility sizes. Plasma wakefields do not suffer from such limitations and allow for the creation of enormous field gradients over shorter distances. Many methods have been proposed for driving such wakefields but laser-based proposals offer very compact, economical and versatile means to this goal. Figure 1 compares current acceleration techniques within an updated Livingston curve. It is apparent that present day laser-driven acceleration is not yet competitive with conventional accelerators, but the technology has still not peaked and shows great promise quickly surpassing the contemporary limits of RF technologies. Current research is seeking to address many of these issues. LWFA may also be considered a novel attempt to increase the frequency of the photon driver: from the DC electric field (such as in the Van de Graaff accelerator) to a microwave in RF accelerators, and now to a laser (optical regime). In this article, we will see potential to further push the driver frequency from an optical laser to an X-ray laser.

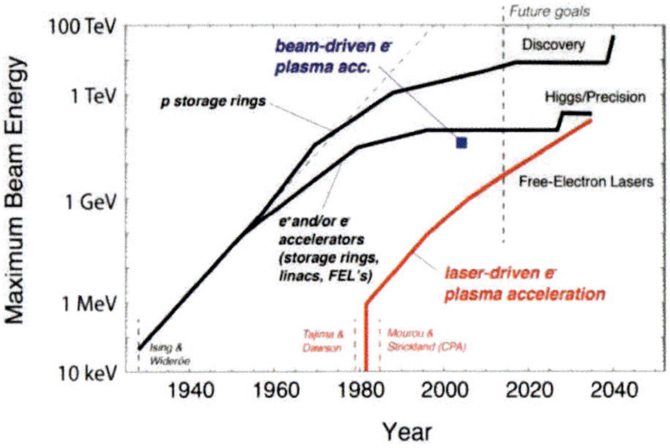

Fig. 1. Updated Livingston chart including both conventional acceleration technology and plasma-based methods. (From Ref. 63.)

2. High Power Laser Systems

With the discovery of chirped pulse amplification (CPA) [2] and later optical parametric CPA (OPCPA) [3], we have seen a considerable jump in peak power of 6–8 orders of magnitude, peaking from 10^{15} W/cm^2 to 10^{22} W/cm^2, or four orders of magnitude above the level where the electron quiver energy equals the rest mass energy, i.e. 10^{18} W/cm^2 of the electron, heralding the relativistic laser–plasma interaction and the subatomic regime, including nuclear and particle physics. The next intensity level leads to a quiver energy equal to the proton rest mass, or 10^{25} W/cm^2. It is these enormous intensities that the large scale, petawatt (PW; 10^{15} W/cm^{-3}) European laser infrastructures such as ELI, LMJ and Apollon, as well as others in Russia, China and Korea, vie to deliver [4]. However, to go directly into the nonlinear quantum-dynamic domain, we must go much higher, with intensities in the 10^{29}/cm^2 range. Based on current laser technology this promises to require either an increase in the scale of future facilities or the introduction of broadly applicable laser techniques that promise to allow an extension of the capabilities of existing facilities by efficiently compressing the pulse durations of the laser systems toward their fundamental, single-cycle wavelength limit. This has benefits both in direct acceleration in the production of secondary X-ray sources capable of yielding extreme acceleration gradients (TeV energy scales over cm length scales) within solid density plasmas. The following sections explore the creation of these sources, and subsequent sections explore their application to acceleration.

2.1. *Compression toward a single cycle*

One effective yet economical method for extending the current laser systems is postcompression of the already short pulses to their wavelength-defined fundamental temporal limit of a single cycle. This postcompression occurs after the amplification and compression steps of a CPA or OPCPA system. Here the damage threshold of the compressor for the laser system often dictates the maximum energy that can be contained with the pulse. At the same time the amplification of the pulse limits the spectral bandwidth of the pulse with typical state-of-the-art systems supporting bandwidth that can produce a 25 fs pulse. Postcompression seeks to introduce additional spectral bandwidth within the pulse — often through a nonlinear material interaction — which can then be further compressed to a shorter duration. If efficiently accomplished, it allows a dramatic increase in the laser pulse's peak intensity along with a corresponding increase in the field gradients. Historically, such compression has been a challenge at high energy. Current laser pulse compression techniques rely on a hollow core capillary filled with noble gases to allow sufficient interaction length to produce the necessary spectral broadening through self-phase modulation (SPM). The postcompression is then

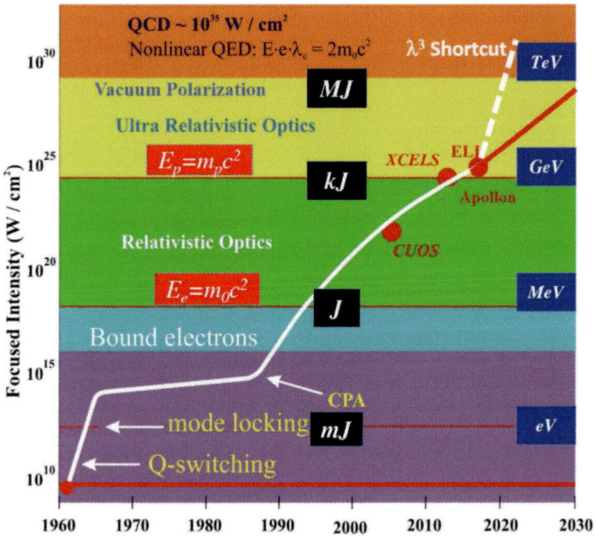

Fig. 2. Laser intensity through the years. The dashed line corresponds to what could be obtained with significant increases in beam size or by increasing the number of beams. The red dashed line corresponds to the "short cut" obtained through the compression technique to sub-attosecond pulse durations. (Adapted from Ref. 64.)

accomplished through subsequent spectral dispersion management using chirped mirrors [5]. However, this technique can accommodate only pulses with low energy, typically up to the mJ level with Gaussian spatial profiles due to the need to efficiently couple within the hollow core capillary. As a consequence this is not easily extended to joule level pulse energies, which typically have beam modes that are super-Gaussian or flat-top. Observation of the sharp transition in the compressible energy as shown in Fig. 3 due to the technological transition from the previously used fused silica fiber [6, 7] to hollow core capillaries suggests that with a new technique a dramatic increase in the compressible energy is feasible. To go higher in energy, bulk compression was attempted by Corkum and Rolland [8]. However, the material nonlinearity is strongly influenced by the Gaussian intensity profile of the beam and makes the pulse impossible to compress uniformly except for the central area near its peak intensity. This non-uniformity has to date greatly diminished the attractiveness of bulk compression.

2.1.1. *Thin film compression*

To solve the uniformity problem, the use of modern PW lasers exhibits pulses with top hat spatial profiles in amplitude and phase in order to maximize the energy extraction within the laser amplification chain. These laser pulses are ideal for a thin film compressor (TFC) [9] which introduces

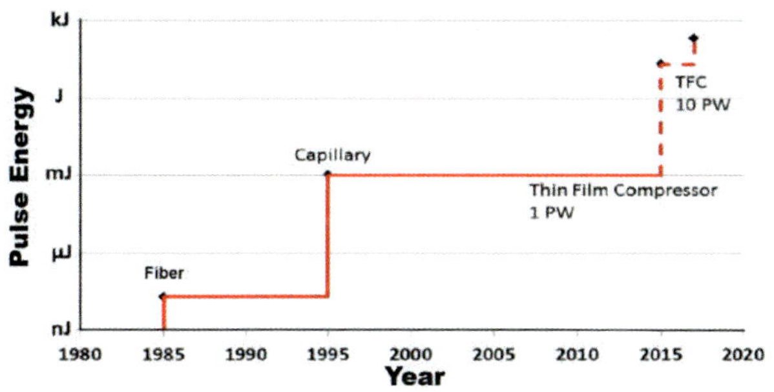

Fig. 3. Evolution of pulse compression in the single-cycle regime. The techniques rely on the nonlinear response to a high intensity pulse within first a single mode fiber and later a noble gas within a hollow core capillary. Each demonstrated that a 20 fs pulse can be further compressed to sub-5 fs within their working energy ranges. The thin film technique is proposed to make the next leap forward in energy for NIR pulse compression. (From Ref. 64.)

inexpensive sheets of plastic or, alternatively, submillimeter glass plates, of uniform thickness — but not necessarily flat — within the PW laser's beam transport. (See the left half of Fig. 4.) The typical transport intensities are sufficient to produce near-constant SPM across the beam while minimizing energy losses in the transparent thin (submillimeter) bulk films. Through the careful balance of the chirp produced during the SPM and dispersion within the thin film and group velocity dispersion of the resultant pulse with chirped mirrors, there is the possibility of producing single-cycle pulses at the tens-of-joules level after several stages of post-compression. Here the chirped mirrors allow for high reflectivity while introducing a second order phase sufficient for the application — typically on the order of tens to hundreds of fs^2. Thus, the expected efficiency for one stage providing a factor of 3 in temporal compression is on the order of 90% transmission in energy. A typical Ti:sapphire 1 PW laser at 30 fs is reducible by a two-stage TFC into a single cycle, 2.5 fs at 10 PW. Through appropriate configuration of thin film and chirped mirror combination, it seems likely that many of the existing laser facilities can be updated to include single-cycle-like capabilities.

2.2. *X-ray generation*

The production of high energy, single-cycle pulses allows another technological leap forward in X-ray pulse production. It has already been demonstrated in gas sources that isolated X-ray pulses can be produced from few-cycle near-infrared (NIR), mJ pulses [10]. The production via the rescattering process in gas targets typically saturates at intensities of around $10^{16}\,\mathrm{W/cm^2}$, but harmonics from solid targets show no such saturation and are producible over a wide range of intensities [11]. The idea of coherent, X-ray generation from a relativistic oscillating mirror (ROM), first proposed by Bulanov *et al.* [12], allows the upconversion of the NIR laser photons to keV or even MeV photons.

2.2.1. *Extreme light generation from solids*

Note that by tightly focusing a single-cycle NIR pulse, it forms an ellipsoid in space with λ^3 dimensions. In order to further compress the NIR single-cycle pulse to the attosecond or zeptosecond regime (corresponding to keV or MeV photon energies,

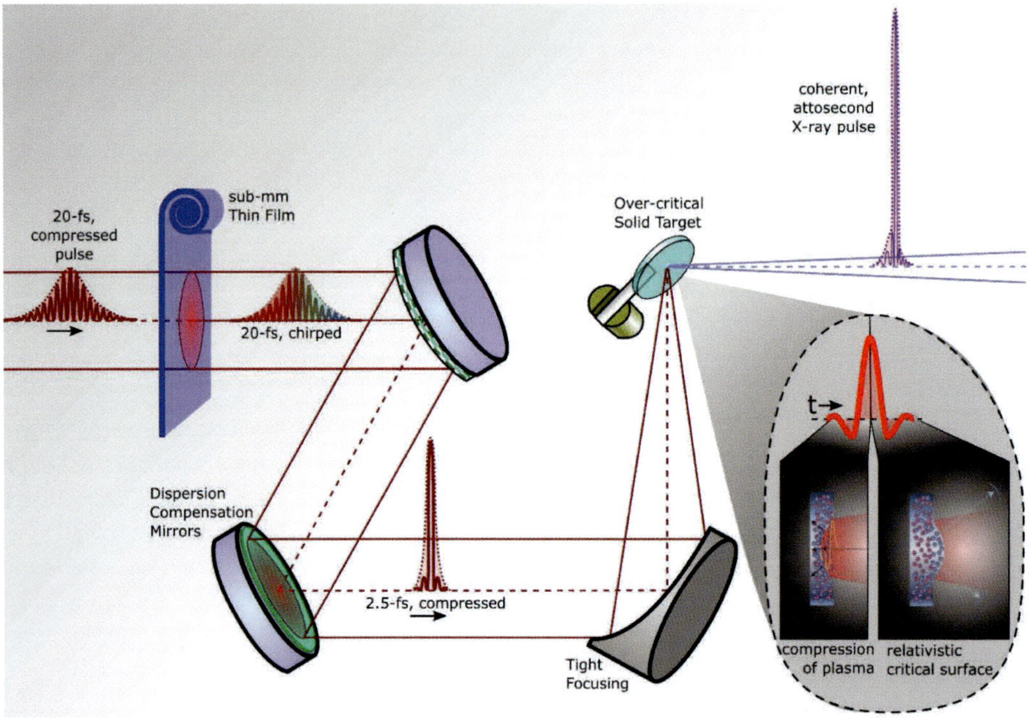

Fig. 4. Two stages of compression. The thin film compression stage relies on the interplay between the spectral broadening produced by SPM and the GVD necessary for stretching the pulse in a large aperture film. The linearly frequency-chirped pulse with increased spectral content compared to the initial pulse can be compressed using dispersive elements such as chirped mirrors. The second stage of compression requires delivery of the single-cycle pulse with appropriate tight focusing to apply a relativistic intensity field to a solid target plasma to upconvert to an X-ray pulse. (From Ref. 64.)

respectively), a laser-driven critical surface within the oversolid density plasma moves in and out under the influence of the very strong incident pulse. At this point the intensity can be in the $10^{23}\,\mathrm{W/cm^2}$ range associated with a formidable light pressure and can move the critical surface at relativistic speeds. In this λ^3 relativistic regime Naumova *et al.* [13] predicts a reflected pulse duration τ_p — compressed by the relativistic motion of the mirror toward the pulse — scaling as

$$\tau_p = 600\,\mathrm{attoseconds}/a_0. \tag{1}$$

Here a_0 is the normalized vector potential of the laser pulse, which is unity at $10^{18}\,\mathrm{W/cm^2}$ and scales as the square root of the intensity. At $10^{23}\,\mathrm{W/cm^2}$, $a_0 \sim 500$, a pulse of 1 as can be produced. Even 1 J of reflected energy delivered by an attosecond pulse is sufficient for exawatt power. This leads to the possibility of zeptosecond scale physics with applications for nuclear photonics. This also provides an unprecedented driver for LWFA, due to the shorter wavelengths and the ability to further compress the beam to even greater peak intensities although it is already at the Schwinger intensity of $10^{29}\,\mathrm{W/cm^2}$.

3. Applications to Acceleration

Over the last decade, rapid progress has been achieved in realizing useful laser plasma accelerators (LPAs) with energies of up to 4.2 GeV, using lasers of one to tens of Joules with tens-of-fs pulse durations in the plasma density range of 10^{17}–$10^{19}\,\mathrm{cm^{-3}}$, demonstrating quasi-monoenergetic electron beams with charge of 1–100 s pC and qualities, e.g. emittances of around 1 mm mrad, energy spreads of 0.3% for 100–200 MeV beams, and bunch durations of the order of 1 fs. Recently there has been a growing interest in laser plasma accelerators driven by PW class lasers, whereby high quality electron beams can be accelerated to multi-GeV energies in a cm scale plasma thanks to the LWFA mechanism, as reported so far, e.g. 1.8 GeV driven by 130 TW at SIOM, 2 GeV driven by 620 TW at TEXAS, 3 GeV driven by 210 TW at GIST, and 4.2 GeV driven by 380 TW at LBNL-BELLA. To date, most of experiments

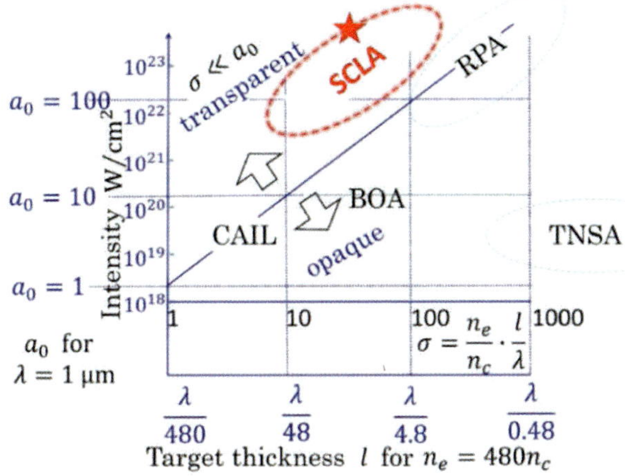

Fig. 5. Proton energy gain in the regimes of single-cycled laser acceleration (SCLA) versus RPA. For the same energy of the laser, SCLA is far more efficient for a far larger energy gain, while the normalized thickness for the optimal acceleration is much smaller. (Adapted from Ref. 39.)

on laser plasma accelerators have been carried out by employing ultrashort pulse lasers with duration $\tau_p = 30-80$ fs, focused onto a short scale plasma target such as a mm scale gas jet and a cm scale plasma channel at plasma densities in the range of $n_e = 10^{18}-10^{19}$ cm^{-3}, where very large amplitude plasma waves of the order of 100 GV/m are excited and trap energetic electrons to be efficiently accelerated in a [illegible] to high energies of the order of several GeV.

Laser-driven ion acceleration has been vigorously investigated both theoretically and experimentally in recent years [12, 13]. The earliest and most-discussed laser ion acceleration regime of target normal sheath acceleration (TNSA) [16–21] has a large divergence of the ion beam and almost a 100% spread in energy. In TNSA the ion energy is tied to the energy of spreading electrons that are heated by the injected laser and, broadly speaking, this leads to a relatively weak laser intensity dependence (such as $I^{1/3}$ to $I^{1/2}$, depending on the exact conditions). Here the laser intensity I is proportional to a_0^2, with $a_0 = eE_0/m_e\omega c$ being the normalized vector potential of the laser, where E_0, ω, c, m_e, and e are the laser electric field, frequency, speed of light in vacuum, electron mass, and charge, respectively. In many modern laser-based ion acceleration experiments, the lasers are in the domain of $a_0 > 1$. Substantial reports have been devoted to decreasing the spread in the ion energy, such as restricting the ions within a small region where the sheath field can be treated as homogenous [22–24] or filtering by a laser-driven plasma lens [25]. These methods rely on complicated target fabrication techniques, and the conversion efficiency is still quite low. The breakout afterburner (BOA) acceleration method can efficiently accelerate the ions with >0.5 GeV carbon ions being demonstrated. However, a laser pulse with ps duration is required and the energy spread of the ion beam is still quite broad [26]. A more sophisticated path is presented by radiation pressure acceleration (RPA), where an ultraintense laser pulse irradiates a thin plane foil [27–32]. The ion energy gain is greater than that of TNSA, and its intensity dependence is more favorable, with a proportionality of I. While the ion energy spectrum is much narrower than that of TNSA, the required laser intensity is huge and is typically on the order of 10^{23} W/cm^2. The coherent acceleration of ions by laser (CAIL) regime, on the other hand, requires less laser power — in line with what is already currently available [33–35]. The ion energy depends on $I^{1/2}$ for $a_0 \gg 1$ and yields a quasi-monoenergetic ion spectrum. Thus, the CAIL mechanism sits between TNSA and RPA. In many ion acceleration experiments the laser contrast requirement remains a challenge for the state of the art in laser technology. In addition, hole-boring effects and transverse effects such as Rayleigh–Taylor instabilities quickly disrupt the ion acceleration process and seriously reduce its efficiency [36–38].

A novel configuration is to apply a single-cycle laser pulse to drive ion acceleration, and it shows great promise as a desirable regime among the conditions for LPA.

3.1. *Single-cycle laser acceleration*

A new highly efficient and instability-free ion acceleration mechanism uses contemporarily available PW class lasers [39]. This regime of acceleration is different and much more efficient, with a far sharper monoenergetic spectrum of ions, unlike the classical TNSA, CAIL, and BOA. In the limit of single-cycle laser pulses, the acceleration becomes more direct and coherent as the ponderomotive acceleration term $\langle \boldsymbol{v} \times \boldsymbol{B} \rangle$ no longer needs averaging. In the case of multicycle laser pulses, the electron acceleration by the ponderomotive force must be averaged over the number of cycles. The former single-cycle situation introduces more coherent electron acceleration and sharper electron layer formation. This single-cycle laser acceleration (SCLA) regime permits a thinner optimal target thickness and leads to a more coherent ion layer following the accelerated electron layer. Because of this our regime takes far smaller laser energy than that required in the known regimes mentioned above. In the present regime, when a single-cycle Gaussian pulse with intensity 10^{23} W/cm^2 is incident on a 50 nm planar CH foil, the ponderomotive force of the laser pulse pushes forward an isolated relativistic electron bunch and, in turn, the resultant longitudinal electrostatic field accelerates the protons (inset of Fig. 6). With a thin target, our mechanism can coherently and stably accelerate ions over a significant distance without suffering from the typical transverse instabilities that arise under previously considered conditions. This uniquely stable acceleration structure is capable of maintaining a highly monoenergetic ultrashort (~fs) GeV proton bunch.

3.2. *Solid density laser wakefield acceleration*

High frequency photons have an advantage in driving wakefields in high density matter. The current approach to increasing the LWFA energy gain is to make the plasma density lower for a fixed optical laser frequency. This is based on the well-known laser wakefield scaling [1]. In LWFA, the energy gain is limited by the critical density of the plasma, with the high intensity LWFA energy gain given by

$$\varepsilon_e = a_0^2 mc^2 \left(\frac{n_c}{n_e}\right),$$

where n_e the electron density. The critical density of a plasma, n_c, is defined by the laser frequency and increases with increasing photon frequency or energy. For 1 eV optical photons, n_c is approximately 10^{21} cm^{-3}, while for X-ray photons of 10 keV n_c is on the order of 10^{29} cm^{-3} — well above solid densities. Therefore, a new path toward higher energy gain (and yet smaller, more compact accelerators) is to increase the photon energies to the level of X-rays [40]. While the ratio n_c/n_e remains large, we need not make n_e lower. Shifting to a higher photon energy permits an impressive energy gain as an important benefit of the higher density plasma which supports a larger acceleration gradient. X-ray-driven wakefield acceleration allows acceleration to larger energies to be achieved over shorter distances. Rather than acceleration distances of meters required to

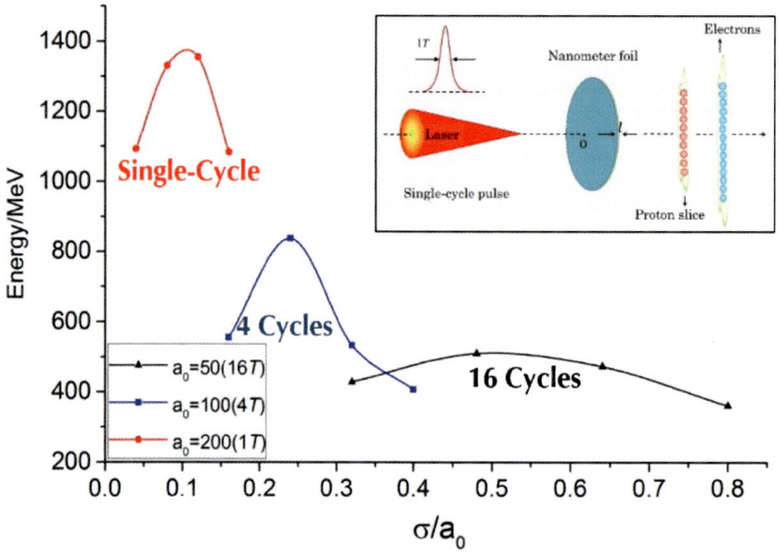

Fig. 6. The resulting proton energies with varying σ/a_0 (the normalized thickness σ divided by the normalized vector potential a_0; see also Fig. 5). The black line indicates a laser pulse with $a_0 = 50$, and pulse duration $\tau_p = 16T$ (T is the laser period); the blue and red lines indicate laser pulses with $a_0 = 100$ ($\tau_p = 4T$) and $a_0 = 200$ ($\tau_p = 1T$), respectively. Inset: Schematic view of the single-cycle pulse ion acceleration regime. The single-cycle laser pulse is focused at point o onto a thin CH foil target of thickness l. The ponderomotive force drives a coherent bunch of electrons in the forward direction beyond the limit of the target. The resulting longitudinal electric field accelerates a coherent slice of protons that trail the electrons leaving from the back surface of the target. (Adapted from Ref. 39.)

reach hundreds-of-GeV energy scales with optical photons, estimates show keV X-rays achieving acceleration of TeV energies over centimeter distances. This is 10^3 times the previous results achieved for optical lasers in gases and 10^6 times that of current RF technology. These are energy levels achieved only at CERN in Geneva, though not over kilometers but over a few-centimeter focus and supported by a laser facility such as the PW level facilities currently being built.

Nanomaterials featuring nanochannels are an ideally suited choice of material in this high density regime [40]. (See Fig. 7.) The solid density walls of the channels support the plasma density required for TeV acceleration gradients, while the nanochannel is of the size scale to act as a low density guide for the accelerated particles and minimizes scattering that would otherwise occur within a uniform high density plasma.

Recent simulations of X-ray driven wakefields for electron acceleration in a nanotube plasma target have been conducted [41]. In this it is shown that the nanostructure helps to radially confine the wakefields, and the expected emittance (longitudinal and transverse) is reduced from the parameters achieved from conventional regimes. While there are many recent works [42–46] on multistage wakefield acceleration in gas plasmas, the shift to an X-ray drive allows for the possibility of a single-stage wakefield acceleration. This is due to the scaling in Eq. (2), the increased value of n_c/n_e in Eq. (2) even for $n_e = 10^{23}\,\mathrm{cm}^{-3}$ (solid density in a nanomaterial).

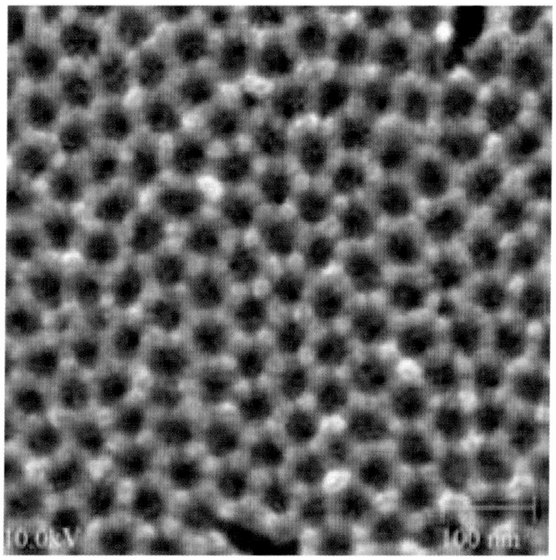

Fig. 7. Porous alumina on a silica substrate. (As presented in Ref. 65.)

4. High Average Power

Sources of ultrashort, high peak power optical pulses show great promise for laser plasma acceleration and high field physics. Considerable progress has been made over the last decade regarding high peak power femtosecond sources reaching up to the PW level.

The architecture for such lasers has been based on flash-lamp-pumped Ti:sapphire and neodymium-doped materials, which are characterized by very low energy efficiency and thermal management requirements that limit the typical repetition rates to sub-Hz. Diode-pumped solid state lasers enable better performance, as the pumping radiation matches with the absorption line of the laser material. Here, the repetition rate exceeds tens of Hz and the energetic efficiency can reach 10%.

However, many applications, such as laser particle acceleration, demand much higher pulse repetition rates in the kHz domain. Hence, the motivation for the development of a high repetition rate, high peak power laser system is to provide such interesting applications and the means to advance from proof-of-principle experiments to real-world implementation. Since the first design of an LWFA-based collider [42], studies have shown that high repetition and high luminosity requirements for a collider demand a laser system to have a high rep rate and high efficiency. The report in 2011 of the joint task force between ICUIL and ICFA shows this point [47]. This prompted the innovation of the high intensity laser to be equipped with the required high rep rate and high efficiency characteristics.

4.1. *The CAN fiber laser system*

Over recent decades, novel architectures of the gain medium have been developed to overcome thermo-optical issues, e.g. the thin disk, slab, and fiber laser concepts [48–50]. Among all these advanced geometries, the fiber laser offers some unique features related to efficiency, gain bandwidth, compactness, simplicity, alignment insensitivity, and average power scalability. State-of-the-art fiber lasers and amplifiers are capable of 10 kW power levels both in continuous wave operation [51] and in average output power for ultrashort pulse operation [50].

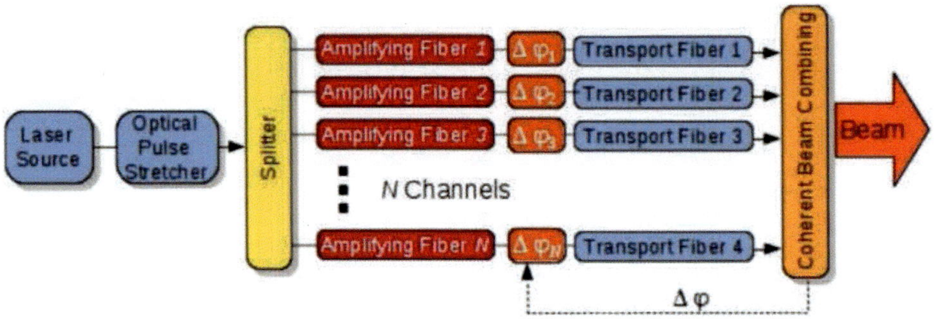

Fig. 8. Schematic of the coherent amplification network (CAN) laser architecture. A femtosecond seed laser pulse is repeatedly split and amplified through this network composed of amplifying and transport fibers. The resulting pulses are then recombined coherently. (Figure from Ref. 60.)

However, the fiber geometry also enhances nonlinear effects by restricting the light to propagate under tight confinement over long interaction lengths. To reduce the impact of this traditional performance limitation, temporal and spatial scaling of the laser pulses is required in order to reduce the peak intensity. The temporal scaling can be achieved by the well-known technique of chirped pulse amplification [2]. In this technique temporally stretched pulses are amplified so that during amplification the peak power of the pulses is considerably reduced and nonlinear effects are mitigated.

Ultrafast fiber laser systems based on such advanced fibers are currently able to deliver mJ pulse energy and GW peak power at high repetition rates of kHz [52]. However, further increases in the performance of single-fiber laser systems become more and more challenging. A promising general approach to overcoming all the limitations of a single emitter system is spatially separated amplification followed by coherent combination of several individual emitters [53, 54]. This is depicted in Fig. 8, where a femtosecond seed laser pulse is demultiplexed into N pulses, which are subsequently amplified in Yb-doped monomode fibers.

This revolutionary laser architecture is known as the coherent amplification network (CAN), originally proposed by Mourou *et al.* [55] as a result of the international ICAN project. It proposes a short pulse high average power and a high peak power laser source composed of several phased combined fibers. Initial work has demonstrated continuous wave coherent combination of 64 individual fibers [56] and 16-fiber [57] amplifiers phased with $\lambda/20$ rms phasing error in a loop time of 1 ms. It is worth mentioning that the combined peak power has reached 5.4 GW [58] and average powers near the kW level [59]. Scaling of a CAN laser to thousands of fibers is needed in order to reach pulse energies of tens of joules and near the petawatt peak power. Such an efficient laser driver would enable many societal applications, including particle accelerators and colliders.

The trend to increase fiber numbers raises several points, as studied in Ref. 60. There is a need to investigate new scalable architectures where heat removal is managed. So far, the geometry of fiber-based lasers has been restricted to planar arrangements, but a proposal has been made for a spherical arrangement that enables removing transport fibers while mitigating thermal effects. (See Fig. 9.) The entire CAN system evolves with increasing scale and this has important implications for stability and noise characteristics, which models suggest are improved with a larger number of contributing fibers.

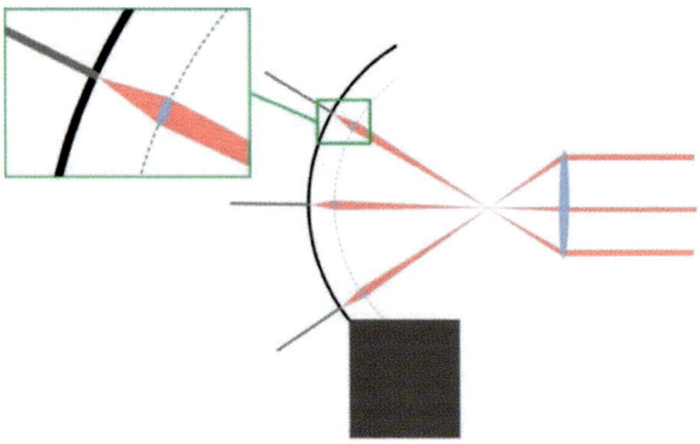

Fig. 9. CAN laser combination. Fiber arrangement on a sphere seen from the side. Beamlets are focused by a spherical microlens array and collimated by a lens. Only three beams are represented here, for clarity. (Figure from Ref. 60.)

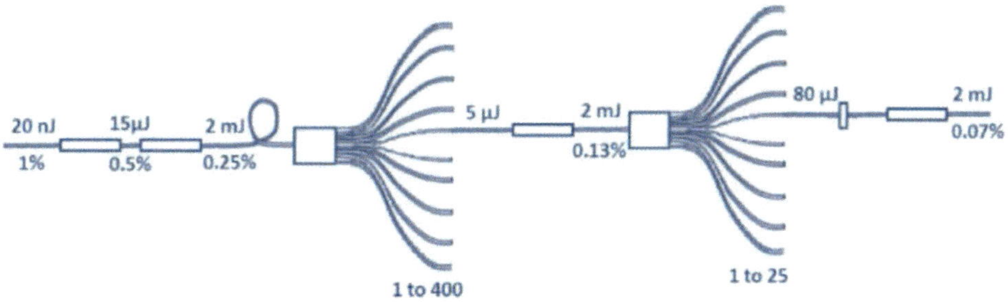

Fig. 10. Scheme of the amplifying silica fiber network. The energy at the output of each amplifier is such that the ratio of damage threshold energy to saturation energy is 1:5. In this condition rms shot-to-shot variation is divided by 2. (Originally from Ref. 60.)

Such a laser system with high average/peak power is capable of exhibiting incomparable amplitude and pointing stability, as well as offering a heuristic capability. The number of fibers employed is important for producing peak and average power, but it also confers on the system an amplitude and pointing stability enhanced by $N^{1/2}$, where N is the number of fibers. As highlighted in Fig. 10, if a single fiber already exhibits an unsurpassed amplitude and pointing stability, for N being of the order of 1000–10,000, a stability improvement of 30–100 is expected. In addition, since each fiber could be addressed and corrected individually with a kHz bandwidth, a feedback for the whole system could seek the best beam characteristics based on the best electron beam or X-ray output. It is paramount for FEL applications that require the most stable output possible.

5. Conclusion

CPA (1985) [2] has answered the demand of the newly posed high intensity (at the relativistic intensity) with an ultrafast (tens of fs) laser pulse to enable the robust and intense wakefield excitation [1]. This was shown by the scientific first proof-of-principle demonstration [61] and later realization of more realistic and/or advanced experiments in the subsequent decade. Beyond these lies an approach of LWFA to societal applications and an eventual embodiment as a collider. For these, the laser must acquire specific features such as high rep rate and high efficiency. Thus, another laser innovation has occurred to meet these demands by way of the CAN fiber laser [55]. This new technology is in the developmental stage via the XCAN project. We envision that such a project will usher in more fully integrated collaboration between the laser community and the accelerator community (particularly toward the realization of an LWFA collider).

Meanwhile, in 2000, the first laser-driven ion acceleration was experimentally shown, as reviewed by Ref. 14. As is clear from the experience of proton acceleration by laser, the laser needs to acquire different and often more demanding properties than LWFA. From this research and demand our TFC technique for intense lasers has emerged [9] to allow the production of even shorter pulses than what CPA alone is capable of, even perhaps to the single-cycle regime. This invention may be characterized by its surprising simplicity. (We note that a similar but distinct TFC technique in a less intense regime has been invented and demonstrated [62].) We demonstrated theoretically that the employment of such a technique would lead to far superior ion acceleration by a laser in the regime of SCLA [39]. We learned that the single-cycle ultrashort laser pulse has such coherence (and getting rid of oscillations in the electron motion, which leads to efficient, higher energy, and more pristine ion beam realization). Thus, a path for laser-driven ion acceleration is extremely attractive for future research.

This leap of technology further ushers in a revolution toward the single-cycle X-ray laser through relativistic compression [13]. This possibility was immediately exploited to lead to the possibility of "TeV on a chip" or acceleration to TeV energies over cm length scales [40]. This opens up the science at the level of the zeptosecond ("zeptoscience"), the X-ray laser at the intensity of exawatts (EW), and extremely compact TeV accelerators in addition to other secondary beams of various new properties.

Taken together, this sets an entirely new horizon of science. In such attributes zeptoscience can directly probe the inner structure of a nucleus by EM photons (nuclear photonics), not to mention the new possibility for particle physics.

Acknowledgments

J. Wheeler acknowledges funding support from Extreme Light Infrastructure — Nuclear Physics (ELI-NP), a project cofinanced by the Romanian Government and European Union through the European Regional Development Fund.

References

[1] T. Tajima and J. M. Dawson, Laser electron accelerator, *Phys. Rev. Lett.* **43**(4), 267–270 (1979).
[2] D. Strickland and G. Mourou, Compression of amplified chirped optical pulses, *Opt. Commun.* **56**(3), 219–221 (1985).
[3] I. Ross, P. Matousek, M. Towrie and A. Langley, The prospects for ultrashort pulse duration and ultrahigh intensity using optical parametric chirped pulse amplifiers, *Optics* **14**, 125 (1997).
[4] M. N. Quinn, J. A. Wheeler, J. L. Miquel, V. Zamfir and A. M. Sergeev, The IZEST Framework, *Eur. Phys. J. Spec. Top.* **223**(6), 985–992 (2014).
[5] M. Nisoli, S. De Silvestri and O. Svelto, Generation of high energy 10 fs pulses by a new pulse compression technique, *Appl. Phys. Lett.* **68**(20), 2793 (1996).
[6] D. Grischkowsky, Optical pulse compression based on enhanced frequency chirping, *Appl. Phys. Lett.* **41**(1), 1 (1982).
[7] R. L. Fork, C. H. Brito Cruz, P. C. Becker and C. V. Shank, Compression of optical pulses to six femtoseconds by using cubic phase compensation, *Opt. Lett.* **12**(7), 483 (1987).
[8] C. Rolland and P. B. Corkum, Compression of high-power optical pulses, **5**(3), 641–647 (1988).
[9] G. Mourou, S. Mironov, E. Khazanov and A. Sergeev, Single cycle thin film compressor opening the door to zeptosecond–exawatt physics, *Eur. Phys. J. Spec. Top.* **223**(6), 1181–1188 (2014).
[10] M. Drescher, X-ray pulses approaching the attosecond frontier, *Science* (80–) **291**(5510), 1923–1927 (2001).
[11] F. Quéré, C. Thaury, H. George, J. P. Geindre, E. Lefebvre, G. Bonnaud, S. Hüller, P. Monot and P. Martin, Basic mechanisms of laser high-order harmonic generation from plasma mirrors, *J. Mod. Opt.* **55**(16), 2711–2721 (2008).
[12] S. V. Bulanov, N. M. Naumova and F. Pegoraro, Interaction of an ultrashort, relativistically strong laser pulse with an overdense plasma, *Phys. Plasmas* **1**(3), 745 (1994).
[13] N. M. Naumova, J. A. Nees, I. V. Sokolov, B. Hou and G. A. Mourou, Relativistic generation of isolated attosecond pulses in a $\lambda\wedge\{3\}$ focal volume, *Phys. Rev. Lett.* **92**(6), 3–6 (2004).
[14] H. Daido, M. Nishiuchi and A. S. Pirozhkov, Review of laser-driven ion sources and their applications, *Rep. Prog. Phys.* **75**(5), 56401 (2012).
[15] A. Macchi, M. Borghesi and M. Passoni, Ion acceleration by superintense laser–plasma interaction, *Rev. Mod. Phys.* **85**(2), 751–793 (2013).
[16] R. A. Snavely, M. H. Key, S. P. Hatchett, I. E. Cowan, M. Roth, T. W. Phillips, M. A. Stoyer, E. A. Henry, T. C. Sangster, M. S. Singh, S. C. Wilks, A. MacKinnon, A. Offenberger, D. M. Pennington, K. Yasuike, A. B. Langdon, B. F. Lasinski, J. Johnson, M. D. Perry and E. M. Campbell, Intense high-energy proton beams from petawatt-laser irradiation of solids, *Phys. Rev. Lett.* **85**(14), 2945–2948 (2000).
[17] A. J. Mackinnon, Y. Sentoku, P. K. Patel, D. W. Price, S. Hatchett, M. H. Key, C. Andersen, R. Snavely and R. R. Freeman, Enhancement of proton acceleration by hot-electron recirculation in thin foils irradiated by ultraintense laser pulses, *Phys. Rev. Lett.* **88**, 215006 (2002).
[18] P. Mora, Plasma expansion into a vacuum, *Phys. Rev. Lett.* **90**(18), 185002 (2003).
[19] R. Hörlein, Y. Nomura, P. Tzallas, S. G. Rykovanov, B. Dromey, J. Osterhoff, Z. Major, S. Karsch, L. Veisz, M. Zepf, D. Charalambidis, F. Krausz and G. D. Tsakiris, Temporal characterization of attosecond pulses emitted from solid-density plasmas, *New J. Phys.* **12**(4), 43020 (2010).
[20] E. L. Clark, K. Krushelnick, M. Zepf, F. N. Beg, M. Tatarakis, A. Machacek, M. I. K. Santala, I. Watts, P. A. Norreys and A. E. Dangor, Energetic heavy-ion and proton generation from ultraintense laser–plasma interactions with solids, *Phys. Rev. Lett.* **85**(8), 1654–1657 (2000).
[21] A. Maksimchuk, S. Gu, K. Flippo and D. Umstadter, Forward ion acceleration in thin films driven by a high-intensity laser, *Phys. Rev. Lett.* **84**(18), 4108–4111 (2000).
[22] H. Schwoerer, S. Pfotenhauer, O. Jäckel, K.-U. Amthor, B. Liesfeld, W. Ziegler, R. Sauerbrey, K. W. D. Ledingham and T. Esirkepov, Laser-plasma acceleration of quasi-monoenergetic protons from microstructured targets, *Nature* **439**(7075), 445–448 (2006).
[23] B. M. Hegelich, B. J. Albright, J. Cobble, K. Flippo, S. Letzring, M. Paffett, H. Ruhl, J. Schreiber, R. K. Schulze and J. C. Fernández, Laser acceleration of quasi-monoenergetic MeV ion beams, *Nature* **439**(7075), 441–444 (2006).
[24] J. Fuchs, P. Antici, E. D'Humières, E. Lefebvre, M. Borghesi, E. Brambrink, C. A. Cecchetti, M. Kaluza, V. Malka, M. Manclossi, S. Meyroneinc, P. Mora, J. Schreiber, T. Toncian, H. Pépin and

P. Audebert, Laser-driven proton scaling laws and new paths towards energy increase, *Nat. Phys.* **2**(1), 48–54 (2006).

[25] T. Toncian, M. Borghesi, J. Fuchs, E. D'Humières, P. Antici, P. Audebert, E. Brambrink, C. A. Cecchetti, A. Pipahl, L. Romagnani and O. Willi, Ultrafast laser-driven microlens to focus and energy-select mega-electron volt protons, *Science (80-)* **312**(5772), 410–413 (2006).

[26] B. M. Hegelich, D. Jung, B. J. Albright, J. C. Fernandez, D. C. Gautier, C. Huang, T. J. Kwan, S. Letzring, S. Palaniyappan, R. C. Shah, H.-C. Wu, L. Yin, A. Henig, R. Hörlein, D. Kiefer, J. Schreiber, X. Q. Yan, T. Tajima, D. Habs, B. Dromey and J. J. Honrubia, Experimental demonstration of particle energy, conversion efficiency and spectral shape required for ion-based fast ignition, *Nucl. Fus.* **51**(8), 83011 (2011).

[27] T. Esirkepov, M. Borghesi, S. V. Bulanov, G. Mourou and T. Tajima, Highly efficient relativistic-ion generation in the laser-piston regime, *Phys. Rev. Lett.* **92**, 175003-1 (2004).

[28] X. Q. Yan, C. Lin, Z. M. Sheng, Z. Y. Guo, B. C. Liu, Y. R. Lu, J. X. Fang and J. E. Chen, Generating high-current monoenergetic proton beams by a circularly polarized laser pulse in the phase-stable acceleration regime, *Phys. Rev. Lett.* **100**(13), 1–4 (2008).

[29] A. Macchi, F. Cattani, T. V. Liseykina and F. Cornolti, Laser acceleration of ion bunches at the front surface of overdense plasmas, *Phys. Rev. Lett.* **94**(16), 2–5 (2005).

[30] M. Chen, A. Pukhov, T. P. Yu and Z. M. Sheng, Enhanced collimated GeV monoenergetic ion acceleration from a shaped foil target irradiated by a circularly polarized laser pulse, *Phys. Rev. Lett.* **103**(2), 24801 (2009).

[31] B. Qiao, M. Zepf, M. Borghesi and M. Geissler, Stable GeV ion-beam acceleration from thin foils by circularly polarized laser pulses, *Phys. Rev. Lett.* **102**(14), 145002 (2009).

[32] B. Qiao, M. Zepf, M. Borghesi, B. Dromey, M. Geissler, A. Karmakar and P. Gibbon, Radiation-pressure acceleration of ion beams from nanofoil targets: The leaky light-sail regime, *Phys. Rev. Lett.* **105**(15), 8–11 (2010).

[33] A. Henig, S. Steinke, M. Schnürer, T. Sokollik, R. Hörlein, D. Kiefer, D. Jung, J. Schreiber, B. M. Hegelich, X. Q. Yan, J. Meyer-Ter-Vehn, T. Tajima, P. V. Nickles, W. Sandner and D. Habs, Radiation-pressure acceleration of ion beams driven by circularly polarized laser pulses, *Phys. Rev. Lett.* **103**(24), 1–5 (2009).

[34] T. Tajima, D. Habs and X. Yan, Laser acceleration of ions for radiation therapy, *Rev. Accel. Sci. Technol.* **2**(1), 201–228 (2009).

[35] X. Q. Yan, T. Tajima, M. Hegelich, L. Yin and D. Habs, Theory of laser ion acceleration from a foil target of nanometer thickness, *Appl. Phys. B Lasers Opt.* **98**(4), 711–721 (2010).

[36] O. Klimo, J. Psikal, J. Limpouch and V. T. Tikhonchuk, Monoenergetic ion beams from ultrathin foils irradiated by ultrahigh-contrast circularly polarized laser pulses, *Phys. Rev. ST Accel. Beams* **11**(3), 31301 (2008).

[37] A. P. L. Robinson, A. Robinson, M. Zepf, M. Zepf, S. Kar, S. Kar, R. Evans, R. Evans, C. Bellei and C. Bellei, Radiation pressure acceleration of thin foils with circularly polarized laser pulses, *New J. Phys* **10**(3), 13021 (2008).

[38] F. Pegoraro and S. V. Bulanov, Photon bubbles and ion acceleration in a plasma dominated by the radiation pressure of an electromagnetic pulse, *Phys. Rev. Lett.* **99**(6), 1–4 (2007).

[39] M. L. Zhou, X. Q. Yan, G. Mourou, J. A. Wheeler, J. H. Bin, J. Schreiber and T. Tajima, Proton acceleration by single-cycle laser pulses offers a novel monoenergetic and stable operating regime, *Phys. Plasmas* **23**(4), 43112 (2016).

[40] T. Tajima, Laser acceleration in novel media, *Eur. Phys. J. Spec. Top.* **223**(6), 1037–1044 (2014).

[41] X. Zhang, T. Tajima, D. Farinella, Y. Shin, G. Mourou, J. Wheeler, P. Taborek, P. Chen, F. Dollar and B. Shen, Particle-in-cell simulation of X-ray wakefield acceleration and betatron radiation in nanotubes, *Phys. Rev. Accel. Beams* **19**(10), 101004 (2016).

[42] M. Xie, T. Tajima, K. Yokoya and S. Chattopadhyay, Studies of laser-driven 5 TeV, in *The Seventh Workshop on Advanced Accelerator Concepts* (1997), Vol. 398(No. 1), pp. 233–242.

[43] J. S. Liu, C. Q. Xia, W. T. Wang, H. Y. Lu, C. Wang, A. H. Deng, W. T. Li, H. Zhang, X. Y. Liang, Y. X. Leng, X. M. Lu, C. Wang, J. Z. Wang, K. Nakajima, R. X. Li and Z. Z. Xu, All-optical cascaded laser wakefield accelerator using ionization-induced injection, *Phys. Rev. Lett.* **107**(3), 2–5 (2011).

[44] B. B. Pollock, C. E. Clayton, J. E. Ralph, F. Albert, A. Davidson, L. Divol, C. Filip, S. H. Glenzer, K. Herpoldt, W. Lu, K. A. Marsh, J. Meinecke, W. B. Mori, A. Pak, T. C. Rensink, J. S. Ross, J. Shaw, G. R. Tynan, C. Joshi and D. H. Froula, Demonstration of a narrow energy spread, ∼0.5 GeV electron beam from a two-stage laser wakefield accelerator, *Phys. Rev. Lett.* **107**(4), 1–4 (2011).

[45] A. J. Gonsalves, K. Nakamura, C. Lin, D. Panasenko, S. Shiraishi, T. Sokollik, C. Benedetti, C. B. Schroeder, C. G. R. Geddes, J. van Tilborg, J. Osterhoff, E. Esarey, C. Toth and W. P. Leemans, Tunable laser plasma accelerator based on longitudinal density tailoring, *Nat. Phys.* **7**(11), 862–866 (2011).

[46] S. Steinke, J. van Tilborg, C. Benedetti, C. G. R. Geddes, C. B. Schroeder, J. Daniels, K. K. Swanson, A. J. Gonsalves, K. Nakamura, N. H. Matlis, B. H.

Shaw, E. Esarey and W. P. Leemans, Multistage coupling of independent laser-plasma accelerators, *Nature* **530**(7589), 190–3 (2016).

[47] W. Leemans, White Paper of the ICFA-ICUIL Joint Task Force — High Power Laser Technology for Accelerators, *ICFA Beam Dyn. Newslett.*, No. 56, pp. 11–88 (2011).

[48] P. Russbueldt, T. Mans, J. Weitenberg, H. D. Hoffmann and R. Poprawe, Compact diode-pumped 11 kW Yb:YAG Innoslab femtosecond amplifier, *Opt. Lett.* **35**(24), 4169 (2010).

[49] C. J. Saraceno, F. Emaury, O. H. Heckl, C. R. E. Baer, M. Hoffmann, C. Schriber, M. Golling, T. Südmeyer and U. Keller, 275 W average output power from a femtosecond thin disk oscillator operated in a vacuum environment, *Opt. Express* **20**(21), 23535 (2012).

[50] T. Eidam, S. Hanf, E. Seise, T. V. Andersen, T. Gabler, C. Wirth, T. Schreiber, J. Limpert and A. Tünnermann, Femtosecond fiber CPA system emitting 830 W average output power, *Opt. Lett.* **35**(2), 94 (2010).

[51] V. Gapontsev, V. Fomin, A. Ferin and M. Abramov, Diffraction limited ultra-high-power fiber lasers, in *Lasers, Sources and Related Photonic Devices* (2010), p. AWA1.

[52] T. Eidam, J. Rothhardt, F. Stutzki, F. Jansen, S. Hädrich, H. Carstens, C. Jauregui, J. Limpert and A. Tünnermann, Fiber chirped-pulse amplification system emitting 38 GW peak power, *Opt. Express* **19**(1), 255 (2011).

[53] E. Seise, A. Klenke, J. Limpert and A. Tünnermann, Coherent addition of fiber-amplified ultrashort laser pulses, *Opt. Express* **18**(26), 27827 (2010).

[54] L. Daniault, M. Hanna, L. Lombard, Y. Zaouter, E. Mottay, D. Goular, P. Bourdon, F. Druon and P. Georges, Coherent beam combining of two femtosecond fiber chirped-pulse amplifiers, *Opt. Lett.* **36**(5), 621 (2011).

[55] G. Mourou, B. Brocklesby, T. Tajima and J. Limpert, The future is fibre accelerators, *Nat. Photonics* **7**, 258–261 (2013).

[56] J. Bourderionnet, C. Bellanger, J. Primot and A. Brignon, Collective coherent phase combining of 64 fibers, *Opt. Express* **19**(18), 17053 (2011).

[57] M. Antier, J. Bourderionnet, C. Larat, E. Lallier, E. Lenormand, J. Primot and A. Brignon, kHz closed loop interferometric technique for coherent fiber beam combining, *IEEE J. Sel. Top. Quantum Electron.* **20**(5), 182–187 (2014).

[58] A. Klenke, E. Seise, S. Demmler, J. Rothhardt, S. Breitkopf, J. Limpert and A. Tünnermann, Coherently-combined two channel femtosecond fiber CPA system producing 3 mJ pulse energy, *Opt. Express* **19**(24), 24280 (2011).

[59] A. Klenke, S. Breitkopf, M. Kienel, T. Gottschall, T. Eidam, S. Hädrich, J. Rothhardt, J. Limpert and A. Tünnermann, 530 W, 13 mJ, four-channel coherently combined femtosecond fiber chirped-pulse amplification system, *Opt. Lett.* **38**(13), 2283 (2013).

[60] R. Soulard, M. N. Quinn and G. Mourou, Design and properties of a coherent amplifying network laser, *Appl. Opt.* **54**(15), 4640 (2015).

[61] K. Nakajima, D. Fisher, T. Kawakubo, H. Nakanishi, A. Ogata, Y. Kato, Y. Kitagawa, R. Kodama, K. Mima, H. Shiraga, K. Suzuki, K. Yamakawa, T. Zhang, Y. Sakawa, T. Shoji, Y. Nishida, N. Yugami, M. Downer and T. Tajima, Observation of ultrahigh gradient electron acceleration by a self-modulated intense short laser pulse, *Phys. Rev. Lett.* **74**(22), 4428–4431 (1995).

[62] M. T. Hassan, T. T. Luu, A. Moulet, O. Raskazovskaya, P. Zhokhov, M. Garg, N. Karpowicz, A. M. Zheltikov, V. Pervak, F. Krausz and E. Goulielmakis, Optical attosecond pulses and tracking the nonlinear response of bound electrons, *Nature* **530**(7588), 66–70 (2016).

[63] R. W. Assmann and J. Grebenyuk, Accelerator physics challenges towards a plasma accelerator with usable beam quality, in *Proc. IPAC2014* (2014), pp. 8–11.

[64] T. Tajima, G. Mourou and J. A. Wheeler, High-energy lasers: Extreme light in zeptoseconds, *Laser Focus World* **52**(4), 27–31 (2016).

[65] N. V Myung, J. Lim, J.-P. Fleurial, M. Yun, W. West and D. Choi, Alumina nanotemplate fabrication on silicon substrate, *Nanotechnology* **15**(7), 833–838 (2004).

Jonathan A. Wheeler is currently a researcher with IZEST based at Ecole Polytechnique. He has studied ultrafast attosecond physics during his postdoctoral work at CEA Saclay, Laboratoire d'Optique Appliquée, and his Ph.D. course at the Ohio State University.

Gérard Mourou is a Professor Haut College at Ecole Polytechnique and Director of IZEST. He is also an Emeritus Professor of the University of Michigan, where in 1991 he created the first NSF Center of Excellence in Optics, known as CUOS (Center for Ultrafast Optical Science). He is active in the field of ultrafast and ultrahigh field science with applications to precision micromachining, eye surgery, and relativistic light–matter interactions, most notably with the invention of the chirped pulse amplification (CPA) technique in 1985. He has also been instrumental in initiating many large scale European laser projects, including Apollon, ICAN (International Coherent Amplification Network), and ELI (Extreme Light Infrastructure). He is the recipient of numerous awards, including the 2009 Charles H. Townes Award from the Optical Society of America.

Toshiki Tajima is Norman Rostoker Chair Professor of the University of California at Irvine. He invented laser wakefield acceleration (with the highest citation among all plasma papers) and leads high field science. Its scientific impact includes worldwide frontier research centers in this field and its applications such as compact accelerators for radiology and collective beam dumps. Prof. Tajima has also served as Blumberg Professor of the University of Texas at Austin (his Ph.D. is from UCI, 1975). His honors include the Enrico Fermi Prize, Nishina Prize, Blaire Pascal Chair, CAS Einstein Professorship, and Fellow of APS. He served as Chairs of ICUIL, ELI-NP Advisory Committee.

Reviews of Accelerator Science and Technology
Vol. 9 (2016) 165–186

DOI: 10.1142/S1793626816300085

Simulations for Plasma and Laser Acceleration

Jean-Luc Vay* and Rémi Lehe†
Lawrence Berkeley National Laboratory
Berkeley, CA 94720, USA
**jlvay@lbl.gov*
†rlehe@lbl.gov

Computer simulations have had a profound impact on the design and understanding of past and present plasma acceleration experiments, and will be a key component for turning plasma accelerators from a promising technology into a mainstream scientific tool. In this article, we present an overview of the numerical techniques used with the most popular approaches to model plasma-based accelerators: electromagnetic particle-in-cell, quasistatic and ponderomotive guiding center. The material that is presented is intended to serve as an introduction to the basics of those approaches, and to advances (some of them very recent) that have pushed the state of the art, such as the optimal Lorentz-boosted frame, advanced laser envelope solvers and the elimination of numerical Cherenkov instability. The particle-in-cell method, which has broader interest and is more standardized, is presented in more depth. Additional topics that are cross-cutting, such as azimuthal Fourier decomposition or filtering, are also discussed, as well as potential challenges and remedies in the initialization of simulations and output of data. Examples of simulations using the techniques that are presented have been left out of this article for conciseness, and because simulation results are best understood when presented together, and contrasted with theoretical and/or experimental results, as in other articles of this volume.

Keywords: Particle-in-cell; plasma simulations; plasma acceleration; relativistic plasmas; particle accelerators.

1. Introduction

Computer simulations have had a profound impact on the design and understanding of past and present plasma acceleration experiments [1–4], with accurate modeling of wake formation, electron self-trapping and acceleration requiring fully kinetic methods (usually particle-in-cell) using large computational resources due to the wide range of space and time scales involved. Numerical modeling complements and guides the design and analysis of advanced accelerators, and can reduce development costs significantly. Despite the recent major experimental successes [5–8], the various advanced acceleration concepts need significant progress to fulfill their potential. To this end, large scale simulations will continue to be a key component toward reaching a detailed understanding of the complex interrelated physics phenomena at play.

For such simulations, the most popular algorithm is the particle-in-cell (PIC) technique, which represents electromagnetic fields on a grid and particles by a sample of macroparticles. However, these simulations are very computationally intensive, due to the need to resolve the evolution of a driver (laser or particle beam) and an accelerated beam into a structure that is orders of magnitude longer and wider than the accelerated beam. Various techniques or reduced models have been developed to allow multidimensional simulations at manageable computational costs: quasistatic approximation [9–13], ponderomotive guiding center (PGC) models [10, 11, 13–15], simulation in an optimal Lorentz-boosted frame [16–28], expansion of the fields into a truncated series of azimuthal modes [29–33], fluid approximation [11, 14, 34] and scaled parameters [35, 36]. Many codes have been developed and are used for the modeling of plasma accelerators. A list of such codes is given in Table 1, with the name of the code, its main characteristics, the website if existing or a reference, and the availability and license, if known.

In Sec. 2 of this article, we review the standard methods employed in relativistic electromagnetic PIC simulations of plasma accelerators, including

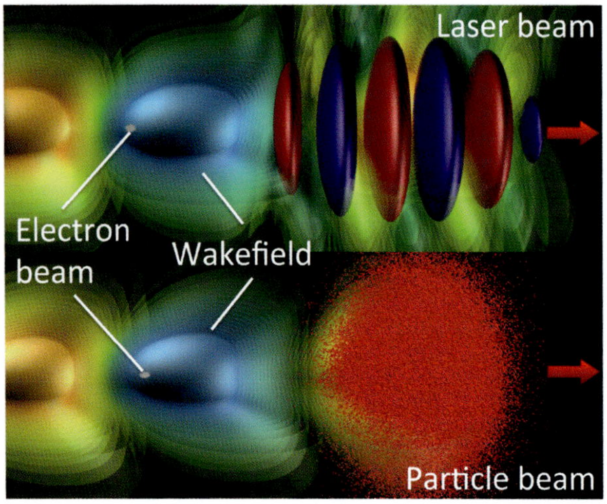

Fig. 1. Plasma laser-driven (top) and charged-particles-driven (bottom) acceleration (rendering from 3D particle-in-cell simulations). A laser beam (red and blue disks in top picture) or a charged particle beam (red dots in bottom picture) propagating (from left to right) through an underdense plasma (not represented) displaces electrons, creating a plasma wakefield that supports very high electric fields (pale blue and yellow). These electric fields, which can be orders of magnitude larger than with conventional techniques, can be used to accelerate a short charged particle beam (white) to high energy over a very short distance.

the core PIC loop steps (particle push, field update, current deposition from the particles to the grid and field gathering from the grid to the particles' positions), the use of a moving window and Lorentz-boosted frame, and the numerical Cherenkov instability and its mitigation. The electromagnetic quasistatic approximation is presented in Sec. 3, the PGC approximation in Sec. 4, and azimuthal Fourier decomposition in Sec. 5. Additional considerations, such as filtering and inputs/outputs, are discussed respectively in Secs. 6 and 7.

2. The Electromagnetic Particle-in-Cell Method

In the electromagnetic PIC method [37], the electromagnetic fields are solved on a grid, usually using Maxwell's equations,

$$\frac{\partial \mathbf{B}}{\partial t} = -\nabla \times \mathbf{E}, \tag{1a}$$

$$\frac{\partial \mathbf{E}}{\partial t} = \nabla \times \mathbf{B} - \mathbf{J}, \tag{1b}$$

$$\nabla \cdot \mathbf{E} = \rho, \tag{1c}$$

$$\nabla \cdot \mathbf{B} = 0, \tag{1d}$$

given here in natural units ($\epsilon_0 = \mu_0 = c = 1$), where t is time, $\mathbf{E}$ and $\mathbf{B}$ are the electric and magnetic field components, and ρ and $\mathbf{J}$ are the charge and current densities. The charged particles are advanced in time using the Newton–Lorentz equations of motion

$$\frac{d\mathbf{x}}{dt} = \mathbf{v}, \tag{2a}$$

$$\frac{d(\gamma \mathbf{v})}{dt} = \frac{q}{m}(\mathbf{E} + \mathbf{v} \times \mathbf{B}), \tag{2b}$$

where $m, q, \mathbf{x}, \mathbf{v}$ and $\gamma = 1/\sqrt{1-v^2}$ are respectively the mass, charge, position, velocity and relativistic factor of the particle given in natural units ($c = 1$). The charge and current densities are interpolated on the grid from the particles' positions and velocities, while the electric and magnetic field components are interpolated from the grid to the particles' positions for the velocity update.

2.1. *Particle push*

A centered finite difference discretization of the Newton–Lorentz equations of motion is given by

$$\frac{\mathbf{x}^{i+1} - \mathbf{x}^{i}}{\Delta t} = \mathbf{v}^{i+1/2}, \tag{3a}$$

$$\frac{\gamma^{i+1/2}\mathbf{v}^{i+1/2} - \gamma^{i-1/2}\mathbf{v}^{i-1/2}}{\Delta t} = \frac{q}{m}(\mathbf{E}^{i} + \bar{\mathbf{v}}^{i} \times \mathbf{B}^{i}). \tag{3b}$$

In order to close the system, $\bar{\mathbf{v}}^i$ must be expressed as a function of the other quantities. The two implementations that have become the most popular are presented below.

2.1.1. *Boris relativistic velocity rotation*

The solution proposed by Boris [38] is given by

$$\bar{\mathbf{v}}^{i} = \frac{\gamma^{i+1/2}\mathbf{v}^{i+1/2} + \gamma^{i-1/2}\mathbf{v}^{i-1/2}}{2\bar{\gamma}^{i}}, \tag{4}$$

where $\bar{\gamma}^i$ is defined by $\bar{\gamma}^i \equiv (\gamma^{i+1/2} + \gamma^{i-1/2})/2$.

The system (3b, 4) is solved very efficiently following Boris' method, where the electric field push is decoupled from the magnetic push. Setting $\mathbf{u} = \gamma\mathbf{v}$,

Table 1. List of simulation PIC codes for the modeling of plasma accelerators.

Code	Type	Website/reference	Availability/license
ALaDyn/PICCANTE	EM-PIC 3D	http://aladyn.github.io/piccante	Open/GPLv3+
Architect	EM-PIC RZ	https://github.com/albz/Architect	Open/GPL
Calder	EM-PIC 3D	http://iopscience.iop.org/article/10.1088/0029-5515/43/7/317	Collaborators/Proprietary
Calder-Circ	EM-PIC RZ^+	http://dx.doi.org/10.1016/j.jcp.2008.11.017	Upon Request/Proprietary
CHIMERA	EM-PIC RZ+	https://github.com/hightower8083/chimera	Open/GPLv3
ELMIS	EM-PIC 3D	http://www.diva-portal.org/smash/record.jsf?pid=diva2%3A681092&dswid=-8610	Collaborators/Proprietary
EPOCH	EM-PIC 3D	http://www.ccpp.ac.uk/codes.html	Collaborators/GPL
FBPIC	EM-PIC RZ^+	https://fbpic.github.io	Open/modified BSD
HiPACE	QS-PIC 3D	http://dx.doi.org/10.1088/0741-3335/56/8/084012	Collaborators/Proprietary
INF&RNO	QS/EM-PIC RZ	http://dx.doi.org/10.1063/1.3520323	Collaborators/Proprietary
LCODE	QS-PIC RZ	http://www.inp.nsk.su/~lotov/lcode	Open/None
LSP	EM-PIC 3D/RZ	http://www.lspsuite.com/LSP/index.html	Commercial/Proprietary
MAGIC	EM-PIC 3D	http://www.mrcwdc.com/magic/index.html	Commercial/Proprietary
Osiris	EM-PIC 3D/RZ^+	http://picksc.idre.ucla.edu/software/software-production-codes/osiris	Collaborators/Proprietary
PHOTON-PLASMA	EM-PIC 3D	https://bitbucket.org/thaugboelle/ppcode	Open/GPLv2
PICADOR	EM-PIC 3D	http://hpc-education.unn.ru/en/research/overview/laser-plasma	Collaborators/Proprietary
PIConGPU	EM-PIC 3D	http://picongpu.hzdr.de	Open/GPLv3+
PICLS	EM-PIC 3D	http://dx.doi.org/10.1016/j.jcp.2008.03.043	Collaborators/Proprietary
PSC	EM-PIC 3D	http://www.sciencedirect.com/science/article/pii/S0021999116301413	Open/GPLv3
QuickPIC	QS-PIC 3D	http://picksc.idre.ucla.edu/software/software-production-codes/quickpic	Collaborators/Proprietary
REMP	EM-PIC 3D	http://dx.doi.org/10.1016/S0010-4655(00)00228-9	Collaborators/Proprietary
Smilei	EM-PIC 2D	http://www.maisondelasimulation.fr/projects/Smilei/html/licence.html	Open/CeCILL
TurboWave	EM-PIC 3D/RZ	http://dx.doi.org/10.1109/27.893300	Collaborators/Proprietary
UPIC-EMMA	EM-PIC 3D	http://picksc.idre.ucla.edu/software/software-production-codes/upic-emma	Collaborators/Proprietary
VLPL	EM/QS-PIC 3D	http://www.tp1.hhu.de/~pukhov/	Collaborators/Proprietary
VPIC	EM-PIC 3D	http://github.com/losalamos/vpic	Open/BSD clause-3 license
VSim (Vorpal)	EM-PIC 3D	https://txcorp.com/vsim	Commercial/Proprietary
Wake	QS-PIC RZ	http://dx.doi.org/10.1063/1.872134	Collaborators/Proprietary
Warp	EM-PIC 3D/RZ^+	http://warp.lbl.gov	Open/modified BSD

EM = electromagnetic; QS = quasistatic; PIC = particle-in-cell; 3D = three-dimensional; RZ = axisymmetric; RZ^+ = axisymmetric with azimuthal Fourier decomposition.

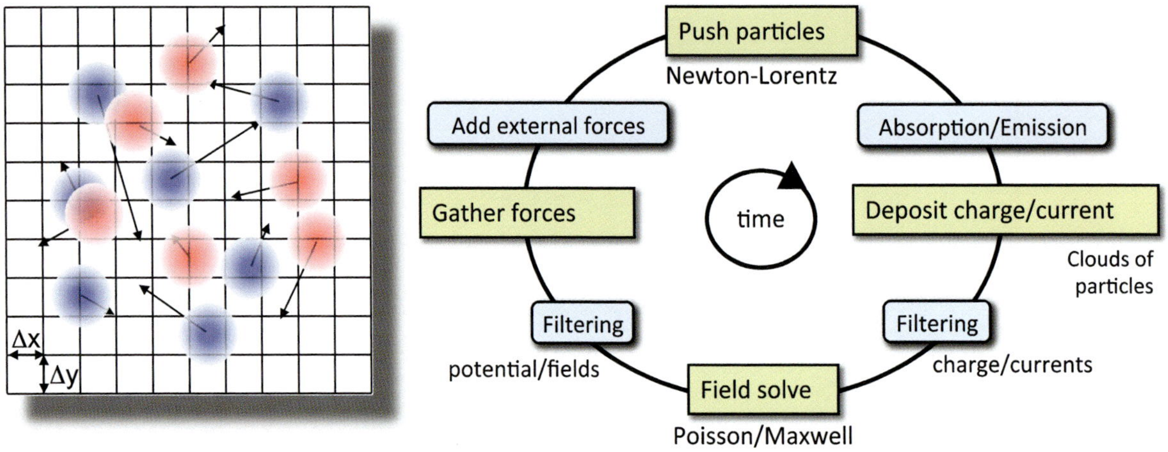

Fig. 2. The particle-in-cell (PIC) method follows the evolution of a collection of charged macroparticles (positively charged in blue in the left plot; negatively charged in red) that evolve self-consistently with their electromagnetic (or electrostatic) fields. The core PIC algorithm involves four operations at each time step: (1) evolve the velocity and position of the particles using the Newton–Lorentz equations, (2) deposit the charge and/or current densities through interpolation from the particles' distributions onto the grid, (3) evolve Maxwell's wave equations (for electromagnetic) or solve Poisson's equation (for electrostatic) on the grid, and (4) interpolate the fields from the grid onto the particles for the next particle push. Additional "add-on" operations are inserted between these core operations to account for additional physics (e.g. absorption/emission of particles, addition of external forces to account for accelerator focusing or accelerating component) or numerical effects (e.g. smoothing/filtering of the charge/current densities and/or fields on the grid).

the velocity is updated using the following sequence:

$$\mathbf{u}^- = \mathbf{u}^{i-1/2} + \left(\frac{q\Delta t}{2m}\right)\mathbf{E}^i, \tag{5a}$$

$$\mathbf{u}' = \mathbf{u}^- + \mathbf{u}^- \times \mathbf{t}, \tag{5b}$$

$$\mathbf{u}^+ = \mathbf{u}^- + \mathbf{u}' \times \frac{2\mathbf{t}}{1+t^2}, \tag{5c}$$

$$\mathbf{u}^{i+1/2} = \mathbf{u}^+ + \left(\frac{q\Delta t}{2m}\right)\mathbf{E}^i, \tag{5d}$$

where $\mathbf{t} = (q\Delta t/2m)\,\mathbf{B}^i/\bar{\gamma}^i$ and where $\bar{\gamma}^i$ can be calculated as $\bar{\gamma}^i = \sqrt{1+(\mathbf{u}^-/c)^2}$.

The Boris implementation is second-order accurate, time-reversible and fast. It is very widespread and used in the vast majority of PIC codes.

2.1.2. *Lorentz-invariant formulation*

It was shown in Ref. 39 that the Boris formulation is not Lorentz-invariant and can lead to significant errors in the treatment of relativistic dynamics. A Lorentz-invariant formulation is obtained by considering the following velocity average

$$\bar{\mathbf{v}}^i = \frac{\mathbf{v}^{i+1/2} + \mathbf{v}^{i-1/2}}{2}. \tag{6}$$

This gives a system that is solvable analytically (see Ref. 39 for a detailed derivation), producing the following velocity update:

$$\mathbf{u}^* = \mathbf{u}^{i-1/2} + \frac{q\Delta t}{m}\left(\mathbf{E}^i + \frac{\mathbf{v}^{i-1/2}}{2} \times \mathbf{B}^i\right), \tag{7a}$$

$$\mathbf{u}^{i+1/2} = \frac{\mathbf{u}^* + (\mathbf{u}^* \cdot \mathbf{t})\mathbf{t} + \mathbf{u}^* \times \mathbf{t}}{1+t^2}, \tag{7b}$$

where $\mathbf{t} = \boldsymbol{\tau}/\gamma^{i+1/2}$, $\boldsymbol{\tau} = (q\Delta t/2m)\mathbf{B}^i$, $\gamma^{i+1/2} = \sqrt{\sigma + \sqrt{\sigma^2 + (\tau^2 + w^2)}}$, $w = \mathbf{u}^* \cdot \boldsymbol{\tau}$, $\sigma = (\gamma'^2 - \tau^2)/2$ and $\gamma' = \sqrt{1+(\mathbf{u}^*/c)^2}$. This Lorentz-invariant formulation is particularly well suited for the modeling of ultrarelativistic charged particle beams, where an accurate account of the cancelation of the self-generated electric and magnetic fields is essential, as shown in Ref. 39.

2.2. *Field solving*

Various methods are available for solving Maxwell's equations on a grid, based on finite difference, finite volume, finite element, spectral or other discretization techniques that apply most commonly on single structured or unstructured meshes and less commonly on multiblock multiresolution grid structures. In this article, we summarize the widespread second order finite difference time domain (FDTD) algorithm, its extension to non standard finite differences, as well as the pseudospectral analytical time domain (PSATD) and pseudospectral time domain (PSTD) algorithms. Extension to multiresolution (or mesh refinement) PIC is described in e.g. Refs. 40 and 41.

2.2.1. *Finite difference time domain (FDTD)*

The most popular algorithm for electromagnetic PIC codes is the finite difference time domain (FDTD) solver:

$$D_t\mathbf{B} = -\nabla \times \mathbf{E}, \tag{8a}$$

$$D_t\mathbf{E} = \nabla \times \mathbf{B} - \mathbf{J}, \tag{8b}$$

$$[\nabla \cdot \mathbf{E} = \rho], \tag{8c}$$

$$[\nabla \cdot \mathbf{B} = 0]. \tag{8d}$$

The differential operator is defined as $\nabla = D_x\hat{\mathbf{x}} + D_y\hat{\mathbf{y}} + D_z\hat{\mathbf{z}}$ and the finite difference operators in time and space are respectively defined as $D_tG|^n_{i,j,k} = (G|^{n+1/2}_{i,j,k} - G|^{n-1/2}_{i,j,k})/\Delta t$ and $D_xG|^n_{i,j,k} = (G|^n_{i+1/2,j,k} - G|^n_{i-1/2,j,k})/\Delta x$, where Δt and Δx are respectively the time step and the grid cell size along x, n is the time index, and i, j and k are the spatial indices along x, y and z, respectively. The difference operators along y and z are obtained by circular permutation. The equations in brackets are given for completeness, as they are often not actually solved, thanks to the use of a so-called charge- conserving algorithm, as explained below. As shown in Fig. 3, the quantities are given on a staggered (or Yee) grid [42], where the electric field components are located between nodes and the magnetic field components are located in the center of the cell faces. Knowing the current densities at half-integer steps, the electric field components are updated alternately with the magnetic field components at integer and half-integer steps, respectively.

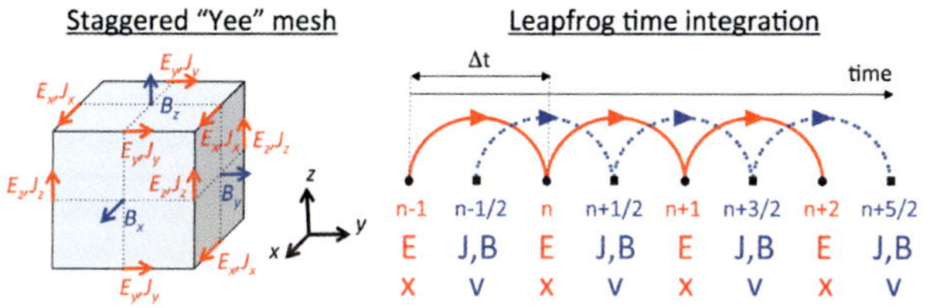

Fig. 3. *Left*: Layout of field components on the staggered Yee grid. Current densities and electric fields are defined on the edges of the cells, and magnetic fields on the faces. *Right*: Time integration using a second order finite difference leapfrog integrator.

2.2.2. *Nonstandard finite difference time domain (NSFDTD)*

In Refs. 43 and 44, Cole introduced an implementation of the source free Maxwells wave equations for narrow band applications based on nonstandard finite differences (NSFD). In Ref. 45, Karkkainen *et al.* adapted it for wideband applications. At the Courant limit for the time step and for a given set of parameters, the stencil proposed in Ref. 45 has no numerical dispersion along the principal axes, provided that the cell size is the same along each dimension (i.e. cubic cells in 3D). The Cole–Karkkainen (CK) solver uses the nonstandard finite difference formulation (based on extended stencils) of the Maxwell–Ampere equation and can be implemented as follows [25]:

$$D_t\mathbf{B} = -\nabla^* \times \mathbf{E}, \tag{9a}$$

$$D_t\mathbf{E} = \nabla \times \mathbf{B} - \mathbf{J}, \tag{9b}$$

$$[\nabla \cdot \mathbf{E} = \rho], \tag{9c}$$

$$[\nabla^* \cdot \mathbf{B} = 0]. \tag{9d}$$

Equations (9c) and (9d) are not being solved explicitly but verified via appropriate initial conditions and the current deposition procedure. The NSFD differential operator is given by $\nabla^* = D^*_x\hat{\mathbf{x}} + D^*_y\hat{\mathbf{y}} + D^*_z\hat{\mathbf{z}}$, where $D^*_x = (\alpha + \beta S^1_x + \xi S^2_x)D_x$ with $S^1_xG|^n_{i,j,k} = G|^n_{i,j+1,k} + G|^n_{i,j-1,k} + G|^n_{i,j,k+1} + G|^n_{i,j,k-1}$, $S^2_xG|^n_{i,j,k} = G|^n_{i,j+1,k+1} + G|^n_{i,j-1,k+1} + G|^n_{i,j+1,k-1} + G|^n_{i,j-1,k-1}$. G is a sample vector component, while α, β and ξ are constant scalars satisfying $\alpha + 4\beta + 4\xi = 1$. As with the FDTD algorithm, the quantities with a half-integer are located between the nodes (electric field components) or in the center of the cell faces (magnetic field components). The operators along y and z, i.e. D_y, D_z, D^*_y, D^*_z, S^1_y, S^1_z, S^2_y and S^2_z, are obtained by circular permutation of the indices.

Assuming cubic cells ($\Delta x = \Delta y = \Delta z$), the coefficients given in Ref. 45 ($\alpha = 7/12$, $\beta = 1/12$ and $\xi = 1/48$) allow the Courant condition to be at $\Delta t = \Delta x$, which equates to having no numerical dispersion along the principal axes. The algorithm reduces to the FDTD algorithm with $\alpha = 1$ and $\beta = \xi = 0$. An extension to noncubic cells is provided by Cowan *et al.* in 3D in Ref. 46 and by Pukhov in

2D in Ref. 47. An alternative NSFDTD implementation that enables superluminous waves is given by Lehe *et al.* in Ref. 48.

As mentioned above, a key feature of the algorithms based on NSFDTD is that some implementations [45, 46] enable the time step $\Delta t = \Delta x$ along one or more axes and no numerical dispersion along those axes. However, as shown in Ref. 25, an instability develops at the Nyquist wavelength at (or very near) such a time step. It is also shown in the same paper that removing the Nyquist component in all the source terms using a bilinear filter (see description of the filter below) suppresses this instability.

2.2.3. *Pseudospectral analytical time domain (PSATD)*

Maxwell's equations in Fourier space are given by

$$\frac{\partial \tilde{\mathbf{E}}}{\partial t} = i\mathbf{k} \times \tilde{\mathbf{B}} - \tilde{\mathbf{J}}, \tag{10a}$$

$$\frac{\partial \tilde{\mathbf{B}}}{\partial t} = -i\mathbf{k} \times \tilde{\mathbf{E}}, \tag{10b}$$

$$[i\mathbf{k} \cdot \tilde{\mathbf{E}} = \tilde{\rho}], \tag{10c}$$

$$[i\mathbf{k} \cdot \tilde{\mathbf{B}} = 0], \tag{10d}$$

where $\tilde{a}$ is the Fourier transform of the quantity a. As with the real space formulation, provided that the continuity equation $\partial\tilde{\rho}/\partial t + i\mathbf{k}\cdot\tilde{\mathbf{J}} = 0$ is satisfied, the last two equations will automatically be satisfied at any time if satisfied initially and do not need to be explicitly integrated.

Decomposing the electric field and current between longitudinal and transverse components $\tilde{\mathbf{E}} = \tilde{\mathbf{E}}_L + \tilde{\mathbf{E}}_T = \hat{\mathbf{k}}(\hat{\mathbf{k}} \cdot \tilde{\mathbf{E}}) - \hat{\mathbf{k}} \times (\hat{\mathbf{k}} \times \tilde{\mathbf{E}})$ and $\tilde{\mathbf{J}} = \tilde{\mathbf{J}}_L + \tilde{\mathbf{J}}_T = \hat{\mathbf{k}}(\hat{\mathbf{k}} \cdot \tilde{\mathbf{J}}) - \hat{\mathbf{k}} \times (\hat{\mathbf{k}} \times \tilde{\mathbf{J}})$ gives

$$\frac{\partial \tilde{\mathbf{E}}_T}{\partial t} = i\mathbf{k} \times \tilde{\mathbf{B}} - \tilde{\mathbf{J}}_\mathbf{T}, \tag{11a}$$

$$\frac{\partial \tilde{\mathbf{E}}_L}{\partial t} = -\tilde{\mathbf{J}}_\mathbf{L}, \tag{11b}$$

$$\frac{\partial \tilde{\mathbf{B}}}{\partial t} = -i\mathbf{k} \times \tilde{\mathbf{E}}, \tag{11c}$$

with $\hat{\mathbf{k}} = \mathbf{k}/k$.

If the sources are assumed to be constant over a time interval Δt, the system of equations is solvable analytically and is given by (see Ref. 49 for the original formulation and Ref. 50 for a more detailed derivation)

$$\tilde{\mathbf{E}}_T^{n+1} = C\tilde{\mathbf{E}}_T^n + iS\hat{\mathbf{k}} \times \tilde{\mathbf{B}}^n - \frac{S}{k}\tilde{\mathbf{J}}_T^{n+1/2}, \tag{12a}$$

$$\tilde{\mathbf{E}}_L^{n+1} = \tilde{\mathbf{E}}_L^n - \Delta t\tilde{\mathbf{J}}_L^{n+1/2}, \tag{12b}$$

$$\tilde{\mathbf{B}}^{n+1} = C\tilde{\mathbf{B}}^n - iS\hat{\mathbf{k}} \times \tilde{\mathbf{E}}^n \tag{12c}$$

$$+ i\frac{1-C}{k}\hat{\mathbf{k}} \times \tilde{\mathbf{J}}^{n+1/2}, \tag{12d}$$

with $C = \cos(k\Delta t)$ and $S = \sin(k\Delta t)$.

Combining the transverse and longitudinal component gives

$$\tilde{\mathbf{E}}^{n+1} = C\tilde{\mathbf{E}}^n + iS\hat{\mathbf{k}} \times \tilde{\mathbf{B}}^n - \frac{S}{k}\tilde{\mathbf{J}}^{n+1/2} \tag{13a}$$

$$+ (1-C)\hat{\mathbf{k}}(\hat{\mathbf{k}} \cdot \tilde{\mathbf{E}}^n) + \hat{\mathbf{k}}(\hat{\mathbf{k}} \cdot \tilde{\mathbf{J}}^{n+1/2})\left(\frac{S}{k} - \Delta t\right), \tag{13b}$$

$$\tilde{\mathbf{B}}^{n+1} = C\tilde{\mathbf{B}}^n - iS\hat{\mathbf{k}} \times \tilde{\mathbf{E}}^n \tag{13c}$$

$$+ i\frac{1-C}{k}\hat{\mathbf{k}} \times \tilde{\mathbf{J}}^{n+1/2}. \tag{13d}$$

For fields generated by the source terms without the self-consistent dynamics of the charged particles, this algorithm is free of numerical dispersion and is not subject to a Courant condition. Furthermore, this solution is exact for any time step size subject to the assumption that the current source is constant over that time step.

As shown in Ref. 50, by expanding the coefficients S_h and C_h in Taylor series and keeping the leading terms, the PSATD formulation reduces to the perhaps-better-known pseudospectral time domain (PSTD) formulation [51, 52]:

$$\tilde{\mathbf{E}}^{n+1} = \tilde{\mathbf{E}}^n + i\Delta t\mathbf{k} \times \tilde{\mathbf{B}}^{n+1/2} - \Delta t\tilde{\mathbf{J}}^{n+1/2}, \tag{14a}$$

$$\tilde{\mathbf{B}}^{n+3/2} = \tilde{\mathbf{B}}^{n+1/2} - i\Delta t\mathbf{k} \times \tilde{\mathbf{E}}^{n+1}. \tag{14b}$$

The dispersion relation of the PSTD solver is given by $\sin(\frac{\omega\Delta t}{2}) = \frac{k\Delta t}{2}$. In contrast to the PSATD solver, the PSTD solver is subject to numerical dispersion for a finite time step and to a Courant condition that is given by $\Delta t \leq \frac{2}{\pi}(\frac{1}{\Delta x^2} + \frac{1}{\Delta y^2} + \frac{1}{\Delta x^2})^{-1/2}$.

The PSATD and PSTD formulations that were just given apply to the field components located at the nodes of the grid. As noted in Ref. 53, they can easily be recast on a staggered Yee grid by multiplying the field components by the appropriate phase factors to shift them from the collocated to the staggered locations. The choice between a collocated and a staggered formulation is application-dependent.

Spectral solvers used to be very popular in the years 1970s to early 1990s, before being replaced by finite difference methods with the advent of parallel supercomputers, which favored local methods. However, it was shown recently that standard domain decomposition with fast Fourier transforms that are local to each subdomain could be used effectively with PIC spectral methods [50], at the cost of truncation errors in the guard cells that could be neglected. A detailed analysis of the effectiveness of the method with exact evaluation of the magnitude of the effect of the truncation error is given in Ref. 54 for stencils of arbitrary order (up to the infinite "spectral" order).

2.3. *Current deposition*

The current densities are deposited on the computational grid from the particle position and velocities, employing splines of various orders [55].

$$\rho = \frac{1}{\Delta x \Delta y \Delta z} \sum_n q_n S_n, \tag{15a}$$

$$\mathbf{J} = \frac{1}{\Delta x \Delta y \Delta z} \sum_n q_n \mathbf{v_n} S_n. \tag{15b}$$

In most applications, it is essential to prevent the accumulation of errors resulting from the violation of the discretized Gauss law. This is accomplished by providing a method for depositing the current from the particles to the grid that preserves the discretized Gauss law, or by providing a mechanism for "divergence cleaning" [37, 56–59]. For the former, schemes that allow a deposition of the current that is exact when combined with the Yee solver are given in Ref. 60 for linear splines and in Ref. 61 for splines of arbitrary order.

The NSFDTD formulations given above and in Refs. 25, 46–48 apply to the Maxwell–Faraday equation, while the discretized Maxwell–Ampere equation uses the FDTD formulation. Consequently, the charge-conserving algorithms developed for current deposition [60, 61] apply readily to those NSFDTD-based formulations. More details concerning those implementations, including the expressions for the numerical dispersion and Courant condition, are given in Refs. 25, 46–48.

In the case of the pseudospectral solvers, the current deposition algorithm generally does not satisfy the discretized continuity equation in Fourier space $\tilde{\rho}^{n+1} = \tilde{\rho}^n - i\Delta t \mathbf{k} \cdot \tilde{\mathbf{J}}^{n+1/2}$. In this case, a Boris correction [37] can be applied in k space in the form $\tilde{\mathbf{E}}_c^{n+1} = \tilde{\mathbf{E}}^{n+1} - (\mathbf{k} \cdot \tilde{\mathbf{E}}^{n+1} + i\tilde{\rho}^{n+1})\hat{\mathbf{k}}/k$, where $\tilde{\mathbf{E}}_c$ is the corrected field. Alternatively, a correction to the current can be applied (with some similarity to the current deposition presented by Morse and Nielson in their potential-based model in Ref. 62) using $\tilde{\mathbf{J}}_c^{n+1/2} = \tilde{\mathbf{J}}^{n+1/2} - [\mathbf{k} \cdot \tilde{\mathbf{J}}^{n+1/2} - i(\tilde{\rho}^{n+1} - \tilde{\rho}^n)/\Delta t]\hat{\mathbf{k}}/k$, where $\tilde{\mathbf{J}}_c$ is the corrected current. In this case, the transverse component of the current is left untouched while the longitudinal component is effectively replaced by the one obtained from integration of the continuity equation, ensuring that the corrected current satisfies the continuity equation. The advantage of correcting the current rather than the electric field is that it is more local and thus more compatible with domain decomposition of the fields for parallel computation [63].

Alternatively, an exact current deposition can be written for the pseudospectral solvers, following the geometrical interpretation of existing methods in real space [60–62], thereby averaging the currents of the paths following grid lines between the positions (x^n, y^n) and (x^{n+1}, y^{n+1}), which is given in 2D (extension to 3D follows readily) for $k \neq 0$ by [63]

$$\tilde{\mathbf{J}}^{k\neq 0} = \frac{i\tilde{\mathbf{D}}}{\mathbf{k}} \tag{16}$$

with

$$D_x = \frac{1}{2\Delta t} \sum_i q_i [\Gamma(x_i^{n+1}, y_i^{n+1}) - \Gamma(x_i^n, y_i^{n+1}) + \Gamma(x_i^{n+1}, y_i^n) - \Gamma(x_i^n, y_i^n)], \tag{17}$$

$$D_y = \frac{1}{2\Delta t} \sum_i q_i [\Gamma(x_i^{n+1}, y_i^{n+1}) - \Gamma(x_i^{n+1}, y_i^n) + \Gamma(x_i^n, y_i^{n+1}) - \Gamma(x_i^n, y_i^n)], \tag{18}$$

where Γ is the macroparticle form factor. The contributions for $k = 0$ are integrated directly in real space [63].

2.4. *Field gathering*

In general, the field is gathered from the mesh onto the macroparticles using splines of the same order as for the current deposition $\mathbf{S} = (S_x, S_y, S_z)$. Three variations are considered:

- *Momentum-conserving*: Fields are interpolated from the grid nodes to the macroparticles using $\mathbf{S} = (S_{nx}, S_{ny}, S_{nz})$ for all field components (if the fields are known at staggered positions, they are first interpolated to the nodes on an auxiliary grid).
- *Energy-conserving* (*or Galerkin*): Fields are interpolated from the staggered Yee grid to the macroparticles using $(S_{nx-1}, S_{ny}, S_{nz})$ for E_x, $(S_{nx}, S_{ny-1}, S_{nz})$ for E_y, $(S_{nx}, S_{ny}, S_{nz-1})$ for E_z, $(S_{nx}, S_{ny-1}, S_{nz-1})$ for B_x, $(S_{nx-1}, S_{ny}, S_{nz-1})$ for B_y and $(S_{nx-1}, S_{ny-1}, S_{nz})$ for B_z (if the fields are known at the nodes, they are first interpolated to the staggered positions on an auxiliary grid).
- *Uniform*: Fields are interpolated directly from the Yee grid to the macroparticles using $\mathbf{S} = (S_{nx}, S_{ny}, S_{nz})$ for all field components (if the fields are known at the nodes, they are first interpolated to the staggered positions on an auxiliary grid).

As shown in Refs. 37, 64 and 65, the momentum- and energy-conserving schemes conserve momentum and energy respectively at the limit of infinitesimal time steps and generally offer better conservation of the respective quantities for a finite time step. The uniform scheme does not conserve momentum or energy in the sense defined for the others but is given for completeness, as it has been shown to offer some interesting properties in the modeling of relativistically drifting plasmas [66].

2.5. *Moving window and optimal Lorentz-boosted frame*

The simulations of plasma accelerators from first principles are very computationally intensive, due to the need to resolve the evolution of a driver (laser or particle beam) and an accelerated particle beam into a plasma structure that is orders of magnitude longer and wider than the accelerated beam. As is customary in the modeling of particle beam dynamics in standard particle accelerators, a moving window is commonly used to follow the driver, the wake and the accelerated beam. This results in huge savings, by avoiding the meshing of the entire plasma that is orders of magnitude longer than the other length scales of interest.

Even when one is using a moving window, however, a full PIC simulation of a plasma accelerator can be extraordinarily demanding computationally, as many time steps are needed to resolve the crossing of the short driver beam with the plasma column. As it turns out, choosing an optimal frame of reference that travels close to the speed of light in the direction of the laser or particle beam (as opposed to the usual choice of the laboratory frame) enables speedups by orders of magnitude [16, 27]. This is a result of the properties of Lorentz contraction and dilation of space and time. In the frame of the laboratory, a very short driver (laser or particle) beam propagates through a much longer plasma column, necessitating millions to tens of millions of time steps for parameters in the range of the BELLA or FACET-II experiments. As sketched in Fig. 4, in a frame moving with the driver beam in the plasma at velocity $v = \beta c$ (c is the speed of light in vacuum), the beam length is now elongated by $\approx (1+\beta)\gamma$, while the plasma contracts by γ ($\gamma = 1/\sqrt{1-\beta^2}$ is the relativistic factor associated with the frame velocity). The number of time steps needed to simulate a "longer" beam through a "shorter" plasma is now reduced by up to $\approx (1+\beta)\gamma^2$ (a detailed derivation of the speedup is given below).

The modeling of a plasma acceleration stage in a boosted frame involves the fully electromagnetic modeling of a plasma propagating at near the speed of light, for which numerical Cerenkov [49, 67] is a potential issue, as explained in more detail below. In addition, for a frame of reference moving in the direction of the accelerated beam (or equivalently, the wake of the laser), waves emitted by the plasma in the forward direction expand while those emitted in the backward direction contract, following the

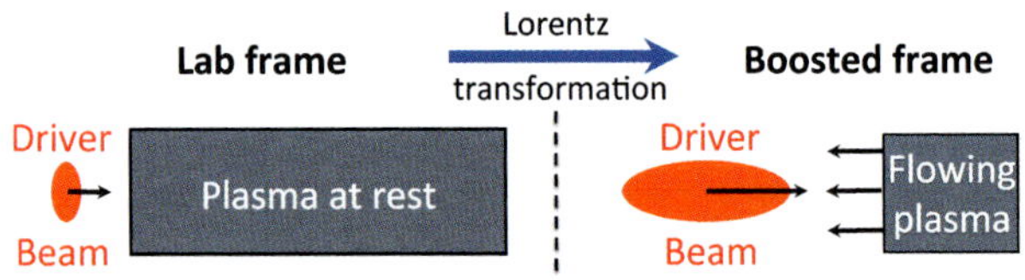

Fig. 4. A first principles simulation of a short driver beam (laser or charged particles) propagating through a plasma that is orders of magnitude longer necessitates a very large number of time steps. Recasting the simulation in a frame of reference that is moving close to the speed of light in the direction of the driver beam leads to simulating a driver beam that appears longer propagating through a plasma that appears shorter than in the laboratory. Thus, this relativistic transformation of space and time reduces the disparity of scales, and thereby the number of time steps to complete the simulation, by orders of magnitude.

properties of the Lorentz transformation. If one had to resolve both forward- and backward-propagating waves emitted from the plasma, there would be no gain in selecting a frame different from the laboratory frame. However, the physics of interest for a laser wakefield is the laser driving the wake, the wake, and the accelerated beam. Backscatter is weak in the short pulse regime, and does not interact as strongly with the beam as do the forward-propagating waves, which stay in phase for a long period. It is thus often assumed that the backward-propagating waves can be neglected in the modeling of plasma accelerator stages. The accuracy of this assumption has been demonstrated by comparison between explicit codes which include both forward and backward waves and envelope or quasistatic codes which neglect backward waves [2, 36, 68].

2.5.1. *Theoretical speedup dependency with the frame boost*

The derivation that is given here reproduces the one given in Ref. 27, where the obtainable speedup is derived as an extension of the formula that was derived earlier [16], taking also into account the group velocity of the laser as it traverses the plasma.

Assuming that the simulation box is a fixed number of plasma periods long, which implies the use (which is standard) of a moving window following the wake and accelerated beam, the speedup is given by the ratio of the time taken by the laser pulse and the plasma to cross each other, divided by the shortest timescale of interest, i.e. the laser period. To first order, the wake velocity v_w is set by the 1D group velocity of the laser driver, which in the linear (low intensity) limit is given by [69]

$$\frac{v_w}{c} = \beta_w = \left(1 - \frac{\omega_p^2}{\omega^2}\right)^{1/2}, \tag{19}$$

where $\omega_p = \sqrt{(n_e e^2)/(\epsilon_0 m_e)}$ is the plasma frequency and $\omega = 2\pi c/\lambda$ is the laser frequency; n_e is the plasma density, λ is the laser wavelength in vacuum, ϵ_0 is the permittivity of vacuum, c is the speed of light in vacuum, and e and m_e are respectively the charge and mass of the electron.

In practice, the runs are typically stopped when the last electron beam macroparticle exits the plasma, and a measure of the total time of the simulation is then given by

$$T = \frac{L + \eta\lambda_p}{v_w - v_p}, \tag{20}$$

where $\lambda_p \approx 2\pi c/\omega_p$ is the wake wavelength, L is the plasma length, v_w and $v_p = \beta_p c$ are respectively the velocity of the wake and of the plasma relative to the frame of reference, and η is an adjustable parameter for taking into account the fraction of the wake which exited the plasma at the end of the simulation. For a beam injected into the nth bucket, η would be set to $n - 1/2$. If positrons were considered, they would be injected half a wake period ahead of the location of the electrons' injection position for a given period, and one would have $\eta = n - 1$. The numerical cost R_t scales as the ratio of the total time to the shortest timescale of interest, which is the inverse of the laser frequency, and is thus given by

$$R_t = \frac{Tc}{\lambda} = \frac{L + \eta\lambda_p}{(\beta_w - \beta_p)\lambda}. \tag{21}$$

In the laboratory, $v_p = 0$ and the expression simplifies to

$$R_{\text{lab}} = \frac{Tc}{\lambda} = \frac{L + \eta\lambda_p}{\beta_w\lambda}. \tag{22}$$

In a frame moving at βc, the quantities become

$$\lambda_p^* = \frac{\lambda_p}{\gamma(1 - \beta_w\beta)}, \tag{23}$$

$$L^* = \frac{L}{\gamma}, \tag{24}$$

$$\lambda^* = \gamma(1 + \beta)\lambda, \tag{25}$$

$$\beta_w^* = \frac{\beta_w - \beta}{1 - \beta_w\beta}, \tag{26}$$

$$v_p^* = -\beta c, \tag{27}$$

$$T^* = \frac{L^* + \eta\lambda_p^*}{v_w^* - v_p^*}, \tag{28}$$

$$R_t^* = \frac{T^* c}{\lambda^*} = \frac{L^* + \eta\lambda_p^*}{(\beta_w^* + \beta)\lambda^*}, \tag{29}$$

where $\gamma = 1/\sqrt{1 - \beta^2}$.

The expected speedup from performing the simulation in a boosted frame is given by the ratio of R_{lab} and R_t^*

$$S = \frac{R_{\text{lab}}}{R_t^*} = \frac{(1 + \beta)(L + \eta\lambda_p)}{(1 - \beta\beta_w)L + \eta\lambda_p}. \tag{30}$$

We note that assuming that $\beta_w \approx 1$ (which is a valid approximation for most practical cases of interest) and that $\gamma << \gamma_w$, this expression is consistent with the expression derived earlier [16] for the laser-plasma acceleration case, which states that $R_t^* = \alpha R_t/(1+\beta)$ with $\alpha = (1-\beta+l/L)/(1+l/L)$, where l is the laser length, which is generally proportional to $\eta\lambda_p$, and $S = R_t/R_T^*$. However, higher values of γ are of interest for maximum speedup, as shown below.

For intense lasers ($a \sim 1$) typically used for acceleration, the energy gain is limited by dephasing [70], which occurs over a scale length $L_d \sim \lambda_p^3/2\lambda^2$. Acceleration is compromised beyond L_d, and, in practice, the plasma length is proportional to the dephasing length, i.e. $L = \xi L_d$. In most cases, $\gamma_w^2 >> 1$, which allows the approximations $\beta_w \approx 1 - \lambda^2/2\lambda_p^2$, and $L = \xi\lambda_p^3/2\lambda^2 \approx \xi\gamma_w^2\lambda_p/2 >> \eta\lambda_p$, so that Eq. (30) becomes

$$S = (1+\beta)^2\gamma^2\frac{\xi\gamma_w^2}{\xi\gamma_w^2 + (1+\beta)\gamma^2(\xi\beta/2 + 2\eta)}. \quad (31)$$

For low values of γ, i.e. when $\gamma << \gamma_w$, Eq. (31) reduces to

$$S_{\gamma<<\gamma_w} = (1+\beta)^2\gamma^2. \quad (32)$$

Conversely, if $\gamma \to \infty$, Eq. (31) becomes

$$S_{\gamma\to\infty} = \frac{4}{1+4\eta/\xi}\gamma_w^2. \quad (33)$$

Finally, in the frame of the wake, i.e. when $\gamma = \gamma_w$, assuming that $\beta_w \approx 1$, Eq. (31) gives

$$S_{\gamma=\gamma_w} \approx \frac{2}{1+2\eta/\xi}\gamma_w^2. \quad (34)$$

Since η and ξ are of order unity, and the practical regimes of most interest satisfy $\gamma_w^2 >> 1$, the speedup that is obtained by using the frame of the wake will be near the maximum obtainable value given by Eq. (33).

Note that without the use of a moving window, the relativistic effects that are at play in the time domain would also be at play in the spatial domain [16], and the γ^2 scaling would transform into γ^4. Hence, it is important to use a moving window even in simulations in a Lorentz-boosted frame. For very high values of the boosted frame, the optimal velocity of the moving window may vanish (i.e. no moving window) or even reverse.

2.6. *Numerical stability and alternate formulation in a Galilean frame*

The numerical Cherenkov instability (NCI) [71] is the most serious numerical instability affecting multidimensional PIC simulations of relativistic particle beams and streaming plasmas [21, 24, 25, 66, 72, 73]. It arises from coupling between possibly numerically distorted electromagnetic modes and spurious beam modes, the latter due to the mismatch between the Lagrangian treatment of particles and the Eulerian treatment of fields [74].

In recent papers, the electromagnetic dispersion relations for the NCI were derived and solved for both FDTD [66, 75] and PSATD [76, 77] algorithms.

Several solutions have been proposed for mitigating the NCI [76–81]. Although they efficiently reduce the numerical instability, they typically introduce either strong smoothing of the currents and fields, or arbitrary numerical corrections, which are tuned specifically against the NCI and go beyond the natural discretization of the underlying physical equation. Therefore, it is sometimes unclear to what extent these added corrections could impact the physics at stake for a given resolution.

For instance, NCI-specific corrections include periodically smoothing the electromagnetic field components [24], using a special time step [21, 25] or applying a wideband smoothing of the current components [21, 25, 26]. Another set of mitigation methods involves scaling the deposited currents by a carefully designed wave-number-dependent factor [75, 77] or slightly modifying the ratio of electric and magnetic fields (E/B) before gathering their value onto the macroparticles [76, 79]. Yet another set of NCI-specific corrections [80, 81] consists in combining a small time step Δt, a sharp low pass spatial filter, and a spectral or high order scheme that is tuned so as to create a small, artificial "bump" in the dispersion relation [80]. While most mitigation methods have only been applied to Cartesian geometry, this last set of methods [80, 81] has the remarkable property that it can be applied [81] to both Cartesian geometry and quasicylindrical geometry (i.e. cylindrical geometry with azimuthal Fourier decomposition [30–32]). However, the use of a small time step proportionally slows down the progress of the simulation, and the artificial "bump" is again an arbitrary correction that departs from the underlying physics.

A new scheme was recently proposed [82, 83], which completely eliminates the NCI for a plasma drifting at a uniform relativistic velocity — with no arbitrary correction — by simply integrating the PIC equations in *Galilean coordinates* (also known as *comoving coordinates*). More precisely, in the new method, the Maxwell equations in *Galilean coordinates* are integrated analytically, using only natural hypotheses, within the PSATD framework (pseudospectral analytical time domain [49, 63]).

The idea of the proposed scheme is to perform a Galilean change of coordinates, and to carry out the simulation in the new coordinates:

$$\boldsymbol{x}' = \boldsymbol{x} - \boldsymbol{v}_{\mathrm{Gal}} t, \tag{35}$$

where $\boldsymbol{x} = x\,\boldsymbol{u}_x + y\,\boldsymbol{u}_y + z\,\boldsymbol{u}_z$ and $\boldsymbol{x}' = x'\,\boldsymbol{u}_x + y'\,\boldsymbol{u}_y + z'\,\boldsymbol{u}_z$ are the position vectors in the standard and Galilean coordinates, respectively.

When choosing $\boldsymbol{v}_{\mathrm{Gal}} = \boldsymbol{v}_0$, where $\boldsymbol{v}_0$ is the speed of the bulk of the relativistic plasma, the plasma does not move with respect to the grid in the Galilean coordinates $\boldsymbol{x}'$ — or, equivalently, in the standard coordinates $\boldsymbol{x}$, the grid moves along with the plasma. The heuristic intuition behind this scheme is that these coordinates should prevent the discrepancy between the Lagrangian and the Eulerian point of view, which gives rise to the NCI [74].

An important remark is that the Galilean change of coordinates (35) is a simple translation. Thus, when used in the context of Lorentz-boosted simulations, it does of course preserve the relativistic dilatation of space and time which gives rise to the characteristic computational speedup of the boosted frame technique.

Another important remark is that the Galilean scheme is *not* equivalent to a moving window (and in fact the Galilean scheme can be independently *combined* with a moving window). Whereas in a moving window grid points are added and removed so as to effectively translate the boundaries, in the Galilean scheme the grid points *themselves* are not only translated but in this case the physical equations are modified accordingly. Most importantly, the assumed time evolution of the current $\boldsymbol{J}$ within one time step is different in a standard PSATD scheme with a moving window and in a Galilean PSATD scheme [83].

In the Galilean coordinates $\boldsymbol{x}'$, the equations of particle motion and the Maxwell equations take the forms

$$\frac{d\boldsymbol{x}'}{dt} = \frac{\boldsymbol{p}}{\gamma m} - \boldsymbol{v}_{\mathrm{Gal}}, \tag{36a}$$

$$\frac{d\boldsymbol{p}}{dt} = q\left(\boldsymbol{E} + \frac{\boldsymbol{p}}{\gamma m} \times \boldsymbol{B}\right), \tag{36b}$$

$$\left(\frac{\partial}{\partial t} - \boldsymbol{v}_{\mathrm{Gal}} \cdot \boldsymbol{\nabla}'\right) \boldsymbol{B} = -\boldsymbol{\nabla}' \times \boldsymbol{E}, \tag{36c}$$

$$\frac{1}{c^2}\left(\frac{\partial}{\partial t} - \boldsymbol{v}_{\mathrm{Gal}} \cdot \boldsymbol{\nabla}'\right) \boldsymbol{E} = \boldsymbol{\nabla}' \times \boldsymbol{B} - \mu_0 \boldsymbol{J}, \tag{36d}$$

where $\boldsymbol{\nabla}'$ denotes a spatial derivative with respect to the Galilean coordinates $\boldsymbol{x}'$.

Integrating these equations from $t = n\Delta t$ to $t = (n+1)\Delta t$ results in the following update equations (see Ref. 83 for the details of the derivation):

$$\begin{aligned}\tilde{\mathbf{B}}^{n+1} &= \theta^2 C \tilde{\mathbf{B}}^n - \frac{\theta^2 S}{ck} i\boldsymbol{k} \times \tilde{\mathbf{E}}^n \\ &\quad + \frac{\theta \chi_1}{\epsilon_0 c^2 k^2}\, i\boldsymbol{k} \times \tilde{\mathbf{J}}^{n+1/2},\end{aligned} \tag{37a}$$

$$\begin{aligned}\tilde{\mathbf{E}}^{n+1} &= \theta^2 C \tilde{\mathbf{E}}^n + \frac{\theta^2 S}{k}\, ci\boldsymbol{k} \times \tilde{\mathbf{B}}^n \\ &\quad + \frac{i\nu\theta\chi_1 - \theta^2 S}{\epsilon_0 c k}\, \tilde{\mathbf{J}}^{n+1/2} \\ &\quad - \frac{1}{\epsilon_0 k^2}(\chi_2\, \hat{\rho}^{n+1} - \theta^2 \chi_3\, \hat{\rho}^n) i\boldsymbol{k},\end{aligned} \tag{37b}$$

where we have used the shorthand notations $\tilde{\mathbf{E}}^n \equiv \tilde{\mathbf{E}}(\boldsymbol{k}, n\Delta t)$ and $\tilde{\mathbf{B}}^n \equiv \tilde{\mathbf{B}}(\boldsymbol{k}, n\Delta t)$, as well as

$$\begin{aligned}C &= \cos(ck\Delta t), \\ S &= \sin(ck\Delta t), \quad k = |\boldsymbol{k}|,\end{aligned} \tag{38a}$$

$$\begin{aligned}\nu &= \frac{\boldsymbol{k} \cdot \boldsymbol{v}_{\mathrm{Gal}}}{ck}, \quad \theta = e^{i\boldsymbol{k}\cdot\boldsymbol{v}_{\mathrm{Gal}}\Delta t/2}, \\ \theta^* &= e^{-i\boldsymbol{k}\cdot\boldsymbol{v}_{\mathrm{Gal}}\Delta t/2},\end{aligned} \tag{38b}$$

$$\chi_1 = \frac{1}{1-\nu^2}(\theta^* - C\theta + i\nu\theta S), \tag{38c}$$

$$\begin{aligned}\chi_2 &= \frac{\chi_1 - \theta(1-C)}{\theta^* - \theta} \\ \chi_3 &= \frac{\chi_1 - \theta^*(1-C)}{\theta^* - \theta}.\end{aligned} \tag{38d}$$

Note that, in the limit $\boldsymbol{v}_{\mathrm{Gal}} = \mathbf{0}$, Eqs. (37a) and (37b) reduce to the standard PSATD equations [49], as expected. As shown in Refs. 82 and 83, the elimination of the NCI with the new Galilean integration is verified empirically via PIC simulations of uniform drifting plasmas and laser-driven plasma acceleration stages, and confirmed by a theoretical analysis of the instability.

3. The Electromagnetic Quasistatic Method

The electromagnetic quasistatic method was developed earlier than the Lorentz-boosted frame method, as a way to tackle the large separation of length and time scales between the plasma and the driver. The quasistatic approximation [9] takes advantage of the fact that (a) the laser or particle beam driver is moving close to the speed of light, and is hence very rigid, with a slow time response, and that (b) the plasma response is extremely fast, in comparison with the driver's. The separation of the driver and plasma time responses enables a separation in the treatment of the two components, as follows.

Assuming the driver at a given time and position, its high rigidity enables the approximation that it is quasistatic during the time that it takes for traversing a transverse slice of the plasma (assumed to be unperturbed by the driver ahead of it). The response of the plasma can thus be computed by following the evolution of the plasma slice as the driver propagates through it (see Fig. 5). The reconstruction of the longitudinal and transverse structure of the wake from the succession of transverse slices gives the full electric and magnetic field map for evolving the beam momenta and positions on a timescale that is commensurate with its rigidity.

Most formulations use the speed-of-light frame, defined as $\zeta = z - ct$, to follow the evolution of the plasma slices. Assuming a slice initialized ahead of the driver, the evolution of the plasma particles inside the slice is given by

$$\frac{d\mathbf{x}_p}{d\zeta} = \frac{d\mathbf{x}_p}{dt}\frac{dt}{d\zeta} = \frac{\mathbf{v}_p}{v_{pz} - c}, \tag{39}$$

$$\frac{d\mathbf{p}_p}{d\zeta} = \frac{q}{v_{pz} - c}(\mathbf{E} + \mathbf{v}_p \times \mathbf{B}). \tag{40}$$

The plasma charge and current densities are computed by accumulating the contributions of each plasma macroparticle i, corrected by the time taken by the particle to cross an interval of ζ:

$$\rho_p = \frac{1}{\delta x \delta y \delta \zeta}\sum_i \frac{q_i}{1 - v_{iz}/c}, \tag{41}$$

$$\mathbf{J}_p = \frac{1}{\delta x \delta y \delta \zeta}\sum_i \frac{q_i \mathbf{v}_\mathbf{i}}{1 - v_{iz}/c}. \tag{42}$$

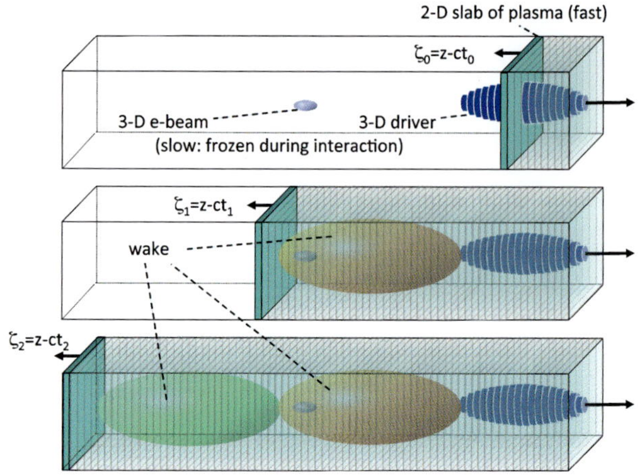

Fig. 5. Schematic of the quasistatic method. The driver beam and the accelerated electron beam are assumed to be static during the time that it takes them to traverse slices of the plasma column that they are propagating through. This enables a separation of scales and the evaluation of the response of the plasma to the driver, forming the wake, and to the accelerated beam, by following the evolution of a 2D slab of the plasma in the coordinates $\zeta = z - ct$.

In contrast, the evolution of a charged particle driver or witness beam (assumed to propagate near the speed of light) is given using the standard equations of motion:

$$\frac{d\mathbf{x}_{d/w}}{dt} = \mathbf{v}_{d/w}, \tag{43}$$

$$\frac{d\mathbf{p}_{d/w}}{dt} = q_{d/w}(\mathbf{E} + \mathbf{v}_{d/w} \times \mathbf{B}), \tag{44}$$

while their contributions to the charge and current densities are

$$\rho_{d/w} = \frac{1}{\delta x \delta y \delta z}\sum_i q_i, \tag{45}$$

$$\mathbf{J}_{d/w} = \frac{1}{\delta x \delta y \delta z}\sum_i q_i \mathbf{v}_\mathbf{i}. \tag{46}$$

The electric and magnetic fields are obtained by either solving the equations of the scalar and vector potentials in the Coulomb or Lorentz gauge [12, 13] or directly the Maxwell equations [84–87] which, under the quasistatic assumption

$$\frac{\partial}{\partial \zeta} = \frac{\partial}{\partial z} = \frac{-\partial}{\partial ct}, \tag{47}$$

become

$$\nabla_\perp \times \mathbf{E}_\perp = \frac{\partial B_z}{\partial \zeta}\hat{\mathbf{z}}, \tag{48}$$

$$\nabla_\perp \times E_z \hat{\mathbf{z}} = \frac{\partial(\mathbf{B}_\perp - \hat{\mathbf{z}} \times \mathbf{E}_\perp)}{\partial \zeta}, \tag{49}$$

$$\nabla_\perp \times \mathbf{B}_\perp - J_z \hat{\mathbf{z}} = -\frac{\partial E_z}{\partial \zeta} \hat{\mathbf{z}}, \tag{50}$$

$$\nabla_\perp \times B_z \hat{\mathbf{z}} - \mathbf{J}_\perp = -\frac{\partial(\mathbf{E}_\perp + \hat{\mathbf{z}} \times \mathbf{B}_\perp)}{\partial \zeta}, \tag{51}$$

$$\nabla_\perp \cdot \mathbf{E}_\perp - \rho = -\frac{\partial E_z}{\partial \zeta}, \tag{52}$$

$$\nabla_\perp \cdot \mathbf{B}_\perp = -\frac{\partial B_z}{\partial \zeta}. \tag{53}$$

The set of equations on the potentials or the fields can then be rearranged in a set of 2D Poisson-like equations that are solved iteratively with the particle motion equations. Unlike the PIC method, there is no single way of marching the set of equations together and the reader should refer to the descriptions of implementations in the various codes for more specific details [12, 13, 84–87].

4. The Ponderomotive Guiding Center Approximation

For laser pulses with envelopes that are long compared to the laser oscillations, it is advantageous to average over the fast laser oscillations and solve the laser evolution with an envelope equation [12, 13, 15, 88, 89]. Assuming a laser pulse in the form of an envelope modulating a plane wave traveling at the speed of light,

$$\tilde{A}_\perp = \hat{A}_\perp(z, \mathbf{x}_\perp, t) \exp ik_0\zeta + \text{c.c.}, \tag{54}$$

the average response of a plasma to the fast laser oscillations can be described by a ponderomotive force that inserts into a modified equation of motion:

$$\frac{d\mathbf{p}}{dt} = q(\mathbf{E} + \mathbf{v} \times \mathbf{B}) - \frac{q^2}{\gamma mc^2} \nabla |\hat{A}_\perp|^2, \tag{55}$$

with

$$\gamma = \sqrt{1 + \frac{|\mathbf{p}|^2}{m^2c^2} + \frac{2|q\hat{A}_\perp|^2}{m^2c^4}}. \tag{56}$$

Most codes [12, 13, 15, 88] solve the approximate envelope equation

$$\left[\frac{2}{c}\frac{\partial}{\partial t}\left(ik_0 + \frac{\partial}{\partial \zeta}\right) + \nabla_\perp^2\right] \hat{A}_\perp = \frac{q^2}{mc^2}\left\langle \frac{n}{\gamma} \right\rangle \hat{A}_\perp, \tag{57}$$

while the more complete envelope equation

$$\left[\frac{2}{c}\frac{\partial}{\partial t}\left(ik_0 + \frac{\partial}{\partial \zeta}\right) + \nabla_\perp^2 - \frac{\partial^2}{\partial t^2}\right] \hat{A}_\perp = \frac{q^2}{mc^2}\left\langle \frac{n}{\gamma} \right\rangle \hat{A}_\perp, \tag{58}$$

which retains the second time derivative is solved in the code INF&RNO [89]. The latter equation is more exact, enabling accurate simulation of the laser depletion into strongly depleted stages. As noted in Ref. 89, in order to avoid numerical inaccuracies, or having to grid the simulation very finely in the longitudinal direction, it is advantageous to use the polar representation of the laser complex field, namely $\hat{A}_\perp(\zeta) = A_\perp(\zeta)\exp[i\theta(\zeta)]$, rather than the more common Cartesian splitting between the real and imaginary parts $\hat{A}_\perp(\zeta) = \Re[A_\perp(\zeta)] + \imath\Im[A_\perp(\zeta)]$. As it turns out, the functions $A_\perp(\zeta)$ and $\theta(\zeta)$ are much smoother functions with respect to ζ than $\Re[A_\perp(\zeta)]$ and $\Im[A_\perp(\zeta)]$, which both exhibit very short wavelength oscillations in ζ, leading to more accurate numerical differentiation along ζ of the polar representation at a given longitudinal resolution.

5. Axisymmetry and Azimuthal Fourier Decomposition

Although full PIC codes are powerful tools, which capture a wide range of physical phenomena, they also require large computational resources. This is partly due to the use of a 3D Cartesian grid, which leads to a very large number of grid cells. (Typical 3D simulations of laser wakefield acceleration require $\sim 10^6$–10^8 grid cells.) For this reason, these algorithms need to be highly parallelized, and high resolution simulations can only be run on costly, large scale computer facilities. However, when the driver is cylindrically symmetric, it is possible to take advantage of the symmetry of the problem to reduce the computational cost of the algorithm [29–32].

5.1. *Azimuthal decomposition*

Let us consider the fields $\boldsymbol{E}$, $\boldsymbol{B}$, $\boldsymbol{J}$ and ρ in cylindral coordinates (r, θ, z), expressed as a Fourier series in θ:

$$F(r, \theta, z) = \text{Re}\left[\sum_{\ell=0}^{\infty} \tilde{F}_\ell(r, z) e^{-i\ell\theta}\right], \tag{59}$$

with

$$\tilde{F}_\ell = C_\ell \int_0^{2\pi} d\theta\, F(r,\theta,z)e^{i\ell\theta} \tag{60}$$

and

$$\begin{cases} C_0 = \dfrac{1}{2\pi}, \\ C_\ell = \dfrac{1}{\pi} \quad \text{for } \ell > 0, \end{cases} \tag{61}$$

where F represents any of the quantities E_r, E_θ, E_z, B_r, B_θ, B_z, J_r, J_θ, J_z and ρ, and $\tilde{F}_\ell$ are the associated Fourier components (ℓ is the index of the corresponding azimuthal mode). In the general case, this azimuthal decomposition does not simplify the problem, since an infinity of modes have to be considered in Eq. (59). However, in the case of a cylindrically symmetric laser pulse, only the very first modes have nonzero components. For instance, the wakefield is represented exclusively by the mode $\ell = 0$. (This is because the quantities E_r, E_θ, E_z, B_r, B_θ, B_z, J_r, J_θ, J_z and ρ associated with the wakefield are independent of θ.) On the other hand, the field of the laser pulse *does* depend on θ, in cylindrical coordinates. For example, for a cylindrically symmetric pulse propagating along z and polarized along $\boldsymbol{e}_\alpha = \cos(\alpha)\boldsymbol{e}_x + \sin(\alpha)\boldsymbol{e}_y$,

$$\boldsymbol{E} = E_0(r,z)\boldsymbol{e}_\alpha \tag{62}$$

$$= E_0(r,z)[\cos(\alpha)(\cos(\theta)\boldsymbol{e}_r - \sin(\theta)\boldsymbol{e}_\theta) + \sin(\alpha)(\sin(\theta)\boldsymbol{e}_r + \cos(\theta)\boldsymbol{e}_\theta)] \tag{63}$$

$$= \mathrm{Re}[E_0(r,z)e^{i\alpha}e^{-i\theta}]\boldsymbol{e}_r + \mathrm{Re}[-iE_0(r,z)e^{i\alpha}e^{-i\theta}]\boldsymbol{e}_\theta. \tag{64}$$

Here the amplitude E_0 does not depend on θ because the pulse was assumed to be cylindrically symmetric. In this case, the above relation shows that the fields E_r and E_θ of the laser are represented exclusively by the mode $\ell = 1$. A similar calculation shows that the same holds for B_r and B_θ. On the whole, only the modes $\ell = 0$ and $\ell = 1$ are *a priori* necessary for modeling laser wakefield acceleration. Under those conditions, the infinite sum in Eq. (59) is truncated at a chosen $\ell_{\max}$. In principle, $\ell_{\max} = 1$ is sufficient for laser wakefield acceleration. However, $\ell_{\max}$ is kept as a free parameter in the algorithm, in order to verify that higher modes are negligible, as well as to allow for less-symmetric configurations. Because codes based on this algorithm are able to take into account the modes with $\ell > 0$, they are said to be "quasicylindrical" (or "quasi-3D" by some authors [31]), in contrast to cylindrical codes, which assume that all fields are independent of θ, and thus consider only the mode $\ell = 0$.

5.2. *Discretized Maxwell equations*

When the Fourier expressions of the fields are injected into the Maxwell equations (written in cylindrical coordinates), the different azimuthal modes decouple. In this case, the Maxwell–Ampère and Maxwell–Faraday equations — which are needed to update the fields in the PIC cycle — can be written separately for each azimuthal mode ℓ:

$$\frac{\partial \tilde{B}_{r,\ell}}{\partial t} = \frac{i\ell}{r}\tilde{E}_{z,\ell} + \frac{\partial \tilde{E}_{\theta,\ell}}{\partial z}, \tag{65a}$$

$$\frac{\partial \tilde{B}_{\theta,\ell}}{\partial t} = -\frac{\partial \tilde{E}_{r,\ell}}{\partial z} + \frac{\partial \tilde{E}_{z,\ell}}{\partial r}, \tag{65b}$$

$$\frac{\partial \tilde{B}_{z,\ell}}{\partial t} = -\frac{1}{r}\frac{\partial (r\tilde{E}_{\theta,\ell})}{\partial r} - \frac{i\ell}{r}\tilde{E}_{r,\ell}, \tag{65c}$$

$$\frac{1}{c^2}\frac{\partial \tilde{E}_{r,\ell}}{\partial t} = -\frac{i\ell}{r}\tilde{B}_{z,\ell} - \frac{\partial \tilde{B}_{\theta,\ell}}{\partial z} - \mu_0 \tilde{J}_{r,\ell}, \tag{65d}$$

$$\frac{1}{c^2}\frac{\partial \tilde{E}_{\theta,\ell}}{\partial t} = \frac{\partial \tilde{B}_{r,\ell}}{\partial z} - \frac{\partial \tilde{B}_{z,\ell}}{\partial r} - \mu_0 \tilde{J}_{\theta,\ell}, \tag{65e}$$

$$\frac{1}{c^2}\frac{\partial \tilde{E}_{z,\ell}}{\partial t} = \frac{1}{r}\frac{\partial (r\tilde{B}_{\theta,\ell})}{\partial r} + \frac{i\ell}{r}\tilde{B}_{r,\ell} - \mu_0 \tilde{J}_{z,\ell}. \tag{65f}$$

In order to discretize these equations, each azimuthal mode is represented on a 2D grid, on which the discretized Maxwell–Ampère and Maxwell–Faraday equations are given by

$$D_t\tilde{B}_r|^n_{j,\ell,k+\frac{1}{2}} = \frac{i\ell}{j\Delta r}\tilde{E}^{\,n}_{z\,j,\ell,k+\frac{1}{2}} + D_z\tilde{E}_\theta|^n_{j,\ell,k+\frac{1}{2}}, \tag{66a}$$

$$D_t\tilde{B}_\theta|^n_{j+\frac{1}{2},\ell,k+\frac{1}{2}} = -D_z\tilde{E}_r|^n_{j+\frac{1}{2},\ell,k+\frac{1}{2}} + D_r\tilde{E}_z|^n_{j+\frac{1}{2},\ell,k+\frac{1}{2}}, \tag{66b}$$

$$D_t\tilde{B}_z|^n_{j+\frac{1}{2},\ell,k} = -\frac{(j+1)\tilde{E}^{\,n}_{\theta\,j+1,\ell,k}}{(j+\frac{1}{2})\Delta r} + \frac{j\tilde{E}^{\,n}_{\theta\,j,\ell,k}}{(j+\frac{1}{2})\Delta r} - \frac{i\ell}{(j+\frac{1}{2})\Delta r}\tilde{E}^{\,n}_{r\,j+\frac{1}{2},\ell,k} \tag{66c}$$

for the magnetic field components, and

$$\frac{1}{c^2} D_t \tilde{E}_r\Big|^{n+\frac{1}{2}}_{j+\frac{1}{2},\ell,k} = -\frac{i\,\ell}{(j+\frac{1}{2})\Delta r} \tilde{B}_{z\,j+\frac{1}{2},\ell,k}^{\;n+\frac{1}{2}} - D_z \tilde{B}_\theta\Big|^{n+\frac{1}{2}}_{j+\frac{1}{2},\ell,k} - \mu_0 \tilde{J}_{r\,j+\frac{1}{2},\ell,k}^{\;n+\frac{1}{2}}, \quad (67a)$$

$$\frac{1}{c^2} D_t \tilde{E}_\theta\Big|^{n+\frac{1}{2}}_{j,\ell,k} = D_z \tilde{B}_r\Big|^{n+\frac{1}{2}}_{j,\ell,k} - D_r \tilde{B}_z\Big|^{n+\frac{1}{2}}_{j,\ell,k} - \mu_0 \tilde{J}_{\theta\,j,\ell,k}^{\;n+\frac{1}{2}}, \quad (67b)$$

$$\frac{1}{c^2} D_t \tilde{E}_z\Big|^{n+\frac{1}{2}}_{j,\ell,k+\frac{1}{2}} = \frac{(j+\frac{1}{2}) \tilde{B}_{\theta\,j+\frac{1}{2},\ell,k+\frac{1}{2}}^{\;n+\frac{1}{2}}}{j\Delta r} - \frac{(j-\frac{1}{2}) \tilde{B}_{\theta\,j-\frac{1}{2},\ell,k+\frac{1}{2}}^{\;n+\frac{1}{2}}}{j\Delta r} + \frac{i\,\ell}{j\Delta r} \tilde{B}_{r\,j,\ell,k+\frac{1}{2}}^{\;n+\frac{1}{2}} - \mu_0 \tilde{J}_{z\,j,\ell,k+\frac{1}{2}}^{\;n+\frac{1}{2}} \quad (67c)$$

for the electric field components.

The numerical operators D_r and D_z are defined by

$$(D_r F)_{j',\ell,k'} = \frac{F_{j'+\frac{1}{2},\ell,k'} - F_{j'-\frac{1}{2},\ell,k'}}{\Delta r},$$

$$(D_z F)_{j',\ell,k'} = \frac{F_{j',\ell,k'+\frac{1}{2}} - F_{j',\ell,k'-\frac{1}{2}}}{\Delta z},$$

where j' and k' can be integers or half-integers. Notice that these discretized Maxwell equations are not valid on-axis (i.e. for $j = 0$), due to singularities in some of the terms. Therefore, on the axis, these equations are replaced by specific boundary conditions, which are based on the symmetry properties of the fields (see Ref. 30 for details).

Compared to a 3D Cartesian calculation with $n_x \times n_y \times n_z$ grid cells, a quasicylindrical calculation with two modes ($l = 0$ and $l = 1$) will require only $3\,n_r \times n_z$ grid cells. Assuming $n_x = n_y = n_r = 100$ as a typical transverse resolution, a quasicylindrical calculation is typically over an order of magnitude less computationally demanding than its 3D Cartesian equivalent.

6. Filtering

It is common practice to apply digital filtering to the charge or current density in PIC simulations as a complement or an alternative to using higher order splines [37]. A commonly used filter in PIC simulations is the three-point filter $\phi_j^f = \alpha\phi_j + (1-\alpha)(\phi_{j-1} + \phi_{j+1})/2$, where ϕ^f is the filtered quantity. This filter is called a bilinear filter when $\alpha = 0.5$. Assuming $\phi = e^{jkx}$ and $\phi^f = g(\alpha, k)e^{jkx}$, the filter gain g is given as a function of the filtering coefficient α and the wave number k by $g(\alpha, k) = \alpha + (1-\alpha)\cos(k\Delta x) \approx 1 - (1-\alpha)\frac{(k\Delta x)^2}{2} + O(k^4)$. The total attenuation G for n successive applications of filters of coefficients $\alpha_1 \ldots \alpha_n$ is given by $G = \prod_{i=1}^n g(\alpha_i, k) \approx 1 - (n - \sum_{i=1}^n \alpha_i)\frac{(k\Delta x)^2}{2} + O(k^4)$. A sharper cutoff in k space is provided by using $\alpha_n = n - \sum_{i=1}^{n-1} \alpha_i$, so that $G \approx 1 + O(k^4)$. Such step is called a "compensation" step [37]. For the bilinear filter ($\alpha = 1/2$), the compensation factor is $\alpha_c = 2 - 1/2 = 3/2$. For a succession of n applications of the bilinear factor, it is $\alpha_c = n/2 + 1$.

It is sometimes necessary to filter on a relatively wide band of wavelength, necessitating the application of a large number of passes of the bilinear filter or the use of filters acting on many points. The former can become very intensive computationally, while the latter is problematic for parallel computations using domain decomposition, as the footprint of the filter may eventually surpass the size of subdomains. A workaround is to use a combination of filters of limited footprint. A solution based on the combination of three point filters with various strides was proposed in Ref. 25, and it operates as follows.

The bilinear filter provides complete suppression of the signal at the grid Nyquist wavelength (twice the grid cell size). Suppression of the signal at integer multiples of the Nyquist wavelength can be obtained by using a stride s in the filter $\phi_j^f = \alpha\phi_j + (1-\alpha)(\phi_{j-s} + \phi_{j+s})/2$, for which the gain is given by $g(\alpha, k) = \alpha + (1-\alpha)\cos(sk\Delta x) \approx 1 - (1-\alpha)\frac{(sk\Delta x)^2}{2} + O(k^4)$. For a given stride, the gain is given by the gain of the bilinear filter shifted in k space, with the pole $g = 0$ shifted from the wavelength $\lambda = 2/\Delta x$ to $\lambda = 2s/\Delta x$, with additional poles, as given by $sk\Delta x = \arccos(\frac{\alpha}{\alpha-1}) \pmod{2\pi}$. The resulting filter is a passband between the poles, but since the poles are spread at different integer values in k space, a wideband low pass filter can be constructed by combining filters using different strides. As shown in Ref. 25, the successive application of 4 passes + compensation of filters with strides 1, 2 and 4 has a nearly equivalent falloff in gain to

80 passes + compensation of a bilinear filter. Yet, the strided filter solution needs only 15 passes of a three-point filter, compared to 81 passes for an equivalent n-pass bilinear filter, yielding a gain of 5.4 in the number of operations in favor of the combination of filters with strides. The width of the filter with stride 4 extends only on 9 points, compared to 81 points for a single pass equivalent filter, hence giving a gain of 9 in compactness for the stride filter combination in comparison with the single pass filter with a large stencil, resulting in more favorable scaling with the number of computational cores for parallel calculations.

7. Inputs and Outputs

Initialization of the plasma columns and drivers (laser or particle beam) is performed via the specification of multidimensional functions that describe the initial state with, if needed, a time dependence, or from reconstruction of distributions based on experimental data. Care is needed when initializing quantities in parallel, to avoid double counting and ensure smoothness of the distributions at the interface of computational domains. When the sum of the initial distributions of charged particles is not charge-neutral, initial fields are computed using generally a static approximation with Poisson solves accompanied by proper relativistic scalings [39, 46].

Outputs include dumps of particle and field quantities at regular intervals, histories of particle distribution moments, spectra, etc., and plots of the various quantities. In parallel simulations, the diagnostic subroutines need to handle additional complexity from the domain decomposition, as well as a large amount of data that may necessitate data reduction in some form before saving to the disk.

Simulations using the quasistatic method or in a Lorentz-boosted frame require additional considerations, as described below.

7.1. *Outputs in parallel quasistatic simulations*

When running in parallel with domain decomposition in the direction of propagation, the slices of the driver and wake are known at various positions (or "stations") in the plasma column, complicating the output of data. The parallelization in the transverse direction (perpendicular to the direction of propagation) uses standard domain decomposition of the particles and fields, and does not add complexity compared to standard PIC simulations. The parallelization in the direction of propagation uses pipelining [90], as follows.

Assuming that the grid has n_z cells along the longitudinal dimension and that N processors are used for a run, n_z/N consecutive slices are assigned to each computational core, as sketched in Fig. 6. During the first iteration, the computational core N evolves the slices at the front of the grid that it contains, then passes the data marching through ζ to the previous core $N-1$ while pushing the driver beam particle to the next station in the plasma column. During the second iteration, both of the cores N and $N-1$ evolve the slices that they contain at two subsequent stations. After N steps, all processors are active and the procedure is repeated until the slices on processor 1 reach the last station at the exit of the plasma column (see Fig. 6). Because each core contains a slice that is at a different station in the plasma column, the output modules must write data at different steps for each computational core.

7.2. *Inputs and outputs in a boosted frame simulation*

The input and output data are often known from, or compared to, experimental data. Thus, calculating in a frame other than the laboratory entails transformations of the data between the calculation frame and the laboratory frame. This section describes the procedures that have been implemented in the PIC framework Warp [91] to handle the input and output

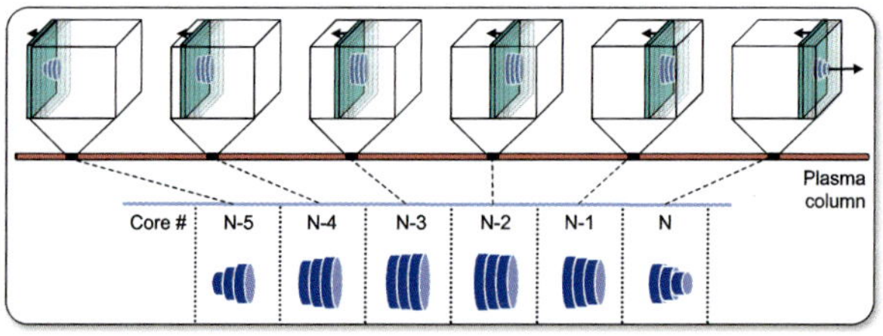

Fig. 6. In the quasistatic mode, the driver is distributed among n slices, which are uniformly spread among N processors. Using a pipelining algorithm, slices on a given processor are pushed from one station to the next, while the slices of the previous processor will be pushed to the same station one time step later. For conciseness, only the portion of the grid that contains the driver is represented in the diagram, but the logic extends to the entire grid.

of data between the frame of calculation and the laboratory frame [27]. Simultaneity of events between two frames is valid only for a plane that is perpendicular to the relative motion of the frame. As a result, the input/output processes involve the input of data (particles or fields) through a plane, as well as output through a series of planes, all of which are perpendicular to the direction of the relative velocity between the frame of calculation and the other frame of choice.

7.2.1. *Input in a boosted frame simulation*

Particles. Particles are launched through a plane using a technique that is generic and applies to Lorentz-boosted frame simulations in general, including plasma acceleration, and is illustrated using the case of a positively charged particle beam propagating through a background of cold electrons in an assumed continuous transverse focusing system, leading to a well-known growing transverse "electron cloud" instability [16]. In the laboratory frame, the electron background is initially at rest and a moving window is used to follow the beam progression. Traditionally, the beam macroparticles are initialized all at once in the window, while background electron macroparticles are created continuously in front of the beam on a plane that is perpendicular to the beam velocity. In a frame moving at some fraction of the beam velocity in the laboratory frame, the beam initial conditions at a given time in the calculation frame are generally unknown and one must initialize the beam differently. However, one can take advantage of the fact that the beam initial conditions are often known for a given plane in the laboratory, either directly or via simple calculation or projection from the conditions at a given time in the laboratory frame. Given the position and velocity $\{x, y, z, v_x, v_y, v_z\}$ for each beam macroparticle at time $t = 0$ for a beam moving at the average velocity $v_b = \beta_b c$ (c is the speed of light) in the laboratory, and using the standard synchronization ($z = z' = 0$ at $t = t' = 0$) between the laboratory and calculation frames, the procedure for transforming the beam quantities for injection in a boosted frame moving at velocity βc in the laboratory is as follows (the superscript $'$ relates to quantities known in the boosted frame, while the superscript $*$ relates to quantities that are known at a given longitudinal position z^* but different times of arrival):

(1) Project positions at $z^* = 0$ assuming ballistic propagation

$$t^* = \frac{z - \bar{z}}{v_z}, \tag{68}$$

$$x^* = x - v_x t^*, \tag{69}$$

$$y^* = y - v_y t^*, \tag{70}$$

$$z^* = 0, \tag{71}$$

with the velocity components being left unchanged.

(2) Apply Lorentz transformation from the laboratory frame to the boosted frame:

$$t'^* = -\gamma t^*, \tag{72}$$

$$x'^* = x^*, \tag{73}$$

$$y'^* = y^*, \tag{74}$$

$$z'^* = \gamma\beta c t^*, \tag{75}$$

$$v'^*_x = \frac{v^*_x}{\gamma(1 - \beta\beta_b)}, \tag{76}$$

$$v'^*_y = \frac{v^*_y}{\gamma(1 - \beta\beta_b)}, \tag{77}$$

$$v'^*_z = \frac{v^*_z - \beta c}{1 - \beta\beta_b}, \tag{78}$$

where $\gamma = 1/\sqrt{1-\beta^2}$. With the knowledge of the time at which each beam macroparticle crosses the plane into consideration, one can inject each beam macroparticle in the simulation at the appropriate location and time.

(3) Synchronize macroparticles in the boosted frame, obtaining their positions at a fixed $t' = 0$ (before any particle is injected)

$$z' = z'^* - \bar{v}'^*_z t'^*. \tag{79}$$

This additional step is needed for setting the electrostatic or electromagnetic fields at the plane of injection. In a PIC code, the 3D fields are calculated by solving the Maxwell equations (or static approximation like Poisson, Darwin or other [39]) on a grid on which the source term is obtained from the macroparticles' distribution. This requires generation of a 3D representation of the beam distribution of macroparticles at a given time before they cross the injection plane at z'^*. This is accomplished by expanding the beam distribution longitudinally such that all macroparticles (so far known at different times of arrival at the injection plane) are synchronized to the same time in the boosted frame. To keep the beam shape constant, the particles are "frozen" until they cross that plane: the three velocity components and the two position components perpendicular to the boosted frame velocity are kept constant, while the remaining position component is advanced at the average beam velocity. As particles cross the plane of injection, they become regular "active" particles with full 6D dynamics.

Figure 7 (top) shows a snapshot of a beam that has passed partly through the injection plane. As the frozen beam macroparticles pass through the injection plane (which moves opposite to the beam in the boosted frame), they are converted into "active" macroparticles. The charge or current density is accumulated from the active and the frozen particles, thus ensuring that the fields at the plane of injection are consistent.

Laser. Similarly to the particle beam, the laser is injected through a plane perpendicular to the axis of propagation of the laser (by default z). The electric field $E_\perp$ that is to be emitted is given by the formula

$$E_\perp(x,y,t) = E_0 f(x,y,t) \sin[\omega t + \phi(x,y,\omega)], \quad (80)$$

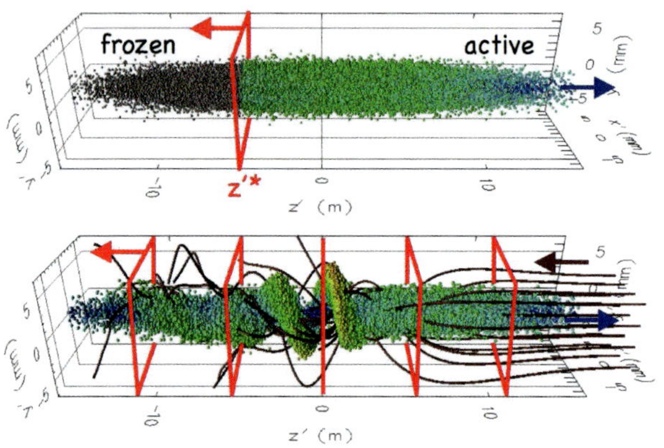

Fig. 7. *Top*: Snapshot of a particle beam showing "frozen" (gray spheres) and "active" (colored spheres) macroparticles traversing the injection plane (red rectangle). *Bottom*: Snapshot of the beam macroparticles (colored spheres) passing through the background of electrons (dark brown streamlines) and the diagnostic stations (red rectangles). The electrons, the injection plane and the diagnostic stations are fixed in the laboratory plane, and are thus counterpropagating to the beam in a boosted frame.

where E_0 is the amplitude of the laser electric field, $f(x,y,t)$ the laser envelope, ω the laser frequency, $\phi(x,y,\omega)$ a phase function to account for focusing, defocusing or injection at an angle, and t time. By default, the laser envelope is a 3D Gaussian of the form

$$f(x,y,t) = e^{-(x^2/2\sigma_x^2 + y^2/2\sigma_y^2 + c^2t^2/2\sigma_z^2)}, \quad (81)$$

where σ_x, σ_y and σ_z are the dimensions of the laser pulse; or it can be defined arbitrarily by the user at run time. If $\phi(x,y,\omega)=0$, the laser is injected at a waist and parallel to the axis z.

If, for convenience, the injection plane is moving at constant velocity $\beta_s c$, the formula is modified to take into account the Doppler effect on frequency and amplitude, and becomes

$$E_\perp(x,y,t) = (1-\beta_s)E_0 f(x,y,t) \times \sin[(1-\beta_s)\omega t + \phi(x,y,\omega)]. \quad (82)$$

The injection of a laser of frequency ω is considered for a simulation using a boosted frame moving at βc with respect to the laboratory. Assuming that the laser is injected at a plane that is fixed in the laboratory, and thus moving at $\beta_s = -\beta$ in the boosted frame, the injection in the boosted frame is given by

$$E_\perp(x',y',t') = (1-\beta_s)E'_0 f(x',y',t') \times \sin[(1-\beta_s)\omega' t' + \phi(x',y',\omega')] \quad (83)$$

$$= \frac{E_0}{\gamma} f(x',y',t') \times \sin\left[\frac{\omega t'}{\gamma} + \phi(x',y',\omega')\right] \quad (84)$$

since $E'_0/E_0 = \omega'/\omega = 1/(1+\beta)\gamma$.

The electric field is then converted into currents that get injected via a 2D array of macroparticles, with one positive and one dual negative macroparticle for each array cell in the plane of injection, whose weights and motion are governed by $E_\perp(x',y',t')$. Injection using this dual array of macroparticles offers the advantage of automatically including the longitudinal component that arises from emission into a boosted frame, and automatically verifing the discrete Gauss law thanks to using the charge-conserving (e.g. Esirkepov) current deposition scheme [61].

7.2.2. *Output in a boosted frame simulation*

Some quantities, such as charge or dimensions perpendicular to the boost velocity, are Lorentz-invariant. They are thus readily available from standard diagnostics in the boosted frame calculations. Quantities that do not fall into this category are recorded at a number of regularly spaced "stations," immobile in the laboratory frame, at a succession of time intervals to record data history, or averaged over time. A visual example is given in Fig. 7 (bottom). Since the space–time locations of the diagnostic grids in the laboratory frame generally do not coincide with the space–time positions of the macroparticles and grid nodes used for the calculation in a boosted frame, some interpolation is performed at run time during the data collection process. As a complement or an alternative, selected particle or field quantities can be dumped at regular intervals and quantities are reconstructed in the laboratory frame during a post-processing phase. The choice of the methods depends on the requirements of the diagnostics and particular implementations.

8. Outlook

The development of plasma-based accelerators depends critically on high performance, high fidelity modeling to capture the full complexity of acceleration processes that develop over a large range of space and time scales. The field will continue to be a driver for pushing the state of the art in the detailed modeling of relativistic plasmas. The modeling of tens of multi-GeV stages, as envisioned for plasma-based high energy physics colliders, will require further advances in algorithmics, coupled with preparation of the codes to take full advantage of the upcoming generation of exascale supercomputers.

Acknowledgments

This work was supported by US-DOE Contract DE-AC02-05CH11231.

This document was prepared as an account of work sponsored in part by the United States Government. While this document is believed to contain correct information, neither the United States Government nor any agency thereof, nor The Regents of the University of California, nor any of their employees, nor the authors makes any warranty, express or implied, or assumes any legal responsibility for the accuracy, completeness, or usefulness of any information, apparatus, product, or process disclosed, or represents that its use would not infringe privately owned rights. Reference herein to any specific commercial product, process, or service by its trade name, trademark, manufacturer, or otherwise, does not necessarily constitute or imply its endorsement, recommendation, or favoring by the United States Government or any agency thereof, or The Regents of the University of California. The views and opinions of authors expressed herein do not necessarily state or reflect those of the United States Government or any agency thereof or The Regents of the University of California.

References

[1] F. Tsung, W. Lu, M. Tzoufras, W. Mori, C. Joshi, J. Vieira, L. Silva and R. Fonseca, *Phys. Plasmas* **13**, 56708 (2006).

[2] C. G. R. Geddes, D. L. Bruhwiler, J. R. Cary, W. B. Mori, J.-L. Vay, S. F. Martins, T. Katsouleas, E. Cormier-Michel, W. M. Fawley, C. Huang, X. Wang, B. Cowan, V. K. Decyk, E. Esarey, R. A. Fonseca, W. Lu, P. Messmer, P. Mullowney, K. Nakamura, K. Paul, G. R. Plateau, C. B. Schroeder, L. O. Silva, C. Toth, F. S. Tsung, M. Tzoufras, T. Antonsen, J. Vieira and W. P. Leemans, Computational Studies and Optimization of Wakefield Accelerators, in *J. Phys. Conf. Ser.* 012002 (2008).

[3] C. Geddes *et al.*, Laser plasma particle accelerators: Large fields for smaller facility sources, in *Scidac Rev.* **13**(12) (2009).

[4] C. Huang, W. An, V. K. Decyk, W. Lu, W. B. Mori, F. S. Tsung, M. Tzoufras, S. Morshed, T. Antonsen, B. Feng, T. Katsouleas, R. A. Fonseca, S. F. Martins, J. Vieira, L. O. Silva, E. Esarey, C. G. R. Geddes, W. P. Leemans, E. Cormier-Michel, J.-L. Vay, D. L. Bruhwiler, B. Cowan, J. R. Cary and K. Paul, *J. Phys. Conf. Ser.* **180**, 012005 (2009).

[5] W. P. Leemans, A. J. Gonsalves, H.-S. Mao, K. Nakamura, C. Benedetti, C. B. Schroeder, C. Tóth, J. Daniels, D. E. Mittelberger, S. S. Bulanov, J.-L. Vay, C. G. R. Geddes and E. Esarey, *Phys. Rev. Lett.* **113**, 245002 (2014).

[6] I. Blumenfeld, C. E. Clayton, F.-J. Decker, M. J. Hogan, C. Huang, R. Ischebeck, R. Iverson, C. Joshi, T. Katsouleas, N. Kirby, W. Lu, K. A. Marsh, W. B. Mori, P. Muggli, E. Oz, R. H. Siemann, D. Walz and M. Zhou, *Nature* **445**, 741 (2007).

[7] S. V. Bulanov, J. J. Wilkens, T. Zh. Esirkepov, G. Korn, G. Kraft, S. D. Kraft, M. Molls and V. S. Khoroshkov, *Phys. Usp.* **57**, 1149 (2014).

[8] S. Steinke, J. van Tilborg, C. Benedetti, C. G. R. Geddes, C. B. Schroeder, J. Daniels, K. K. Swanson,

A. J. Gonsalves, K. Nakamura, N. H. Matlis, B. H. Shaw, E. Esarey and W. P. Leemans, *Nature* **530**, 190 (2016).
[9] P. Sprangle, E. Esarey and A. Ting, *Phys. Rev. Lett.* **64**, 2011 (1990).
[10] T. M. Antonsen and P. Mora, *Phys. Rev. Lett.* **69**, 2204 (1992).
[11] J. Krall, A. Ting, E. Esarey and P. Sprangle, *Phys. Rev. E* **48**, 2157 (1993).
[12] P. Mora and T. Antonsen, *Phys. Plasmas* **4**, 217 (1997).
[13] C. Huang, V. K. Decyk, C. Ren, M. Zhou, W. Lu, W. B. Mori, J. H. Cooley, T. M. Antonsen Jr. and T. Katsouleas, *J. Comput. Phys.* **217**, 658 (2006).
[14] C. Benedetti, C. B. Schroeder, E. Esarey, C. G. R. Geddes and W. P. Leemans, *AIP Conf. Proc.* **1299**, 250 (2010).
[15] B. M. Cowan, D. L. Bruhwiler, E. Cormier-Michel, E. Esarey, C. G. R. Geddes, P. Messmer and K. M. Paul, *J. Comput. Phys.* **230**, 61 (2011).
[16] J.-L. Vay, *Phys. Rev. Lett.* **98**, 130405/1 (2007).
[17] D. L. Bruhwiler, J. R. Cary, B. M. Cowan, K. Paul, C. G. R. Geddes, P. J. Mullowney, P. Messmer, E. Esarey, E. Cormier-Michel, W. Leemans and J.-L. Vay, New developments in the simulation of advanced accelerator concepts, in *AIP Conf. Proc.* (2009), pp. 29–37.
[18] J.-L. Vay, D. L. Bruhwiler, C. G. R. Geddes, W. M. Fawley, S. F. Martins, J. R. Cary, E. Cormier-Michel, B. Cowan, R. A. Fonseca, M. A. Furman, W. Lu, W. B. Mori and L. O. Silva, *J. Phys. Conf. Ser.* **180**, 012006 (2009).
[19] J.-L. Vay *et al.*, Application of the reduction of scale range in a Lorentz-boosted frame to the numerical simulation of particle acceleration devices, in *Proc. Particle Accelerator Conference* (Vancouver, Canada, 2009).
[20] S. F. Martins *et al.*, Boosted frame PIC simulations of LWFA: Towards the energy frontier, in *Proc. Particle Accelerator Conference* (Vancouver, Canada, 2009).
[21] J. L. Vay, C. G. R. Geddes, C. Benedetti, D. L. Bruhwiler, E. Cormier-Michel, B. M. Cowan, J. R. Cary and D. P. Grote, *AIP Conf. Proc.* **1299**, 244 (2010).
[22] S. F. Martins, R. A. Fonseca, W. Lu, W. B. Mori and L. O. Silva, *Nat. Phys.* **6**, 311 (2010).
[23] S. F. Martins, R. A. Fonseca, J. Vieira, L. O. Silva, W. Lu and W. B. Mori, *Phys. Plasmas* **17**, 56705 (2010).
[24] S. F. Martins, R. A. Fonseca, L. O. Silva, W. Lu and W. B. Mori, *Comput. Phys. Commun.* **181**, 869 (2010).
[25] J. L. Vay, C. G. R. Geddes, E. Cormier-Michel and D. P. Grote, *J. Comput. Phys.* **230**, 5908 (2011).
[26] J. Vay, C. G. R. Geddes, E. Cormier-Michel and D. P. Grote, *Phys. Plasmas* **18**, 30701 (2011).
[27] J. L. Vay, C. G. R. Geddes, E. Esarey, C. B. Schroeder, W. P. Leemans, E. Cormier-Michel and D. P. Grote, *Phys. Plasmas* **18**, 123103 (2011).
[28] P. Yu, X. Xu, A. Davidson, A. Tableman, T. Dalichaouch, F. Li, M. D. Meyers, W. An, F. S. Tsung, V. K. Decyk, F. Fiuza, J. Vieira, R. A. Fonseca, W. Lu, L. O. Silva and W. B. Mori, *J. Comput. Phys.*, 747 (2016).
[29] B. Godfrey, *The IPROP Three-Dimensional Beam Propagation Code* (Defense Technical Information Center, 1985).
[30] A. F. Lifschitz, X. Davoine, E. Lefebvre, J. Faure, C. Rechatin and V. Malka, *J. Comput. Phys.* **228**, 1803 (2009).
[31] A. Davidson, A. Tableman, W. An, F. Tsung, W. Lu, J. Vieira, R. Fonseca, L. Silva and W. Mori, *J. Comput. Phys.* **281**, 1063 (2015).
[32] R. Lehe, M. Kirchen, I. A. Andriyash, B. B. Godfrey and J.-L. Vay, *Comput. Phys. Commun.* **203**, 66 (2016).
[33] I. A. Andriyash, R. Lehe and A. Lifschitz, *Phys. Plasmas* **23**, 033110 (2016).
[34] B. A. Shadwick, C. B. Schroeder and E. Esarey, *Phys. Plasmas* **16**, 56704 (2009).
[35] E. Cormier-Michel, C. G. R. Geddes, E. Esarey, C. B. Schroeder, D. L. Bruhwiler, K. Paul, B. Cowan and W. P. Leemans, Scaled simulations of a 10 GeV accelerator, in *AIP Conf. Proc.* 297–302 (2009).
[36] C. G. R. Geddes *et al.*, Scaled simulation design of high quality laser wakefield accelerator stages, in *Proc. Particle Accelerator Conference* (Vancouver, Canada, 2009).
[37] C. K. Birdsall and A. B. Langdon, *Plasma Physics via Computer Simulation* (Adam-Hilger, 1991).
[38] J. Boris, Relativistic plasma simulation-optimization of a hybrid code, in *Proc. Fourth Conf. Num. Sim. Plasmas* (Naval Res. Lab., Washington, D.C., 1970), pp. 3–67.
[39] J. L. Vay, *Phys. Plasmas* **15**, 56701 (2008).
[40] J.-L. Vay, D. P. Grote, R. H. Cohen and A. Friedman, *Comput. Sci. Disc.* **5**, 014019 (2012).
[41] J.-L. Vay, J.-C. Adam and A. Heron, *Comput. Phys. Commun.* **164**, 171 (2004).
[42] K. Yee, *IEEE Trans. Antennas Propag.* **Ap14**, 302 (1966).
[43] J. Cole, *IEEE Trans. Microw. Theory Tech.* **45**, 991 (1997).
[44] J. Cole, *IEEE Trans. Antennas Propag.* **50**, 1185 (2002).
[45] M. Karkkainen, E. Gjonaj, T. Lau and T. Weiland, Low-dispersion wakefield calculation tools, in *Proc. Int. Computational Accelerator Physics Conference* (Chamonix, France, 2006), pp. 35–40.
[46] B. M. Cowan, D. L. Bruhwiler, J. R. Cary, E. Cormier-Michel and C. G. R. Geddes, *Phys. Rev. ST Accel. Beams* **16** (2013).
[47] A. Pukhov, *J. Plasma Phys.* **61**, 425 (1999).

[48] R. Lehe, A. Lifschitz, C. Thaury, V. Malka and X. Davoine, *Phys. Rev. ST Accel. Beams* **16**, 021301 (2013).
[49] I. Haber, R. Lee, H. Klein and J. Boris, Advances in electromagnetic simulation techniques, in *Proc. Sixth Conf. Num. Sim. Plasmas* (Berkeley, CA, 1973), pp. 46–48.
[50] J.-L. Vay, I. Haber and B. B. Godfrey, *J. Comput. Phys.* **243**, 260 (2013).
[51] J. M. Dawson, *Rev. Mod. Phys.* **55**, 403 (1983).
[52] Q. Liu, *Microw. Opt. Technol. Lett.* **15**, 158 (1997).
[53] Y. Ohmura and Y. Okamura, *Piers Online* **6**, 632 (2010).
[54] H. Vincenti and J.-L. Vay, *Comput. Phys. Commun.* **200**, 147 (2016).
[55] H. Abe, N. Sakairi, R. Itatani and H. Okuda, *J. Comput. Phys.* **63**, 247 (1986).
[56] A. B. Langdon, *Comput. Phys. Commun.* **70**, 447 (1992).
[57] B. Marder, *J. Comput. Phys.* **68**, 48 (1987).
[58] J.-L. Vay and C. Deutsch, *Phys. Plasmas* **5**, 1190 (1998).
[59] C. Munz, P. Omnes, R. Schneider, E. Sonnendrucker and U. Voss, *J. Comput. Phys.* **161**, 484 (2000).
[60] J. Villasenor and O. Buneman, *Comput. Phys. Commun.* **69**, 306 (1992).
[61] T. Esirkepov, *Comput. Phys. Commun.* **135**, 144 (2001).
[62] R. Morse and C. Nielson, *Phys. Fluids* **14**, 830 (1971).
[63] J. L. Vay, I. Haber and B. B. Godfrey, *J. Comput. Phys.* **243**, 260 (2013).
[64] R. W. Hockney and J. W. Eastwood, *Computer Simulation Using Particles* (1988).
[65] H. Lewis, *J. Comput. Phys.* **10**, 400 (1972).
[66] B. B. Godfrey and J.-L. Vay, *J. Comput. Phys.* **248**, 33 (2013).
[67] J. Boris and R. Lee, *J. Comput. Phys.* **12**, 131 (1973).
[68] B. Cowan, D. Bruhwiler, E. Cormier-Michel, E. Esarey, C. G. R. Geddes, P. Messmer and K. Paul, Laser wakefield simulation using a speed-of-light frame envelope model, in *AIP Conf. Proc.* (2009), pp. 309–314.
[69] E. Esarey, C. B. Schroeder and W. P. Leemans, *Rev. Mod. Phys.* **81**, 1229 (2009).
[70] C. B. Schroeder, C. Benedetti, E. Esarey and W. P. Leemans, *Phys. Rev. Lett.* **106**, 135002 (2011).
[71] B. Godfrey, *J. Comput. Phys.* **15**, 504 (1974).
[72] L. Sironi and A. Spitkovsky, private communication (2011).
[73] X. Xu, P. Yu, S. F. Martins, F. S. Tsung, V. K. Decyk, J. Vieira, R. A. Fonseca, W. Lu, L. O. Silva and W. B. Mori, *Comput. Phys. Commun.* **184**, 2503 (2013).
[74] B. Godfrey, *J. Comput. Phys.* **19**, 58 (1975).
[75] B. B. Godfrey and J.-L. Vay, *J. Comput. Phys.* **267**, 1 (2014).
[76] B. B. Godfrey, J.-L. Vay and I. Haber, *J. Comput. Phys.* **258**, 689 (2014).
[77] B. B. Godfrey, J.-L. Vay and I. Haber, *IEEE Trans. Plasma Sci.* **42**, 1339 (2014).
[78] B. B. Godfrey, J.-L. Vay and I. Haber, *J. Comput. Phys.* **258**, 689 (2014).
[79] B. B. Godfrey and J.-L. Vay, *Comput. Phys. Commun.* **196**, 221 (2015).
[80] P. Yu, X. Xu, V. K. Decyk, F. Fiuza, J. Vieira, F. S. Tsung, R. A. Fonseca, W. Lu, L. O. Silva and W. B. Mori, *Comput. Phys. Commun.* **192**, 32 (2015).
[81] P. Yu, X. Xu, A. Tableman, V. K. Decyk, F. S. Tsung, F. Fiuza, A. Davidson, J. Vieira, R. A. Fonseca, W. Lu, L. O. Silva and W. B. Mori, *Comput. Phys. Commun.* **197**, 144 (2015).
[82] M. Kirchen, R. Lehe, B. B. Godfrey, I. Dornmair, S. Jalas, K. Peters, J.-L. Vay and A. R. Maier, *Phys. Plasmas* **23**, 100704 (2016).
[83] R. Lehe, M. Kirchen, B. B. Godfrey, A. R. Maier and J.-L. Vay, *Phys. Rev. E* **94**, 053305 (2016).
[84] K. V. Lotov, *Phys. Plasmas* **5**, 785 (1998).
[85] K. V. Lotov, *Phys. Rev. ST Accel. Beams* **6**, 061301 (2003).
[86] T. Mehrling, C. Benedetti, C. B. Schroeder and J. Osterhoff, *Plasma Phys. Control. Fus.* **56**, 084012 (2014).
[87] W. An, V. K. Decyk, W. B. Mori and T. M. Antonsen, *J. Comput. Phys.* **250**, 165 (2013).
[88] D. F. Gordon, W. B. Mori and T. M. Antonsen, *IEEE Trans. Plasma Sci.* **28**, 1135 (2000).
[89] C. Benedetti, C. B. Schroeder, E. Esarey and W. P. Leemans, Efficient modeling of laser-plasma accelerators using the ponderomotive-based code INF&RNO, in *ICAP* (Jacow, Rostock-Warnemunde, Germany, 2012), p. THAAI2.
[90] B. Feng, C. Huang, V. Decyk, W. Mori, P. Muggli and T. Katsouleas, *J. Comput. Phys.* **228**, 5340 (2009).
[91] D. P. Grote, A. Friedman, J.-L. Vay and I. Haber, The warp code: modeling high intensity ion beams, in *AIP Conf. Proc.* (749) (2005), pp. 55–58.

Jean-Luc Vay received his M.S. and Ph.D. degrees in Physics from the University of Paris, France, in 1993 and 1996, respectively. He is currently a senior research scientist at Lawrence Berkeley National Laboratory, where he leads the accelerator modeling effort in the Accelerator Technology and Applied Physics Division. Dr. Vay's research interests center on the development and usage of advanced numerical techniques in application to the modeling of conventional particle accelerators and advanced concepts such as plasma accelerators. He is the author of over 150 papers and reports.

Rémi Lehe received his M.S. and Ph.D. degrees in Physics from Ecole Normale Supérieure, France, and Ecole Polytechnique, France, in 2011 and 2014, respectively. He is currently a postdoctoral fellow at Lawrence Berkeley National Laboratory, where his research focuses on high accuracy spectral PIC algorithms, and on various methods to speed up large scale simulations of laser-plasma accelerators. Dr. Lehe also actively contributes to the PIC code Warp, and is the original developer of the code FBPIC.

Reviews of Accelerator Science and Technology
Vol. 9 (2016) 187–207

DOI: 10.1142/S1793626816300097

Luminosity Limitations of Linear Colliders Based on Plasma Acceleration*

Valeri Lebedev†, Alexey Burov and Sergei Nagaitsev
*Fermi National Accelerator Laboratory,
PO Box 500, Batavia, IL 60510, USA*
†*val@fnal.gov*

Particle acceleration in plasma creates a possibility of exceptionally high accelerating gradients and appears to be a very attractive option for future linear electron–positron and/or γ–γ colliders. These high accelerating gradients have already been demonstrated in a number of experiments. However, a linear collider requires exceptionally high beam brightness which still needs to be demonstrated. In this article we discuss major phenomena which limit the beam brightness of accelerated beam and, consequently, the collider luminosity.

Keywords: Plasma acceleration; collider; luminosity.

1. Introduction

In recent years two basic concepts for a linear electron–positron collider based on acceleration in plasma have been proposed and developed. They differ by the method of exciting the plasma wave. The first concept, developed by the SLAC/UCLA group [1], is based on plasma wave excitation by a high energy electron beam; the second one, developed by LBNL [2–4], is based on plasma excitation by a short pulse of laser radiation. In the case of plasma excitation by an electron bunch, the length of a single accelerating section is much longer due to the much higher energy stored in the electron beam (compared to the energy stored in a laser pulse) and essentially negligible RF dephasing with length. In the case of plasma excitation with a laser pulse, the RF dephasing is associated with the speed of a laser pulse being below the speed of light due to plasma permeability. Together dephasing and the energy stored in the laser pulse limit the length of acceleration to about 10 cm and the energy gain to about 10 GeV. Recently, a proposal to excite the plasma by a proton beam has also been actively discussed; and an experiment to validate such a concept is currently under construction at CERN [5]. The plasma excitation with a proton beam allows one to achieve the largest length of a single accelerating section, which potentially could greatly simplify the accelerator. However, to our knowledge, nobody has published a concept for a collider based on proton-excited plasma acceleration. In the following, we mostly concentrate on the acceleration of colliding beams. In this case, the method of plasma excitation does not play a significant role and, with minor exceptions, will not be discussed.

Plasma acceleration can be produced in two regimes. The first one is the quasilinear regime, where the amplitude of the plasma electron density perturbation is smaller than the initial plasma electron density. The second one is called the bubble regime, where the plasma electrons are completely blown away from the axis by the drive pulse. In this case, for a given plasma density the maximum accelerating gradient is achieved. As will be seen below, only the latter regime looks practical. In the case of electron bunch acceleration, this regime does not have any plasma electrons inside of the accelerated (witness) bunch and, consequently, does not suffer from expulsion of plasma electrons by the field of the witness bunch, which degrades beam focusing in the bunch tail and results in emittance growth. As will be seen below, acceleration of positrons with brightness required for a collider is presently unfeasible for both regimes.

*Work supported by the US DOE under contract #DE-AC02-07CH11359.

Both the SLAC/UCLA and LBNL concepts suggest that they can achieve a collider operation with luminosity $10^{34}\,\mathrm{cm}^{-2}\mathrm{s}^{-1}$ or above (see Table 1). However, a close examination of the presented parameters shows that they are not completely consistent with limitations coming from fundamental principles of accelerator physics. To demonstrate that, the following limitations are considered:

- Limitations on the beam emittance and momentum spread coming from the synchrotron radiation in the final focus lenses/quads;
- Emittance growth due to multiple scattering in the plasma and/or residual gas;
- Pinching/expulsion of plasma electrons by the electric field of the positron/electron bunch;
- Pinching/expulsion of plasma ions (in the bubble regime) by the electric field of the electron/positron bunch;
- Bremsstrahlung on ions and impact ionization of ions by the witness bunch electric field, if heavy ions are used to mitigate ion pinching by an electron bunch in the bubble regime;
- Transverse beam breakup (BBU) instability due to the transverse wakefield excited in plasma.

Separately, each of the above limitations can, in principle, be mitigated to an acceptable level. However, they cannot be satisfied simultaneously within a set of ideas discussed up to the present time. We knowingly do not consider technical limitations, which are very tight but can potentially be overcome with future developments in technology.

2. Final Focus and Beam Emittance Limitations

To be competitive with high energy proton colliders and, in particular, with the LHC, a lepton collider needs to have the luminosity $\geq 10^{34}\,\mathrm{cm}^{-2}\mathrm{s}^{-1}$. It is exceptionally challenging requirement, demanding very high beam brightness. As will be seen in the following sections, an increase in the accelerated beam emittance would significantly simplify problems associated with acceleration of high brightness beam. However, a compensation for luminosity loss by tighter focusing at the interaction point (IP) is limited by many effects. In this section we consider the beam emittance and the momentum spread limitations caused by the final focus (FF) chromaticity and synchrotron radiation (SR) in the FF lenses. Note that other effects, which are not considered here, may require even smaller emittances. However, the obtained values are already sufficiently small to switch on a number of other limitations discussed in the following sections. A more detailed discussion on final focusing limitations is presented in Ref. 6.

An increase in the beam emittance can be potentially compensated for by an increase in the focusing strength of the FF lenses resulting in a reduction of the FF chromaticity and, consequently, a smaller beam size at the IP. Using plasma lenses instead of quadrupoles allows one to achieve focusing gradients that are orders of magnitude above what could be achieved using conventional technology. This could potentially greatly reduce the effects of the FF chromaticity. However, an increase in the focusing gradient results in SR from the FF lens. It limits the increase in the focusing strength.

To make a rough estimate, we assume that there is only one axially symmetric focusing lens[a] in the FF; the focusing length is equal to the lens length, $F = L_{\mathrm{lens}}$; and the focusing chromaticity is suppressed by factor $\eta_F \gg 1$ with help of upstream optical elements. We also assume that in the case of flat beams, the vertical emittance and the IP beta-function are smaller than their values for the horizontal plane; and in the case of round beams, the horizontal and vertical emittances and the IP beta-functions are equal, respectively. Consequently, the vertical focusing chromaticity is larger than the horizontal one for flat beams, and the horizontal and vertical chromaticities are equal for round beams. Then, in the absence of chromaticity suppression, a particle with a momentum deviation of $\Delta p/p$ is misfocused and its vertical offset in the IP is equal to

$$\Delta y \approx F\sqrt{\frac{\varepsilon_y}{\beta_y^*}}\frac{\Delta p}{p}. \tag{1}$$

Here ε_y is the vertical single particle emittance (twice the particle action), and β_y^* is the vertical beta function at the IP. We also assume that the betatron phase is chosen to obtain the maximum particle displacement in the lens. Requiring this offset to be smaller than the maximum particle offset at the IP,

[a]Axial symmetry implies that it is a plasma lens. Although with minor corrections this estimate is justified for quadrupole focusing.

$\sigma_y = \sqrt{\varepsilon_y \beta_y^*}$ (i.e. the offset of a particle with the same action but the betatron phase shifted by 90°), one obtains a limitation on the momentum deviation:

$$\frac{\Delta p}{p} < \eta_F \frac{\beta_y^*}{F}, \tag{2}$$

where we have also accounted for the suppression of focusing chromaticity of the final lens by a factor η_F. A finite value of η_F is typically related to the higher order corrections and practical limitations of the correction scheme accuracy, which we characterize by a single number. As one can see from Eq. (2), the limitation on the momentum deviation does not depend on ε_y.

The momentum change due to SR in the FF lens is[b]

$$\left.\frac{\Delta p}{p}\right|_{\mathrm{SR}} \approx \frac{2}{3}\frac{e^4 B^2 \gamma L_{\mathrm{lens}}}{m_e^3 c^6} \approx \frac{2}{3}\frac{r_e \gamma^3}{F}\left(\frac{\varepsilon_x}{\beta_x^*} + \frac{\varepsilon_y}{\beta_y^*}\right), \tag{3}$$

where γ is the particle Lorentz factor, e and m_e are its charge and mass, r_e is its classical radius, and B is the lens magnetic field at radius $r = \sqrt{x^2+y^2}$ at the particle location in the lens so that $B = Gr \approx GF\sqrt{\varepsilon_x/\beta_x^* + \varepsilon_y/\beta_y^*}$. In transition to the second half of the equation, we also assumed that $1/F \approx eGL_{\mathrm{lens}}/(mc^2\gamma)$ and $F = L_{\mathrm{lens}}$. Equaling $\Delta p/p$ of Eqs. (2) and (3) for particles with 2σ amplitudes, one obtains limitations on the normalized RMS beam emittance,

$$\varepsilon_{ny}|_{\max} \equiv \varepsilon_y \gamma \le \frac{1}{4}\sqrt[3]{\frac{3S^{*2}\eta_F}{4\pi^2 r_e R^2 R_\varepsilon}}, \tag{4}$$

and the relative RMS momentum spread,

$$\left.\frac{\Delta p}{p}\right|_{\max} \le \frac{\gamma}{F}\sqrt[3]{\frac{4 r_e S^* \eta_F^2 R_\varepsilon}{3\pi R}}. \tag{5}$$

Here $S^* = \pi\sigma_x^*\sigma_y^*$ is the effective beam cross-section at the IP, $R = \sigma_x^*/\sigma_y^* \ge 1$ is the ratio of beam sizes at the IP, and $R_\varepsilon = (1+\varepsilon_x\beta_y^*/\varepsilon_y\beta_x^*)/2$ is the parameter characterizing the ratio of beam sizes in the lens. Taking into account that S^* is determined by the luminosity, one can see from Eq. (4) that the limitation on the beam emittance does not depend on the lens focusing strength and has weak dependence on η_F, i.e. stronger focusing does not help to increase the beam emittance. However, as one can see from Eq. (5), it allows one to accept a larger momentum spread.

[b]Gaussian units are used throughout this article.

For the luminosity of $2\cdot 10^{34}\,\mathrm{cm^{-2}s^{-1}}$, the bunch frequency of 15 kHz and the beam energy of 0.5 TeV, one obtains $S^* = 314\,\mathrm{nm}^2$. Assuming round beam ($R = 1, R_\varepsilon = 1$), a factor of 100 times suppression of the FF chromaticity ($\eta_F = 100$) and the focal distance of $F = 1$ m, one obtains from Eqs. (4) and (5) $\varepsilon_{ny} \le 1.6\,\mu$m and $\Delta p/p \le 1.5\cdot 10^{-2}$. Note that this estimate is still quite optimistic, and it will be significantly smaller if quadrupoles are used instead of an axial symmetric lens.

Table 1 presents parameters for plasma-based colliders [7] suggested by SLAC/UCLA (PWFA — plasma wake field accelerator) [1] and LBNL (LPA—laser plasma accelerator) [2–4]. Parameters for much more mature ILC and CLIC projects are also presented [7, 8] One can see that the beam emittance for an LPA is approximately an order of magnitude smaller than in the above estimate. At first glance it appears to be a significant margin. However, repulsion of plasma electrons or pinching of plasma ions by the electron bunch electric field, discussed in the following sections, exclude possible use of a plasma lens at the required beam brightness. The use of quadrupoles leaves no margin for possible emittance increase for the LPA. The last line of Table 1 presents the ratio of the emittance limitation of Eq. (4) to the RMS emittance presented in the same table. As one can see, the margins for different machines are not significantly different.

Table 1. RMS emittances and momentum spreads for various linear collider proposals.

	ILC	CLIC	LPA	PWFA
Beam energy, TeV	0.25	1.5	0.5	1.5
Luminosity, $10^{34}\,\mathrm{cm^{-2}s^{-1}}$	1.8	6	2	6.3
Particles per bunch, 10^{10}	2	0.37	0.4	1
Bunch rep. rate, kHz	6.5	15.6	15	10
σ_x, nm	474	40	10	194
σ_y, nm	6	1	10	1.1
RMS norm.h.emit., ε_x, μm	10	0.66	0.1	10
RMS norm.v.emit., ε_y, μm	0.035	0.020	0.1	0.035
RMS mom. spread, %	0.1*	0.35	N/C	N/C
RMS bunch length, μm	300	44	1	20
IP size ratio, $R = \sigma_x/\sigma_y$	79	40	1	176
$R_\varepsilon = (1+\varepsilon_x\beta_y^*/\varepsilon_y\beta_x^*)/2$	7	0.84	1	1.8
Emit. margin, $\varepsilon_{ny}\vert_{\max}/\varepsilon_y$	12	4	16	2

*Taken from Ref. 8.
N/C: not cited.

3. Particle Focusing and Acceleration in a Plasma

In plasma-based accelerators, the thermal velocities of plasma electrons are much smaller than the velocities excited by a plasma wave and the electron temperature and the motion of ions can be neglected. In this case a plasma wave can be described by a hydrodynamical approximation, where only electron motion is taken into account:

$$\begin{cases} \dfrac{\partial n_e}{\partial t} + \mathrm{div}(n_e \mathbf{v}_e) = 0, \\ \mathrm{div}(\mathbf{E}) = -4\pi e(n_e - n_0), \\ \dfrac{\partial \mathbf{v}_e}{\partial t} + (\mathbf{v}_e \cdot \nabla)\mathbf{v}_e = -\dfrac{e}{m_e}\left(\mathbf{E} + \dfrac{1}{c}[\mathbf{v}_e \mathbf{B}]\right); \end{cases} \tag{6}$$

where n_0 is the unperturbed electron density, and the sign minus takes into account the negative charge of plasma electrons. Linearizing these equations, one obtains

$$\begin{cases} \dfrac{\partial \tilde{n}}{\partial t} + n_0\, \mathrm{div}(\mathbf{v}_e) = 0, \\ \mathrm{div}(\mathbf{E}) = -4\pi e\tilde{n}, \\ \dfrac{\partial \mathbf{v}_e}{\partial t} = -\dfrac{e}{m_e}\mathbf{E}, \end{cases} \tag{7}$$

where $\tilde{n} = n_e - n_0$ is the plasma density perturbation. Excluding $\mathbf{v}$ and $\mathbf{E}$ from the above equations, one obtains:

$$\frac{\partial^2 \tilde{n}}{\partial t^2} + \omega_p^2 \tilde{n} = 0, \tag{8}$$

where $\omega_p = \sqrt{4\pi n_0 e^2/m}$ is the plasma frequency. Equation (8) shows that in the linear regime the plasma oscillates at the plasma frequency, independently of the initial density perturbation.

To find conditions of applicability of the linearized Eqs. (7), we consider a one-dimensional solution with all values changing as $e^{i(\omega t - kz)}$. Then, we obtain from Eqs. (7) the following relationships:

$$\begin{cases} v_{e_z} = c\tilde{n}/n_0, \\ E = -E_0 \tilde{n}/n_0. \end{cases} \tag{9}$$

Here, we have taken into account that the plasma wave is propagating with the speed of light, $\omega = kc$, $\omega = \omega_p$, and

$$E_0 = \frac{4\pi e n_0}{k} \tag{10}$$

is the electric field amplitude for a harmonic perturbation of plasma electron density with 100% amplitude ($\tilde{n}/n_0 = 1$). As one can see, a density perturbation comparable to the initial plasma density results in velocities of plasma electrons comparable to the speed of light. Consequently, if the density perturbation is comparable to the initial density, the relativistic equations for plasma electron motion have to be used in Eqs. (6).

Performing a Fourier transform in time for the Maxwell equation,

$$\mathrm{rot}\mathbf{B} = \frac{4\pi \mathbf{j}}{c} + \frac{1}{c}\frac{\partial \mathbf{E}}{\partial t}, \tag{11}$$

and obtaining from the bottom equation of Eqs. (7) a relationship between the harmonics of the electric field and the current $\mathbf{j}_\omega = n_e e^2 \mathbf{E}_\omega / i\omega m_e$, one can write

$$\mathrm{rot}\mathbf{B}_\omega = \frac{i\omega}{c}\left(1 - \frac{\omega_p^2}{\omega^2}\right)\mathbf{E}_\omega. \tag{12}$$

For infinite plasma in a linear regime, $\omega = \omega_p$, which results in rot $\mathbf{B} = 0$. In an axially symmetric case, this shows that plasma oscillations do not create the azimuthal component of the magnetic field. Note that this conclusion is not correct for a plasma with a boundary where a surface plasma wave is excited, and for high amplitude plasma oscillations. In both cases $\omega < \omega_p$, and the magnetic field penetrates into the plasma but exponentially decays with the penetration depth.

To find the beam focusing force, we assume an azimuthal symmetry of the accelerating field but do not assume that the plasma wave needs to have a small amplitude. Rewriting the central equation of Eqs. (6) in the cylindrical coordinates, we obtain

$$\frac{1}{r}\frac{\partial (rE_r)}{\partial r} + \frac{\partial E_z}{\partial z} = -4\pi e(n_e - n_0). \tag{13}$$

Taking into account that $E_r|_{r=0} = 0$ and solving Eq. (13) near the axis, where one can consider the plasma density and longitudinal electric field to be constant, we obtain the plasma electric field near the axis:

$$E_r = -\left[2\pi e(n_e - n_0) + \frac{1}{2}\frac{\partial E_z}{\partial z}\right] r. \tag{14}$$

Similarly, using Eq. (11), we get the magnetic field near the axis:

$$B_\theta = \frac{r}{2c}\left(\frac{\partial E_z}{\partial t} - 4\pi e n_e v_{e_z}\right), \tag{15}$$

where v_{e_z} is the longitudinal velocity of plasma electrons near the axis. Consequently, the focusing force is

$$F_r = e\left(E + \frac{[\mathbf{v}_b\mathbf{B}]}{c}\right)$$
$$= \mp\frac{er}{2}\left(\frac{\partial E_z}{\partial z} + \frac{v_b}{c^2}\frac{\partial E_z}{\partial t} + 4\pi e\Big(n_e\left(1 - \frac{v_b v_{e_z}}{c^2}\right) - n_0\Big)\right), \quad (16)$$

where v_b is the bunch velocity, the top and bottom signs should be taken for accelerated positrons and electrons respectively, and the positive longitudinal electric field accelerates positrons. For the wave propagating with the velocity of the driving beam v_d, one can write $\mathbf{E}(\mathbf{r}_\perp, z, t) = \mathbf{E}(\mathbf{r}_\perp, z - v_d t)$. It results in

$$\partial E_z/\partial z + (v_b/c^2)\partial E_z/\partial t = (1 - v_b v_d/c^2)\partial E_z/\partial z.$$

Consequently, if both the driving and the accelerated beams move with velocities close to the speed of light, the first two terms of Eq. (16) are canceled and the focusing force is determined by the particle density at the axis.

$$F_r = \mp 2\pi e^2 r[n_e(1 - v_{e_z}/c) - n_0]. \quad (17)$$

As one can see, the motion of plasma electrons in the beam direction reduces their contribution to the focusing force. In the bubble regime, when electrons are absent on the axis and the ion density changes inside the bubble can be neglected, the focusing force is constant inside the bubble.

4. Emittance Growth Due to Multiple Scattering in a Plasma

There are two major limitations which require very strong focusing of bunches, which are accelerated in a plasma. The first one is related to multiple scattering of accelerated particles on the plasma ions and electrons (if present); and the second one is related to the BBU instability associated with the transverse impedance of the plasma channel. In this section we consider the first limitation.

The normalized emittance growth in a singly ionized plasma due to multiple Coulomb scattering is given by (in the ultrarelativistic case) [9]:

$$\frac{d\varepsilon_n}{d\gamma} \approx \frac{2\pi(Z(Z+1) + n_e/n_0 - 1)r_e^2 n_0 \Lambda_c}{\gamma(d\gamma/ds)}\beta_f(\gamma), \quad (18)$$

where Z is the plasma ion charge, $\Lambda_c \approx 18$ the Coulomb logarithm, and $\beta_f(\gamma)$ the beta function of beam focusing.

Below we consider the emittance growth for the most optimistic case of acceleration of electrons in a bubble regime in a fully ionized hydrogen plasma ($n_e = 0, Z = 1$). To simplify an estimate we will assume that a harmonic wave with $E = E_0\cos(\omega t - kz)$ is excited.[c] Here we also assume 100% density modulation resulting in E_0 being determined by Eq. (10), and $k = \omega_p/c$. Then, the acceleration rate is equal to:

$$\frac{d\gamma}{ds} \approx \frac{eE_0}{mc^2}\cos\phi_{\text{acc}} = \sqrt{4\pi n_0 r_e}\cos\phi_{\text{acc}}, \quad (19)$$

where ϕ_{acc} is the accelerating phase. Assuming that the beam focusing is supported by the plasma itself, we obtain from Eq. (17) the corresponding beta function:

$$\beta_f(\gamma) = \frac{c\sqrt{2\gamma}}{\omega_p}. \quad (20)$$

Substituting Eqs. (19) and (20) into Eq. (18) and integrating to the final energy with the γ-factor of γ_f, we obtain the normalized emittance increase due to multiple scattering in the course of the entire acceleration:

$$\Delta\varepsilon_n = \frac{\sqrt{2\gamma_f}r_e\Lambda_c}{\cos\phi_{\text{acc}}}. \quad (21)$$

Here, we have neglected a contribution to the integral of its lower limit corresponding to the initial beam energy. As we can see from Eq. (21), the emittance growth does not depend on the plasma density.

For the beam acceleration to 0.5 TeV with the accelerating phase of 60° one obtains $\Delta\varepsilon_n = 0.14$ nm. This value is almost three orders of magnitude smaller than the value for the LPA emittance presented in Table 1. However, it has been achieved due to extremely strong plasma focusing. For the plasma density of $10^{17}\,\text{cm}^{-3}$ suggested in Ref. 2, one obtains the effective magnetic field gradient of plasma focusing of $3\cdot 10^4$ T/cm. It is more than three orders of magnitude larger than for the state-of-the-art superconducting quadrupoles. This implies that only plasma focusing is capable of suppressing the emittance growth due to multiple scattering on the plasma ions. Figure 1 presents the RMS beam size and the beta function for beam acceleration with LPA parameters. As one can see, the RMS beam size

[c] It is straightforward to extend this calculation for a more accurate description of beam acceleration in the bubble regime considered in detail in Secs. 6 and 7.

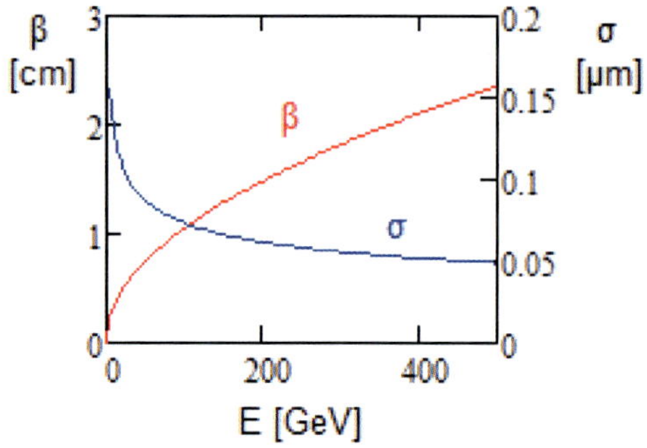

Fig. 1. The beta function and beam size for acceleration in the bubble regime; $n_0 = 10^{17}\ \mathrm{cm}^{-3}$.

is well below 1 μm for the major part of the acceleration. Finally, we need to note that multiple scattering also limits the use of heavy plasma ions. In the case of LPA it requires $Z \leq 10$ if singly charged ions are used.

5. Bunch Deceleration in a Uniform Plasma

In this section we consider the deceleration of a short bunch in a uniform plasma. We assume that the transverse beam size is equal to zero. The consequences and the accuracy of this approximation will be considered at the end of this section. To find the decelerating force along the bunch, we will follow the standard procedure presented in Ref. 10. The approach consists of three steps: (1) finding a collective plasma response at large impact parameters, where the plasma perturbation theory is justified; (2) computing a contribution of small impact parameters, where the plasma perturbation theory is not justified but the collective response of the plasma can be neglected; and (3) combining two contributions at some medium impact parameter.

To find the collective plasma response, we follow Ref. 11. We consider a pointlike charge q moving with velocity v_b along axis z through a plasma. Its charge density can be presented as

$$\rho(\mathbf{r}_\perp, z, t) = q\delta(\mathbf{r}_\perp)\delta(z - v_b t). \tag{22}$$

To describe a plasma response we add this external charge to the middle equation of Eqs. (7) so that

$$\mathrm{div}(\mathbf{E}) = 4\pi(q\delta(\mathbf{r}_\perp)\delta(z - v_b t) - e\tilde{n}). \tag{23}$$

Excluding $\mathbf{E}$ and $\mathbf{v}$ from the modified Eqs. (7), we obtain

$$\frac{\partial^2 \tilde{n}}{\partial t^2} + \omega_p^2 \tilde{n} = \frac{4\pi n_0 e q}{m}\delta(\mathbf{r}_\perp)\delta(z - v_b t). \tag{24}$$

The solution to Eq. (24) is the Green function of a harmonic oscillator, namely

$$\tilde{n} = \frac{\omega_p q}{e v_b}\sin(\omega_p t - k_p z)\theta(t - z/v_b)\delta(\mathbf{r}_\perp). \tag{25}$$

Here $\theta(t)$ is the step function, which is 1 or 0 for positive or negative values of its argument, respectively; we have also taken into account that $\delta(z - v_b t) = \delta(t - z/v_b)/v_b$, and $k_p = \omega_p/v_b$. Transitioning to the bunch frame, we cancel the dependence on time. For $|z - v_b t| \gg r_\perp/\gamma$, a solution to the resulting Poisson equation is straightforward. Coming back to the lab frame, we obtain the resulting scalar potential,

$$\varphi = \begin{cases} -2qk_p K_0(r_\perp k_p)\sin(\omega_p t - k_p z), & -s \gg r_\perp/\gamma, \\ 0, & s \gg r_\perp/\gamma, \end{cases} \tag{26}$$

where $s = z - v_b t$ and $K_0(x)$ is the modified Bessel function, and we have also taken into account that the negative sign of plasma electron charge introduces a sign change in Eq. (26). In the ultrarelativistic case $r_\perp/\gamma \to 0$, that results in a replacement of inequalities in Eq. (26) by the θ-function. Setting $v_b = c$, we obtain the corresponding electric field for the ultrarelativistic case ($k_p = \omega_p/c$):

$$\begin{aligned} E_\parallel &= -2qk_p^2 K_0(r_\perp k_p)\cos(\omega_p t - k_p z)\theta(t - z/c), \\ E_\perp &= -2qk_p^2 K_1(r_\perp k_p)\sin(\omega_p t - k_p z)\theta(t - z/c). \end{aligned} \tag{27}$$

Here $K_1(x)$ is the modified Bessel function. The evaluation of rot($\mathbf{E}$) demonstrates that the magnetic field in the trailing wave is equal to zero. It is related to the fact that the magnetic field, excited by the longitudinal component of the plasma current, is compensated for by the magnetic field excited by changing longitudinal electric field, as can be seen from Eq. (12). As we can see from Eq. (25), the density perturbation is equal to zero everywhere except for the axis. Consequently, everywhere except for the axis, $\mathrm{div}_{\mathbf{v}_e} = \mathrm{div}\,\mathbf{E} = 0$.

The above-obtained equations diverge at small $\mathbf{r}_\perp$. It is related to a violation of applicability of the linearized plasma equations. Therefore, a contribution of small impact parameters (small $\mathbf{r}_\perp$) needs to be evaluated separately. First, we consider how the wakefield will be changed if the plasma is removed from small impact parameters. To do so, we consider a problem where a charge moves along axis z in a plasma with a round channel of radius b around the particle trajectory. This problem was solved in Ref. 12 (see also Ref. 13[d]). The obtained longitudinal electric field inside the channel is constant across it and is equal to:

$$E_{\|} = -2qk_p^2 K_0(bk_p)\kappa_0(k_p b) \times \cos[\Omega(k_p b)\omega_p t - k_p z]\theta(t - z/c), \quad r_\perp < b. \tag{28}$$

The corresponding radial electric and azimuthal magnetic fields are:

$$E_r = H_\theta = qk_p^3 r_\perp K_0(bk_p)\kappa_0(k_p b) \times \sin[\Omega_0(k_p b)\omega_p t - k_p z]\theta(t - z/c), \quad r_\perp < b, \tag{29}$$

where the correcting factors are:

$$\Omega_0(x) = \sqrt{\frac{2K_1(x)}{2K_1(x) + xK_0(x)}},$$

$$\kappa_0(x) = \frac{1}{x(K_1(x) + xK_0(x)/2)}. \tag{30}$$

Figure 2 shows plots of these functions. As we can see, for $x \ll 1$ both functions are close to 1. Consequently, for $b \ll k_p^{-1}$ we can put them equal to 1. In this case the longitudinal electric fields of Eqs. (27) and (28) coincide if $r_\perp = b$.

Before proceeding to the bunch with finite length, we consider a pointlike bunch. The decelerating electric field acting on a charged particle (pointlike bunch) traveling through plasma is well known

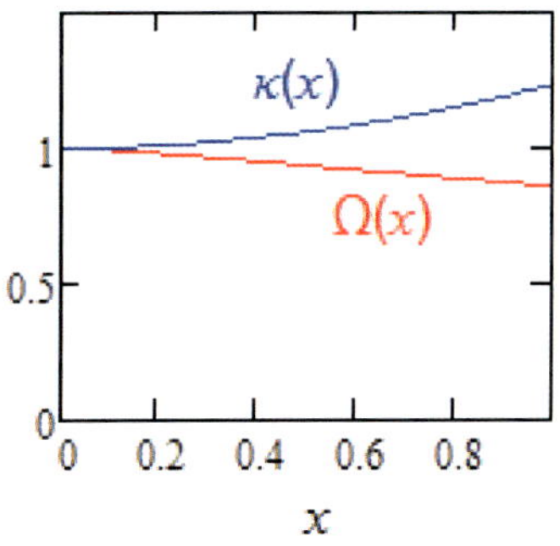

Fig. 2. Plots of the functions $\Omega_w(x)$ and $\lambda_w(x)$.

[d]It is straightforward to obtain the equations describing bunch interaction with plasma from the theory developed in Ref. 13, where the resistive wall impedances of a round channel were found. The replacement of the wall material conductivity σ, by the plasma conductivity $\sigma_p = -e^2 n_0/im\omega$ addresses the problem.

and is derived in a number of textbooks. In particular, it can be found in Ref. 10. For a particle moving with the speed of light, its result can be presented in the following form:

$$E_p = -qk_p^2 \ln\left(\frac{2}{e^{\gamma_E}}\frac{\rho_{\max}}{\rho_{\min}}\right),$$

$$\rho_{\min} = r_e q/e, \quad \rho_{\max} = 1/k_p, \tag{31}$$

where $\gamma_E \approx 0.577$ is the Euler constant [$\exp(\gamma_E) \approx 1.781$]. Note that the decelerating electric field immediately behind the particle is twice larger than the decelerating field of Eq. (31). Expanding the Bessel function in Eq. (28) one obtains

$$E_{\|}(\xi) = -2qk_p^2 \ln\left(\frac{2}{e^{\gamma_E}}\frac{1}{k_p b}\right)\cos(k_p\xi), \quad k_p b \ll 1, \tag{32}$$

where $\xi = ct - z$. Comparing Eqs. (31) and (32) we can see that the results coincide if $b = r_{\min}$. This means that the contribution of impact parameters lower than $\rho_{\min}$ is greatly reduced relative to the prediction of the linearized plasma theory, i.e. accurate accounting of small impact parameters should remove the divergence at small $r_\perp$ in Eqs. (27)–(29).

For a bunch with finite length, the contribution of small impact parameters depends on the longitudinal particle distribution inside the bunch and, consequently, on the bunch length. As will be seen below, the contribution of small impact parameters also depends on the sign of the charge. If the bunch is much shorter than k_p^{-1}, we can neglect collective effects in the plasma and consider the bunch interaction with noninteracting electrons. To further simplify this problem, we consider a bunch with a uniform longitudinal density with the bunch length L_b and the number of particles N_b so that $q = N_b e$. To eliminate the dependence on time, we consider the problem in the bunch frame. In this case the plasma with density $n' = \gamma n_0$ moves along the z' axis toward the bunch with the bunch length, $L' = \gamma L$, which is much larger than any transverse dimension in the problem. Here and below we use the sign $'$ to denote values in the bunch frame if they are different from the lab frame. In the ultrarelativistic case the equation of motion for a plasma electron in the bunch frame is:

$$\frac{d^2 r_\perp}{ds'^2} = \pm\frac{2e^2 N_b}{mc^2\gamma^2 L_b}\frac{1}{r_\perp}, \tag{33}$$

where the plus and minus signs are taken for electron and positron bunches, respectively. Note that this equation is justified even for the case where the transverse motion of plasma electrons is relativistic in the plasma (lab) frame. The only condition of Eq. (33) applicability is that the transverse momentum of plasma electrons is smaller than its initial momentum in the bunch frame $p_\perp \ll mc\gamma$. This is always the case for a major part of beam acceleration in a plasma-based collider, with possible exceptions for very small beam energy at the accelerator's beginning. Note also that a description of electron motion in the lab frame for a relativistic motion of a plasma electron would require accounting for the electron longitudinal motion excited by the bunch axial magnetic field and a solution to relativistic equations of the electron motion.

First, we consider plasma interaction with an electron bunch. The right hand side of Eq. (33) does not depend on s', and its integration is straightforward. The first integration yields

$$\frac{dr_\perp}{ds'} = \sqrt{\frac{4e^2 N_b}{mc^2\gamma^2 L_b}\ln\left(\frac{r_\perp}{r_0}\right)}, \tag{34}$$

where r_0 is the impact parameter (electron initial radius). The next integration yields

$$2r_m\frac{s'}{\gamma L_b} = r_0\int_1^{r_\perp/r_0}\frac{dx}{\sqrt{\ln(x)}}, \tag{35}$$

where

$$r_m = \sqrt{N_b r_e L_b} = \sqrt{\rho_{\min} L_b}. \tag{36}$$

Let us introduce the function

$$\mathrm{R}_e(x) = \int_1^x \frac{dx'}{\sqrt{\ln(x')}} = -i\sqrt{\pi}\,\mathrm{erf}(i\sqrt{\ln(x)}), \tag{37}$$

where $\mathrm{erf}(x) = (2/\sqrt{\pi})\int_0^x \exp(-t^2)dt$ is the error function. Results of numerical inversion of $\mathrm{R}_e(x)$ can be presented by a fitting formula:

$$\mathrm{R}_e^{-1}(x) \approx 1 + \frac{x^2}{4(1+0.123x)} + \frac{0.055x^{10/3}}{10^4 + x^2}, \tag{38}$$

which has a few-percent accuracy in the range $x \in [0, 4000]$ — the range sufficient for most applications. Then, a particle trajectory can be described by the following equation:

$$r_\perp(s', r) = r_0\mathrm{R}_e^{-1}\left(\frac{2r_m}{r_0}\frac{s'}{\gamma L_b}\right). \tag{39}$$

To find the contribution of small impact parameters to the decelerating force, we first, find a scalar potential at the axis. After passing distance s' the plasma electrons contained before collision in a thin cylinder with the radius r and the thickness dr are separated from the ions (those deflections can be neglected) and, consequently, the ions stay at the same radius. The summation of contributions of all radii yields the scalar potential on axis:

$$\varphi(s') = 4\pi\gamma n_0 e\int_0^b \ln(r_\perp(s', r)/r)r\,dr. \tag{40}$$

Consequently, the longitudinal electric field excited at the axis is

$$E_\parallel(s') = -\frac{d\varphi}{ds} = -4\pi\gamma n_0 e\int_0^b \frac{dr_\perp}{ds'}\frac{r}{r_\perp(s', r)}dr. \tag{41}$$

Substituting $r_\perp(s', r)$ from Eq. (39), transiting to a dimensionless variable in the integration and, finally, returning to the lab frame, we obtain the contribution to the longitudinal electric field at the axis produced by collisions with small impact parameters:

$$E_\parallel(\xi) = 2eN_b k_p^2\frac{\xi}{L_b}\Phi_e\left(\frac{L_b b}{2r_m\xi}\right), \quad \xi = s'/\gamma, \tag{42}$$

where the sign is changed because s and s' are directed in opposite directions, and

$$\Phi_e(x) = 2\int_0^x \frac{\mathrm{D}_e(1/x')}{\mathrm{R}_e^{-1}(1/x')}dx', \quad \mathrm{D}_e(x) = \frac{d}{dx}\mathrm{R}_e^{-1}(x). \tag{43}$$

The function $\Phi(x)$ was calculated by a numerical integration, whose results can be approximated by the following equation:

$$\Phi_e(x) \approx \ln\left(1 + \frac{2.93x^2}{1+2x}\right). \tag{44}$$

To find the electric field coming from large impact parameters, one needs to sum contributions for all particles in the bunch. Integrating Eq. (32) so that only contributions for upstream particles are taken into account, one obtains:

$$E_\parallel(\xi) = 2eN_b k_p^2\frac{\xi}{L_b}\ln\left(\frac{2}{e^{\gamma_E}}\frac{1}{k_p b}\right), \quad \begin{cases}\xi k_p \ll 1, \\ b k_p \ll 1.\end{cases} \tag{45}$$

Leaving only the leading term in Eq. (42) and summing it with Eq. (45), one finally obtains, in logarithmic approximation, the total decelerating electric field along the electron bunch with uniform longitudinal distribution:

$$E_{\parallel}(\xi) \approx 2eN_b k_p^2 \frac{\xi}{L_b} \ln\left(\frac{2.93}{e^{\gamma_E}} \frac{1}{r_m k_p} \frac{L_b}{\xi}\right),$$

$$\begin{cases} \xi k_p \ll 1, \\ r_m k_p \ll 1. \end{cases} \quad (46)$$

As one can see, a dependence on b disappeared. That supports an applicability of the method where collisions at small and large impact parameters are taken into account separately. Figure 3 shows the electric field along the electron bunch for different values of $r_m k_p$.

Now, we consider plasma interaction with the positron bunch, where we will follow the procedure used above for the electron bunch. In this case the solution Eq. (33) is

$$2r_m \frac{s'}{\gamma L_b} = r_0 \int_1^{r_\perp/r_0} \frac{dx}{\sqrt{-\ln(x)}}. \quad (47)$$

Introducing the function

$$\mathrm{R}_p(x) = -\int_1^x \frac{dx'}{\sqrt{-\ln(x')}} = \sqrt{\pi}\mathrm{erf}(\sqrt{-\ln(x)}), \quad (48)$$

we obtain the particle trajectory,

$$r_\perp(s', r) = r_0 \mathrm{R}_p^{-1}\left(\frac{2r_m}{r_0}\frac{s'}{\gamma L_b}\right), \quad (49)$$

where $\mathrm{R}_p^{-1}(x)$ is the inverse of function $\mathrm{R}_p(x)$, and it can be approximated as follows:

$$\mathrm{R}_p^{-1}(x) \approx \left[\left(\frac{9}{8} - \frac{1}{3\pi}\right)\cos\left(\frac{\sqrt{\pi}}{2}x\right) + \left(\frac{1}{3\pi} - \frac{1}{8}\right)\cos\left(\frac{3\sqrt{\pi}}{2}x\right)\right]^{3/4}. \quad (50)$$

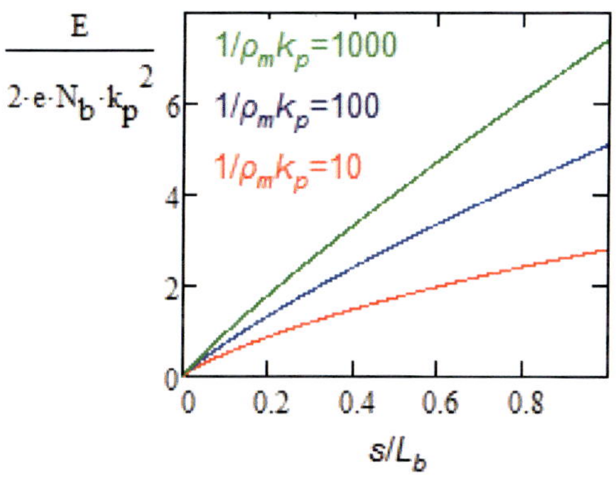

Fig. 3. Dependence of the decelerating field along the bunch for different values of $\rho_m k_p$.

Similar to the case electron bunch where Eq. (42) describes contribution of small impact parameters to the electric field at the axis one can use Eq. (42) for the positron bunch where the function $\Phi(x)$ is replaced by the function

$$\Phi_p(x) = \int_0^x F\left(\frac{1}{x'}\right)dx',$$
$$F(x) = 2\frac{D_p(x)}{R_p^{-1}(x)}, \quad D_p(x) = \frac{d}{dx}R_p^{-1}(x). \quad (51)$$

To eliminate the divergence in the integrand, we replace the variable so that $x \to 1/x$, the split area of integration at pieces equal to a half period, and group addends of opposite signs together:

$$\Phi_p(x) = \int_{1/x}^{\infty} \frac{F(x')}{x'^2}dx' = \int_{1/x}^{\sqrt{\pi}/2} \frac{F(x')}{x'^2}dx' + \int_{\sqrt{\pi}/2}^{2\sqrt{\pi}} \frac{F(x')}{x'^2}dx' + \int_0^{\sqrt{\pi}/2}\left(\frac{F(\sqrt{\pi}+x')}{(\sqrt{\pi}+x')^2} + \frac{F(\sqrt{\pi}-x')}{(\sqrt{\pi}-x')^2}\right)dx' + \sum_{k=1}^{\infty}\int_0^{\sqrt{\pi}}\left(\frac{F(\sqrt{\pi}+x')}{((1+2k)\sqrt{\pi}+x')^2} + \frac{F(\sqrt{\pi}-x')}{((1+2k)\sqrt{\pi}-x')^2}\right)dx', \quad 1/x \le \sqrt{\pi}/2. \quad (52)$$

Here we have used the periodicity of $\mathrm{F}(x)$ with period $2\sqrt{\pi}$ and $\mathrm{F}((1+2k)\sqrt{\pi}+x) = -\mathrm{F}((1+2k)\sqrt{\pi}-x)$. The results of numerical integration can be approximated as

$$\Phi_p(x) \approx -\ln\left(1 + \frac{3.02x^3}{x^2 + 1.05}\right). \quad (53)$$

Finally, we obtain the contribution of small impact parameters,

$$E_{\parallel}(\xi) = 2eN_b k_p^2 \frac{\xi}{L_b}\Phi_p\left(\frac{L_b b}{2r_m \xi}\right), \quad (54)$$

and the total decelerating field for the positron bunch in the logarithmic approximation:

$$E_{\parallel}(\xi) \approx -2eN_b k_p^2 \frac{\xi}{L_b} \ln\left(\frac{3.02}{e^{\gamma_E}} \frac{1}{r_m k_p} \frac{L_b}{\xi}\right),$$

$$\begin{cases} \xi k_p \ll 1, \\ r_m k_p \ll 1. \end{cases} \quad (55)$$

The same as for electrons, the major contribution to the decelerating force comes from plasma electrons with impact parameters larger than r_m.

We need to note that an assumption that an interaction of plasma electrons can be neglected is not quite accurate in the case of positron beam. Plasma electrons are attracted to the axis and create very large charge density in its close vicinity. For electrons with sufficiently small impact parameters, this results in their repulsion from the axis. However, it has little effect on the decelerating force, due to the overall small contribution of small impact parameters, $r < r_m$. It is important to note that the large density of plasma electrons destroys the linearity of plasma focusing (discussed in Sec. 2) and greatly amplifies multiple scattering (discussed in Sec. 3). Each of these effects makes impossible an acceleration of the collider quality positron bunch in a plasma-based accelerator.

The decelerating forces for positrons and electrons are quite similar for a small bunch population when $k_p r_m$ is small. However, with increase in the intensity, the decelerating force for a positron bunch is up to ~1.5 times larger, as shown by the ratio $\Phi_p(1/e^{\gamma_E} r_m k_p)/\Phi_e(1/e^{\gamma_E} r_m k_p)$ in Fig. 4. Note that the above approximations are only justified for $x = 1/r_m k_p \gg 1$ and one needs to ignore behavior of $\mathrm{R}(x)$ for $x \approx 1$.

The r_m represents the impact parameter where the radial displacement of a plasma electron during its passage through the bunch is about its initial value. It is demonstrated in Fig. 5 presenting

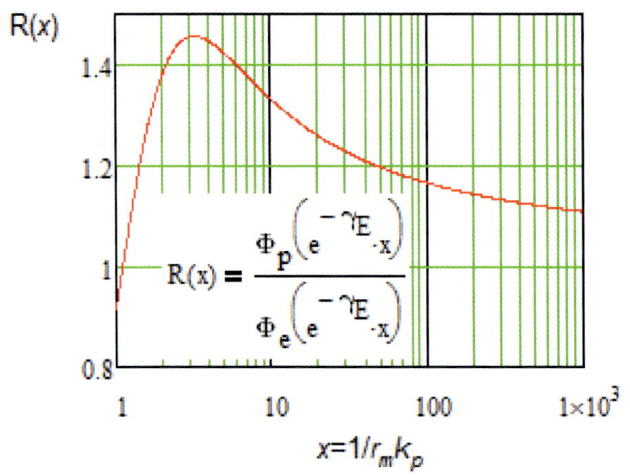

Fig. 4. The ratio of deceleration forces for positron and electron bunches at the bunch tail as a function of $r_m k_p$.

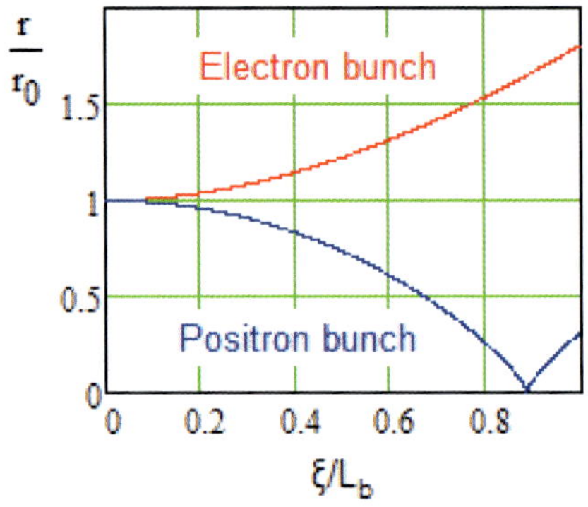

Fig. 5. The trajectory of a plasma electron for collisions with electron and proton bunches for the impact parameter equal to r_m, i.e. $r_0 = r_m$.

a plasma electron trajectory in the course of collision with electron and positron bunches. For both cases, the major contribution to the decelerating field comes from large impact parameters, $r > r_m$. Therefore, if the transverse bunch size is significantly smaller than r_m, its value does not affect the decelerating field. Note that for a bunch brightness required for a collider this condition is always satisfied; and one can consider that, in finding the deceleration force, a bunch has zero transverse size.

As one can see from Eqs. (46) and (55), there is a logarithmic dependence of decelerating force on ξ. This means that a contribution to the decelerating force left behind by a specific particle depends on its longitudinal position in the bunch. Such a dependence is related to plasma density perturbation produced by previous particles. As one can see from Fig. 3, it becomes less significant with reducing $r_m k_p \ll 1$. Note also that this effect is much less significant for a bunch with Gaussian distribution.

The decelerating forces of Eqs. (46) and (55) are justified for $r_m k_p \ll 1$. As one can see, they change sign when the argument in the logarithm becomes equal to 1. This happens because the maximum impact parameter, k_p^{-1}, becomes smaller than the minimum impact parameter, r_m. An addition of r_m to the maximum impact parameter — in other words, an addition of 1 to the arguments in the logarithms — corrects this problem and yields reasonably good estimates of the decelerating field even for the case of the bubble regime where $r_m k_p \geq 1$. A comparison of numerical simulations presented in Ref. 14 with calculations in the present article shows reasonable coincidence. In the case of $r_m k_p \gg 1$, the logarithm can be approximated by a linear function. Then, using Eq. (46), one obtains the decelerating force for electron beam:

$$E_{\|}(\xi) \approx 3.2ek_p\sqrt{\frac{N_b}{r_eL_b}}, \quad \frac{2\cdot 2.93}{\mathrm{e}^{\gamma_E}} \to 3.2,$$

$$\begin{cases} \xi k_p \ll 1, \\ r_m k_p \gg 1. \end{cases} \tag{56}$$

As one can see, in the strong bubble regime the decelerating field grows proportionally to $\sqrt{N_b}$. We will see this dependence in Sec. 7, where the bunch acceleration in the strong bubble regime is considered.

The above calculations describe the decelerating force for the case of $L_bk_p \ll 1$. In linear approximation a plasma response results in an oscillation of this field with distance from the bunch at the space frequency of k_p. In most of the practical cases such approximation is not quite valid. Even a small charge electron bunch moving in a plasma pushes plasma electrons out of its way, thus creating a cavity, void of plasma electrons. An evaluation of the radial size and length of the cavity created by a pointlike bunch is considered in Ref. 15 for the case where the motion of plasma electrons is nonrelativistic: $N_bk_p^3/(4\pi n_0) \ll 1$. The paper concludes that the length of the cavity does not depend on the number of particles in the bunch and is equal to $3.8/k_p$, and its radial size is

$$r_{\text{cav}} \approx 2.82\sqrt{N_b}\sqrt[4]{\frac{r_e}{4\pi n_0}} = 2.82\frac{r_m}{\sqrt{k_pL_b}}. \tag{57}$$

Although this result was obtained for zero bunch length, it is actually justified for $L_bk_p \leq 0.4$. In the range $0 \leq L_bk_p \leq 0.4$ the cavity radius is decreasing approximately linearly with the bunch length and achieves $\sim 80\%$ of its zero length value at $L_bk_p = 0.4$. From the second half of Eq. (57) one can see that $r_{\text{cav}} > r_m$ for $L_b < k_p^{-1}$.

The minimum impact parameter, r_m, depends on the longitudinal density distribution in the bunch. To obtain the decelerating force for the Gaussian distribution, we followed the above described procedure. Finding particle trajectories in the first order perturbation theory, substituting them into Eq. (41), performing integration over the impact parameter, and adding the contribution of large impact parameters, one obtains the longitudinal field with logarithmic accuracy:

$$E_{\|}(\xi) \approx -eN_bk_p^2[1+\mathrm{erf}(\hat{\xi})] \times \ln\left(\frac{2\sqrt{2}/(e^{\gamma_E}r_{mg}k_p)}{\sqrt{\sqrt{\pi}\hat{\xi}[1+\mathrm{erf}(\hat{\xi})]+\exp(-\hat{\xi}^2)}}\right), \tag{58}$$

$$\hat{\xi} = \frac{\xi}{\sqrt{2}\sigma_s},$$

where $r_{mg} = \sqrt{2\sqrt{2/\pi}N_br_e\sigma_s}$ is the minimum impact parameter for beam with Gaussian distribution, σ_s is the bunch RMS length, and we assume that the bunch center is at $\xi = 0$. We can see that, similar to the rectangular particle distribution considered above, there is a logarithmic dependence of single particle response on its longitudinal coordinate.

6. Longitudinal and Transverse Wakes in Plasma

To create an accelerating field in plasma one has to excite it by a short electron bunch or a short pulse of laser radiation. In the context of this section the method of excitation is not important. The excitation creates a cavity moving with the velocity close to the speed of light. The cavity's electric field accelerates and focuses particles of the accelerated (witness) bunch. The longitudinal electric field experiences a half period oscillation along the cavity, changing from deceleration of electrons in its first half to acceleration in the second half, so that the maximum acceleration is achieved at the cavity end. In this section we consider an interaction of a bright electron bunch with a plasma surrounding the cavity.

First, we ignore the fact that the cavity walls are moving and replace them by a plasma channel considered at the beginning of the previous section. Then it is straightforward to obtain the longitudinal wake function from Eq. (28). At small distances ($\xi \ll k_p^{-1}$) it is constant and is equal to

$$W_{\|} \equiv -\frac{E_{\|}(0^+)}{q} = \frac{2k_p^2K_0(R)}{R[K_1(R)+RK_0(R)/2]}, \quad 0 < \xi \ll k_p^{-1}, \tag{59}$$

where $R = k_p\tilde{b}$, with $\tilde{b} = \max(b, r_m)$ being the maximum between the channel radius and the minimum

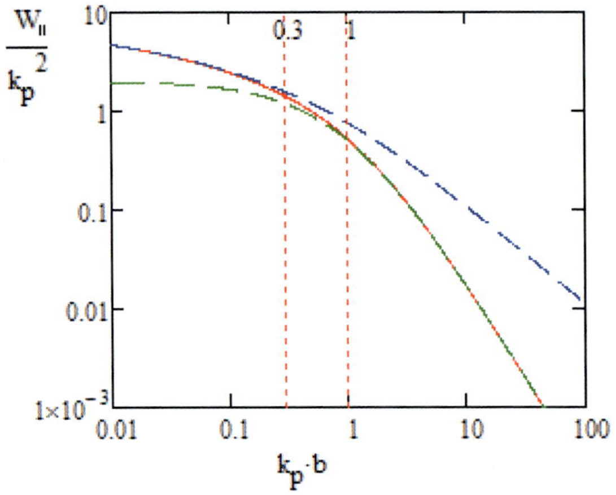

Fig. 6. Dependence of the longitudinal wake function on parameter $R = \tilde{b}k_p$ (red line); the blue and green lines represent the asymptotes at low and high values of R, respectively.

impact parameter introduced in Eq. (36). Figure 6 presents the plot of this function and its asymptotes:

$$W_{\parallel} \approx 2k_p^2 \begin{cases} \ln(1 + 2/e^{\gamma_E} k_p \tilde{b}), & 0 \le \tilde{b}k_p \le 0.3, \\ 2/(1 + k_p\tilde{b})^2, & \tilde{b}k_p \ge 1. \end{cases} \tag{60}$$

For sufficiently large $R \ge 10$ the dependence of the wake on k_p becomes insignificant:

$$W_{\parallel} \approx 4/b^2. \tag{61}$$

In the case of the bubble regime the electron density near the bubble boundary is significantly higher than in the plasma channel considered here. It increases the local k_p and extends the applicability of Eq. (61) to smaller values of R. The same result is obtained in Ref. 16 where the longitudinal wake is obtained for the large-size bubble, where the motion of plasma electrons is relativistic.

To find the transverse wake we will split the problem into two steps — separately considering collisions with small and large impact parameters following the recipe of the previous section. However, in the evaluation of small impact parameter's contribution, we will keep only linear terms in the plasma response so as to keep the obtained result in a simple form. If required, it is straightforward to include nonlinear terms using the perturbation theory.

In the evaluation of small impact parameters contribution, as in the previous section, we consider collisions in the bunch frame and neglect the reaction of plasma to the motion of plasma electrons. In this case a plasma with density distribution $\gamma n(r)$ encounters a pointlike charge q displaced from the axis by distance a in the horizontal plane. The electrons of plasma are scattered on the charge, and transverse motion of ions is neglected. In the first order of perturbation theory the scattering angle of the electron located at radius $r \gg a$ is:

$$\theta' \approx \frac{2eq}{mc^2\gamma r}\left(1 + \frac{a}{r}\cos\phi\right), \tag{62}$$

where ϕ is the angle between the directions of a and r, and we take into account that a and r are Lorentz-invariant. The first and second terms of Eq. (62) are responsible for creating longitudinal and transverse wakes, respectively. To find the transverse electric field excited near the axis by plasma electrons, we split the incoming plasma beam into thin cylinders with thickness dr. Taking into account that the ions in the cylinder are not scattered, we obtain the transverse electric field coming from one cylinder:

$$\begin{aligned} \Delta E'_{\perp} &= 2\pi e\gamma \frac{dn}{dr} s'\theta'_1 \Delta r \\ &= 2\pi\gamma e \frac{dn}{dr} s'\Delta r \left(\frac{2eqa}{mc^2\gamma r^2}\right), \end{aligned} \tag{63}$$

where $\theta'_1 = 2eqa/mc^2\gamma r^2$ is the maximum of the second order deflecting angle. Integrating over impact parameters and returning to the lab frame, we obtain the transverse impedance:

$$W_{\perp} = \frac{4\pi e^2 \xi}{mc^2} \int \frac{1}{r^2}\frac{dn}{dr} dr, \tag{64}$$

where we have taken into account that $W_{\perp} = \gamma \Delta E'_{\perp}/qa$ and $\xi = s'/\gamma$. In the case of a plasma channel of radius b with a rigid boundary $dn/dr = n_0\delta(r - b)$ and the integral evaluation is straightforward:

$$W_{\perp} = \frac{k_p^2 \xi}{b^2}, \quad \begin{array}{l} k_p\xi \ll 1, \\ k_p b \ll 1. \end{array} \tag{65}$$

This result is justified only for the case where the reaction of plasma can be neglected, i.e. $k_p\xi \ll 1, k_p b \ll 1$.

The transverse impedance of a plasma channel was found in Ref. 12. The result can be presented in the following form:

$$\begin{aligned} W_{\perp} &= \frac{k_p^2}{b^2}\kappa_1(k_p b)\frac{\sin[\Omega_1(k_p b)k_p\xi/\sqrt{2}]}{\Omega_1(k_p b)k_p/\sqrt{2}}, \\ \kappa_1(x) &= \frac{8K_1(x)}{R[4K_2(x) + RK_1(x)]}, \\ \Omega_1(x) &= \sqrt{\frac{4K_2(x)}{4K_2(x) + RK_1(x)}}. \end{aligned} \tag{66}$$

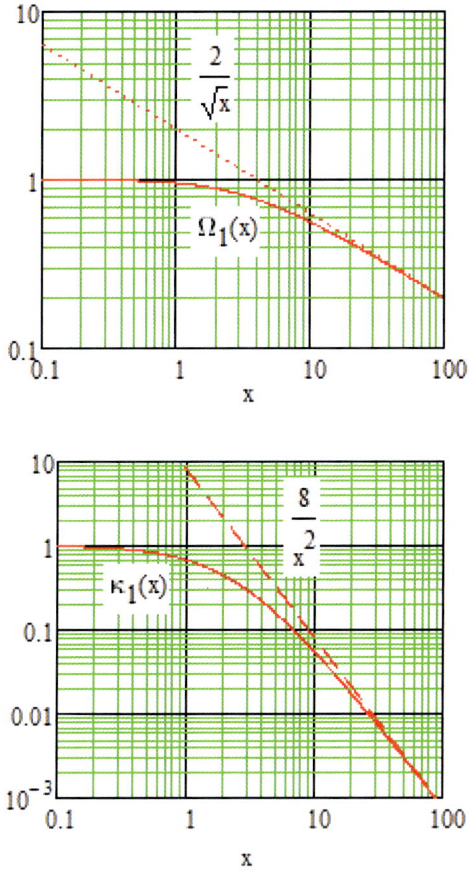

Fig. 7. Plots of functions $\Omega_l(x)$ (top) and $\kappa_1(x)$ (bottom).

Here $K_2(x)$ is the modified Bessel function. Plots of the functions $\kappa_1(x)$ and $\Omega_1(x)$ and their asymptotes at large x are shown in Fig. 7. One can see that Eqs. (65) and (66) are identical in the parameter range where Eq. (65) is applicable, $k_p\xi \ll 1, k_p b \ll 1$. This proves that for a plasma channel of sufficiently small radius, $b \ll k_p^{-1}$, one can separately consider collisions at small and large impact parameters and then combine the obtained results. In particular, it creates a straightforward procedure for making an accurate evaluation of the transverse impedance for more complicated cases like if the plasma boundary is not rigid or the motion of plasma electrons is excited at sufficiently small radii. However, such a procedure is applicable only for channels of sufficiently small radius, $r \ll k_p^{-1}$, and intensity $r_m \ll k_p^{-1}$.

Finally, we should note that for a large bubble size, $b \gg k_p^{-1}$, the transverse wake is determined by a collective plasma response and is equal to

$$W_\perp \approx 8\xi/b^4, \quad bk_p \gg 1, \quad \xi k_p \ll 1. \tag{67}$$

Similarly to the longitudinal wake, it does not depend on plasma parameters. It is important to note that for $b \gg k_p^{-1}$ both the longitudinal and transverse wakes are strongly suppressed relative to the predictions of a binary collision model, and that this model is not applicable for a large size bubble (plasma channel). For $b \gg k_p^{-1}$, all currents in plasma are concentrated in a thin layer near channel boundary, b. This brings us to the well-known relationship between impedances, $W_\perp \approx 2W_\parallel\xi/b^2$, which follows from the Panofsky–Wenzel theorem.

7. Strong Bubble Regime

As one can see from the previous sections, acceleration of a collider quality electron bunch in the quasi-linear regime and acceleration of a bright positron bunch are not presently feasible. Therefore, in this section, we will focus on the limitations of electron bunch acceleration in the strong bubble regime ($b \gg k_p^{-1}$). In this case, an accurate quantitative description can only be obtained with numerical simulations. However, a qualitative description based on the above described approaches and Ref. 14 delivers reasonably accurate results and, what is even more important, allows one to look into a relationship of different limitations and shows possible methods for optimization of plasma acceleration.

As will be seen below, the main parameter determining the beam transverse stability is the ratio of transverse and longitudinal impedances. It decreases fast with increasing size of the plasma channel, b, and thus it is preferable for beam acceleration. In the strong bubble regime the motion of plasma electrons is relativistic and, consequently, the maximum transverse size of the bubble, R_b, is much larger than k_p^{-1}. The corresponding equation, which describes the dependence of the bubble radius, r_b, on the longitudinal coordinate was developed in Ref. 14:

$$r_b\frac{d^2 r_b}{d\xi^2} + 2\frac{dr_b}{d\xi}^2 + 1 = \frac{2}{\pi n_e r_b^2}\frac{dN_b}{d\xi}, \tag{68}$$

where $dN_b/d\xi$ is the longitudinal particle density. Below we will call this equation the Lu equation.

In the absence of particles the right-hand side of Eq. (68) is equal to zero. An introduction of variable $\hat{p} = dr_b/d\xi$ allows one to rewrite it in the following form:

$$r_b\hat{p}\frac{d\hat{p}}{dr_b} + 2\hat{p}^2 + 1 = 0. \tag{69}$$

Its integration yields

$$\frac{dr_b}{d\xi} \equiv \hat{p} = \pm\sqrt{\frac{1}{2}\left(\frac{R_b^4}{r_b^4} - 1\right)}, \tag{70}$$

where we have taken into account that $dr_b/d\xi = 0$ at $r_b = R_b$, and the signs + and − are used for the upstream and downstream halves of the bubble. Consequently, the halfbubble length is equal to:

$$\xi_b = \sqrt{2}\int_0^{R_b}\left(\frac{R_b^4}{r^4} - 1\right)^{-1/2} dr = \frac{1}{\sqrt{\pi}}\Gamma\left(\frac{3}{4}\right)^2 \approx 0.847R_b. \tag{71}$$

The shape of the bubble and its fitting with the approximation

$$r_b \approx R_b\sqrt[3]{1 - \left(\frac{\xi}{\xi_b}\right)^2} \tag{72}$$

are shown in Fig. 8. Here we assume that the bubble center is located at $\xi = 0$.

Reference 14 also suggests the equation for the evaluation of the longitudinal electric field in the bubble:

$$E_{\|} = 2\pi e n_0 r_b \frac{dr_b}{d\xi}. \tag{73}$$

It is straightforward to obtain this equation in the bunch frame (see details in Sec. 4) assuming that all plasma electrons pushed out by the beam are located in the thin layer near the bubble boundary. Substituting $dr_b/d\xi$ of Eq. (70), one obtains the field in the bubble:

$$E_{\|} = \pm\pi e n_0 r_b\sqrt{2\left(\frac{R_b^4}{r_b^4} - 1\right)}$$
$$\approx \pm\pi e n_0 R_b\sqrt{2\left[1 - \left(\frac{\xi}{\xi_b}\right)^2\right]^{-\frac{2}{3}} - \left[1 - \left(\frac{\xi}{\xi_b}\right)^2\right]^{\frac{2}{3}}}, \tag{74}$$

where the signs ± are related to the first and the second half of the bubble, and we have used Eq. (72) to obtain the second half of Eq. (74). Figure 9 presents a numerically computed dependence of this electric field on the coordinate within the bubble. Equation (68) yields $d^2r_b/d\xi^2 = 1/R_b$ in the bubble center. Using this result in the top part of Eq. (74), one obtains the field gradient in the bubble center:

$$\frac{dE_{\|}}{d\xi} = -2\pi e n_0. \tag{75}$$

As one can see, it does not depend on the bubble radius. Note that the focusing part of the electromagnetic field is determined by Eq. (17) with $n_e = 0$.

To obtain the longitudinal wake at short distances we substitute the particle density, as $q\delta(\xi - \xi_1)$ into Eq. (68), integrate it in the close vicinity of ξ_1 and substitute the obtained result into Eq. (73). The longitudinal wake obtained this way coincides with the wake of Eq. (61) with replacement b by r_b, i.e. $W_{\|} = 4/r_b^2$. To obtain the dependence of the wake on the coordinate within the bubble, one needs to integrate Eq. (70). Figure 10 shows wakes numerically calculated for "leading" particles

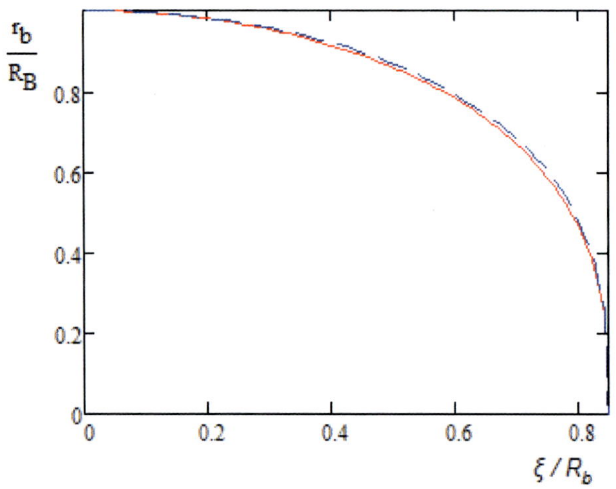

Fig. 8. The half bubble profile of Eq. (70) and its fitting by Eq. (72) are shown by solid and dashed lines, respectively.

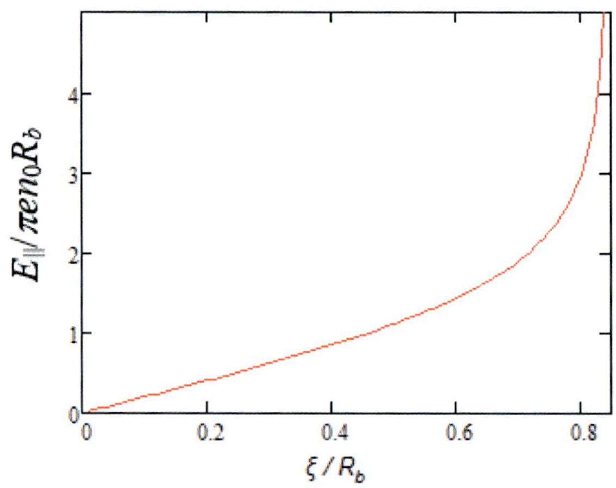

Fig. 9. Dependence of the dimensionless longitudinal electric field, $E_{\|}/\pi e n_0 R_b$, on the longitudinal coordinate for the bubble trailing half; $\xi = 0$ corresponds to the bubble center.

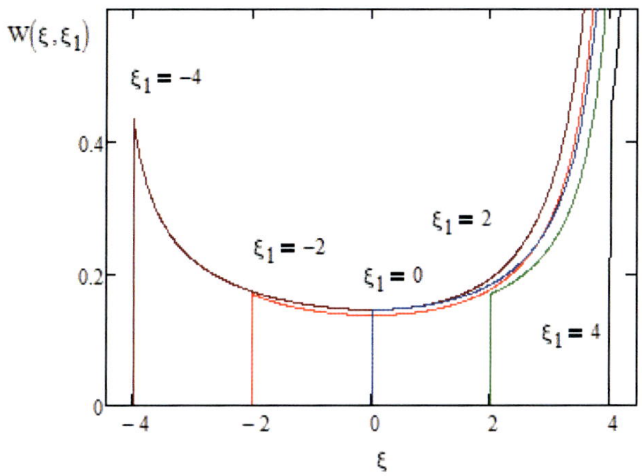

Fig. 10. Dependence of longitudinal wake on the coordinate within the bubble for different points of excitation, ξ_1; $R_b = 5$, and the bubble center is at $\xi = 0$.

located at different coordinates, ξ_1. The results can be roughly approximated as

$$W_{\|}(\xi,\xi_1) \approx \frac{4}{r_b(\xi)^2}\theta(\xi-\xi_1). \tag{76}$$

As one can see, the wake does not change its sign to the end of the bubble and diverges at its end. It is completely different from the weak bubble case where the motion of plasma electrons is nonrelativistic and the bubble length does not depend on its radius. In the strong bubble case the duration of the bubble is increasing with additional excitation. Combined with independence of the maximum gradient from the excitation strength, it makes the wake positive for the entire length of the unperturbed bubble. Also note that Eq. (68) is nonlinear and, consequently, an introduction of the longitudinal wake is justified for small perturbations only.

8. Beam Acceleration in the Strong Bubble Regime

Now we consider a plasma excitation and beam acceleration in the strong bubble regime. First, we require that all particles in the driving bunch be decelerated at the same rate, E_d. In this case an integration of Eq. (73) binds up the bubble radius and electric field:

$$E_d\xi = \pi n_0 e(r_b^2 - r_{b0}^2), \tag{77}$$

where r_{b0} is the initial bubble radius. Taking into account that at the bunch head $\xi = 0$ and $r_{b0} = 0$, expressing r_b from the obtained equation, substituting it into Eq. (68) and performing simple calculations, one obtains the longitudinal particle distribution,

$$\frac{dN_b}{d\xi} = \frac{E_d}{8e}\left(\frac{E_d}{\pi n_0 e} + 4\xi\right), \tag{78}$$

and the total number of particles in the bunch,

$$N_b = \frac{E_d}{8e}L_b^2\left(\frac{E_d}{\pi n_0 e L_b} + 2\right). \tag{79}$$

Equation (77) results in

$$L_b = \left.\frac{r_b}{2dr_b/d\xi}\right|_{r_b=r_{bd}}, \quad E_d L_b = \pi n_0 e r_{bd}^2,$$

where r_{bd} is the bubble radius at the end of the driving bunch. Using these equations one can rewrite Eq. (79) in the form

$$N_b = \frac{\pi n_0 r_{bd}^3}{8}\left(2\frac{dr_{bd}}{d\xi} + \frac{1}{dr_{bd}/d\xi}\right), \tag{80}$$

which results in the total power being transferred from the beam to the plasma:

$$P = eN_bE_dc = \frac{\pi^2}{4}e^2n_0^2cR_b^4. \tag{81}$$

Here we have expressed the electric field of through r_{bd} and $dr_{bd}/d\xi$ using Eq. (73), and $dr_{bd}/d\xi$ through R_b and r_b using Eq. (70). As we can see, the power is uniquely determined by the maximum bubble radius.

One can see from Eq. (68) that if the density distributions in the driving and accelerated bunches are mirror-symmetric relative to the bunch center, the accelerating bunch is accelerated at the same rate as the deceleration of the driving bunch and 100% of its energy will be transferred to the accelerated (witness) bunch. Actually, the ratio of the acceleration to the deceleration forces can be arbitrary as long as the particle distributions and bunch lengths are determined by Eqs. (78) and (79). In this case, if the accelerated bunch extends to the cavity end, 100% energy transfer efficiency is achieved.

It is important to note that Eq. (68) is approximate and one has to be cautious about the accuracy and meaning of the obtained results. There is a striking difference between the statement in this section that the decelerating force is constant along the bunch with a trapezoidal distribution and the conclusions of Sec. 4 [see Eq. (46)] that the decelerating force is zero for the very head of a bunch. This difference originates from the poor description of the plasma reaction by the Lu equation for small r_b, for which the longitudinal wake diverges as $1/r_b^2$.

Note that in a more accurate model the divergence will be limited by the minimum impact parameter. Therefore, Eq. (68) works well only for $L_b \geq k_p^{-1}$. In this case the absence of deceleration for the bunch head becomes insignificant for the bubble formation and particle acceleration for the rest of the bunch. We also need to stress that Eq. (68) conserves energy exactly. In a real plasma, the accuracy of these statements improves with increase of R_b, however, even for large R_b ($R_b \gg k_p^{-1}$), it is not perfectly accurate. Numerical simulations carried out in Ref. 17 show the efficiency of acceleration of $\sim 90\%$ for $k_p R_b \approx 5$.

For a given number of particles the decelerating electric field can be obtained from Eq. (79):

$$E_d = \sqrt{(\pi e n_0 L_b)^2 + \frac{8\pi e^2 n_0 N_b}{L_b}} - \pi e n_0 L_b. \tag{82}$$

For $N_b \geq n_0 L_b^3$ the second addend in the square root dominates and the decelerating field grows proportionally to the square root of the bunch population [see Eq. (56)]. With the use of above-derived relationships one, after comparatively simple calculations, can express the bubble radius through the bunch length and population:

$$R_b = \frac{L_b}{\sqrt[4]{2}} \sqrt[4]{\frac{8N_b}{\pi n_0 L_b^3}\left(\sqrt{\frac{8N_b}{\pi n_0 L_b^3}+1}-1\right)}. \tag{83}$$

Here we assume that the particle distribution is chosen so as to keep constant the decelerating field along the entire bunch. One can see that $R_b \to \infty$ for $L_b \to 0$, which is obviously an incorrect result related to a violation of Eq. (68) applicability ($L_b \geq k_p^{-1}$).

Now we consider limitations on the acceleration of the accelerated (witness) bunch. High efficiency of acceleration requires this bunch to extend to the very end of the bubble, but it creates a problem with BBU instability because the transverse impedance diverges as $1/r_b^4$ and becomes very large at the cavity end. Therefore, the tail of accelerated bunch should be located sufficiently far from the bubble end.

Similarly, let us assume that the particle density is chosen so that all particles are accelerated at the same rate. In this case, similar to Eq. (78), the particle distribution is trapezoidal but with the particle density linearly decreasing to the bunch tail. Expressing coordinates of the bunch head and tail through the bubble radii at their locations with the use of Eq. (77), one obtains the total number of particles in the accelerated bunch:

$$N_a = \frac{E_a}{8e}\left[r_{b2}^2 - r_{b1}^2 + 2\left(\frac{\pi n_0 e}{E_a}\right)^2 (r_{b2}^4 - r_{b1}^4)\right]. \tag{84}$$

Here E_a is the accelerating field, r_{b2} and r_{b1} are the bubble radii in the locations of the bunch head and tail, respectively. Then, the power transferred to the accelerated bunch is

$$P_a = \frac{\pi^2 e^2 n_0^2 c}{4}(r_{b2}^2 - r_{b1}^2)\left(\frac{R_b^4}{r_{b2}^2} + r_{b1}^2\right). \tag{85}$$

To obtain this equation we expressed the accelerating field in Eq. (84) through the bubble radius using Eq. (74). Consequently, the efficiency of power transfer from the driving to the accelerated bunch is

$$\eta_P = \frac{P_a}{P} = \frac{r_{b2}^2 - r_{b1}^2}{R_b^2}\left(\frac{R_b^2}{r_{b2}^2} + \frac{r_{b1}^2}{R_b^2}\right). \tag{86}$$

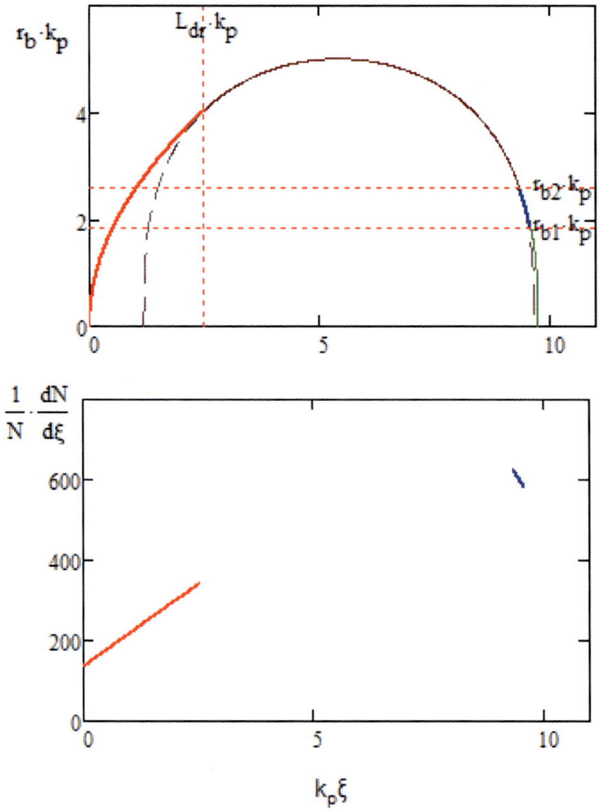

Fig. 11. The bubble shape (top) and particle distributions for driving and accelerated bunches (bottom). The red and blue lines show the driving and accelerated bunches, respectively. The dashed brown line shows the bubble shape in the absence of particles inside the bubble. Accelerating and decelerating fields are constant along the bunches: $E_d/E_0 = 1.64$; $E_a/E_0 = 3.28$.

Figure 11 shows an example illustrating the bubble shape and the particle distributions of the driving and accelerated bunches for the power transfer efficiency of 50% and the transformer ratio of 2. For $n_0 = 10^{17}\,\mathrm{cm}^{-3}$ the driving bunch parameters are chosen to be $R_b k_p = 5$, $L_b k_p = 2.5$, yielding the decelerating field of $E_d = 50\,\mathrm{GV/m}$ and $N_b = 3.55 \cdot 10^{10}$. The accelerated bunch parameters are $r_{b2} = 0.518 R_b$, $r_{b1} = 0.373 R_b$, $E_a = 100\,\mathrm{GV/m}$, $N_a = 8.86 \cdot 10^9$.

9. Beam Breakup Instability in the Strong Bubble Regime

In this section we consider a beam breakup (BBU) instability for an accelerated (witness) bunch. The main parameter determining the development of BBU is the ratio of the bunch deflection force to the focusing force. Focusing is completely dominated by plasma and is described by Eq. (17), where in the bubble regime $n_e = 0$. This yields

$$F_r = -2\pi e^2 n_0 r.$$

Defocusing is determined by the transverse wake. In the bubble regime, where all currents are localized in a thin layer near the bubble boundary, the transverse and longitudinal wakes are related by the Panofsky–Wenzel theorem. Using Eqs. (76) and (67) one can write

$$W_\perp(\xi, \xi_1) \approx \frac{8(\xi - \xi_1)}{r_b(\xi)^4}\theta(\xi - \xi_1). \quad (87)$$

Assuming that initially all particles of the bunch are deflected by the same amount r, one obtains the wake-induced force acting on particles in the bunch tail:

$$F_w \approx \frac{4e^2 N_a L_a}{r_{b1}^4} r, \quad (88)$$

where L_a is the length of the accelerated bunch. Relatively simple calculations yield the following result for the ratio of deflecting to focusing strengths:

$$\eta_w = \frac{F_w}{F_r} = \frac{(r_{b2}^2 - r_{b1}^2)^2}{4 r_{b1}^4} \frac{R_b^4 + r_{b1}^2 r_{b2}^2}{R_b^4 - r_{b2}^4}. \quad (89)$$

Expressing r_{b1} from Eq. (86), one obtains η_w as a function of η_P and r_{b2}/R_b. In the area of interest $r_{b2}/R_b \le 0.5$, where acceleration is reasonably fast, η_w can be presented by the approximate formula

$$\eta_w(\eta_P) \approx \frac{\eta_P^2}{4(1-\eta_P)^2}, \quad \frac{r_{b2}}{R_b} \le 0.5. \quad (90)$$

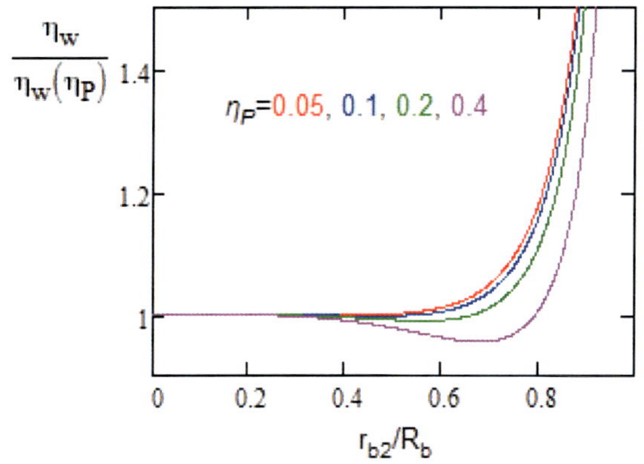

Fig. 12. The ratio of $\eta_w/\eta_w(\eta_P)$ as a function of r_{b2}/R_b.

Figure 12 presents the dependence of ratios of Eq. (89) to Eq. (90) on r_{b2}/R_b for different values of η_P. One can see good coincidence for $r_{b2}/R_b \le 0.5$ and that η_w grows fast for $r_{b2}/R_b \ge 0.5$, thus making this range of parameters uninteresting for plasma acceleration. Also, one can see from Eq. (90) that the normalized deflection force does not depend on details of plasma acceleration and is uniquely determined by η_P in the parameter range useful for plasma acceleration.

Strong focusing in the bubble results in a very large number of betatron oscillations in the course of beam acceleration. The total betatron phase advance can be estimated as

$$\mu_{\mathrm{tot}} \approx \sqrt{2\gamma_f} E_0/E_a, \quad E_0 = 4\pi n_0 e/k_p,$$

where γ_f is the final value of the particle Lorentz factor. For a 1 TeV linac with $E_a/E_0 = 2$ it yields $\mu_{\mathrm{tot}}/2\pi \approx 160$. In this case oscillations of the bunch head resonantly drive particles in the tail, resulting in the increase of the effective transverse emittance.

To describe the head–tail motion we assume a rectangular longitudinal particle distribution and a linear dependence of the transverse wake on ξ. Then, transiting to the variables natural for betatron motion description,

$$X = \frac{x}{\sqrt{\beta_f}}\sqrt{\frac{p}{p_0}}, \quad d\mu = \frac{dz}{\beta_f}, \quad (91)$$

we obtain the equation of betatron motion:

$$\frac{d^2 X}{d\mu^2} + \frac{X}{1 + \frac{\Delta p}{p}} = \frac{2\eta_w}{\left(1 + \frac{\Delta p}{p}\right) L_a^2} \int_0^\xi X(\xi')(\xi - \xi')d\xi'. \quad (92)$$

Here β_f is the beta-function determined by Eq. (20), p_0 is the momentum at the acceleration start, and we have taken into account a possible momentum deviation $\Delta p/p$ dependent on the particle longitudinal position in the bunch and assumed being constant in the course of acceleration. From a practical point of view the most interesting case is when all particles have the same initial offset—for example, excited by an error of bunch transfer between accelerating sections. For $\Delta p/p = 0$ Ref. 12 presents an asymptotic solution to this problem for large μ. However, a solution for small μ is more important in practical estimates. Therefore, we solved the problem numerically. Figure 13 shows the dependence of growth for the tail particle amplitude on $\eta_w\mu$ for different strengths of the transverse wake. As one can see, all solutions presented this way coincide so well that the red points on the plot are covered by the blue and green ones, and the blue points are covered by the green ones, respectively, making them invisible for low $\eta_w\mu$. The obtained results suggest a simple approximate parametrization for the amplitude of the tail particle,

$$\frac{A}{A_0} = \exp\left[\frac{(\mu\eta_w)^2}{12 + 1.416(\mu\eta_w)^{1.57}}\right], \quad \begin{array}{l}\mu\eta_w \le 100,\\ \eta_w \le 0.1,\end{array} \tag{93}$$

and the RMS amplitude of particle motion in the bunch,

$$\frac{\sigma_A}{A_0} \equiv \frac{1}{A_0}\sqrt{\frac{1}{L_a}\int_0^{L_a} A(\xi)^2 d\xi} = \exp\left[\frac{(\mu\eta_w)^2}{56 + 2.24(\mu\eta_w)^{1.57}}\right], \quad \eta_w \le 0.1, \quad \mu\eta_w \le 100, \tag{94}$$

where A_0 is the initial amplitude being equal for all particles. The requirement for the instability threshold depends on the requirements for the shot-to-shot machine stability. In the case of a single small kick this requirement is more forgiving. However, in real life there are many perturbations to the machine alignment coming from ground motion, jitter in the driving beam position or positions of laser beams in the case of LPA, etc. Therefore, to be on the safe side, one needs to require that a single kick should not increase the RMS motion by more than a factor of 2 compared to the case where instability is absent. That yields $\mu\eta_w \le 10$. For the above mentioned

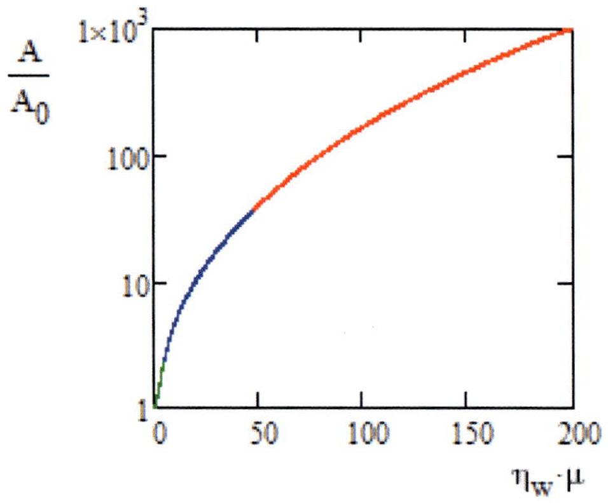

Fig. 13. Dependence of the tail particle amplitude on the betatron phase advance normalized by wake strength ($\eta_w\mu$) for different strengths of the wake. Data for $\eta_w = 0.1, 0.01, 0.001$ are shown in red, blue and green.

1 TeV linac we have $\mu \approx 10^3$, which yields $\eta_w \le 0.01$. Using Eq. (90) we obtain a limitation on the energy efficiency $\eta_P < 17\%$.

An effective way to suppress the BBU instability was suggested by Balakin, Novokhatsky and Smirnov [18]. It is called BNS damping. The idea is to introduce a dependence of particle momentum on the longitudinal coordinate in the bunch so that it would compensate for frequency detuning due to the transverse wake. In this case, from Eq. (92) one obtains

$$\frac{1}{1+\frac{\Delta p}{p}} - \frac{2\eta_w}{\left(1+\frac{\Delta p}{p}\right)L_a^2}\int_0^{\xi}(\xi-\xi')d\xi' = 1. \tag{95}$$

An integration and simple transformations result in

$$\frac{\Delta p}{p} = -\eta_w\frac{\xi^2}{L_a^2}. \tag{96}$$

A requirement to have the total momentum spread of 1% yields the same value, $\eta_w \le 0.01$, and subsequently the same, limitation on the energy efficiency as for the case where BNS damping is absent. Note also that it is unclear how the quadratic dependence of momentum deviation on ξ can be created on the entire length of acceleration. However, one can expect some improvement in beam stability with an introduction of the energy droop along the bunch and minor improvements in the energy transfer efficiency from plasma to accelerated beam.

10. Other Limitations

It has been already mentioned that the collapse of plasma electrons to the positron bunch axis greatly affects the beam focusing coming from the plasma. The focusing becomes strongly nonlinear and dependent on the particle longitudinal position in the bunch. All electrons which are located at radii smaller than r_m [introduced in Eq. (36)] are pulled into the positron bunch. Even for a modest number of positrons in a bunch, r_m is orders of magnitude larger than the transverse beam size. Note that the beam emittance is fixed by the luminosity and the only way to increase the beam size is an increase in the beta function, but it is limited by the necessity to prevent the emittance growth due to multiple scattering and by suppression of the BBU instability. For a bunch population of $4 \cdot 10^9$ and a bunch length of $10\,\mu$m, one obtains $r_m = 10.6\,\mu$m. The frequency of plasma electrons oscillations increases inversely proportionally to their impact parameters. This results in good mixing of plasma electrons near the axis, making variations of their density along the axis smooth. In the absence of electron interaction this density would diverge logarithmically with the radius near the axis. A repulsion of electrons prevents divergence but still cannot suppress a very large electron density variation and a plasma focusing nonlinearity related to it. Such nonlinear focusing would not create tremendous problems with emittance growth in the case of a single accelerating section and an absence of the final focus. But this is not the case for a plasma-based collider. Another way to characterize the problem is the phase advance of small amplitude oscillations of plasma electrons in the field of the positron bunch:

$$\mu_e = \frac{\sqrt{2r_e N_a L_a}}{\sigma_\perp} = \sqrt{2}\frac{r_m}{\sigma_\perp}. \qquad (97)$$

Here $\sigma_\perp$ is the RMS transverse size of the positron bunch and a round beam is assumed. For the above considered parameters ($N_a = 4 \cdot 10^9$, $L_a = 10\,\mu$m) and $\sigma_\perp = 0.15\,\mu$m (see Fig. 1), one obtains $\mu_e = 100$. To have reasonably linear focusing this phase advance need to be reduced by at least two orders of magnitude. That requires a drastic decrease for the number of positrons per bunch.

In the case of electron bunch acceleration we have a problem of collapsing plasma ions in the field of the electron bunch. In this case, in the expression for r_m one needs to use the ion mass instead of the electron one. For the above-considered case of electron beam acceleration with the bunch population of $8.86 \cdot 10^9$ and the bunch length of $4.2\,\mu$m, one obtains $r_m = 0.2\,\mu$m for proton plasma. This size is still larger than the electron beam radius, varying in the range of 0.05–0.15 μm (see Fig. 1). This means that a problem of ion collapse in the field of the electron bunch is also quite severe and will be an important limitation on the collider parameters. As in the case of positrons, we introduce the phase advance of small amplitude oscillations of plasma ions in the field of the electron bunch:

$$\mu_{\text{ion}} = \frac{\sqrt{2r_p N_a L_a}}{\sqrt{M_i}\sigma_\perp},$$

where M_i is the ratio of the ion mass to the proton mass. For the above discussed parameters of the accelerated electron bunch and proton plasma, the phase advance grows with beam energy and achieves $\sim 360°$ at 500 GeV. To avoid problems this value needs to be reduced by at least an order of magnitude.

Note that although the use of heavy ions looks to be a possible means of mitigating the problem of ion collapse, it does not look to be a real possibility due to the impact ionization of the ions. For the bunch parameters mentioned in the previous paragraph, one obtains the electric field at the bunch boundary exceeding 10^3 GV/cm. It is more than two orders of magnitude larger than the electric field in a hydrogen atom of ~ 6 GV/cm. Using ions stripped to the level sufficient to avoid impact ionization looks to be unrealistic. Heavy ions also increase the effects of bremsstrahlung, which are not completely negligible even for proton plasma.

11. Discussion

There are a number of problems which need to be solved before a credible concept of a plasma-based e^+–e^- or γ–γ collider can be put forward. Above we have intentionally discussed only beam physics limitations, leaving aside multiple outstanding challenges in technology and engineering. As far as we can presently judge, there is still no viable path which could lead to a high luminosity collider concept within the present paradigm.

The most outstanding problem is the acceleration of positrons with bunch brightness, required for a linear collider. One needs to have plasma electrons at the system axis to have sufficiently strong focusing required to suppress the transverse BBU (hosing) instability. The plasma transverse impedance (transverse wake) is very large and the suppression of this instability cannot be done by means other than plasma focusing. Any other existing focusing is orders of magnitude weaker than the plasma focusing and is incapable of suppressing the instability. But the problem is that the presence of plasma electrons at the positron path results in their pinching by the bunch head. It creates a very nonlinear focusing field in the tail. This field is driven by the high density of plasma electrons at the axis, which is orders of magnitude higher than the initial electron density in the plasma. This high density of electrons also greatly increases multiple scattering on the plasma electrons, resulting in unacceptably large emittance growth. The plasma channel was suggested to mitigate pinching and multiple scattering. In this case there are no charged and neutral particles in the close vicinity of the system axis. Consequently, such an arrangement solves problems of multiple scattering and pinching. However, it brings another problem — the absence of a mechanism capable of suppressing the BBU instability due to the large transverse impedance of a channel. To address it, one needs external focusing orders of magnitude higher than can be achieved with conventional means.

The situation is much better in the case of electrons. Potentially, one can consider a γ–γ or e^-–e^- collider, but its luminosity will still be significantly lower than the luminosity of the ILC or CLIC. There are several main phenomena which limit the luminosity. All of them, one way or another, are related to the efficiency of energy transfer to the beam. Studies of luminosity limitations carried out for the ILC and CLIC show that in the case of a global machine optimization, the collider luminosity depends only on the beam power; and that an achievement of luminosity comparable to the LHC luminosity requires the beam power to be above tens of megawatts. The operation at this power level requires a very high efficiency of energy transfer to the accelerated beam. In the ILC and CLIC it is achieved by acceleration of a very large number of bunches in a single linac pulse. Such an operation is supported by high Q values of accelerating cavities. In this case, each bunch after passing a cavity receives a small percentage of energy stored in this cavity. The lost energy is replenished by the RF power source before the arrival of the next bunch. As a result, in the case of the ILC, after passing many bunches, about 20% of the RF energy is transferred to the beam. Consequently, the overall energy efficiency is about 10%.

Achieving such an efficiency in the plasma-based accelerator represents a great challenge. The problem originates from a low Q value of plasma oscillations (especially in the bubble regime), with the result that only one bunch can be accelerated in a single pulse. Contrary to conventional accelerating structures, a plasma-based accelerating structure is excited at a single frequency which is close to the plasma frequency. Consequently, an absence of higher order modes suggests that the efficiency of plasma acceleration for a single bunch can be significantly higher than for a conventional structure. Although this statement is supported by a number of simulations carried out in recent years, we are still far away from demonstrating that it could work for acceleration of a collider quality beam. The above-mentioned BBU instability, driven by transverse impedance of the plasma bubble, is one of the major problems. It limits the number of particles in the bunch and, consequently, limits the efficiency of acceleration. In optimistic scenarios, the instability does not lead to unacceptably large emittance growth in the course of acceleration; however, it still greatly amplifies the emittance growth due to errors of the relative alignment of different accelerating sections. Note that presently the required alignment accuracy does not look achievable even in the absence of the BBU instability. The BNS damping, which could potentially help, requires too large energy spread, which is unacceptable from the final focus point of view.

For present proposals, pinching of plasma ions by a bright electron beam limits the luminosity of the e^-–e^- or γ–γ collider to well below $10^{34}\,\mathrm{cm}^{-2}\mathrm{s}^{-1}$ if a plasma with light ions is used. The use of heavy ions, which cannot be completely stripped due to the required energy efficiency, is prohibited by their impact ionization by an electron bunch field. It also greatly amplifies multiple scattering and bremsstrahlung.

In conclusion, we should note that there are many potential applications for plasma-based accelerators. However, it is presently unclear how the above-mentioned limitations could be overcome for the high luminosity linear collider.

References

[1] E. Adli *et al.*, SLAC-PUB-15426, arXiv:1308.1145.
[2] C. B. Schroeder, E. Esarey and W. P. Leemans, *Phys. Rev. ST Accel. Beams* **15**, 051301 (2012).
[3] *ICFA Beam Dynamics Newsletter*, No. 56, Dec. 2011.
[4] C. B. Schroeder *et al.*, *Phys. Rev. ST Accel. Beams* **13**, 101301 (2010).
[5] E. Gschwendtner *et al.*, arXiv:1512.05498.
[6] K. Oide, *Phys. Rev. Lett.* **61**, 1713 (1988); K. Hirata *et al.*, *Phys. Lett. B* **224**, 437 (1989).
[7] D. Schulte, "Future linear collider requirements", Presentation at LCWS-2015, (Whistler, Nov. 2015), http://agenda.linearcollider.org/event/6662/other-view?view=standard#2015-11-04.
[8] ILC Technical Design Report/Accelerator, https://www.linearcollider.org/ILC/Publications/Technical-Design-Report.
[9] V. Lebedev and S. Nagaitsev, *Phys. Rev. ST Accel. Beams* **16**, 108001 (2013).
[10] A. I. Akhiezer *et al.*, *Plasma Electrodynamics: Linear Theory* (Pergamon, 2013).
[11] T. Katsouleas *et al.*, Beam loading in plasma accelerators, *Part. Accel.* **22**, 81–99 (1987).
[12] C. B. Schroeder, D. H. Whittum and J. S. Wurtele, Multimode analysis of the hollow plasma channel wakefield accelerator, *Phys. Rev. Lett.* **82**, 1177–1180 (1999).
[13] A. Chao, *Physics of Collective Beam Instabilities in High Energy Accelerators*, (John Wiley & Sons, 1993).
[14] W. Lu *et al.*, Limits of linear plasma wakefield theory for electron or positron beams, *Phys. Plasmas* **12**, 063101 (2005).
[15] G. Stupakov *et al.*, Wake excited in plasma by an ultra-relativistic point-like bunch, arXiv:1606.02336.
[16] A. Burov, V. Lebedev and S. Nagaitsev, Instability of a witness bunch in a plasma bubble, http://arxiv.org/abs/1602.05260.
[17] M. Tzoufras *et al.*, Beam loading in the nonlinear regime of plasma-based acceleration, *Phys. Rev. Lett.* **101**, 145002 (2008).
[18] V. Balakin, A. Novokhatsky and V. Smirnov, in *Proc. 12th Int. Conf. High Energy Accelerators* (Fermilab, Batavia, IL, 1983), p. 119.

Valeri Lebedev graduated Novosibirsk State University in 1978 and earned his Ph.D. in electron cooling at BINP, Novosibirsk, in 1986. He worked at BINP from 1978 to 1994: at the single pass electron cooling and the Novosibirsk B-factory project; he was the BINP representative in the SSC in 1992-1993. In addition, he worked at Aarhus University, Denmark, from 1994 to 1995; and at TJNAF from 1995 to 2001, regarding commissioning and operational improvements of CEBAF. Since 2001 Dr. Lebedov has been at Fermilab, dealing with Tevatron Run II commissioning and luminosity improvements. Presently he is PIP-II Project Scientist.

Alexey Burov graduated from Novosibirsk State University and received his Ph.D. in Theoretical and Mathematical Physics at Budker Institute of Nuclear Physics (Novosibirsk, Russia), where he worked on various problems of beam physics. His main teachers were Eugene Perevedentsev and Slava Derbenev. Since 1997 he has been at Fermilab. He has authored many papers on beam dynamics.

Sergei Nagaitsev received a University Diploma in Plasma Physics from Novosibirsk University in 1989. He then completed his Ph.D. in Accelerator Physics at Indiana University Bloomington in 1995. He joined the Fermi National Accelerator Laboratory (Fermilab) in 1995 and led the effort to design, install and commission in 2005 the world's first relativistic electron cooling system of antiprotons at Fermilab. In 2006, Dr. Nagaitsev joined in the International Linear Collider (ILC) collaboration effort at Fermilab. He has since remained at Fermilab as an accelerator scientist and as Head of the Accelerator Division. He is currently Fermilab's Chief Accelerator Officer. In 2006, Dr. Nagaitsev was elected APS Fellow (Division of Physics of Beams). In 2007, he received the US Particle Accelerator School Prize for Achievement in Accelerator Physics and Technology.

Reviews of Accelerator Science and Technology
Vol. 9 (2016) 209–233
© World Scientific Publishing Company
DOI: 10.1142/S1793626816300103

Application of Advanced Accelerator Concepts for Colliders

Daniel Schulte
CERN, CH-1211, Geneva 23, Switzerland
daniel.schulte@cern.ch

In this article the potential application of advanced accelerator concepts to future high energy colliders is discussed. The emphasis is on lepton colliders, since the advanced technologies appear to be most applicable in this field.

Keywords: Plasma acceleration; linear colliders.

1. Introduction

Advanced acceleration technologies, such as plasma-based acceleration [1] and dielectric accelerating structures, allow the reaching of very high accelerating gradients, well in excess of what can be achieved with conventional normal-conducting accelerating structures. It is therefore timely to consider which role they could play in future colliders for particle physics.

1.1. *Future high energy collider goals*

The main collider that is currently operated for high energy physics is the LHC [2]. It collides proton beams with the ultimate center-of-mass energy of 14 TeV. The LHC has discovered the Higgs and is exploring the physics at higher energy scales.

Two main directions are being explored for the next generation of high energy frontier colliders. One direction aims at high energy hadron colliders, and the other at high energy lepton colliders. The main reason for this two-pronged approach is that the two options offer different advantages. A decision will be based on further input from the LHC experiments.

With current technologies, pushed further in performance, proton colliders have a larger energy reach than lepton colliders. This is beneficial for direct searches of new physics phenomena. Lepton colliders offer the potential of high precision complementing the hadron collider experiments. They can explore high energy scales by indirect searches made possible by the high precision. There are two main reasons for this difference. First, protons are not fundamental particles like leptons; rather, they consist of different partons, which carry variable fractions of the proton energy. This makes the reconstruction of the physics event difficult, since the initial conditions are not well known. In contrast, the initial particles in lepton colliders are known and have relatively well-defined energies. Second, in the proton–proton collisions a large number of background events are produced in each collision, which crowd the detector. This makes the reconstruction of the interesting part of the physics significantly more difficult. It is one of the main concerns of the current LHC experiments and the foreseen luminosity upgrade.

The hadron collider direction is addressed by two main studies. Within the framework of the Future Circular Study (FCC [3]), a hadron–hadron collider (FCC-hh) is studied for implementation at CERN. It should provide proton–proton collisions with a center-of-mass energy of 100 TeV. An electron–positron collider in the same tunnel with an energy of up to 400 GeV is also studied as a potential first stage; it is called FCC-ee. A second study is the Chinese Proton–Proton Collider (SPPC [4]), which is directed toward a proton–proton collider with a center-of-mass energy 50–70 TeV, to be implemented in China. The main focus of this study is on a circular electron–positron collider, CepC, with a center-of-mass energy of 240 GeV, which could later be upgraded with the proton collider in the same tunnel.

The main studies for high energy lepton colliders are on two linear electron–positron colliders.

The International Linear Collider (ILC) [5] is developing a collider for potential implementation in Japan. It should provide electron–positron collisions with a center-of-mass energy of 500 GeV. In the initial phase, the energy may be lower, and an upgrade to 1 TeV is also considered. The Compact Linear Collider (CLIC) [6] studies the implementation of a linear collider next to CERN. It will be implemented in three energy stages, the first at 380 GeV to cover Higgs and top physics, and the final stage at about 3 TeV.

1.2. *Use of advanced acceleration concepts*

The energy reach of proton colliders is mainly limited by their size and the field of the beam-guiding dipole magnets. In contrast, linear colliders mainly require accelerating the beam over a short distance to full energy. Hence, advanced technologies that can provide high accelerating gradients, such as plasma acceleration or dielectric RF structures, can play an important role in these machines. They can provide an alternative technology for the main components of linear colliders.

The advanced technologies are by far not as mature as ILC or CLIC. An important R&D program will be required to establish the feasibility of a collider based on advanced technologies, in particular the plasma-based cases. It therefore appears reasonable to assume that the technology could be used for a higher energy stage of a linear collider but that it would not be available for the first stage or stages. Also, the benefit of this solution is limited, since the linac represents a limited fraction of the project cost at these energies. We will consider colliders at 3 TeV for comparison with the third energy stage of CLIC.

It should be noted that the goal of advanced concepts must be not only to reduce the size of the machine. More importantly, it should reduce the cost of the project and increase the efficiency of converting electrical power from the grid to luminosity. These latter two requirements have critical importance for the application of the technology, as detailed below. For CLIC a systematic optimization of the collider parameters has been performed to achieve the lowest cost for a fixed performance. The optimum design that has been found is at a lower gradient than the maximum that one could achieve.

The design of linear colliders has reached a high level of maturity. One collider, the Stanford Linear Collider (SLC [7]), has operated successfully. In addition, several decades have been invested in further R&D on linear colliders. The performances of the different systems have been pushed as much as possible. The most promising path to profit from advanced acceleration technology appears therefore to integrate them it into linear collider designs and to limit the changes to the necessary minimum.

Two additional options for the use of plasma technology will be mentioned at the end. First, the use of plasma technology will be discussed to upgrade an existing electron–positron collider. While this is the most likely scenario, it cannot be treated sufficiently in this article. Second, a very different approach is to use protons to generate the plasma. Currently no proposal for using this technology for an electron–positron collider exists. However, one can consider using the proton beam to accelerate electrons, which can then be collided with a proton beam.

In the following, the design and beam requirements of linear colliders are discussed first. This will be the basis for the discussion of plasma-based colliders.

2. Linear Collider Overview

In circular electron–positron colliders, synchrotron radiation becomes a severe issue as the beam energy increases, since the radiation power increases with the fourth power of the energy. This is a fundamental limitation on the beam energy in FCC-ee and CepC. The limitation can be avoided in linear colliders, where the beam trajectory is not bent.

A generic example of a linear collider is shown in Fig. 1. The beams are produced in an electron and a positron source, respectively. They are accelerated to several GeV and transported to a damping ring. Here, their emittance is reduced to very small values, especially in the vertical plane. Then the beams are transported from the rings to the main linac. During this transport they are accelerated somewhat more and compressed longitudinally. In the main linac they are accelerated to full energy. The beams are then focused to very small sizes at the collision point in the beam delivery system. Finally, they are disposed of in beam dumps.

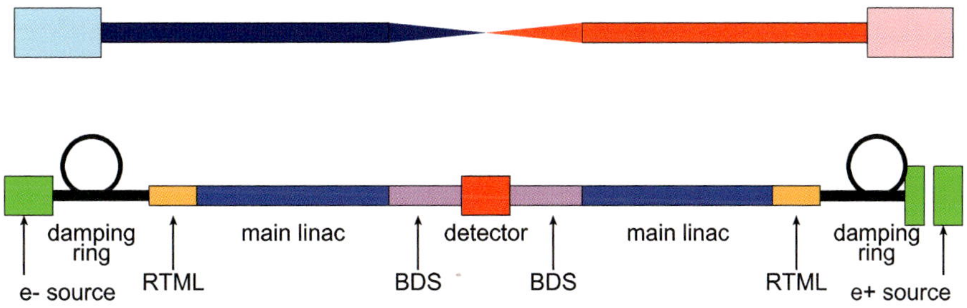

Fig. 1. The concept of a linear collider.

Linear colliders face two main challenges. First, the high colliding beam energy E must be achieved in a single pass through the main linac. This requires a very high accelerating gradient. Plasma and other advanced acceleration technology can obviously provide high gradients. Second, the high luminosity of the colliding beams must also be achieved in a single pass, since the beams have to be dumped afterward. This equally critical goal requires excellent beam quality and control.

Some key parameters of ILC and CLIC are given in Table 1. For the ILC the parameters are for 500 GeV, the main thrust of the proposal. For CLIC the parameters of the third energy stage are given. They are very useful for a direct comparison with colliders based on advanced technologies at high energies. As will be discussed below, a number of the considerations made in the CLIC design are also very relevant to colliders based on advanced technologies.

Significant differences exist between ILC and CLIC. They arise from the difference in acceleration technology used in the main linac. In the ILC, superconducting accelerating cavities are used. Because they are superconducting, almost none of the power that is fed into the cavities is lost into the walls. This allows one to use a very long, low current beam pulse and still reach high efficiency. The main drawback of the technology is that the gradient that one can achieve in the cavities is limited. The ILC goal is about 31.5 MV/m.

In contrast, CLIC is based on normal-conducting accelerating structures. These can hold much higher gradients; the CLIC goal is 100 MV/m. However, in these structures the RF power is rapidly lost into the walls. Therefore, one has to use short, high current beam pulses to be efficient.

This difference of acceleration technology between ILC and CLIC gives rise to another difference between the two projects. The low current beam

Table 1. Key collider parameters taken from Refs. 5, 6, 8, 10 and 13. The luminosity in brackets was calculated by the author for the DLA. The emittance values and the RMS bunch length for the LPA are private communications [9] and the beta function was calculated based on this. It should be noted that the ILC and CLIC parameters are supported by the results of an extensive R&D program. The other parameter sets contain important uncertainties concerning the feasibility; this is most true for the DLA.

Parameter	Symbol [unit]	ILC	CLIC	LPA	PWFA	DLA
CMS energy	$E_{\rm cm}$ [GeV]	500	3000	3000	3000	3000
Luminosity	L [$10^{34}\,{\rm cm}^{-2}{\rm s}^{-1}$]	1.8	6	10	6.3	10.7 (4.4)
Luminosity in peak	$L_{0.01}$ [$10^{34}\,{\rm cm}^{-2}{\rm s}^{-1}$]	1	2	?	2.5	(3.8)
Total beam power	[MW]	10.5	28	48	48	68.8
Loaded gradient	G [MV/m]	31.5	100	3000	7600	1000
Particles per bunch	N [10^9]	20	3.72	1.19	10	$3 \cdot 10^{-5}$
Bunch length	σ_z [μm]	300	44	8	20	0.0028
Interaction point beam size	σ_x/σ_y [nm/nm]	474/6	40/1	18/0.5	194/1.1	0.75/0.75
Normalized emittances	ϵ_x/ϵ_y [nm]	$10^4/35$	660/20	50/5	$10^4/35$	0.1/0.1
Beta functions	β_x/β_y [mm]	10/0.4	7/0.07	−/−	11/0.1	16.5/16.5
Initial beam energy spread	σ_E [%]	O(0.1)	0.35	—	—	—
Bunches per train	n_b	1312	312	1	1	159
Bunch distance	Δz [ns]	554	0.5	$11.9 \cdot 10^3$	10^5	$6.7 \cdot 10^{-6}$
Repetition rate	f_r [Hz]	5	50	$84 \cdot 10^3$	10^4	$3 \cdot 10^7$

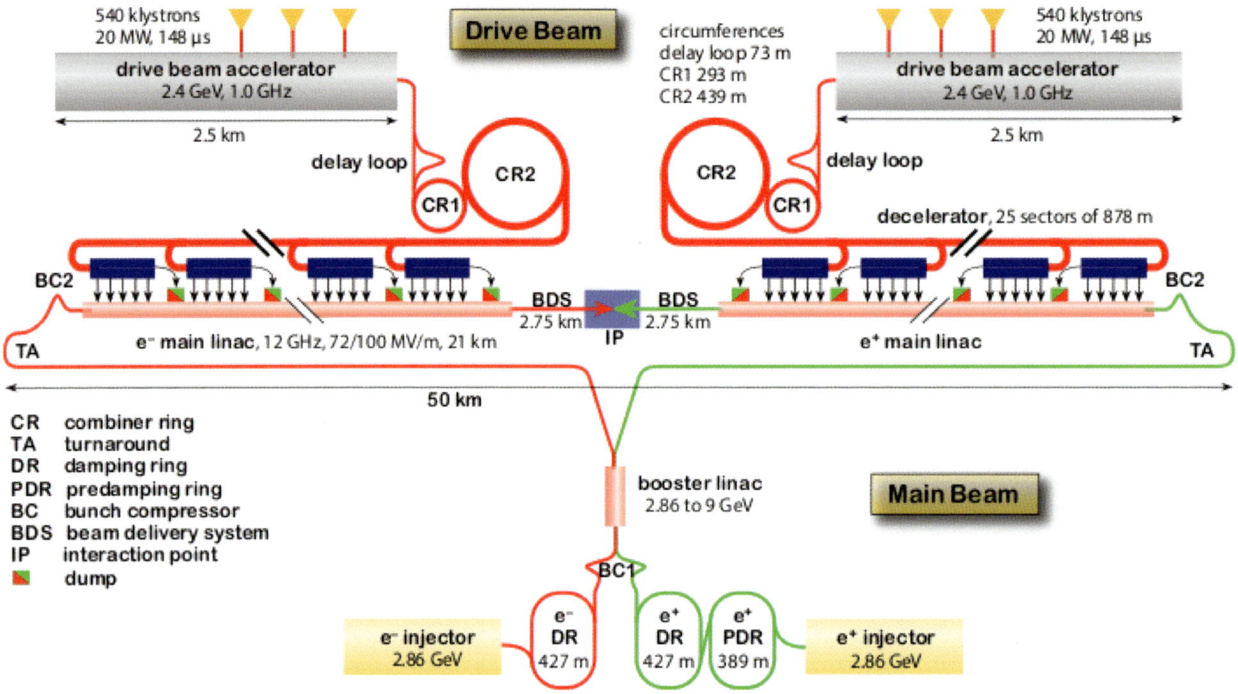

Fig. 2. CLIC design with the drive beam complex.

pulse requires only relatively low power but long RF pulses. These can be generated efficiently with conventional klystrons. In contrast, CLIC requires very high power, short RF pulses. These cannot efficiently be generated with klystrons. Hence, a unique method to produce the power is used. This is illustrated in the layout of CLIC, which can be seen in Fig. 2.

In a drive beam accelerator a very long train of drive beam bunches is accelerated. The beam is then sent into a system of delay loop and combiner ring which forms a sequence of short, very intense beam pulses. These pulses are then sent into the main linac, where each of them is used to power some part of the length. To this end, the drive beam is sent to fly in parallel with the main beam. The drive beam travels through power extraction and transfer structures (PETS), where it is decelerated and generates RF power. This power is then transferred to the main beam accelerating structures. The scheme allows one to compress a 140 μs beam pulse in the drive beam accelerator to a 244 ns RF pulse for the main beam, increasing the power by the corresponding factor.

In this need to use high power, short pulses, CLIC shares some similarity with the advanced acceleration technologies discussed here. Some still consider it advanced and it certainly was an advanced and novel technology some decades ago. CLIC requires one to maximize the main beam current for efficiency and the beam is powered by another beam; both features are quite similar to the advanced technologies discussed below. Hence, in the following, we will use the CLIC design as a reference point for conventional colliders.

3. Linear Collider Based on Advanced Acceleration Technology

Two different advanced acceleration concepts based on the use of plasma are being studied for high energy electron–positron colliders. One is based on the use of plasma that is driven by a laser beam, and the other is based on plasma that is driven by an electron beam. The study of these concepts has not reached a maturity of the design that is anywhere comparable to that of ILC and CLIC. In particular, no conceptual design exists but rather a set of ideas. We will, however, use some sample parameter sets and a layout proposed in different references; see Table 1. While it is not certain that they can be reached, they allow one to discuss the issues more specifically.

The example for a laser-driven plasma is called the LPA (laser plasma accelerator [8]). The beam-driven approach is called PWFA (plasma wakefield acceleration [10]). Since the use of the plasma-based technologies seems most beneficial at very high energies, parameter sets for a center-of-mass energy of 3 TeV have been chosen. It should be noted that the parameters are only illustrations of what one might consider. They will have to evolve, as the understanding of the collider is advancing. This is witnessed by a recent change of the parameters for the LPA, which assumed drastically different beam parameters from Ref. 11 to Ref. 8. One can expect more changes of parameters as the physics and technology driving the choices is better understood.

Dielectric acceleration can be considered to be in between conventional and plasma-based acceleration. It can be applied in a number of different forms. In the most simple case it replaces conventional normal-conducting RF structures [12], the main difference being that one can hope to achieve higher gradients in the dielectric structures before they spark. More aggressive schemes propose to power the structure with a laser. An example is DLA (dielectric laser acceleration [13]). Example parameters for this are also found in Table 1 based on Ref. 13 with some modification [14]. These parameters are very different from the other proposals and the concept is less mature than the plasma-based ones. In this article the discussion of this technology is rather limited, in order to maintain a reasonable scope.

In the following, the focus will be on the plasma-based electron–positron collider, where the plasma will be used from the very beginning.

3.1. *Beam-driven plasma-based collider*

The potential layout of a beam-driven plasma collider is shown in Fig. 3; the design is actually for a center-of-mass energy of 1 TeV. The key parameters for a design at 3 TeV are given in Table 1 to ease comparison with CLIC.

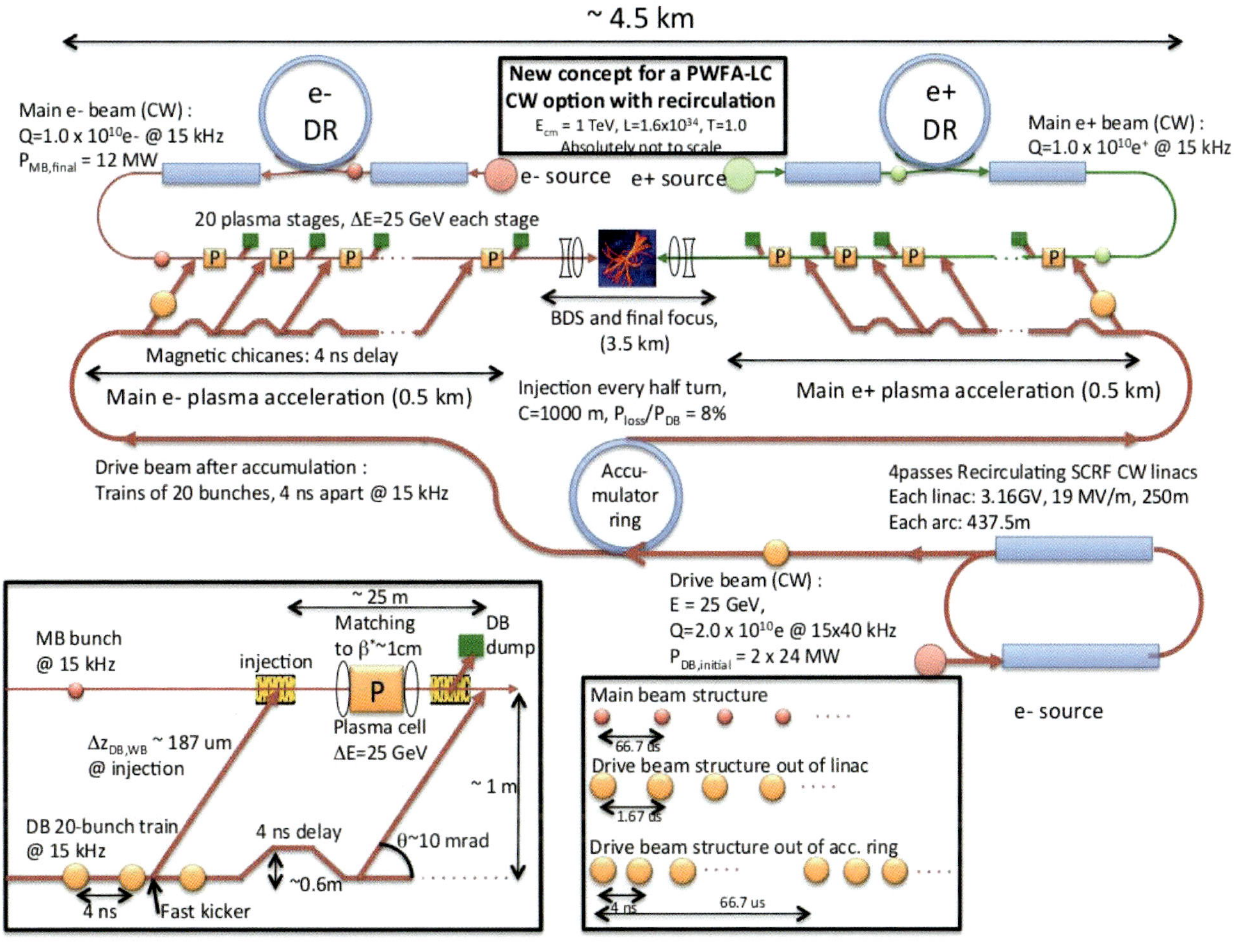

Fig. 3. The concept of linear collider based on beam-driven plasma. The layout is from Ref. 10.

The design is similar to that of CLIC or ILC. The electron and the positron main beams are generated in sources, cooled in damping rings and accelerated further before they are injected into the main linacs. The main linac consists of a sequence of 60 plasma cells; each accelerates the beam by 25 GeV. After the acceleration the beams are focused to collide in the detector.

The plasma cells will be discussed in more detail below. In each of the plasma cells, the accelerating field is generated by a drive beam bunch consisting of $2 \cdot 10^{10}$ electrons with an energy of 25 GeV, which is about a factor of 10 higher than the drive beam energy in CLIC. The plasma is operated in the blowout regime.

In the proposal, the drive beam bunches are generated and accelerated in a central complex at a constant rate. A recirculating superconducting linac is used for this purpose. The generated bunches are injected into a accumulator ring, where they circulate. The size of the ring is chosen such that the circulating bunches are very close to the point of injection when the next bunch arrives. This allows one to form two bunch trains — one for the electron linac and one for the positron linac. In Ref. 10 the distance between bunches in the train was chosen to be 4 ns; currently a shorter distance of 2 ns is preferred [15]. The distance is defined by the minimum rise time that a fast kicker can have; see below.

Once the trains are complete they are extracted and travel to the beginning of the main linacs. In each main linac the last bunch of the train is used to power the first plasma cell. To this end it is extracted from the train with a very fast kicker. Hence, the minimum rise time of this kicker defines the minimum distance between the bunches. The other bunches are then sent into a chicane, in order to delay them by one bunch distance. Now the last bunch is again in the proper longitudinal position to power the next plasma cell. The plasma cells are 3.3 m long and are spaced by 25 m. Hence, the main beam is accelerated over a fraction of the main linac only.

The drive beam energy is very high, which is challenging, as will be discussed later. The reason for the high energy is the wish to limit the number of plasma cells. A higher number of cells would most likely result in an increased length of the machine if one cannot reduce the spacing between the cells. It would certainly result in an increase in the number of technical components, which could lead to higher cost. The drive beam energy and the main beam energy gain are coupled through beam physics. The drive beam generates the longitudinal fields which accelerate the main beam. These fields, however, decelerate the drive beam. If the drive beam energy were too low, particles would be lost in the cell. The ratio between the drive beam energy and the energy gain by the main beam is called the transformer ratio. It is difficult to increase it significantly above unity. However, as will be discussed in the section on the drive beam complex, a reduced drive beam energy would potentially be very beneficial. An increase in the transformer ratio should be one important line of research.

Many of the challenges that need to be met for this scheme will not be discussed here. For example, one has to have very fast kickers and the optics for the handling of the drive beam and the main beam are far from trivial. In the following, the focus will be on some key issues in the plasma cell and the generation of the drive beam.

3.2. *Laser-driven plasma acceleration*

The potential key parameters of a laser-driven plasma collider at 3 TeV are given in Table 1. A simple sketch can be found in Ref. 16, but it shows a very limited amount of details.

Also in the LPA, the main beams will be accelerated in a series of plasma cells. In this case the accelerating field is generated by laser pulses. Each cell has a length of 1.7 m and a loaded gradient of 3 GV/m. The plasma cell should be hollow, i.e. it has a central cylinder that does not contain any plasma; its radius is 22 μm. The optics for the main beam between the plasma cells is not specified and might be quite challenging. As for the PWFA, it will not be discussed here.

3.3. *Configurations of plasma*

The configuration of the plasma is a fundamental choice for a collider. It has important consequences for the overall collider design, in particular on the acceleration of positrons. Different options exist. One can work in the quasilinear or in the blowout [17] regime. Also, the concept of hollow plasmas exists. The choice of regime is particularly important for the acceleration of positrons. The key parameters of

Table 2. The key parameters of the plasma cells in the LPA and PWFA.

Parameter	Symbol [unit]	PWFA	LPA
Cell length	L_{cell} [m]	3.3	1.7
Number of cells per linac		60	300
Peak gradient	G_0 [GV/m]	14	6
Used gradient	G_{eff} [GV/m]	7.6	3
Plasma density	n_0 [10^{17}cm^{-3}]	0.2	1.0
Plasma wavelength	λ_p [μm]	215	100
Hollow channel radius	[μm]	—	22

the currently proposed values of the PWFA and LPA plasma cells are given in Table 2.

In the blowout regime, the driver is pushing the plasma electrons completely to the side [17], generating an electron free cavity or bubble behind it. When the electrons start to rush back into the center, they produce a longitudinal field together with the ions. An electron beam can be nicely accelerated in the second half of the bubble before the electrons meet at the central axis. The ions in the bubble also generate a transverse field that focuses the electrons. A positron beam will be accelerated just behind the driver, when the plasma electrons are still rushing out. However, they will also experience a transverse field from the ions, which is defocusing and will destroy the beam rapidly. The next longitudinal position that can accelerate positrons is just behind the position where the plasma electrons start to meet at the center. However, this region is quite short. In addition, one can only use the fraction of this region in which the electron density at the axis is higher than the ion density, in order to avoid a defocusing force on the positrons.

One could drive the bubble with a positron (or proton) beam. In this case the electrons of the plasma would accelerate toward the center of the driver; this has been used to accelerate positrons [18]. While this is an interesting option, the region available for positron acceleration remains very limited.

In conclusion, positron acceleration appears possible in the blowout regime but it will be very difficult to accelerate a significant amount of charge. Efficient positron acceleration is, however, instrumental for an electron–positron collider. Since an antimatter plasma does not currently appear practical, another solution has to be found.

Another option is the quasilinear regime, where the beam density is smaller than the plasma density. The driver (electron beam or laser) pushes the electrons only slightly away from the beam axis. This causes a harmonic oscillation of the electron density in the plasma without removing the electrons completely. This density variation generates the longitudinal field. The individual electrons are moving only a little transversely.

An electron beam can be accelerated in this configuration in the same way as in the bubble regime. It needs to be placed behind the driver when the electron density is increasing and the ion density is still higher, providing the focusing force. A positron bunch can be accelerated behind this location when the electron density is decreasing again but is still higher than the ion density.

However, to achieve a good plasma-to-beam efficiency, the bunch charge has to be large enough. As Ref. 19 has demonstrated, it is not possible to have a high efficiency without violating the requirement that the beam density be smaller than the plasma density for the quasilinear regime.

This leaves another option for accelerating electrons and positrons; the use of hollow plasma channels. This will allow one to avoid transverse forces in the main beam. While this solves the problem of defocusing forces, it also suppresses the benefit of focusing forces. The beam will therefore be much more sensitive to instabilities than in the bubble regime. In addition, the properties of the plasma channel in the highly loaded regime remain to be studied.

For the electrons, one could even consider having some ions in the channel to provide a calibrated focusing. This still would not solve the problem for the positrons. Also, the generation of such a hollow channel seems a formidable challenge, particularly since very high accuracy is likely required.

An alternative approach is to generate a hollow laser pulse [20]. This is sent into the uniform gas and forms a hollow plasma; the inner part of this channel remains filled with neutral gas. Then a driver is used to produce the fields in the plasma. This scheme has recently been demonstrated [21]. This approach is very useful for experimental purposes. However, a number of limitations arise from the fact that gas remains inside of the plasma channel. Most importantly, the fields generated by the beam have to remain small enough to not ionize the gas. In this respect it is useful to note that in CLIC the beam density is indeed high enough to perform field ionization of the rest gas in the vacuum at the end of the main linac in the 3 TeV design.

In summary, the acceleration of positrons is more challenging than the acceleration of electrons in all types of plasma.

If the difficulty of accelerating positrons cannot be overcome, the only solution appears to be to consider a gamma–gamma collider. In such a collider one would need two high energy electron beams. Just before the collision point each electron beam would collide with a laser beam. The electrons would kick some of the laser photons very hard, transferring most of their energy to them. The relative importance of such a collider for the physics community remains to be established and will depend on findings at existing facilities such as the LHC.

4. CLIC Parameter Choice Strategy

The derivation of the CLIC parameters gives a valuable insight into the considerations relevant to advanced acceleration technologies.

The CLIC parameters have been optimized in a systematic way, taking into account the beam dynamics and constraints of all the different subsystems. In this approach the main linac accelerating structure defines the single bunch charge and length. In order to optimize the RF to beam efficiency, one aims to maximize the single bunch charge. The bunch charge and length are coupled by the goal of achieving an acceptable energy spread at the end of the linac; the larger the charge is, the longer the bunch has to be. A maximum acceptable bunch charge results from the requirement that the beam should remain stable in the presence of self-generated transverse fields. This charge limit can be improved by strong focusing of the beam, which is indeed pushed to the maximum reasonable value. As a consequence, tight tolerance has to be met for the alignment and stability of the components in the main linac.

The possible performance of the main beam injection and the focusing system, such as the damping ring and beam delivery system, are then used to determine the other beam parameters. This allows one to estimate the luminosity that can be achieved with a given power. With the help of a cost model the overall project cost can be estimated. This systematic approach allows one to define one optimum parameter set for each accelerating structure design. By scanning many options of different structure designs, one can then determine the overall best collider design.

In the following, first the basic considerations of luminosity will be presented, in particular the limitations and requirements arising from the collision itself. Then longitudinal beam parameters and the efficiency of the main linac will be discussed. Finally, the beam physics that constrains the parameter choice in the main linac will be touched on.

5. Beam Parameter Considerations

In the following, we will discuss the beam parameters that are required in order to achieve the luminosity in a collider. Also, the generation of the beam will be discussed. Subsequently, the impact of these parameters on the use of advanced acceleration in the main linac will be highlighted.

5.1. *Luminosity and beam parameters*

Based on the requests of the high energy physics community, the luminosity target for linear colliders is in the range of $\mathcal{L} \approx 10^{34}\,\mathrm{cm^{-2}s^{-1}}$. From Table 1, we can see that ILC and CLIC use flat beams to achieve this ambitious luminosity target and that the vertical beam size is only of the order of nm. The reason will be detailed below. For the PWFA similar parameters have been used from the beginning, and the LPA has recently adopted similar parameters, while the DLA uses round beams.

The luminosity in a linear collider is given by the following formula:

$$\mathcal{L} = H_D \frac{N^2 n_b f_r}{4\pi\sigma_x\sigma_y}. \tag{1}$$

Here N is the number of particles per bunch, σ_x and σ_y are the horizontal and vertical beam sizes at the collision point, n_b is the number of bunches per train, f_r is the rate of trains per second and H_D is a factor that contains the impact of beam–beam forces and other relevant effects. H_D is typically in the order of 1.5–2, with the exception of the DLA. For simplicity we ignore this factor in the following discussion.

It is useful to rewrite the luminosity formula in the following form:

$$\mathcal{L} \propto \frac{N}{\sigma_x}\frac{1}{\sigma_y} N n_b f_r. \tag{2}$$

The term Nn_bf_r represents the beam current. Its upper limit arises from the power consumption P of the collider and the efficiency η to turn this power into beam power. As can be seen from Table 1, the beam powers of all projects are quite significant. It is therefore important to maximize the efficiency.

In addition, one has to optimize the luminosity per beam current, i.e. the factors N/σ_x and $1/\sigma_y$. However, a lower limit to the horizontal beam size arises from the beam–beam effects, and to the vertical beam size from the beam quality.

5.2. *Beam–beam effect*

At collision, the beams are so dense that they generate very strong electromagnetic fields. In an electron–positron collider they focus each other; Fig. 4 illustrates this. In circular colliders this deflection is quite small and can be understood as a thin lens kick. In a linear collider the beams are so dense that the particles move strongly during the collision. This so-called pinch effect reduces the effective beam size and leads to an increase in luminosity. It has been experimentally verified in the SLC [22], where under some conditions the luminosity was more than doubled. But this motion also leads to the emission of radiation, the so-called beamstrahlung, which alters the energy of the colliding beam particles. As a consequence, the luminosity is provided not only at full beam energy but also at lower energies; a luminosity spectrum forms — see Fig. 5.

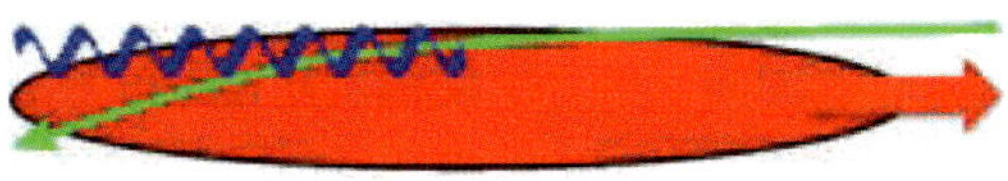

Fig. 4. The beam fields focus the other particles of the other beam, which leads to the emission of beamstrahlung.

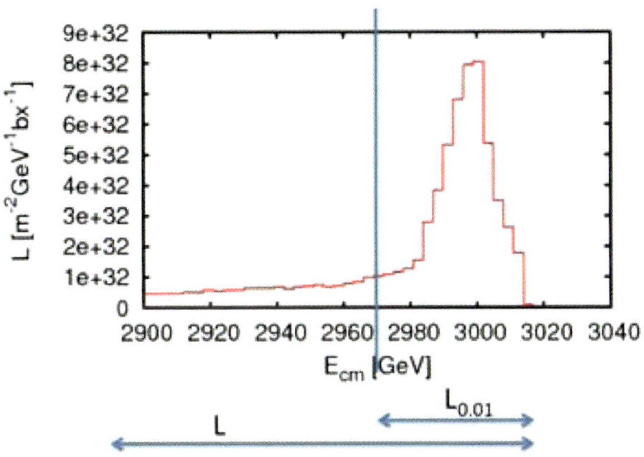

Fig. 5. The luminosity spectrum in CLIC at 3 TeV.

The physics experiments require a certain quality of the luminosity spectrum. In the case of CLIC the goal is to have more than 30% of the luminosity above 99% of the nominal center-of-mass energy. This places a constraint on the beam parameters. The radiation increases as $N/(\sigma_x + \sigma_y)$ increases. Since the luminosity is proportional to $N/\sigma_x\sigma_y$, one chooses $\sigma_x \gg \sigma_y$. Hence, the value of σ_x is fixed by the level of beamstrahlung that is allowed by the experiment. At these high energies, the acceptable value also depends on the bunch length, since shorter bunches reduce the beamstrahlung. The actual dependence is given by

$$\mathcal{L} \propto \frac{n_\gamma^{3/2}}{\sqrt{\sigma_z}} \frac{1}{\sigma_y} N n_b f_r. \qquad (3)$$

Here n_γ is the number of photons emitted per beam particle. The bunch length is not a free parameter. A reduction in the vertical beam size will improve the luminosity; the same is true for an increase in the beam current.

As one can see from Table 1, all proposed colliders use flat beams with the exception of the DLA. In the case of the latter, the charge is very small in the bunch train, leading to little beamstrahlung. The spot size is defined by other considerations than beamstrahlung.

5.3. *Implications for beam parameters*

In a simplified way, one can calculate the beam size at the collision point as

$$\sigma_{x,y} = \sqrt{\frac{\epsilon_{x,y}\beta_{x,y}}{\gamma}}. \qquad (4)$$

Here $\epsilon_{x,y}$ is the normalized beam emittance in x or y, the beta function $\beta_{x,y}$ describes the focusing at the interaction point, and γ is the relativistic factor. The emittance is a property of the beam.

In linear colliders, the beams are focused using a system of magnets. The final component is a doublet of quadrupoles that focuses the beam to have a waist at the collision point. An important R&D effort has been invested in the design of the focusing systems. This includes very sophisticated computer-based optimizations and the experimental verification at two test facilities — the FFTB [23] and ATF2 [24]. An international collaboration has constructed and operated the latter for several years.

However, the performance is constrained by a number of effects. The final doublet creates a high chromaticity, i.e. a beam energy spread will lead to a much-increased beam size at the collision point. This has to be corrected with nonlinear lenses. These generate geometric aberrations. In addition, the beam particles can radiate in the field of the magnets, leading to an energy mismatch.

These different effects result in lower limits for the beta functions. Since one wants to achieve flat beams, one tries to minimize the beta function in the vertical plane as much as is reasonably possible and accepts a larger value in the horizontal plane. If we take the CLIC system as an example, it has been possible to reach nominal beta functions of $\beta_x \approx 7\,\text{mm}$ and $\beta_y \approx 70\,\mu\text{m}$ and an RMS energy spread acceptance of about 0.35% at 3 TeV. In this design the effective beam size is already larger than the nominal one calculated with Eq. (4). In particular, in the vertical plane the beam size is a factor of 1.4 larger than in the simple model.

If one investigates Table 1, one can make a comparison with the beam delivery system design achieved for CLIC:

- A simulation shows that the LPA beam size is about compatible with the use of the CLIC beam delivery system, provided that the beam has essentially no incoming energy spread and the emittance target is reached [25].
- The PWFA beta functions' vertical beam size is similar to that of CLIC, but with almost twice the emittance. The horizontal beta function is larger than in CLIC, which will make the system design slightly easier. However, it may still be necessary to decrease the emittances, particularly in the vertical plane, in order to reach the spot sizes.
- The DLA has very relaxed beta functions. This is the result of the extremely small emittances that are assumed for this collider.

The beam sizes in the table seem roughly consistent with what could be expected provided that the energy spread of the beam can be kept very small and that the emittance targets can be reached. In the case of the PWFA some reduction of the emittance target could be required. Certainly, a full design of the focusing system will have to be made for the advanced projects in order to ensure the quantitative correctness of the predictions.

5.4. *Availability of the main beam sources*

The availability of sources that can provide a beam of sufficient quality is essential. The sources are challenging, since the emittance targets are very small. Also, the bunches need to be short and potentially have a specific longitudinal profile, as will be discussed below. In ILC and CLIC the design of the beam generation system has been one of the most demanding challenges.

The emittance produced in the injector has to be smaller than in Table 1, since some emittance growth will occur during the transport from the source to the collision point. In the case of CLIC the required emittances from the injector complex are $\epsilon_x = 500\,\text{nm}$ and $\epsilon_y = 5\,\text{nm}$. This allows one to have emittance growth budgets of $\Delta\epsilon_x = 160\,\text{nm}$ and $\Delta\epsilon_y = 15\,\text{nm}$.

As one can see, the absolute emittance growth budget is larger in the horizontal plane than in the vertical. However, the relative emittance growth is more important in the vertical plane; the final value is mainly defined by the growth. This is a result of the effects in the main linac. The growth depends on the beam intensity. If one would reduce the budget for emittance growth, the beam current would have to be reduced, which would lead to a loss of luminosity.

Based on the CLIC considerations and since the vertical emittance is much smaller than the horizontal, one can make the working assumption that the source should be able to provide a horizontal emittance slightly smaller than in the table. In the vertical plane, however, effects that will increase the emittance will be quite visible, because efficiency will likely demand this. So one should assume that the emittance will grow by at least a factor of 2 in the vertical plane.

The generation of the high quality beams for colliders has been subject to an extended R&D program. For electrons, the production of the main beam is relatively straightforward; typically, one would use a photoinjector. However, with the current state of the art, the transverse emittances of the produced bunches are too large. This is even more important in the case of positrons. These are usually produced by sending a photon beam into the target. The showers on the other side then contain positrons, which one can collect. These beams typically have very large emittances. In ILC and CLIC, this is overcome by sending the beams into a damping ring, where the emittance is reduced. In these rings the beam particles radiate photons, which remove some of the transverse momentum, and are then reaccelerated, thus reducing the transverse emittance.

A damping ring can achieve very small emittances in the vertical plane; the limit is imposed by collective effects of the beam and imperfections of the machine. However, a lower limit exists for the horizontal emittance, e.g. 500 nm in the case of CLIC. This is because in the bending plane of the beam, the fact that the radiation consists of individual photons leads to an emittance growth, which at some level balances the decrease. In addition, the beam brilliance is limited by collective effects such as intra-beam scattering. However, for the positron beam, the author is not aware of a better scheme than the use of damping rings to produce high quality beams, which has been demonstrated.

The assumptions that have been made for the various proposals in Table 1 differ significantly. The PWFA is quite in line with ILC and CLIC transverse emittances. As discussed above, some reduction of the emittance might be needed to reach the luminosity target but this would still be consistent with CLIC. The LPA would require a very significant improvement of the emittances compared to CLIC in both planes. Whether it is possible to reach these small values remains to be studied. This will require the development of improved damping rings or other solutions. If such an effort is successful, this will also be beneficial to ILC and CLIC.

The DLA has emittances far below that of any other project. It is not clear how one can produce such a positron beam.

In a damping ring the bunch length will become quite long (typically of the mm scale). Therefore, the produced bunches need to be compressed longitudinally. The bunch compressors typically consist of an RF station that induces an energy chirp along the bunch. This is followed by a chicane or an equivalent, in which the path length is energy-dependent. In the case of CLIC, which requires a 44 μm bunch length in the main linac, the beam has to be compressed in two stages, which require some acceleration in between them. To achieve the very short bunch lengths in the advanced colliders might turn out to be challenging, particularly since a very specific longitudinal profile might be required, as discussed below. The machines will likely be very sensitive to small errors in the profile.

If the beams cannot be generated with damping rings, novel schemes for generating them will have to be devised. If one can come up with a good scheme, a demonstration will be required.

Side note: The operation of the damping rings in a plasma collider requires individual bunches, while in ILC and CLIC trains of bunches are injected, cooled and extracted together. However, it should be possible to operate the ring in a mode where many bunches are cooled together and only one is extracted and then replaced with a fresh one for each pulse.

5.5. *Impact on beam parameters*

Based on the considerations at the interaction point, the beam parameters that are required at the end of the linac have been determined. The horizontal emittance of the beam is a function of its charge (and length), while the vertical emittance should be as small as possible and the RMS energy spread should remain in the final focus acceptance.

The production of the beams for the PWFA appears to be possible using the experience from ILC and CLIC; the vertical emittance is similar with 40 nm. For the LPA a rigorous R&D program will be required to understand if beams of sufficiently small emittance can be produced. If such an effort is successful it will also benefit the other projects. In particular, a reduction in the vertical emittance will directly benefit in the form of an increased luminosity. The preservation of the beam emittance in the main linac will also be more challenging, since a smaller budget for the growth exists. The production of the positron beam for the DLA will require completely novel approaches. However, in all cases a more detailed design will be required in order to also understand the longitudinal beam dynamics that will be crucial for the main linac.

The most important challenges for the main linac are to limit the increase in the vertical emittance to a modest level and to obtain a beam RMS energy spread of the order of 0.3% or less.

6. Main Linac Efficiency

As shown in Table 1, the beam powers in the different colliders are quite high; this is mainly because the luminosity is proportional to the beam power. High efficiency to transform the power from the grid into beam power is therefore important.

The efficiency is the product of three main factors in plasma colliders:

- The efficiency to generate the drive beam (electrons or photons);
- The fraction of the drive beam energy that can be transferred to the plasma;
- The fraction of the energy that the main beam can extract from the plasma.

In the following we will first point out the target efficiencies, before we discuss the impact on the design.

6.1. *Efficiencies in CLIC*

As a point of reference, in CLIC the power from the grid can be transformed into drive beam power with an efficiency of 58%. The efficiency of transforming this drive beam power into RF power is 78%. Both of these efficiencies are dominated by technology, such as the efficiency of klystrons and modulators. The efficiency of turning the RF power into beam power is 28%. This efficiency is strongly dependent on the parameters of the main beam, as will be discussed below. The total efficiency of transforming the drive beam into main beam power is thus 22% and the overall efficiency from the grid to the main beam is 12.7%. This is the fundamental efficiency of the technology.

The overall efficiency of the collider is further impacted by other power consumers. Additional overheads are required for operational considerations. If one includes them the overall efficiency is 11%. However, not only the drive beam acceleration requires power. Many other systems need to be considered, including the beam-guiding magnets, cooling and ventilation, the generation of the main beam, and others. Including all of these in the calculation reduces the overall efficiency to 5%. It is therefore important to consider the full design in order to give a realistic estimate of the power consumption. In the following we will, however, discuss only the efficiency of the main beam acceleration since the power consumption of the other systems cannot be reliably estimated without a design. Also, the plasma-to-main-beam efficiency is most relevant to the beam physics.

6.2. *Efficiency of plasma acceleration*

For PWFA the efficiency to produce the drive beam could be similar to that in CLIC, as discussed below. Simulation studies [10] have found that 77% of the drive beam energy can be transferred to the plasma and 65% of this energy can be absorbed by the main beam. These simulations assume Gaussian bunches. The total efficiency of transferring drive beam power into main beam power is thus 50%. The direct comparison with CLIC shows that the main improvement in efficiency is expected from the plasma-to-main-beam efficiency, which is assumed to be about twice as high.

The power that the drive beam transfers to the plasma but that the main beam does not extract has to be extracted in a different way. For PWFA with the current design about $77\%*(1-0.65) = 27\%$ of the drive beam power is lost in the plasma; this is about 160 kW per plasma cell. The cooling of the plasma cell will thus be important and it has to be ensured that this is done efficiently so as not to increase the power consumption of the collider too much.

For laser-based plasma acceleration, similar numbers are assumed. The target efficiency to generate laser power is of the order of 50% including the combination of laser pulses from different sources into the full pulse required for the plasma cell. The efficiency of transfer from the laser into the plasma is only 20% but most of the remaining laser power can be recovered. An efficiency of 75% from the plasma to the beam is assumed. This leads to a total efficiency of 12%.

It has been proposed to use a second laser pulse in each cell to extract the power from the plasma that has not been transferred to the beam [26]. This should reduce the heat load in the gas.

In summary, if one considers only the wall plug power required for the main beam acceleration in the main linac, the LPA might be able to reach the same efficiency of 12% as CLIC. The beam-driven PWFA might hope for an efficiency that is about twice as high. Detailed studies and technological R&D will be required to consolidate the numbers for the plasma-based colliders. In all projects the efficiency of transferring power from the RF (in the case of CLIC) or the plasma (in the other cases) to the beam is critical.

6.3. *RF-to-beam efficiency in CLIC*

In CLIC, originally single bunch operation had been foreseen and the parameters chosen accordingly —

hence a situation close to the one for plasma-based colliders. However, it has been found that the efficiency that can be achieved with good beams in this way is limited.

To generate the accelerating field in a normal-conducting structure, it has to be filled with electromagnetic energy. The passing bunch extracts some of this energy as it is accelerated. In order to obtain high efficiency it has to extract as large a fraction of the stored energy as possible. This is called high beam-loading. It requires a high bunch charge to maximize the extracted energy and a small transverse size of the accelerating structure to minimize the stored energy required to provide the gradient. However, both requirements reduce the beam stability.

This problem can be reduced by accelerating trains of bunches. In this case each bunch needs only to extract a small fraction of the stored energy. RF power is continuously flowing into the structure. When the target gradient is reached the first bunch will be sent through the structure and extract some energy. When the inflowing RF has replaced the extracted energy the next bunch arrives, and so on. This allows one to accelerate all bunches at the same gradient. A train of bunches has also the advantage that the unwanted transverse wakefields that can lead to an instability can be removed between bunches. The design of the main beam accelerating structure has therefore been changed accordingly by adding damping waveguides.

While multibunch operation greatly enhances the RF-to-beam efficiency, still larger bunch charges lead to better efficiency. Hence, in CLIC the bunch charge is still at the limit of what can be transported in a stable way, including some safety margin.

6.4. *Bunch length and energy spread*

Because a bunch that passes through an accelerating structure (or a plasma cell) extracts a fraction of the stored energy, the field amplitude is reduced. Hence, the tail of the beam will be accelerated less, which leads to a large energy spread. As discussed above, the energy spread of the beam has to remain small in order to allow efficient focusing at the interaction point. In the case of CLIC an RMS energy spread of 0.35% is the goal and this appears an adequate goal for plasma-based colliders. If it were possible to increase this acceptance the efficiency of CLIC would be improved. However, so far no better system design has been found in spite of considerable effort.

Let us consider the simplest possible model: an accelerating structure in which only a single longitudinal mode can exist. In this case all the energy of the structure has to be stored in this mode. One can calculate the energy spread induced in a short bunch if it is accelerated when the RF field reaches its maximum. If the bunch extracts 2% of the energy of the structure, the field G is reduced by 1%, i.e. the last particle sees only 99% of the gradient that is seen by the first. For a bunch with a uniform charge profile, the RMS energy spread that will develop is 0.3%, i.e. very close to the limit. This will set a rather severe limit on the RF-to-beam efficiency of only 2%.

The energy spread in the beam can be reduced by accelerating the bunches slightly before the maximum field is reached. The increase in the accelerating field and the additional decelerating field produced by the head of the bunch can compensate. In this case the acceleration of the tail is the same as for the head.

In principle one can suppress the gradient spread completely by giving the bunch a particular triangular longitudinal profile [27]. The compensation of the beam-loading is illustrated in Fig. 6. A harmonic variation of the accelerating field with time is assumed. The decelerating field induced by the beam is also assumed to only excite the fundamental mode of the structure. Based on this concept for any given charge, the bunch length is fixed. For small efficiencies, the bunch length scales as $\sigma_z \propto \sqrt{N}$. The

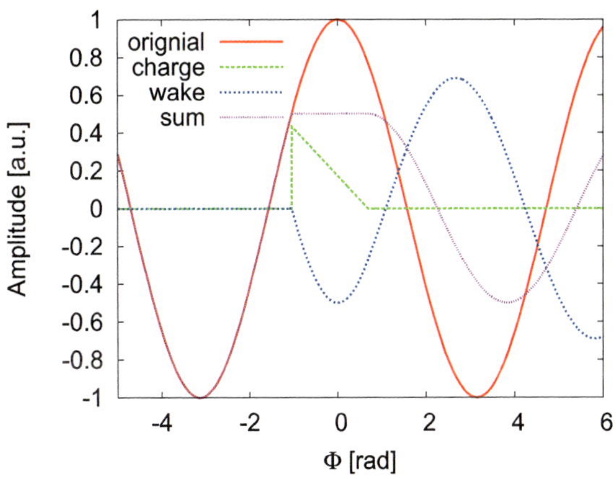

Fig. 6. The beam-loading by a beam in an accelerating structure and the compensation by accelerating it off-crest.

actual limit on the bunch charge will then arise from beam physics and tolerances, as dicsussed in the next section.

A note: In CLIC this principle is applied assuming a Gaussian longitudinal beam profile. It also has to be considered that the field induced by the bunch not only consists of the fundamental mode but also has higher frequency modes, which has been neglected in the above model. In this case a full suppression of the gradient variation is not possible. However, the principles hold true.

6.5. *Plasma-to-beam efficiency*

In the considered plasma-based colliders, only a single main beam bunch is accelerated. It therefore has to extract a large fraction of the energy in the plasma in order to achieve high efficiency.

In the case of the LPA, a triangular longitudinal bunch profile is suggested, which should eliminate the energy spread completely following the simple model above. The accuracy of this model for this case, particularly given the high efficiency of 75% of energy transfer from the plasma to the beam, has to be determined.

In the case of the PWFA with a 65% plasma-to-beam efficiency, an RMS spread in the gradient of 3% has been found for Gaussian bunches [10]. Since this gradient variation is systematic as a function of the longitudinal particle position in the bunch, it will lead to a final RMS energy spread of the beam of almost the same size. This is much larger than the typical energy acceptance of a final focus system. An improved reduction in the energy spread is thus required. A longitudinal profile of the bunch could achieve that. Because of the field profile in the bubble it would need to have a shape different from a simple triangle.

In principle a small energy spread can be achieved in plasma-based colliders by giving the main beam bunches a specific longitudinal profile. To achieve such a profile is challenging and requires careful development and study of the main beam generation system. A careful and detailed study of the main beam generation system and the plasma cells with high beam-loading is required to determine what is ultimately possible. In addition, the limit from the beam quality preservation has to be considered, as will be discussed next.

7. Beam Quality Preservation

In the above, only the efficiency to transfer power into the beam has been considered. For plasma-based colliders considerations of tolerances, emittance growth and beam stability seem to have played a minor or no role in the definition of the parameters. In the case of CLIC, the actual limits for the beam parameters, in particular also the efficiency and luminosity, come from beam quality considerations. In plasma-based colliders, beam physics and tolerances will also determine the performance. Hence, a systematic assessment of all the relevant effects is mandatory before any claim of achievable luminosity can be made.

7.1. *General considerations*

The beam quality can be affected by the design. A good example is beam gas scattering in the case of a nonhollow plasma, which leads to emittance growth of the main beam. However, for linear colliders and other accelerators it has been found that static and dynamic imperfections are highly important. They arise from the fact that the machine can only be built with certain tolerances and that it is not completely stable. The consideration of tolerances is fundamental for the application of a technology in a collider.

In CLIC, different limits exist for the bunch charge due to transverse effects. The first arises from transverse wakefields. A bunch that has a transverse offset in an accelerating structure induces a transverse field in the structure, which will kick the tail toward the structure. If this field is too strong, it will render the beam unstable. The strength of these fields depends mainly on the transverse size of the accelerating structure.

The limit for the beam stability is coupled to the strength of the focusing. The stronger the focusing, the more stable the beam. That is to say, the faster the beam oscillates in the guiding fields, i.e. the smaller the beta functions, the more charge can be stably transported. In CLIC the strength of the focusing is maximized to the level that can be reasonably achieved. About 10% of the linac is filled with focusing magnets that have the maximum strength that one can achieve with the normal-conducting technology and good field quality. Any further increase would compromise the effective gradient significantly.

However, the strong focusing creates new challenges. The components of the collider can only be placed with limited accuracy in the tunnel. The tolerances for these alignment errors of the magnets become tighter as their number and strength increase. In the case of CLIC the tolerance has been well below what state-of-the-art alignment could deliver a decade ago. A very intense R&D program has been carried out in the past few years in order to develop a novel alignment system. This system uses overlapping wires to define the reference positions along the tunnel and actively aligns all components to this reference. This improves the alignment quality by about an order of magnitude. In addition, a similarly intense program has been carried out to develop and test the beam-based alignment algorithms that are instrumental in preserving the beam quality.

The strong focusing also leads to very tight tolerances for the stability of the magnets. In the case of CLIC, even the natural motion of the ground would destroy the luminosity in many sites. This has been addressed with another R&D program. An active stabilization system has been developed using inertial sensors that can detect the magnet motion and actuators that can correct it.

In the following a few examples of tolerances are highlighted. Simple models are used to allow some very rough first order quantitative assessment. This is meant to point out the importance of the work that is required to arrive at a realistic design and parameters. A substantial effort is required to arrive at the mandatory complete understanding of the relevant effects. Based on this, one can identify R&D needs and make informed design choices.

7.2. *Longitudinal effects*

The energy of the colliding beams has to be well controlled in order to be able to efficiently focus them. The typical tolerance for existing beam delivery system designs is an RMS energy jitter of about 0.1% in order to keep the luminosity stable at the 1% level [28]. An RMS jitter of 0.3% would already lead to a luminosity loss by 10% and larger values would reduce the luminosity quickly to zero.

It should be noted that the tolerances in a collider are based on a luminosity loss of at most a few percent. This is because several effects that reduce the luminosity exist and one has to limit their combined effect. A very large luminosity loss due to dynamic effects also implies a strong fluctuation of the luminosity. This makes the tuning of the machine in operation impossible.

If the timing between the drive beam and the main beam is changing slightly, the main beam will be accelerated at a different gradient. Let us again consider the example of a perfectly harmonic field oscillation given in the section on efficiency. In order to keep the acceleration error below 0.1%, one finds that the relative timing error has to remain below about 0.1° for a plasma-to-beam efficiency of 50% and 0.07° for an efficiency of 75%.

If the main beam has the wrong timing, it will see the same error in all plasma cells. The final energy error will thus be the same as the gradient error in each cell. For a plasma wavelength of around 200 μm, as in the PWFA this would translate roughly into a timing tolerance of better than 0.2 fs; this corresponds to about 50 nm. Similarly for the LPA, with a wavelength of 100 μm the tolerance is twice tighter. This is two or three orders of magnitude tighter than the tolerance in CLIC, which is already very tight. Hence, it calls for an important R&D program.

One should note that the exact tolerance depends on the plasma-to-beam efficiency. The higher the efficiency is, the more sensitive the acceleration will be to phase jitter. Actually, this tolerance is more relaxed in CLIC and ILC, because the bunches are accelerated closer to the maximum of the field, since the single bunch beam-loading is smaller, and because the RF wavelength is longer. Also, the mode of the plasma acceleration, i.e. linear versus blowout, is important. Hence, a full simulation is required to establish a precise value. The tolerances can also be relaxed by increasing the plasma wavelength (and consequently the bunch length).

It is worth noting that if the drive beam has a phase jitter that varies from cell to cell, the tolerance is increased since the main beam averages over the errors. For the 3 TeV PWFA this relaxes the tolerance by a factor of about 8.

Other variations can also cause energy errors. For example, a main beam bunch charge variation of 0.1% will cause a similar energy error. The dependence of the field error on the drive beam variation has to be established using full simulations. The same is true for a variation of the plasma density. However, one can expect that the relative acceleration error is of the same magnitude as the relative drive beam charge error and the relative plasma density error. Also, these errors will benefit from an averaging over the different cells of the accelerator, which increases the tolerance.

7.3. *Transverse drive beam jitter*

In the following we consider the transverse jitter tolerance for the drive beam in the case of a plasma in the linear or blowout regime. Again, the consideration is valid for an electron or a laser drive beam. We assume a constant ion density around the path of the main beam. In addition, we consider the limit of small main beam bunch charges, such that we can ignore the impact of the main beam on the plasma. The latter effects will be discussed below.

The main electron beam is focused by the ions in the plasma. The strength of this focusing can easily be calculated from the electric field that the ions produce. One can assume a cylinder with uniform charge density. In this field, the main beam electrons will oscillate with a wavelength $2\pi\beta$, where β is given by

$$\beta \approx 3.3\,\mathrm{mm}\sqrt{\frac{10^{16}\,\mathrm{cm}^{-3}}{n_0}\frac{E}{\mathrm{GeV}}}. \qquad (5)$$

Let us use the PWFA parameters at the end of the linac as an example. In this case we find a wavelength of $2\pi\beta \approx 0.57\,\mathrm{m}$. The cell has a length of more than 3 m. Hence, the beam particles will perform several oscillations in the plasma cell.

In order to be matched to this focusing the beam size has to be

$$\sigma_{x,y} \approx 41\,\mathrm{nm}\left(\frac{10^{16}\,\mathrm{cm}^{-3}}{n_0}\frac{\mathrm{GeV}}{E}\right)^{\frac{1}{4}}\sqrt{\frac{\epsilon_{x,y}}{\mathrm{nm}}}. \qquad (6)$$

At the end of the PWFA the main beam size is about 33 nm in the vertical plane for an emittance of 35 nm. This is the smallest size in the main linac and the location where the beam is most sensitive to the effect described below.

One can easily estimate the required stability of the drive beam center. The main beam will be focused toward the center of the drive beam. If the drive beam has a small fluctuation of the transverse offset with respect to the nominal position, the main beam will be focused around a different axis; see Fig. 7. This means that the main beam will jitter from pulse to pulse.

For simplicity let us ignore the acceleration of the main beam and the deceleration of the drive beam,

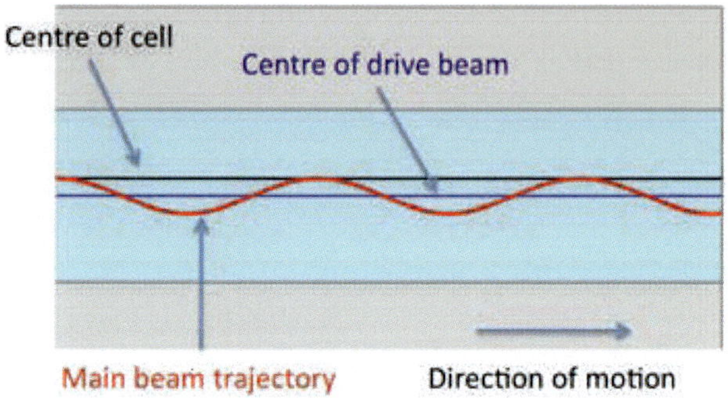

Fig. 7. Scheme showing how an offset drive beam can lead to a wrong main beam position at the end of the plasma cell.

and assume that the only error present is a transverse position error $\Delta y_{\mathrm{drive}}$ of the drive beam when it enters the plasma cell. In this case the center of the focusing has an offset with respect to the main beam position (which we assume to be in the nominal position). The main beam will hence oscillate around this center. At the end of the cell, the offset Δy_{main} will be

$$\Delta y_{\mathrm{main}} = -\Delta y_{\mathrm{drive}}\left[1 - \cos\left(\frac{L}{\beta_{\mathrm{main}}}\right)\right]. \qquad (7)$$

Here β_{main} is the main beam beta function and L the length of the plasma cell. The drive beam jitter will hence induce a jitter of similar size in the main beam.

One can estimate the expectation value σ_{jitt} for the main beam jitter at a later point in the beam line. Let us ignore the phase factor L/β_{main}. One wants to limit the transverse jitter of the main beam to a fraction of its size; let us assume 20% to be specific. Such a jitter in the two colliding beams corresponds to a luminosity loss of 2%.

If one just considers the last cell, the transverse jitter of the drive beam (be it a laser or an electron beam pulse) has therefore to be limited to about 7 nm. The tolerance in this last cell is the tightest, since the tolerance scales with the main beam size and the beam size is decreasing with energy. However, all of the cells will add some jitter. If we assume that this is uncorrelated from cell to cell and take into account the variation of the main beam size along the main linac, we find a tolerance of less than 1.4 nm. This is a challenging goal, particularly since the drive beam will have relatively large transverse dimensions.

For this specific case, the problem could in principle be eased by adjusting the length of each plasma cell to bring the factor $L/\beta_{\rm main}$ close to a multiple of 2π. If the beam has an energy spread, the effectiveness of this scheme is reduced. More importantly, angular jitters of the drive beam are also of concern. The position error of the focusing axis at the end of the cell is given by the product of incoming angular error and cell length. In the PWFA the cells are about 3 m long. Hence, the transverse position jitter tolerance of 1.4 nm is equivalent to an angular jitter tolerance of 0.5 nrad.

Achieving such a tight tolerance is quite challenging. This can be easily seen by comparing the required angular stability to the angular spread, the divergence, in the drive beam. Let us assume that the drive beam has a small emittance of only 1 μm; this is actually less than is assumed in Ref. 30, for example. In this case the drive beam will have a beta function of less than 12 mm at the entrance of the cell. This yields a beam size of 490 nm at the entrance of the plasma cell and a divergence of 42 μrad. Hence, it needs to be stabilized to less than 0.28% of the beam size and a fraction of $1.2 \cdot 10^{-5}$ of its divergence. Assuming a perfect Gaussian transverse distribution, the fact that one has only $2 \cdot 10^{10}$ particles in the drive beam bunch will already lead to a jitter of the center of $0.7 \cdot 10^{-5}$ times the divergence just because of the finite number of particles.

Different methods to improve this tolerance might be considered. One of them is to slowly increase the plasma density at the entrance of the cell and slowly decrease it at the exit. This leads to larger beta functions at the entrance and exit. The beam sizes will increase accordingly and the jitter tolerance will increase in proportion. It will also be easier to meet the angular jitter tolerance, since the divergence of the beams is reduced. The angular tolerance can further be improved by using shorter cells, which reduces the drive beam position error at the end of the cell. However, the tolerance appears very demanding and it remains to be estimated if these measures will be sufficient and if the tolerances can be met. Obviously, a possibility of avoiding this effect is to use a hollow plasma.

7.4. *Transverse beam alignment*

We discussed above the tolerance for a drive beam jitter. One can use similar considerations to estimate the tolerance for static offsets between the two beams. In this case the relevant quantity is the growth of the emittance of the beam, while the total transverse deflection in each cell will be corrected with other machine elements. We will again ignore the impact of the main beam on the plasma.

When the main beam oscillates around the trajectory of the drive beam, it will filament. The particles in the main beam have different energies and hence will oscillate at different speeds. Thus, their motion will decohere. Intuitively, it is obvious that in this case the beams need to be aligned to a fraction of the main beam size. Quantitative estimates have been performed [29, 30]. A tolerance in the range of tens of nm has been found. This is three orders of magnitude tighter than for CLIC. However, it may be possible to devise beam-based tuning strategies to resolve this issue.

7.5. *Hose instability*

As the beam passes through the plasma, it modifies the plasma. For example, in a bubble, the head of the beam will move the center of the bubble by pushing the electrons in the sheath surrounding the bubble slightly to one side. The plasma in turn will impact the tail of the beam. In particular, it can deflect it. This can lead to a hosing instability. For small perturbations the effect is similar to wakefield effects in a linear collider. The main difference is that the accelerating structures will not be moved transversely by the field of the beam, unlike the plasma.

The hosing instability will limit the amount of charge that one can transport in a bunch. This is equivalent to the limit obtained in CLIC from the transverse wakefields. The effect exists in any plasma. It is, however, most important in the absence of the strong focusing force from the plasma ions, i.e. in the case of a hollow plasma. In the definition of the target parameters for a collider, the hose instability should play the main role.

In CLIC, the single bunch transverse wakefield is given by the following equation for small distances z of the witness particle after the driving particle:

$$W_\perp(z) \approx 2\frac{Z_0 c}{\pi a^4} z. \tag{8}$$

Here, $Z_0 \approx 377\Omega$ is the vacuum impedance, a the radius of the accelerating structure irises, i.e. 2.75 mm, and c the speed of light. Hence, one finds that $W_\perp(z) \approx 1.3 \cdot 10^6\,\mathrm{kV/(pCm^3)}z$.

For a plasma with a hollow channel of radius b, a similar calculation can be performed [31]. It yields

$$W_\perp(z) \approx 2\frac{Z_0 c}{4\pi}\frac{(2\pi)^2}{b^2\lambda^2}\frac{K_1(R)}{RK_2(R)}\frac{1}{1+\frac{RK_1(R)}{4K_2(R)}}. \quad (9)$$

Here, K are the Bessel functions and $R = 2\pi b/\lambda$ is the radius of the channel in the plasma divided by the wavelength. One can insert the number from Table 1. Using the LPA bunch length and gradient, one finds that $W_\perp(z) \approx 4.2 \cdot 10^{13}\,\mathrm{kV/(pCm^3)}z$. This is about $3 \cdot 10^7$ times larger than in CLIC. The wakefield kick on the bunch tail will also depend on the bunch length and charge, so one has to compare $W_\perp(\sigma_z)N$. Here the LPA is a factor of $1.7 \cdot 10^6$ worse than CLIC.

The instability induced by the coupling to the wall of the hollow channel can be partially mitigated by strongly focusing the beam. Since the beam is not focused by the plasma in the hollow channel, the beta function of the drive and main beams has to be adapted to the length of the plasma cell. It would hence be much larger in the case of a solid plasma and comparable to the length of the plasma cell. An obvious way to improve this is to have a limited ion density in the central channel. However, the control of the focusing will be challenging. Also, part of the issue with the transfer of the drive beam jitter to the main beam will exist.

A hollow channel also has to deal with additional imperfections. If the laser beam that creates the hollow channel is misaligned with respect to the nominal trajectory of the beam, the hollow cylinder is created with a slight offset, which can give rise to further wakefield effects. In addition, if the drive beam has an offset it will induce a transverse wakefield which will deflect the main beam. The tolerances for both effects have to be evaluated.

8. Drive Beam

The production of the drive beam is critical for the performance of the PWFA. The drive beam bunches have to be short and, as described above, one likely wants to have a specific optimized longitudinal profile. Also, the longitudinal and transverse stability of the beam has to be excellent.

Currently no design for the longitudinal manipulations exists that allows the generation of the short bunches. It is essential to address this in order to identify the technologies that have to be developed to produce these bunches with the required stability.

However, concepts of how such short drive beam bunches can be accelerated and distributed to the main linac plasma cells exist and will be discussed here.

8.1. *Drive beam production*

For the PWFA, the generation of the drive beam is an interesting challenge. It has been proposed to use a CW recirculating superconducting linac to produce the drive beam bunches [10]. The bunches would be produced and accelerated with large distances and sent into a combiner ring. Here trains of bunches are accumulated. Once a sufficient number of bunches has been collected to feed each main linac plasma cell, the train is extracted and sent to the main linac. The collection of the next train then starts.

This scheme has similarities to the CLIC drive beam scheme. It has some potential shortcomings. First, the circulating beams radiate strongly because their energy is an order of magnitude higher than in CLIC and because they have to perform more turns. The energy of the first bunch would vary strongly between the first and the last turn. Experience with the CLIC combiner ring complex in the test facility CTF3 indicates that this might be very challenging. A detailed study of the full beam dynamics would be required to establish the feasibility of this scheme.

However, recently, the author proposed using a CW superconducting linac but with a pulsed beam; more information can be found in Ref. 32. Each beam pulse would contain one bunch per plasma cell that needs to be powered. The spacing between bunches would correspond to the one that is used in the main linac, e.g. 2 ns. The beam pulses therefore can directly be sent to the main linacs with no storage.

One point that has to be considered with this scheme is the beam-loading. Each bunch of the pulse will extract energy out of the cavity, thus reducing the field amplitude. This can lead to a decrease in the acceleration from one bunch to the next, which results in a variation of the final energy. This variation has to be limited. As an example we assume a maximum difference of 1%, but detailed studies are required to arrive at a more precise value.

The reduction of the amplitude due to beam-loading depends on the stored energy in the cavity. If we accelerate the beam on the crest of the RF wave, the stored energy in the cavity needs to be 50 times larger than the energy extracted by the beam pulse.

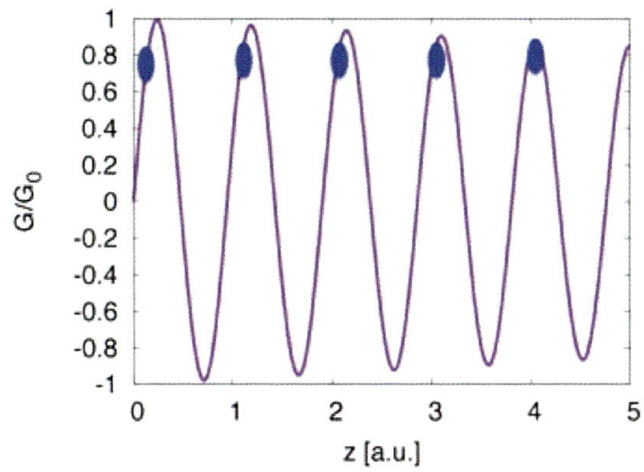

Fig. 8. Compensation of the beam-loading by frequency detuning. The phase of the bunches changes with respect to the RF to compensate in first order for the amplitude reduction.

Unfortunately, the stored energy is slowly dissipated in the form of heat into the cavity walls, even in a superconductor. The cavity has to be kept cold, i.e. around 2 or 4 K, to remain superconducting. The removal of power dissipated at this temperature is very inefficient. If it is dissipated at 2 K the cryogenic system typically consumes 700 times the power that one removes.

However, we can use a compensation scheme, as sketched in Fig. 8. The frequency of the cavity is slightly detuned with respect to the frequency of the bunches. The first bunch is accelerated before the field in the cavity reaches its maximum. The subsequent bunches are then accelerated closer and closer to the maximum of the field, thus compensating for the decrease in the amplitude. In this case we need to store only about five times the extracted energy; the effective gradient will be 80% of the maximum. Note: If the energy spread within each bunch is too high with this scheme, it can be compensated for by detuning half of the cavities to have a lower frequency than the train and to accelerate the bunches after the maximum of the RF has been reached.

For the 3 TeV collider the drive beam pulse contains 9.6 kJ, so about 50 kJ have to be stored in the cavities. If one would use 1 GHz cavities, one would need an active length of about 4200 m and a gradient of about 7.5 MV/m to achieve the drive beam energy of 25 GeV and have enough stored energy. These cavities would lose 16 kW into the wall if we assume a quality factor of $Q_0 = 2 \cdot 10^{10}$. So the cooling would consume about 11 MW. If one would double the gradient, the linac length could be halved, reducing the cost. The stored energy would double, increasing the cooling power to 22 MW, which would still be a correction to the overall power consumption. Certainly, a full optimization of the linac for cost and power should be performed.

The main issue to address for such a design is the stability of the accelerated drive beam train. Multibunch wakefields can render the beam unstable if the focusing is not strong enough.

8.2. *Drive beam distribution*

A scheme that has been proposed for distributing the drive beam bunches is illustrated in Fig. 3. The train of drive beam bunches is sent parallel to the main beam. At the first cell the last bunch is extracted and used to power this cell. All the other bunches in the train are delayed by a chicane by exactly the distance between the bunches. Now the new last bunch is sent into the next plasma cell to power it. This scheme is continued until the last bunch has been used for the last cell.

In principle, this scheme is very compact. However, to delay the high energy drive beam (E_D = 25 GeV) by 2 ns over 25 m would require four 6.25-m-long bending magnets with a field of about 5 T, even neglecting the space needs for other beam line components. These strong superconducting magnets would lead to the emission of a high level of synchrotron radiation. In fact, in the first chicane the drive beam would lose 2% of its energy. The total power of 1 MW leads to a load of 40 kW/m. In a superconducting magnet this is impossible to sustain.

In order to use such a scheme one would need to reduce the drive beam energy drastically. This would, however, imply using many more plasma cells for the main linac, or increasing the transformer ratio.

An alternative scheme that could be used is illustrated in Fig. 9. In this scheme the drive beam is sent through separate tunnels that have an angle with respect to the main linac. This angle is chosen to produce the correct delay with respect to the main beam, i.e. for a delay of 2 ns equivalent to 0.6 m over 25 m it has an angle θ defined by $1 - \cos\theta = 0.6/25$. In this specific example the last four bunches will be kicked down in the first kicker (A) to power the first

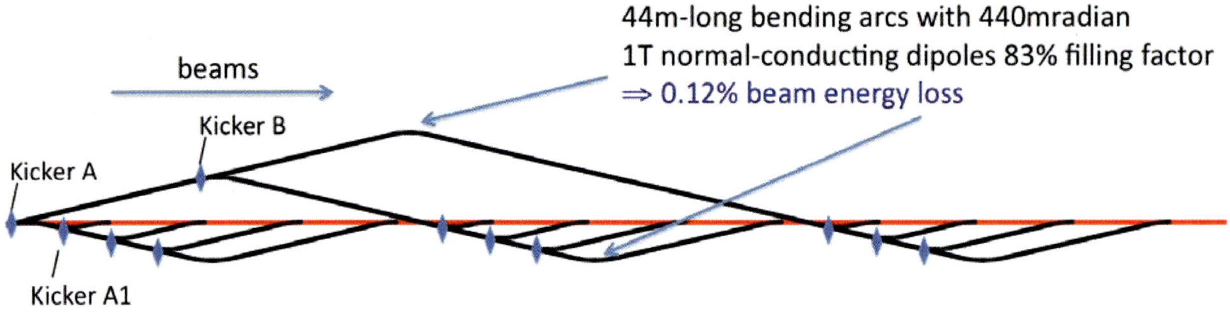

Fig. 9. An alternative drive beam distribution scheme proposed by the author.

four cells. The rest of the train will continue to kicker B, where again the last four bunches are deflected, and so on. The last of the four bunches deflected in A will be deflected by kicker A1 to power the first cell. The third bunch (now the last) will be kicked at the next kicker. Here it will be in time to accelerate the main beam.

For the 3 TeV PWFA one would use 64 cells per linac. This would allow the use of three stages with four exits each. The total additional tunnel for the drive beam distribution would add up to 5.4 km for the straight tunnels. For the bends a radius of 100 m has been assumed, with a filling factor of 83% for the dipoles. This allows one to use normal-conducting magnets and reduces the maximum synchrotron radiation power to 200 W/m. The energy loss per bend is 0.12% of the beam energy. The total energy lost by the bunches is 0.42%. The total length of the tunnel with bends is 5.1 km. The total bending strength is 4250 Tm, which is less than in the solution with the chicanes. While this option may be viable, it certainly leaves plenty of room for improvements and fundamentally better ideas.

Another scheme has been suggested earlier for the PWFA [33]. The system uses return arcs in a similar fashion to CLIC; see Fig. 10.

In that proposal the drive beam accelerator is normal-conducting. For efficiency reasons this requires very small distances between the drive beam bunches. As a consequence trains of drive beam bunches have to be kicked into each return arc in order to accelerate trains of main beam bunches. The feasibility of accelerating such trains with very small bunch distances in the plasma remains to be investigated. With the here-proposed superconducting linac with pulsed bunch trains, the scheme could be used. The drive beam bunches in the train can have the

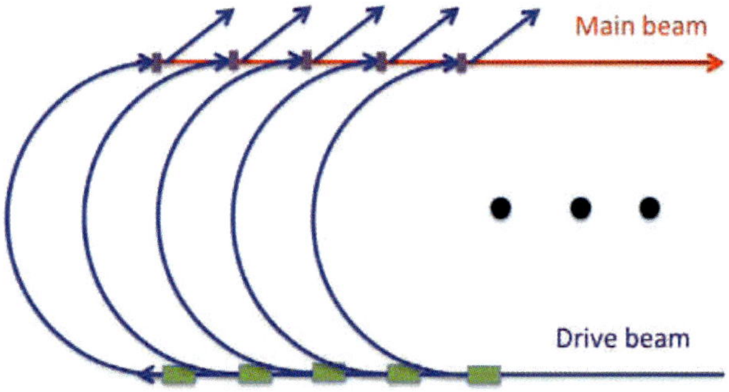

Fig. 10. A drive beam distribution system similar to the CLIC concept.

required spacing of 50 m to match the return arc spacing of 25 m. This allows one to use single main beam bunches.

However, if one assumes that the dipoles in the return arcs are limited at 1 T, as in the above scheme, and that the filling factor in the arcs is 83%, the return arcs will have a radius of 100 m and a length of about 314 m. For each linac of a 3 TeV collider this adds up to almost 19 km of additional tunnels filled with arcs. This is more than in the first alternative described.

8.3. *Drive beam summary*

Different ideas for the drive beam generation and distribution have been presented. For all of the proposals a more detailed conceptual design has to be made to ensure that they are feasible.

9. Use of Plasma-Based Technology for a Linear Collider Upgrade

In the case that ILC or CLIC has already been constructed and operated, one can consider upgrading the collider with the help of plasma technology. First, one can consider adding a plasma cell after the main linac that will be driven by the main beam itself. Second, one can add plasma cells at the end of the main linac that are powered by a laser or external drive beam. Third, one can consider rebuilding the main linac using plasma technology. In this case the ILC main linac RF components or the CLIC drive beam components can be reused to generate the drive beam. For ILC, the first two scenarios have been considered in Ref. 10.

9.1. *Main-beam-driven energy booster*

In this scenario one adds a plasma cell at the end of each linac. The main beam bunches have to be split longitudinally into two parts. The first part serves as a drive beam and the second as a main beam. In order to double the main beam energy, one will roughly need to place two-thirds of the charge in the drive beam and one-third in the main beam. This concept has actually been demonstrated at SLAC [34].

This concept has a number of advantages. Only one stage of plasma acceleration is required. The drive beam quality is very good and the drive beam can be very stable due to its high energy. One can, for example, expect a very high jitter stability. However, the charge of the colliding bunches will be reduced by a factor of 3. As a consequence the luminosity at the collision point will be limited.

While the principle has been demonstrated, a number of critical points remain to be addressed for a realistic design of this scheme. Most important is the production of the bunch with the proper longitudinal profile. The energy spread has to remain small in the accelerated part of the bunch after the energy booster. Therefore, a specific longitudinal profile is required for both the driving and the accelerated part of the bunch. This may be difficult to achieve. In addition, the acceleration of the very short bunches may lead to higher order mode losses in the superconducting main linac.

In ILC and CLIC the power deposited and remaining in the plasma will be much larger during the bunch train than on average. This may make the heating problem for the plasma cell even more severe than in a collider that uses single bunches.

As a limited upgrade at a limited cost, this still appears to be an interesting option. Except for the plasma cell, one would mainly rebuild the beam delivery system so as to be able to focus the higher energy beam.

9.2. *Additional plasma acceleration*

The addition of plasma cells at the end of the linac requires that they accelerate a bunch train. This leads to some modifications for the plasma driver and for the cell itself.

First, the plasma has to recover between bunches. In ILC the time between bunches in the train is relatively large. So one might expect that the plasma has recovered from the previous bunch that passed. In CLIC the spacing between bunches is very short, so one would need to investigate whether the plasma can sustain this rate. In any case the very high heating of the plasma during the beam pulse has to be considered as in the first scheme.

In this scenario the drive beam source has to be pulsed. This practically excludes the drive beam generation in a CW superconducting linac, which we have discussed before. The stored energy required for a whole train appears probibitive. For ILC one could use a pulsed superconducting linac. For CLIC (and maybe ILC) one could consider using a pulsed, fully loaded, normal-conducting linac to generate the drive beam — similar to the CLIC drive beam accelerator.

If one drives the plasma with lasers, the pulsed operation may also create some issues. During the bunch train the average laser power has to be much larger than if the bunches were equally distributed in time.

9.3. *Replacement of the main linac with plasma technology*

The other option one can consider is to replace the main linac with plasma technology. This will allow one to reuse the injectors and infrastructure, which represent a large fraction of the overall project cost. One can also consider reusing a part of the main linac complex in the case of ILC or the drive beam complex in the case of CLIC. This will lead to a further cost saving, since these components can provide most of the drive beam generation complex. It should be noted that here just some idea of what might be possible is given to illustrate the concept. Much more detailed studies will be required to show the feasibility of the concept. Also, one will have to work on a serious optimization of the parameters, while here we just use the ones from the PWFA.

In the case of the ILC one could use a part of the existing main linac cavities and RF to produce the drive beam and also to accelerate the main beam to a high-enough energy. Two scenarios are possible. First, one could use the existing, pulsed modulators and klystrons and continue to pulse the cavities. This would create a scenario similar to the one discussed below for CLIC. Second, one could replace the RF generation in order to operate the drive beam linac in CW mode. This corresponds to the case mentioned above. For the 3 TeV collider based on the PWFA parameters, 2500 cavities would be required with a gradient of 12.6 MV/m to achieve the drive beam energy of 25 GeV and have enough stored energy. These cavities would lose 41 kW into the wall, if we assume a quality factor of $Q_0 = 10^{10}$. So the cooling would consume about 29 MW. Since these cavities have an RF of 1.3 GHz, the drive beam bunch spacing would have to be adjusted accordingly. For example, a spacing of 2.3 ns could be used (every third bucket filled). Another part of the ILC cavities could be used to construct a CW injector linac that provides a main beam with sufficient energy to be injected into the main linac.

If CLIC has been constructed, one can also consider using its drive beam generation complex for the plasma collider. This complex is normal-conducting, so it has to be operated in pulsed mode. The design of this option is well beyond the scope of this article. So only the adjustment required to match the drive beam for the PWFA is disucssed. The CLIC drive beam energy is 2.5 GeV, the bunch charge 5×10^{10} and the bunch spacing 2 ns. For the PWFA one needs a charge of 2×10^{10} with 2 ns spacing. This reduction of the current by a factor of 2/5 allows one to use 5/2-times-longer drive beam accelerating structures. The drive beam accelerating structures represent a relatively small fraction of the drive beam accelerator cost. This modification allows one to reach a drive beam energy of about 6.25 GeV in a single pass.

The linac can be configured in a recirculating mode, as described above. This will, however, raise similar challenges. If the beam performs four turns it can reach a total energy of 25 GeV. The length of the recirculating complex will be adjusted to the total pulse length of the drive beam, i.e. it will have a circumference of 10.5 km, which will allow one to store a 35-μs-long pulse. The klystrons and modulators can deliver a 140-μs-long pulse, which is sufficient to power the four turns of the beam. This scheme can produce a total of 125 pairs of drive beam pulses with 64 bunches each. The repetition rate of 50 Hz allows for a total collision rate of 6 kHz.

However, one should remain conscious that these are just ideas and that no studies have been performed to validate them.

10. Dielectric Acceleration

Conventional normal-conducting acceleration structures achieve the longitudinal accelerating field by periodic variations of the structure radius. The gradient that can be achieved is limited by breakdowns. Another method to achieve the field is to use a smooth pipe with a dielectric surface. With such a structure it should be possible to reach significantly higher fields than with a copper surface. A dielectric structure might also be easier to fabricate, in particular if one aims for very high frequencies of the accelerating field, which corresponds to small dimensions of the accelerating structure. The goal is to achieve high gradients with such structures; 690 MV/m has been demonstrated [35].

In principle, dielectric structures can replace copper structures in the whole collider complex. However, if one wants to use them in order to achieve a higher gradient, the energy density in the accelerating structures increases. Since the stored energy increases with the square of the field, while the length can only be reduced inversely with the field, the total energy in the main linac will increase linearly with the accelerating gradient. That will make the production of this energy more costly. It may therefore be better to also increase the RF, which will lead to smaller volumes that are filled with the gradient, and hence smaller energy per pulse. A full optimization of the design using the same approach as for CLIC will thus be useful.

An extreme proposal for the use of dielectric acceleration is the DLA. In this case an RF wavelength of only 2 μm is chosen — about four orders of magnitude smaller than in CLIC. The structures are not powered by conventional RF but rather by a laser. Still, trains of bunches will be accelerated. However, these trains consist of very short bunches and are themselves very short — a mere 318 μm. At the interaction point the trains will collide as if they were single bunches. The charge per train is also very small — only 159 bunches of $3 \cdot 10^4$ electrons or positrons each. The transverse emittance goal is far smaller than in any other proposal. To compensate for the small charge, a high train repetition rate of 30 MHz is used. In principle this is advantageous for the detector.

At the collision point the small charge per bunch does not require the use of flat beams. A very small beam emittance of 0.1 nm is targeted. Therefore, the beam size of less than 1 nm can be achieved with a comparatively large beta function of 16.5 mm. In combination with the small emittance the particles enter the collision point with almost no angular spread. The beams are therefore focused to a very small size, leading to a very high luminosity enhancement factor. This works very nicely for straight beams, where the transverse distribution is independent of the longitudinal position.

The production of such small emittance beams will be very difficult; in particular, for positrons no solution exists. Also, the preservation of the very small emittances poses a formidable challenge. An exploration of key tolerances is mandatory in order to develop an informed option about the feasibility and required R&D for this option.

A serious consideration of the use of dielectric accelerating structures at more conventional frequencies could be very beneficial for CLIC.

11. Proton Plasma Acceleration

In principle one can use protons to generate the plasma for the main beam acceleration. This remains to be demonstrated in AWAKE [36] at CERN, and the associated technology will therefore require even more R&D than the other options.

Proton bunches tend to be quite long, which is not consistent with driving a dense plasma, which has a short wavelength. Therefore, it will be necessary to transform such a long bunch into a train of short bunches. These short bunches then drive the plasma in a resonant fashion, increasing the amplitude of the field from plasma wavelength to plasma wavelength. Indeed, in AWAKE one about 10-cm-long proton bunch will be split into a train of about 100 short bunches to drive a plasma with a wavelength of 1 mm.

A proton drive beam has an important advantage compared to an electron drive beam. It can be circulated in a ring for a long time without losing a significant amount of synchrotron radiation. Hence, it can be efficiently accelerated to very high energies. As a result of the high energy of the proton driver, a single acceleration stage can reach very high main beam energies, which simplifies the collider design.

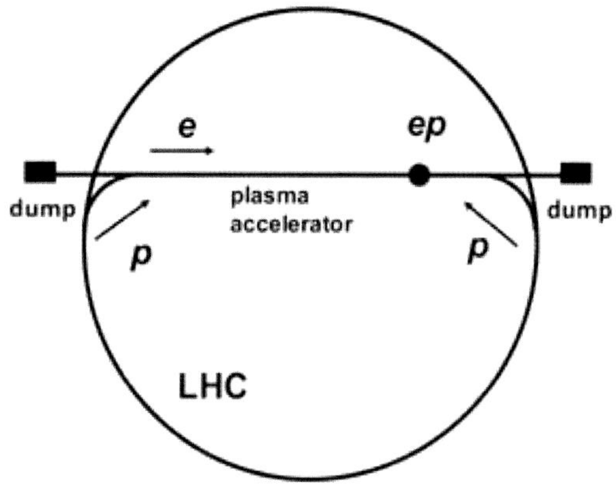

Fig. 11. Sketch of a potential electron–proton collider from Ref. 37.

A potential use of proton driven plasma could be in an upgrade of the LHC for electron–proton collisions [37]. In this scenario, one of the two proton beams is used to accelerate an electron beam to about 3 TeV. The other proton beam is then collided with the produced electron beams; see Fig. 11.

The reference assumes 3000 LHC bunches per fill with a 30 min cycle time, $1 \cdot 10^{11}$ electrons and $4 \cdot 10^{11}$ protons per bunch, as well as a transverse RMS beam size of 4 μm, to reach a luminosity of $4 \cdot 10^{28}$ cm^{-2}s^{-1}. The number of protons is ambitious — about twice the value foreseen for the LHC luminosity upgrade. In contrast, the electron emittance required is quite relaxed; for a beta function of 10 cm, $\epsilon_{x,y} \approx 1$ mm will be sufficient. In the example layout a significant amount of tunnels would be needed. The magnets for the arcs that connect the ring to the collider would have to use magnets that are much stronger than in the LHC. Other layout options might be considered.

12. Conclusion

Advanced acceleration technologies are interesting options that may become useful for the second generation of future high energy lepton colliders. The plasma and the dielectric technology have made great progress in demonstrating high gradients. With this progress it becomes important to start to also address the even more demanding challenges of high efficiency and beam quality that are instrumental for a collider. This will require the development of other technologies, as it did for linear colliders.

A fundamental problem that has to be addressed in plasma-based colliders is the stable acceleration of positrons. The use of hollow plasma has been suggested as a solution, but lacks the focusing that stabilizes the electron beam in a full plasma.

A plasma-based collider faces very demanding tolerances. The timing has to be accurately adjusted and kept stable between the main beam and the laser or electron beam that drives the plasma. The transverse position of both beams has to be well controlled and the driver position has to be kept extremely stable. The collective interaction of the main beam with the plasma has to be studied. This is expected to put severe constraints on the efficiency and hence the luminosity performance of the collider.

In summary, in order to make the discussed advanced technologies available for future colliders, an important R&D program has to be carried out by the advanced technology experts in close collaboration with collider designers. This program should identify and address the challenges that need to be met in order to make the promising technologies available for a collider. It should culminate in the construction and operation of a demonstrator facility that can achieve simultaneously high gradient, high efficiency and excellent beam quality.

Acknowledgments

The author is pleased to acknowledge the very helpful discussions that he had with E. Adli, P. Muggli, E. Marin Lacoma and J. Pfingstner, as well as with R. J. England, E. Esarey, C. B. Schroeder, M. Hogan, A. Petrenko, V. K. Berglyd Olsen, M. Wing and A. Caldwell.

References

[1] T. Tajima and J. Dawson, *Phys. Rev. Lett.* **43**, 267 (1979).
[2] L. Evans and P. Bryant, LHC machine, *JINST* **3**, S08001 (2008).
[3] M. Benedikt, B. Goddard, D. Schulte, F. Zimmermann and M. J. Syphers, FCC-hh hadron collider — parameter scenarios and staging options. IPAC-2015-TUPTY062.
[4] CEPC-SPPC Study Group, CEPC-SPPC Preliminary Conceptual Design Report. 2. Accelerator. IHEP-CEPC-DR-2015-01, IHEP-AC-2015-01.
[5] T. Behnke *et al.*, The International Linear Collider Technical Design Report — Volume 1: Executive Summary, arXiv:1306.6327 [physics.acc-ph]; H. Baer *et al.*, The International Linear Collider Technical Design Report — Volume 2: Physics, arXiv:1306.6352 [hep-ph]; C. Adolphsen *et al.*, The International Linear Collider Technical Design Report — Volume 3.I: Accelerator & in the Technical Design Phase, arXiv:1306.6353 [physics.acc-ph]; C. Adolphsen *et al.*, The International Linear Collider Technical Design Report — Volume 3.II: Accelerator Baseline Design, arXiv:1306.6328 [physics.acc-ph]; T. Behnke *et al.*, The International Linear Collider Technical Design Report — Volume 4: Detectors, arXiv:1306.6329 [physics.ins-det].
[6] M. Aicheler *et al.*, A multi-TeV Linear Collider Based on CLIC Technology: CLIC Conceptual Design Report, doi:10.5170/CERN-2012-007; L. Linssen, A. Miyamoto, M. Stanitzki and H. Weerts, Physics and Detectors at CLIC: CLIC Conceptual Design Report, doi:10.5170/CERN-2012-003, arXiv:1202.5940 [physics.ins-det].
[7] One can use the following as a starting point: N. Phinney, SLC final performance and lessons, eConf C **00082** (2000) MO102 [physics/0010008 [physics.acc-ph]].
[8] C. B. Schroeder *et al.*, *Nucl. Instrum. Methods A* (2016).
[9] C. B. Schroeder, private communication.
[10] E. Adli *et al.*, A beam driven plasma-wakefield linear collider: from Higgs factory to multi-TeV. SLAC-PUB-15426.
[11] C. B. Schroeder, E. Esarey and W. P. Leemans, *Phys. Rev. ST Accel. Beams* **15**, 051301 (2012).
[12] W. Gai, J. G. Power and C. Jing, Short-pulse dielectric two-beam acceleration, *J. Plasma Phys.* **78**/Special Issue 04/Aug. 2012, pp. 339–345; M. E. Conde, W. Gai, J. G. Power and C.-J. Jing, Staging in two beam dielectric wakefield accelerators, *Conf. Proc. C* **110904**, 2802 (2011); M. E. Conde, W. Gai, J. G. Power, S. P. Antipov, C.-J. Jing, A. Kanareykin and P. Schoessow, Study of a TeV level linear collider using short rf pulse (20 ns) two beam accelerator concept, *Conf. Proc. C* **110328**, 2279 (2011).
[13] R. J. England, R. J. Noble, Z. Wu and M. Qi, Dielectric laser acceleration, *Rev. Mod. Phys.* **86**, 1337 (2014).
[14] R. J. England, private communication.
[15] J.-P. Delahaye, private communication.
[16] W. Leemans and E. Esarey, Laser-driven plasma-wave electron accelerators, *Phys. Today* **62**(3), 44 (2009).
[17] J. B. Rosenzweig *et al.*, *Phys. Rev. A* **44**, R6189 (1991).
[18] S. Corde *et al.*, Multi-gigaelectronvolt acceleration of positrons in a self-loaded plasma wakefield, *Nature* **524**(7566), 442 (2015).

[19] V. Lebedev and S. Nagaitsev, Comment on beamstrahlung considerations in laser-plasma-accelerator–based linear colliders. FERMILAB-PUB-13-093-AD.
[20] W. D. Kimura *et al.*, *Phys. Rev. ST Accel. Beams* **14**, 041301 (2011).
[21] Gessner *et al.*, to be published in *Nature.*
[22] T. Barklow *et al.*, Experimental evidence for beam–beam disruption at the SLC. SLAC-PUB-8043.
[23] M. Berndt *et al.*, Final focus test beam: Project design report. SLAC-0376, SLAC-376, SLAC-R-0376, SLAC-R-376.
[24] B. I. Grishanov *et al.* [ATF2 Collaboration], SLAC-R-771, CERN-AB-2005-035, CLIC-NOTE-636, DESY-05-148, ILC-ASIA-2005-22, JAI-2005-002, KEK-REPORT-2005-2, UT-ICEPP-05-02.
[25] E. M. Lacoma, private communication.
[26] C. B. Schroeder, E. Esarey, C. Benedetti and W. P. Leemans, Efficiency considerations for high-energy physics applications of laser-plasma accelerators, in *Advanced Accelerator Concepts* (AIP, San Jose, 2014).
[27] S. van der Meer, CERN/PS/85-65 (AA) (1985).
[28] D. Schulte and R. Tomas, Dynamic effects in the new CLIC main linac, in *PAC 2009* (Vancouver, 2009).
[29] R. Assmann and K. Yokoya, Transverse beam dynamics in plasma based linacs, *Nucl. Instrum. Methods A* **410**, 544 (1998).
[30] C. A. Lindstrom *et al.*, in *IPAC 2016* (2016).
[31] C. B. Schroeder, D. H. Whittum and J. S. Wurtele, *Phys. Rev. Lett.* **82**, 1187 (1999).
[32] J. Pfingstner *et al.*, in *IPAC 2016.*
[33] S. Pei, M. J. Hogan, T. O. Raubenheimer, A. Seryi, H. H. Braun, R. Corsini and J. P. Delahaye, Conceptual Design of the Drive Beam for a PWFA-LC. SLAC-PUB-13723.
[34] I. Blumenfeld *et al.*, *Nature* **445**, 741 (2007).
[35] K. P. Wootton *et al.*, Observation of 690 MV/m Electron Accelerating Gradient with a Laser-Driven Dielectric Microstructure. SLAC-PUB-16442.
[36] http://awake.web.cern.ch/awake/A. Caldwell *et al.*, AWAKE Design Report: A Proton–Driven Plasma Wakefield Acceleration Experiment at CERN. CERN-SPSC-2013-013; SPSC-TDR-003.
[37] A. Caldwell and M. Wing, VHEeP: A very high energy electron–proton collider, arXiv:1606.00783 [hep-ex].

Daniel Schulte studied Physics in Hamburg and finished his Ph.D. thesis at DESY in 1996 on detector backgrounds in future linear colliders. Since then he has been working at CERN on the beam physics of different projects. These include the two linear collider studies CLIC and ILC, the LHC, and the study for the future very large hadron collider FCC-hh. Currently, he is leading the parameter and design activities for CLIC and the FCC-hh collider design.

Reviews of Accelerator Science and Technology
Vol. 9 (2016) 235–260

DOI: 10.1142/S1793626816300115

Advanced Accelerators for Medical Applications

Mitsuru Uesaka

Nuclear Professional School, University of Tokyo
Tokai-mura, Ibaraki-ken, 319-1188, Japan
uesaka@nuclear.jp

Kazuyoshi Koyama

High Energy Accelerator Research Organization, KEK
Tsukuba-shi, Ibaraki-ken, 305-0801, Japan
Nuclear Professional School, University of Tokyo
Tokai-mura, Ibaraki-ken, 319-1188, Japan
koyamakz@post.kek.jp
koyama@nuclear.jp

We review advanced accelerators for medical applications with respect to the following key technologies: (i) higher RF electron linear accelerator (hereafter "linac"); (ii) optimization of alignment for the proton linac, cyclotron and synchrotron; (iii) superconducting magnet; (iv) laser technology. Advanced accelerators for medical applications are categorized into two groups. The first group consists of compact medical linacs with high RF, cyclotrons and synchrotrons downsized by optimization of alignment and superconducting magnets. The second group comprises laser-based acceleration systems aimed of medical applications in the future. Laser plasma electron/ion accelerating systems for cancer therapy and laser dielectric accelerating systems for radiation biology are mentioned. Since the second group has important potential for a compact system, the current status of the established energy and intensity and of the required stability are given.

Keywords: Medical applications; higher RF electron linac; compact synchrotron; superconducting cyclotron; laser plasma acceleration; laser dielectric acceleration.

1. Introduction

In this review article, the following key technologies of advanced accelerators for medical applications are considered:

(i) Higher RF electron linac;
(ii) Optimization of alignment for the proton linac, cyclotron and synchrotron;
(iii) Superconducting magnet;
(iv) Laser technology.

Based on the key technologies, we define two groups. In the first group are compact medical linacs with high RF, cyclotrons and synchrotrons downsized by optimization of alignment, and superconducting magnets. The advanced accelerators in the second group are driven by intense lasers. This article will introduce a laser-based acceleration system aimed at medical applications in the near future, laser plasma electron/ion accelerating systems for cancer therapy, and a laser dielectric accelerating system for radiation biology.

We are actually developing and experimenting with S-band (2.856 GHz) and X-band (9.3 GHz) electron linacs, laser plasma electron/ion acceleration and laser dielectric acceleration. Thus, we understand the requirement for medical accelerators in the first group and the current status of the development and stability of the second group. Although the maturity is different between the two groups, we will try to review their current statues in series.

Among the medical applications of accelerators including imaging and radionuclide production, the therapeutic use of accelerators has been growing sharply for external beam radiation therapy (EBRT), in which ionizing radiation breaks strands of DNA in tumor cells to cure cancer. Radiation therapy with X-rays is the most popular form of EBRT, in which an MV-class electron linac is used to generate therapeutic X-rays. Particle therapy using proton

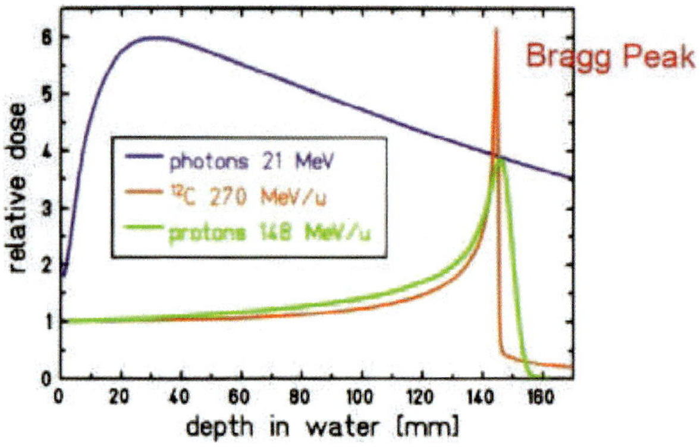

Fig. 1. Depth–dose distribution for photons, protons and carbon ions. The Bragg peak can be observed for protons and carbon ions [2].

and heavier ions (mainly carbon ions), also called hadron therapy, has been growing rapidly worldwide since the 1990s because of its better dose conformity which comes from the physical characteristics of energy deposition. MV photons deposit their energy throughout the body along their path, and the energy deposition gently peaks several centimeters from the entrance. On the contrary, ions deposit their maximum energy at a steep peak just in front of their stopping depth, called the Bragg peak (Fig. 1). The better dose conformity can be realized by adjusting the beam positon and ion energy so that the Bragg peak is delivered at the tumor position. Thus, the accelerator for particle therapy requires an energy high enough to deliver the ions to tumors passing through body tissue — typically 250 MeV for proton therapy and 400 MeV/u for carbon ion therapy. For proton therapy, the cyclotron and synchrotron are commercially available, and the proton linac is under development by a UK-based developer [1]. For carbon ion therapy, only the synchrotron is commercially available, while a superconducting cyclotron for carbon therapy has been proposed but has not been realized yet.

Section 2 (on the linac), Sec. 3 (on the synchrotron) and Sec. 4 (on the cyclotron) will deal with the first group, while Sec. 5 (on laser dielectric acceleration), Sec. 6 (on laser plasma electron acceleration) and Sec. 7 (on laser-driven ion acceleration) will deal with the second group.

2. Linac for Cancer Therapy [3]

The downsizing of accelerating structure by adopting higher accelerating electromagnetic frequency is schematically depicted in Fig. 2. Now, the smallest medical electron linacs use X-band (9.3 GHz) RF. The laser dielectric acceleration system uses an ~100 THz wave, namely an ~1 μm infrared wave.

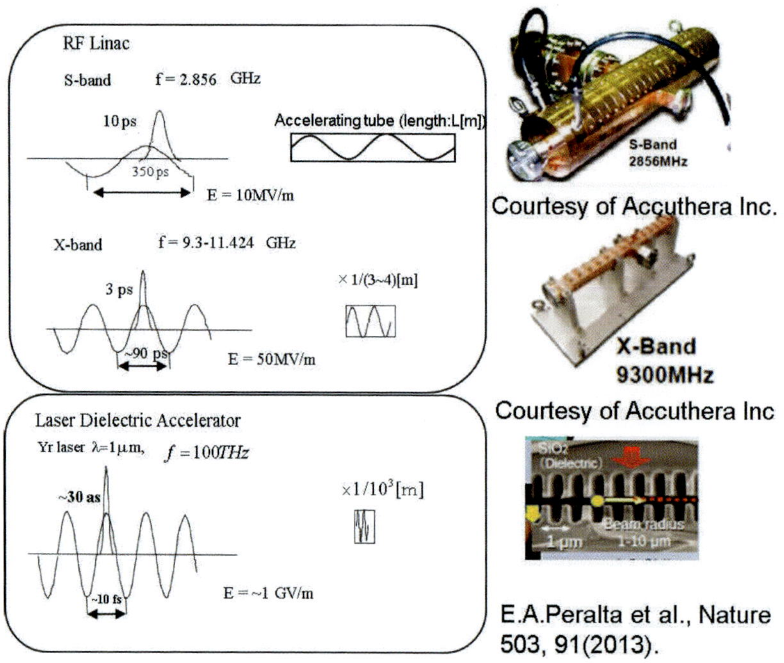

Fig. 2. Downsizing of an electron linac by higher accelerating electromagnetic wave frequency.

As for RF, the S-band [2.856 GHz, 105 mm (wavelength)] is most common for medical linacs in the world. The X-band (9.3 GHz, 32 mm) was adopted for the stereotactic therapy system. Recently, the C-band (5.712 GHz, 53 mm) electron linac and klystron were developed for scientific uses, such as a linear collider for high energy physics, and an X-ray free electron laser. The parameters of typical S-, C- and X-band accelerating structures are summarized in Table 1. It is obvious that the size becomes smaller for higher frequency. In addition, the shunt impedance becomes higher for smaller structure, namely higher frequency. The maximum electron energy and beam current depend on the design and power of the RF source. So far, the maximum electron energy of the X- and C-band linacs is 6 MeV.

The most common gantry-type medical linac system is shown by Fig. 3. Its inner structure is shown in (ii) of the figure. It consists of an electron gun, accelerator, bending magnet, X-ray target, collimator, klystron, waveguide and rotary joint. S-band RF is used for this type of medical electron linac. Conformational radiation therapy and IMRT (intensity-modulated radiation therapy) are performed by rotation of the gantry

Table 1. Comparison of S-, C- and X-band accelerating structures [3].

Frequency	2856 or 2998 MHz (S-band)	5712 MHz (C-band)	9300 MHz (X-band)
Shunt impedance	90 MΩ/m	100 MΩ/m	150 MΩ/m
Energy	10 MeV	6 MeV	6 MeV
Beam current	30 mA	100 mA	100 mA
Pulse width	3 μs	4 μs	4 μs

(i) Overview(http://www.siemens.com/entry/cc/en/)

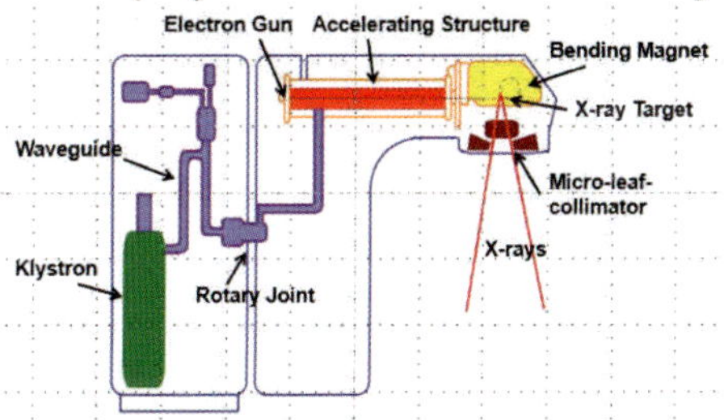

(ii) Inner structure

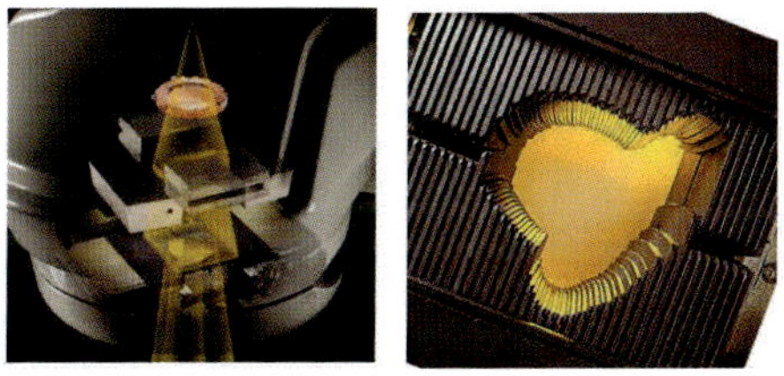

(iii) Microleaf collimator for IMRT

Fig. 3. A gantry-type electron linac.

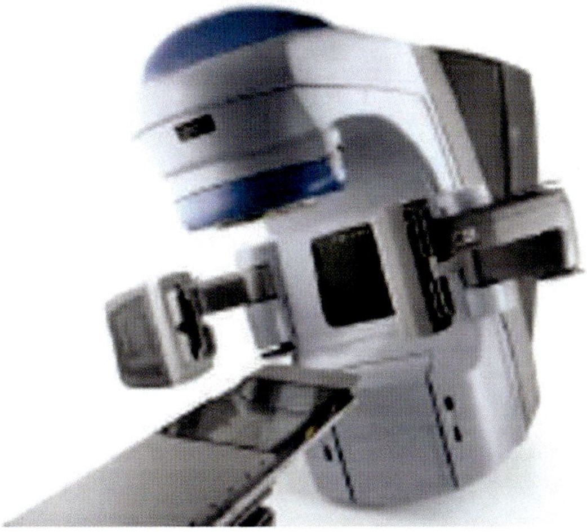

http://www.varian.com/

http://www.elekta.com/

Fig. 4. A medical electron linac with cone beam X-ray CT.

and multi-directional irradiations with transversely intensity-modulated X-ray beams. Figure 4 shows an S-band electron linac X-ray therapy system with a cone beam X-ray CT system. IGRT (image-guided) is available by *in situ* X-ray diagnosis for checking the position of moving cancer. An electron linac X-ray therapy system installed in an X-ray CT (computed tomography) system is shown in Fig. 5. The systems in (i) and (ii) of the figure adopt the S- and C-band linacs. Due to the optimization of component alignment and the choice of compact higher RF, the linacs are installed in the X-ray CT devices, where the accelerator systems do not appear directly. A robot-armed system with an X-band electron linac X-ray source is shown in Fig. 6. The compact size of the X-band linac enables this configuration. Stereotactic X-ray therapy by multidirectional irradiations with a small (~5 mm or less) X-ray beam can be conducted. Even the dynamic tracking RT aiming at moving lung cancer becomes possible in order to reduce the irradiation dose for surrounding normal organs.

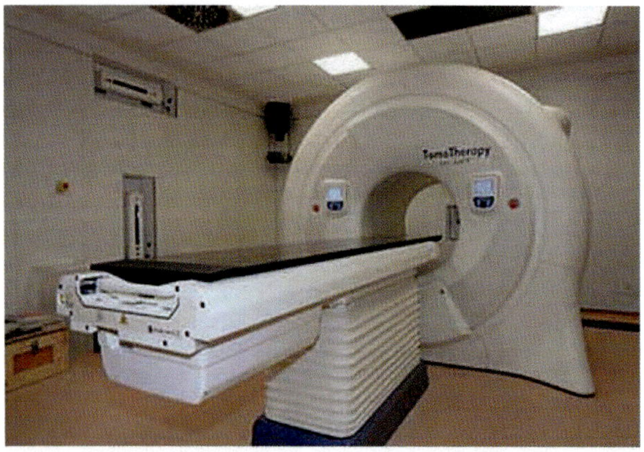

http://www.accuray.com/

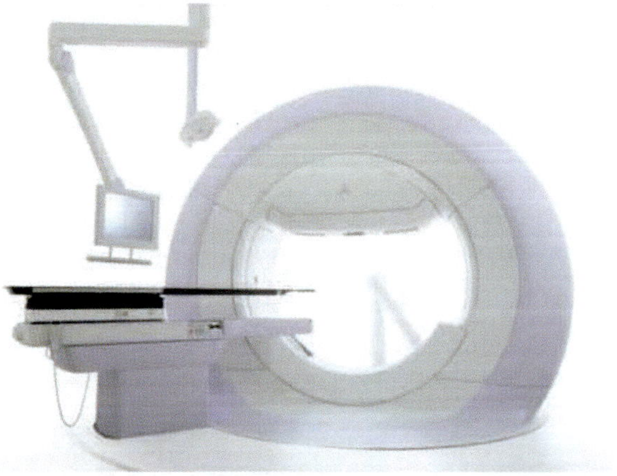

http://www.mhi-global.com/index.html

Fig. 5. A medical electron linac installed in the X-ray CT system.

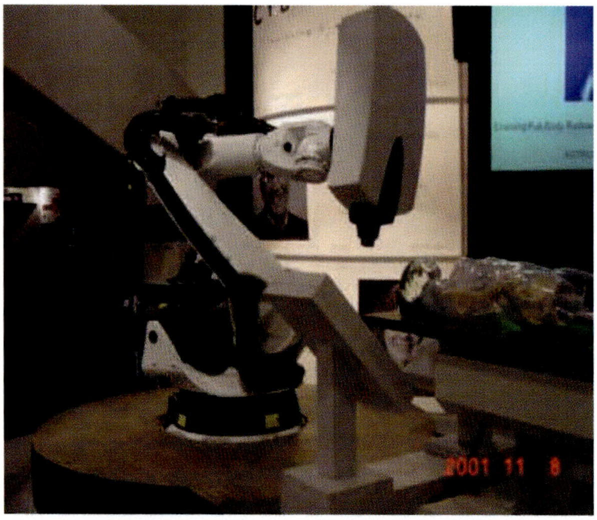

http://www.accuray.com/

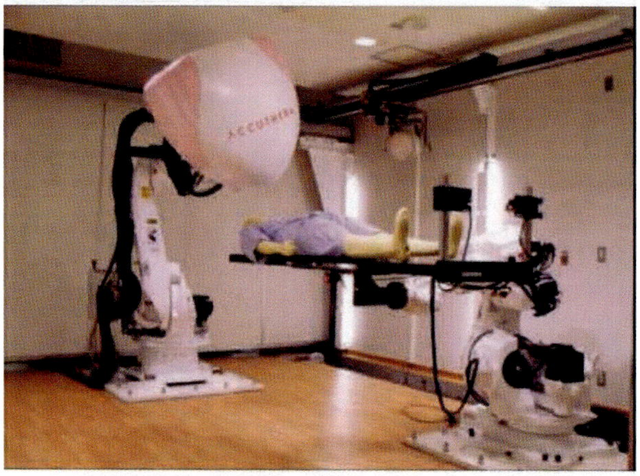

http://www.accurthera.com/

Fig. 6. A robot-arm-type medical electron linac.

Here we explain a proton linac for BNCT (boron neutron capture therapy) for the first group. There are three types of accelerators for BNCT, depending on the proton energy and neutron generation target. The requirement for the BNCT accelerator, the available energy of the neutron, is 0.5 eV–10 keV, and a neutron intensity of more than 1×10^9/s/cm^2 is desirable. From the characteristics of the target materials and the efficiency of moderation, the beam energy of proton beams for the lithium target is selected as 2–3 MeV. The pronounced resonance of 2.25 MeV is used for neutron production. For the purpose, both single-ended and tandem electrostatic accelerators and RFQ accelerators are candidates. On the other hand, the proton beam in the range of 7–13 MeV is used for the beryllium target. A proton linac with RFQ (radio frequency quadrupole) and a drift tube linac (DTL) are a candidate for this case. An 8 MeV, 10 mA (50 mA peak, 20% duty factor) proton linac and of a beryllium-based neutron production target for the BNCT is under construction in Japan [4]. A bird's-eye view of the BNCT facility in Ibaraki, Japan and photographs of the RFQ and DTL of J-PARC appear in Fig. 7. The high current RFQ and a front part of the DTL at the J-PARC provide a good start for the accelerator design. The beam dynamics in the high current region would be quite similar to that of J-PARC. But the intensity required by the BNCT is quite high compared with that of J-PARC. In order to produce a high intensity of epithermal neutrons, a 30 MeV proton cyclotron and beryllium target is chosen, as explained in Sec. 4.3.

The details of recent intense accelerator-based neutron sources are reviewed in Ref. 5.

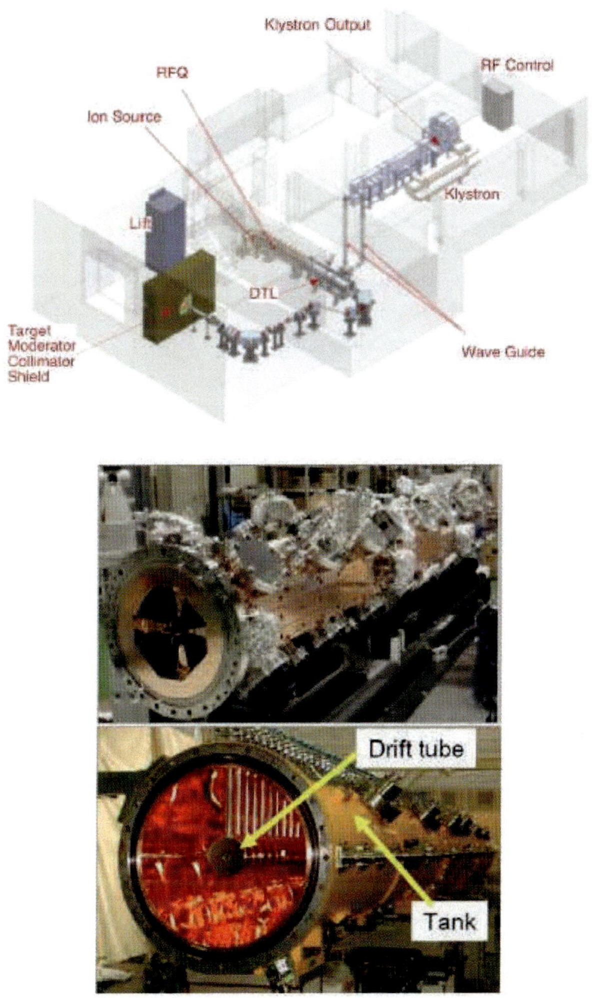

Fig. 7. Bird's-eye view of the BNCT facility in Ibaraki, Japan and photographs of the RFQ and DTL of J-PARC.

3. Synchrotron for Proton and Carbon Cancer Therapy

3.1. *Bevalac* [6]

In the 70-year history of particle therapy since Wilson's suggestion [7], the first synchrotron for therapeutic use was Bevatron, a weak focusing proton accelerator at Lawrence Berkeley National Laboratory, USA. It was originally constructed in 1954 for high energy physics (not including therapeutic use), then improved with the SuperHILAC installation as a heavy ion injector in 1971 called Bevalac. A dedicated biomedical experimental area with three irradiation stations was developed to begin PHASE I/II clinical trials with the University of California, San Francisco Medical Center in 1975. A total of 433 patients were irradiated at Bevalac with helium, carbon, neon, silicon and argon. For the heavy ion use, a wobbler beam delivery system was developed to reduce the material in the beam. In addition, a raster scanning system was developed, but soon it reached the termination of the facility operation in 1993.

Although Bevalac did not treat a large number of patients, the obtained results looked promising for ion therapy in some aspects. Following them, two dedicated ion therapy facilities were constructed in Japan (HIMAC) and Germany (GSI) in the 1990s (see Sec. 3.3).

3.2. *Synchrotron for proton therapy*

Proton therapy was initiated with a cyclotron in the 1950s, and was developed with cyclotrons until the 1980s to clarify its effectiveness. Cyclotrons form the majority of worldwide proton therapy machines even now; however, their disadvantage of fixed exit energy remains. Synchrotrons have the advantage of

Fig. 8. The proton synchrotron at Hokkaido University, supplied by Hitachi [10].

variable exit energy to reduce in-beam materials substantially. Lower radioactive waste is another advantage of synchrotrons.

The first synchrotron dedicated to proton therapy was constructed at Loma Linda University Medical Center (LLUMC), USA in 1990 [8]. This was also the world's first hospital-based proton therapy facility. It consists of an ion source, linac-based injector, main ring, beam transport, and beam delivery system. (This structure was subsequently common among medical synchrotrons.) It also has three isocentric 360° rotating gantries. The gantry has been a standard for proton therapy since LLUMC.

In the 2010s, the demand for commercial proton therapy systems increased explosively, partly because proton therapy had become financially profitable. About one-third of the running and planned proton therapy centers use synchrotrons worldwide [9]. Three vendors are supplying therapeutic proton synchrotrons (Hitachi, Mitsubishi Electric and Protom). Hitachi and Mitsubishi Electric provide very compact synchrotron (~6 m in diameter). Hitachi has successfully reduced the circumference to 18 m with combined function magnets. Protom's Radiance 330 produces 330 MeV protons, 80 MeV higher than the general requirement of proton therapy, for the future imaging, such as proton CT.

Recently, Danfysik proposed a unique idea of a rotating proton synchrotron at the 55th Annual Conference of the Particle Therapy Co-operative Group (PTCOG) in Prague: a synchrotron and an injector are located at the end of a cylindrical rotator approximately 8 m in length and 8 m in diameter, to rotate with it.

3.3. *Synchrotron for heavier ions*

Following the ion therapy research at Bevalac, Japan and Germany continued its legacy. The essential insight they had in common was that the carbon ion is more promising than the neon ion, which had been strongly investigated at Bevalac. The main reason is the coincidence of the dose–peak depth and the relative biological effectiveness (RBE) peak depth for the carbon ion [11, 12].

In Japan, the first synchrotron facility dedicated to ion therapy was constructed — the Heavy-Ion Medical Accelerator in Chiba (HIMAC), which has been operational since 1994. In Germany, an experimental heavy ion therapy facility was constructed by Gesellschaft fur Schwerionenforschung (GSI) in Darmstadt, and was operational from 1997 to 2008, using an existing heavy ion synchrotron, SIS-18.

3.3.1. *HIMAC*

HIMAC is the first dedicated ion therapy research center in the world, and is operated by the National Institute of Radiological Sciences (NIRS). Since not only clinical research but also basic research in various fields is targeted, it has a high specification and occupies an area larger than a football field. It has vertically configured two independent synchrotrons 130 m in circumference, enabling flexible operation. It can produce He to Fe ions with the maximum energy of 800 MeV/u. HIMAC has treated more than 9000 patients since 1994 — more than half of carbon ion therapy patients worldwide.

Two major upgrades have been ongoing at HIMAC in the 2010s, in a new research building. One is a fast 3D scanning system, installed in 2011. The hybrid scanning method is applied, where the depth is controlled coarsely (11 levels) by the accelerator and precisely by the range shifter. The scanning speed is so fast $[(v_x, v_y) = (100, 50)\,\mathrm{mm/ms}]$ that the fast repainting can be combined with the respiratory gating system [13]. The respiratory gating treatment with 3D scanning has been undergoing a clinical trial since 2015.

The other upgrade is a 360° isocentric gantry. This is the second gantry for carbon ion therapy after HIT, but its size is reduced by the application of superconducting magnets. Ten combined function superconducting magnets are serially located at the rotator [14] (Fig. 9). The energy scanning method is applied to the gantry instead of the hybrid scanning system. A clinical trial will begin in 2016.

These two upgrades are collaborations between NIRS and Toshiba. They are also collaborating on the development of a superconducting synchrotron for carbon ion therapy. It may reduce the footprint of the carbon ion therapy facility with one gantry to no more than $20 \times 10\,\mathrm{m}^2$ [15].

3.3.2. *SIS*-18 *at GSI and HIT*

SIS-18 is a synchrotron constructed for physics research at GSI in Darmstadt in the late 1980s. Particle therapy research was proposed jointly by GSI and Heidelberg University Radiology, and the German Cancer Research Centre, in 1993. Since they are going to employ the raster-scanning method for beam delivery, the accelerator control system has been upgraded for fine energy control of 255 steps from 80 to 430 MeV/u. The energy can be varied pulse by pulse. Heavy ion therapy was carried out at GSI from 1997 to 2008 to treat 440 patients [12].

The valuable experience of ion therapy at GSI has been continued by the Heidelberg Ion-Beam Therapy Center (HIT), the first dedicated proton and carbon therapy center in Europe. The synchrotron of HIT is specially designed for ion therapy with the raster-scanning delivery method. Its circumference is about 65 m. It can accelerate protons and carbon up to 430 MeV/u. The penetration depth can be controlled from 20 to 300 mm with 255 steps. It has two horizontal beam ports, and the world's first 360° rotating gantry for carbon ion therapy [17]. The gantry has two 45° bending magnets before a

Fig. 9. The isocentric rotating gantry with superconducting magnets (blue) installed at NIRS. All the magnets, as well as the profile monitors, are mounted on the gantry structure [16].

90° final bending magnet, all of which are conventional magnets using magnetic poles made of steel. Although it weighs as heavy as 670 tons in total, the precision is below ±0.5 mm [17]. Patient treatment began in 2009 by horizontal beam, and in 2012 by the gantry. More than 2500 patients were treated at HIT until the end of 2014, about one-third by protons and two-thirds by carbon ions.

The basic design of the HIT synchrotron was followed by the Marburg Ion Therapy Center in Germany and the Shanghai Proton and Carbon Ion Therapy Center in China.

3.3.3. *CNAO and MedAustron*

The National Center for Oncological Hadrontherapy (CNAO) in Pavia, Italy and MedAustron in Wiener Neustadt, Austria are similar proton and carbon ion therapy facilities, whose accelerator was based on a design study at CERN from 1995 to 2000. The

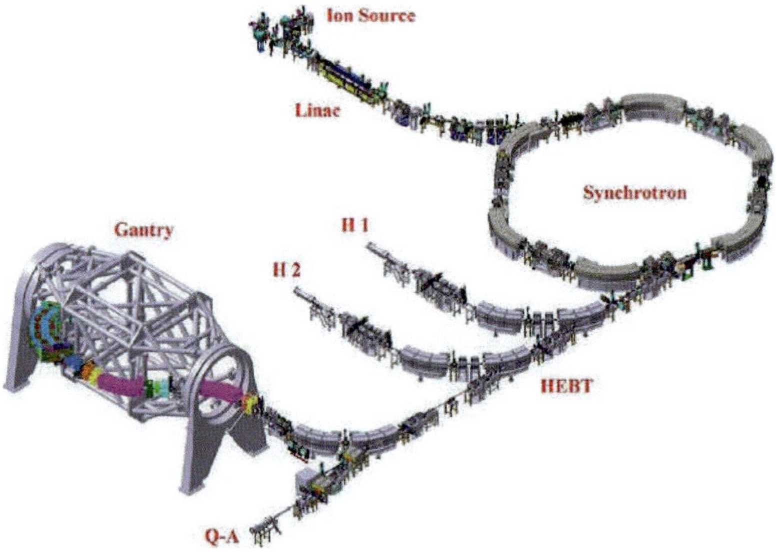

Fig. 10. Overview of the accelerator at HIT [17].

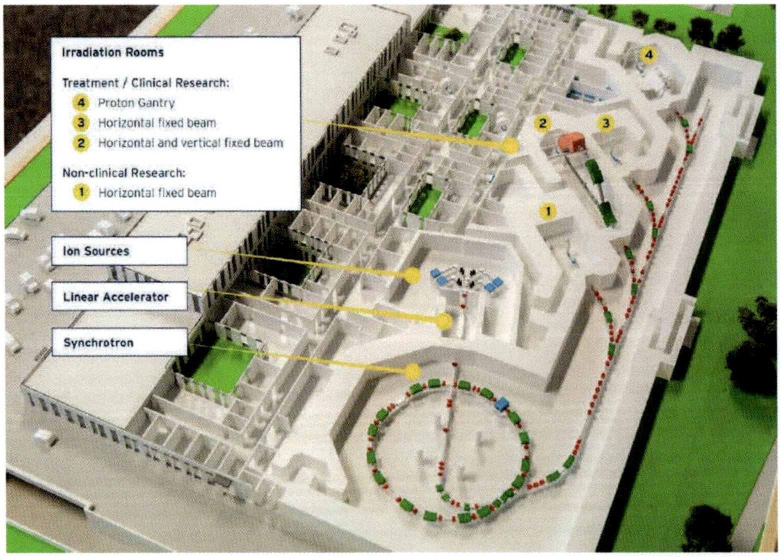

Fig. 11. Overview of MedAustron [20].

synchrotron, about 80 m in circumference, can accelerate protons to carbon ions up to 400 MeV/u [18]. CNAO does not have a gantry; however, MedAustron will have a gantry not for carbon ions but for protons [19]. CNAO began treatment in 2012, while MedAustron will begin proton therapy in 2016 and carbon ion therapy in 2017.

3.3.4. *Carbon-dedicated compact synchrotron*

Following the significant success at HIMAC, several dedicated ion therapy facilities were constructed in Japan. The first post-HIMAC facility in Japan was the Hyogo Ion Beam Medical Center (HIBMC), which has been operational since 2003. The synchrotron can accelerate both protons and carbon up to 230 and 375 MeV/u, respectively. Its circumference is about 90 m. More than 7000 patients have been treated with protons or carbon-ions [21].

After HIBMC, a number of proton and carbon ion therapy facilities were separately constructed in Japan. The first dedicated carbon ion therapy center was launched at Gunma University, which has been operational since 2010. Its compact synchrotron was designed by NIRS; the circumference was reduced to approximately 63 m. Accelerators, three treatment rooms (horizontal, horizontal/vertical, vertical) and the necessary modalities are located within an area of 3140 m^2 [22]. The SAGA Heavy Ion Medical Accelerator in Tosu (SAGA HIMAT) and the Ion-Beam Radiation Oncology Center in Kanagawa (i-ROCK) followed Gunma University, and have been operational since 2013 and 2015, respectively. As they were therapy facilities rather than basic research facilities, the overall system was supplied by a vendor like other medical modalities. Two more carbon ion therapy centers are under construction in Japan: Osaka University and Yamagata University. The former is located at the center of downtown Osaka, whose population is 8.8 million, and its synchrotron, designed by Hitachi, is the most compact for carbon therapy, with a circumference that is no more than approximately 56 m. The latter has configured the main accelerator and treatment rooms vertically for the first time as a carbon ion therapy facility to reduce the facility footprint to 2000 m^2. A 360° rotating gantry with superconducting magnets will be installed by Toshiba based on the HIMAC gantry, but it will be a somewhat smaller and lighter one [23].

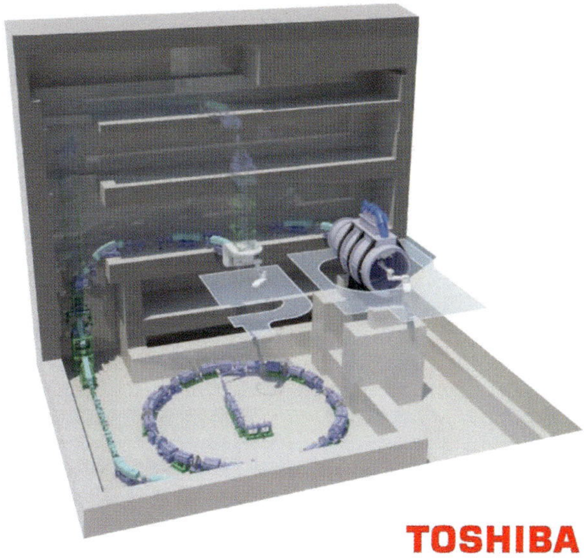

Fig. 12. Vertical configuration of the main components in the carbon ion therapy facility at Yamagata University [24].

China is another country which has great interest in carbon ion therapy. The first carbon ion therapy in China was conducted at the Heavy Ion Research Facility in Lanzhou (HIRFL), operated by the Institute of Modern Physics, Chinese Academy of Sciences. Carbon is accelerated through two cyclotrons and one storage ring called the Cooling Storage Ring (CSR). A clinical trial for skin tumor therapy began in 2006 with two cyclotrons, and then a clinical trial for deep-seated tumors began in 2009 with the two cyclotrons and the CSR [25].

Based on the experience at HIRFL, two carbon ion treatment facilities are under construction in Lanzhou city and Wuwei city, both in Gansu province. The unique point of the facilities is the use of a cyclotron as injector. The synchrotron will be a compact carbon synchrotron with a circumference of approximately 56 m [25].

4. Cyclotron for Proton Cancer Therapy, BNCT and RI Production for PET

Apart from special AVF and separate-sector ring cyclotrons for fundamental physics research, several types of compact cyclotrons have been developed for medical fields. In particular, because of growing interest in medical applications of accelerators, several manufacturers have developed commercial cyclotrons for both diagnostic and treatment purposes.

Compact, negative ion cyclotrons for the PET system are the main products for diagnostic purposes. PET cyclotrons consist of wide energy range machines, from 7 MeV to 20 MeV, in order to meet the requirements of different positron emitters, such as ^{11}C, ^{15}O and ^{18}F, for producing various PET medicines. GE Healthcare (GE), Ion Beam Applications, SA (IBA), Siemens Healthcare and Sumitomo Heavy Industries, Ltd. (SHI) are the main suppliers of these cyclotrons.

Cancer therapy by protons or heavy ions is one of the major trends of accelerator application for treatment purposes. SHI and IBA have jointly developed a compact, high field and normal-conducting AVF cyclotron (C235) dedicated for proton therapy, which can accelerate protons up to 230 MeV. The first machine was installed at National Cancer Center Hospital East (NCCHE) in Japan in 1998. Early in 2001, the system was approved as a medical device. Recently, the new systems P235 (by SHI) and Proteus 235 (by IBA), which are updated versions of C235, have been developed and installed in many facilities all over the world.

Although C235-based cyclotrons are very compact for normal-conducing machines (4.3 m in diameter and 220 tons in weight), a more compact machine is required for the hospital-based treatment system. Varian Medical Systems, Inc. (VMS), Mevion Medical Systems (Mevion) and IBA have developed compact superconducting cyclotrons.

BNCT is a combined therapy with boron compounds and neutrons but uses high intensity proton or deuterium beams in order to produce epithermal neutrons. SHI has developed a negative-ion, 30 MeV cyclotron (HM-30) dedicated for BNCT, which can produce very high intensity (2 mA) protons. HM-30 was developed by employing many years' experience with PET cyclotrons and also using the axial injection technique acquired from manufacturing other big cyclotrons for research purposes.

4.1. *Cyclotron for proton therapy*

P235/Proteus 235 (here after called P235) is a fixed-energy isochronous cyclotron dedicated for medical treatment purposes. Developed initially in collaboration between SHI and IBA [26, 27], it features a high magnetic field to minimize size, weight and cost while avoiding superconducting coils.

P235 comprises an internal ion source, poles with a spiral sector, two dee cavities and extraction equipment (electrostatic deflector, gradient corrector and permanent quadrupoles). A schematic is shown in Fig. 13 and the main specifications are presented in Table 2.

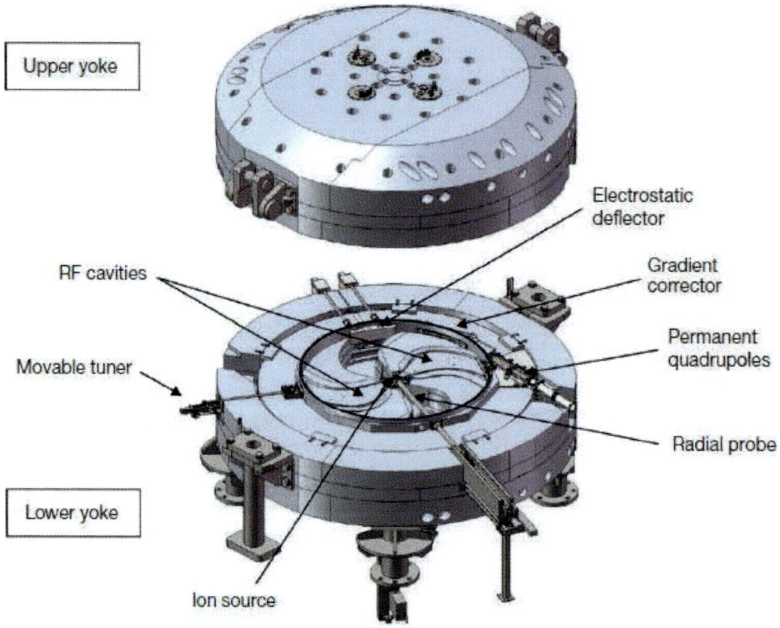

Fig. 13. Internal view of the cyclotron p235.

Table 2. Specifications of the cyclotron p235.

Type	AVF cyclotron
Accelerating ions	Proton
Extraction energy	230 MeV
Extraction method	Electrostatic deflector
Beam current	1–300 nA
Magnet size	ϕ4.34 m × 1.2
Weight	220 tons
Electrical power consumption	
Full beam extracted	330 kW
Vacuum in standby condition	8 kW

In order to reduce the magnet size with normal conduction, the elliptical profile of the pole gap is used, which gives a high magnetic field up to the outermost region of the pole (average 2.1 T). This can reduce the pole diameter. The elliptical profile also produces a rapid field decrease beyond the pole edge and greatly facilitates extraction. With such a gap shape, if the iron is well saturated, then its contribution to the magnetic field is uniform and the radial gradient for an isochronous field depends mainly on the coil field.

The isochronous field required for accelerating protons up to 230 MeV causes a very strong vertical defocusing force [$\nu_Z^2 = -(\gamma/B)(dB/dr) \sim -0.55$]. To compensate for this effect and to obtain sufficient vertical focusing ($\nu_Z \sim 0.2$), the flutter is enhanced by a deep valley and the spiral angle is increased.

Because of the elliptical profile of the pole gap, the RF cavity is installed in the deep valley. In this case, the span angle of the dee electrode is limited to about 30°. The harmonic mode 4 is selected for better acceleration efficiency and consistency with the cavity resonant frequency.

The RF system requires 60 kW at a frequency of 106 MHz. Acceleration is performed with a pair of cavities located in opposite valleys and mutually connected at the center. The dee electrode of each cavity is supported by two vertical pillars located symmetrically with respect to the median plane. Optimization of the diameter of each pair of opposed pillars allows shaping of the amplitude of the dee voltage as a function of the radius to obtain a voltage varying from 60 kV peak dee voltage at the center (to avoid sparking problems in the central region) to 150 kV at extraction (to improve turn separation).

The protons are produced at the center of the cyclotron by a hot filament PIG ion source. The ion source is introduced axially through a vacuum lock system, so that vacuum ventilation of the cyclotron is not necessary for maintenance of the ion source.

The extraction system includes three main components: an electrostatic deflector, a magnetostatic gradient corrector channel and a doublet of permanent magnet quadrupoles. The main element of the system is a short (58-cm-long) electrostatic deflector located in a valley, taking advantage of wide-vertical dimensions.

The technique of beam current stability and fast beam switching is required for scanning irradiation. The beam current stability control system detects the beam current from the dose monitor in the nozzle and controls the arc current of the ion source to compensate for differences between the preset and the actual beam current. Using this feedback system, a beam current stability of $< 1\%$ (2σ) has been achieved. A fast beam chopper system is installed in the central region of P235 for this purpose. A pair of parallel electrodes gives vertical kicks to the second-to-fifth turns of beams in a period of beam off, and a vertical slit completely stops the beams. The response time of chopper electronics is $< 1\,\mu$s. The beams can be turned on and off in $< 50\,\mu$s. This delay results from the acceleration time of P235.

4.2. *Superconducting cyclotron for proton therapy*

The use of superconducting (SC) coils in a cyclotron is attractive because a high magnetic field by high current density in the coils with small consumption power allows the cyclotron to be compact and lightweight.

The first isochronous SC cyclotron for proton therapy was originally designed by H. G. Blosser in 1993 [28], and manufactured by ACCEL/Varian in 2003. The weight of the iron yoke is 90 tons, which is about half of that of the normal-conducting cyclotron P235. The first beam was extracted in 2005 at the Paul Scherrer Institute. The maximum beam current is 800 nA CW.

The approximate vertical stability condition of beam motion in an isochronous cyclotron,

$$0 < \nu_z^2 \cong 1 - \gamma^2 + F^2(1 + \tan^2 \xi), \qquad (1)$$

means that the sector spiral angle should be large, especially at the extraction energy of 230–250 MeV (= 1.25–1.27). In P235, it is around 60°, and it is

Fig. 14. The Varian 250 MeV isochronous SC cyclotron for proton therapy [29]. The yoke diameter is 3.1 m, and the weight is 90 tons.

Fig. 15. CAD cutaway of the NbTi coils of the Varian SC cyclotron [29].

hard to increase it. The flutter,

$$F^2 = \{\langle B^2\rangle - \langle B\rangle^2\}/\langle B\rangle^2 \propto (B_H - B_V)^2/\langle B\rangle^2, \tag{2}$$

decreases as the magnetic field increases. So, it is difficult to achieve a more compact isochronous cyclotron with an iron pole.

Fig. 16. The Mevion 250 MeV SC synchrocyclotron [32]. The yoke diameter is 1.8 m, and weight is 20 tons.

One way to avoid this limitation is to abandon isochronism. In a synchrocyclotron [30], the vertical beam stability is obtained by gradually reducing the magnetic field (and accordingly the RF) with the beam orbit radius. Typically, a several-microsecond width, several-hundred-Hz repetition rate pulsed beam is extracted.

The first SC synchrocyclotron for proton therapy [31] using Nb3Sn coils was designed by T. Antaya in 2007, and manufactured by Mevion [32]. The magnetic field is around 9 T in the central region. The first patient was treated in 2013.

Another SC synchrocyclotron using NbTi coils was constructed by IBA in 2013 [33].

4.3. *Cyclotron for BNCT*

Boron neutron capture therapy (BNCT) is a cancer treatment using boron compounds and neutron irradiation [34]. First, the compounds are infused into the patient. They are concentrated on cancer cells. Second, neutrons are irradiated to the patient. The boron (especially ^{10}B) has a large cross-section for absorbing thermal neutrons. Immediately after capturing a thermal neutron, it disintegrates into two particles. The range of the particles is almost the

Fig. 17. The IBA 230 MeV SC synchrocyclotron [8]. The yoke diameter is 2.5 m, and the weight is 50 tons.

same as the cell size. Finally, a much large fraction of cancer cells are killed by the particles.

Sumitomo Heavy Industries developed the cyclotron-based neutron source for BNCT jointly with the Kyoto University Research Reactor Institute [35]. The neutron source is optimized for BNCT, which means that the energy range of neutrons is from 0.5 eV to 40 keV. Neutron flux is measured to be 1.2×10^9 neutrons/cm^2/s at 1 mA of proton beam. The typical treatment time is 30–45 min.

In order to produce a high intensity of epithermal neutron flux, a 30 MeV proton beam and a beryllium target are adopted. The cyclotron was developed for the neutron source for BNCT by Sumitomo Heavy Industries in Japan. The specifications of the cyclotron are shown in Table 3.

Table 3. The main specifications of the cyclotron HM-30 for BNCT.

Type	AVF cyclotron
Accelerating ions	Negative hydrogen
Extraction energy	30 MeV
Extraction method	Foil stripping
Beam current	
Designed max	2 mA
Nominal operation	1 mA
Magnet size	3.0 m × 1.6 m × 1.7 m
Weight	60 tons

Figure 18 shows a photo of the cyclotron HM-30 for the neutron source of BNCT. This cyclotron was designed as an upgraded version of PET cyclotrons. However, the required beam current is much higher than for PET cyclotrons, and a high current external ion source was adopted instead of an internal source. The ion source is installed under the cyclotron. The ion beam is injected upward into the center of the cyclotron through the injection line. The injected beam is bent by spiral inflectors to match the accelerating orbit.

At the extraction radius, the negative hydrogen ion is stripped of its electrons by thin carbon foil. This extraction method achieves with lower beam losses at the extraction device, compared with the

Fig. 18. The Sumitomo cyclotron HM-30 for BNCT.

method using an electric deflector. Low beam loss at the extraction device means low activation and a low heat load to the cyclotrons.

HM-30 was developed for the BNCT neutron source. The cyclotron is compact but can accelerate a 30 MeV, 1 mA proton beam by using the external negative hydrogen ion source and foil stripping extraction. It has good electricity efficiency compared with other types of accelerators at 30 MeV.

4.4. *Cyclotron for PET*

The emergence of positron emission tomography (PET) as a diagnostic technique of metabolic function in 1980 promoted the development of cyclotrons for the production of PET radiotracers. As shown in Table 4, the radionuclides used in PET are relatively light elements emitting positron particles that are annihilated by surrounding electrons. Protons (p) of 10–15 MeV or deuterons (d) of around 8 MeV, at 10–50 μA beam currents, are used for the production of these radionuclides.

The first generation of cyclotrons, developed in the 1980s, accelerated positively ions (p or d) that were electrostatically extracted in deflectors.

In order to make small cyclotrons it was necessary to decrease the extraction diameters and it was a trend among the manufacturers to design high magnetic fields. Therefore, the magnet average gap became smaller and the magnetic field flutter (the intensity in the valley/hill) was decreased. Also, an angled spiral in the sector was the norm. However, instability due to the following design limitations was observed:

(1) Since the pole gap was small, discharge of the high frequency electrode (about 35 kV) was common.
(2) The high magnetic fields called for high intensity electric fields in the electrostatic extraction deflector, leading to frequent discharges.
(3) The high magnetic field and low acceleration voltage made the extraction turn separation small, with the extraction efficiency as low as 50%. For this reason, problems involving melting of the septum electrode were also frequent.

Table 4. PET radionuclides and their production routes.

Radionuclide	Proton reactions	Deuteron reactions
^{11}C	^{14}N(p, α)^{11}C	
^{13}N	^{16}O(p, α)^{13}N	
^{15}O	^{15}N(p, n)^{15}O	^{14}N(d, n)^{15}O
^{18}F	^{18}O(p, n)^{18}F	^{20}Ne(d, α)^{18}F

In the 1990s, the acceleration of negative ions was introduced, and these PET cyclotrons were commercialized. The main features of negative ion cyclotrons are presented below. All the problems of the first machines were solved and a "dream cyclotron" came true.

(1) Extraction efficiency of 100%, with negligible radioactivation;
(2) Dual irradiation through two ports became available (Fig. 19);
(3) Stable operation, since a deflector was not needed;
(4) Small beam sizes that allowed reduction of the volume in ^{18}O water targets below 2 mL;
(5) Self-shielded machines that allowed installation in common buildings (the cyclotron was surrounding by a shield that allowed a dramatic reduction of the wall thickness — initially 2 m).

The advantages of negative cyclotrons were appreciated by the users, and positive cyclotrons went off the market.

In the 21st century, PET has gained wide recognition and the demand for cyclotrons has sharply increased. Furthermore, in order to increase the radionuclide production yield, high beam intensity has been necessary and cyclotron manufacturers have developed high current PET cyclotrons.

5. Laser Dielectric Acceleration System for Radiation Biology

5.1. *Microbeams of ionization radiation*

In order to estimate the health risk associated with a low radiation dose, basic radiobiological processes must be clarified by irradiating a cell with a well-defined microbeam of ionization radiation at a precisely defined location. The use of microbeams of X-rays, electrons, and light and heavy ions for radiobiological applications and microfabrications has substantially increased since the early 1990s, and some review papers have reported the features of the main facilities [36–42]. Despite the high cost, the majority of the ion microbeam facilities use electrostatic or magnetic lenses to focus MeV-class ion beams to a submicron size. A much simpler and low cost device called a tapered capillary is also used as an adjustable collimator of the ion beam. Nowadays, the tapered capillary collimator is made of glass and can deliver submicron ion beams [43]. The majority of ion accelerators used at microbeam facilities are Van de Graaf accelerators or cyclotrons. X-ray and electron microbeams have also been developed in order to provide useful tools for quantitative analysis in radiation biology research which is complementary to the ion microbeam. X-ray microbeams with energies up to a few tens of keV are delivered by a synchrotron, and are focused to a spot diameter of 1 μm using a grazing incidence mirror system such as Kirkpatrick–Baez mirrors [44]. Electron microbeams are easily obtained by using electron microscopes. However, the low energy electron beam has a troublesome scattering problem. In order to restrict the expansion of the electron beam to less than 1 μm when passing through a biological cell, the energy of the electron beam must be at least 0.5 MeV [45].

The dimensions of beam sources such as MeV-class ion accelerators, synchrotrons for hard X-ray radiation and MeV-class electron microscopes are extremely large, such that only a limited number of facilities can install them. If the size of an accelerators is reduced to be an order of magnitude smaller than that of present accelerators, many radiobiological studies will be performed at small laboratories. Since a laser dielectric acceleration system (LDA) has the potential to realize an on-chip accelerator, the research and development of LDAs has been conducted at various institutes.

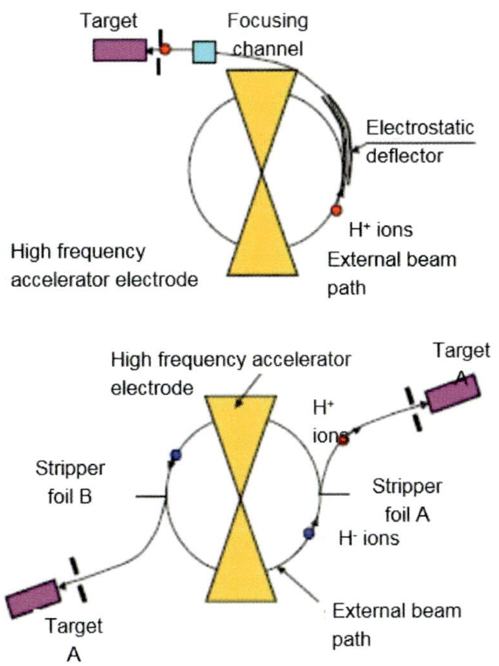

Fig. 19. The Sumitomo cyclotron HM-30 for BNCT.

5.2. *Laser dielectric acceleration system*

A high frequency electric field and a material with a high withstand voltage are indispensable for overcoming the limitation of the acceleration gradient. A combination of an intense laser pulse and the periodic structure of a dielectric is capable of increasing the acceleration gradient by an order of magnitude compared to that of a conventional accelerator. The first idea for an accelerator based on phase modulation near the grating surface was the acceleration of electrons by the modulated laser beam near the surface of the reflection-type grating. Evidence of acceleration was not observed because of the low damage threshold of the metal grating [46]. A phase-modulation-type laser accelerator is also possible using a transparent dielectric. The accelerator proposed by Plettner [47] had a one-dimensional periodic structure (a phase-modulation-masked structure), which is much simpler than the two-dimensional [48] and three-dimensional waveguide structures [49]. The typical shapes of LDAs are shown in Fig. 20.

5.2.1. *Physics of LDAs*

A linearly polarized wavefront of laser light passing through a rectangular wave grating is modulated owing to the optical path difference between two paths in the dielectric and vacuum. When the optical path difference is the same as a half cycle, π of the laser light, an alternating electric field similar to a standing wave is produced, as shown in Fig. 21.

The standing wave consists of the counterpropagation of waves with a wavelength equal to the grating constant L_G. The propagation velocity of the waves is $\pm\omega L_G/2\pi$, where ω is the angular frequency of the laser light. Electrons synchronized with one of the wave components are trapped by the wave and continuously accelerated. The synchronization condition is expressed as $v/c = L_G/\lambda_0$, where v, c

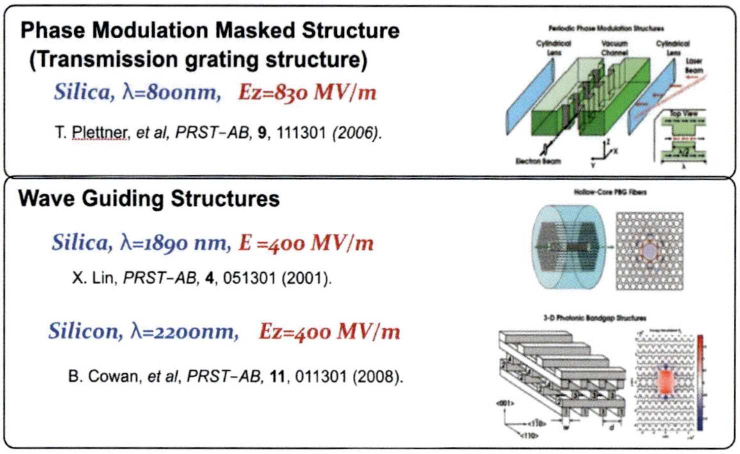

Fig. 20. Typical structures of LDAs.

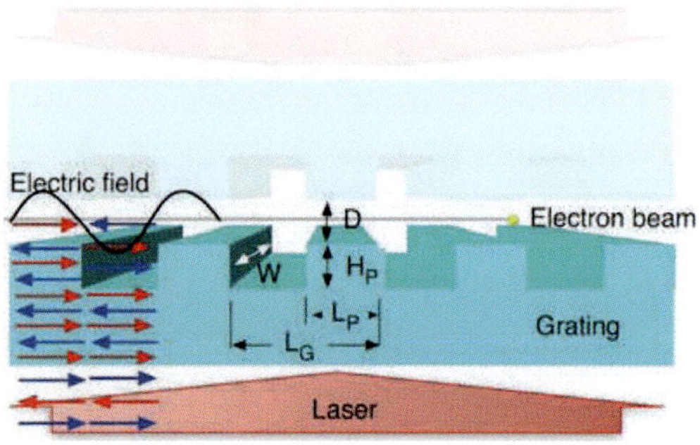

Fig. 21. Parameters defining an LDA. Both single- and dual-grating structures are possible.

and λ_0 are the velocity of an electron, the velocity of light in vacuum, and the wavelength of the laser light, respectively. The height of the grating pillar H_P, which is determined by the conditions for producing the optical path difference of π, is expressed by $H_P/\lambda_0 = 1/2(n-1)$, where n is the index of refraction of the dielectric. Since the standing-wave-like acceleration field appears only in the proximity of the grating surface, the thickness of the acceleration channel D is approximated as $D/\lambda_0 \leq L_P/\lambda_0 = L_G/2\lambda_0$, where $L_P = L_G/2$ is the pillar width. In short, the grating constant, the pillar height and the width of the acceleration channel are almost the same as the laser wavelength. The synchronization condition means that the electric field produced by a smaller grating constant or a higher order component of the spatial distribution of the field must be used in order to accelerate nonrelativistic electrons. However, the acceleration gradient for a nonrelativistic electron is smaller than that of a relativistic electron.

The laser intensity is limited by the optical breakdown of the dielectric [50], and the nonlinear optical effect is $10^{11}\,\mathrm{W\,cm^{-2}}$ for silica (SiO_2) [45]. According to numerical simulation results for the dual-grating structure, the acceleration gradient for relativistic electrons is 600–800 $\mathrm{MeV\,m^{-2}}$.

5.3. *Recent experiments of LDA*

The rapid progress in microfabrication and laser technology makes it possible to perform acceleration experiments after the long dormancy of successful experiments. Electron acceleration by an LDA was successfully demonstrated at two institutes [51, 52]. An acceleration experiment at a relativistic energy of 60 MeV was performed by using the output electron beam from a conventional linear accelerator. The highest acceleration gradient was 310 $\mathrm{MeV\,m^{-1}}$ at the peak incident electric field of the laser of 3.4 $\mathrm{GV\,m^{-1}}$ in the dual-grating structure. The grating and experimental setup are shown in Fig. 22.

An acceleration experiment for nonrelativistic electrons was performed by another institute [52]. The third spatial harmonic ($n = 3$) was used to accelerate 27.9 keV electrons. The maximum acceleration gradient was 25.0 $\mathrm{MeV\,m^{-1}}$ in the proximity (100 nm) of a surface of the single grating at an incident electric field of the laser of 2.85 $\mathrm{GV\,m^{-1}}$.

5.4. *Application of LDAs and problems*

5.4.1. *Applications*

Since the cross-section of the acceleration channel of an LDA is smaller than 1 μm and the bunch charge of

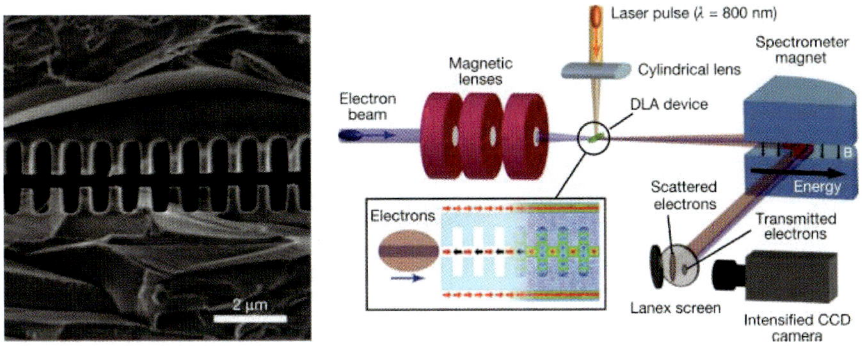

(a) (b)

Fig. 22. (a) Cross-section of the grating; (b) experimental setup. The figure is from: Petralta, *Nature* **503** (2013) [51].

the beam is less than 1 fC, the required laser power and pulse energy are much smaller than those for a laser wakefield accelerator. The estimated values for accelerating electrons from 30 keV to 1 MeV are about 10 MW and 20 pulses of 50 μJ/pulse [45]. Some research groups are developing on-chip high energy accelerators for future energy frontier colliders and on-chip X-ray free electron lasers.

5.4.2. *Future problems*

In order to accomplish a practical LDA, the following crucial problems must be solved: an increase in the energy transfer efficiency from the laser to an electron, attosecond (as) synchronization of pump pulses, and the as-bunching of electron beam. These problems will be solved by applying the physics of a photonic crystal for designing the LDA structure.

6. Laser Plasma Electron Acceleration System for Cancer Therapy

A superintense laser pulse ionizes atoms within a very short time and propagates in the plasma, followed by a laser-excited wake. A laser wakefield accelerator (LWFA), which consists of an ionized gas (plasma), possesses the ability to downsize a high energy particle accelerator the size of a tabletop, because the electric field strength of a laser-excited plasma wave, i.e. the laser wakefield, is more than a few hundred times higher than that of conventional accelerators. Many proof-of-principle experiments for LWFAs have demonstrated the ability to realize tabletop GeV-class accelerators, as shown in Fig. 23. A current top priority issue in LWFA

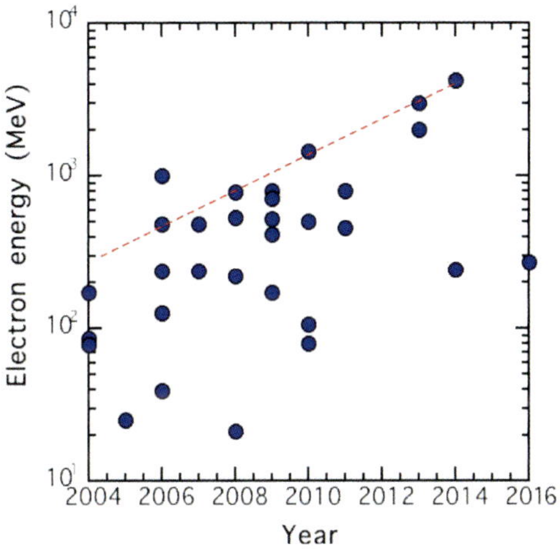

Fig. 23. Progression of the electron energies of LWFA experiments.

research is the achievement of the same beam parameters as for a state-of-the-art accelerator. All of these endeavors will produce practical uses of LWFAs in various fields. In the field of medicine, a cancer therapy with a very high energy electron (VHE) beam delivered by the LWFA will provide an opportunity to install a therapy system for a small scale hospital [53].

Stabilities of output, energy and charge, of the present LWFAs are insufficient for medical applications, because the present LWFAs are operated in the nonlinear mode (bubble regime) [54]. In order to obtain stable output, the LWFA must be separated into the electron source (injection part) and an acceleration part, which are operated in the nonlinear regime and the linear regime, respectively. The physics and the technology of LWFAs for VHE are the same as for GeV-class accelerators.

6.1. *Physics of LWFAs*

Since a charged particle gains energy from the parallel component of an electric field for particle movement, the charged particle oscillates in the plane electromagnetic wave (transverse wave) and does not gain a net energy. In order to accelerate a particle continuously, the longitudinal component of the electric field must be produced by introducing a boundary that is made of a metal/dielectric or converting a mode from a transverse wave into a longitudinal wave (plasma wave). The former is known as a radio frequency (RF) accelerator or dielectric accelerator. An LWFA is in the latter category. The typical setup of an LWFA experiment is very simple, consisting of an ultraintense laser, an off-axis parabolic mirror to focus the laser, and a gas jet/cell, as shown in Fig. 24.

A superintense laser pulse above 10^{18} W cm^{-2} ionizes atoms at the foot of the pulse and propagates in the plasma. The laser pulse repels electrons in the plasma by the strong ponderomotive force while propagating through the plasma. The electron plasma wave (wake) is excited behind the laser pulse, which propagates at the speed of light, as shown in the lower picture of Fig. 25. The wave forms of the linear wakefield are similar to a sinusoidal wave. The longitudinal electric field associated with the fast wakefield is able to accelerate trapped electrons into the plasma. By surfing on the fast electrostatic wave, electrons can be boosted to high energies over very short distances [55]. In order to increase the energy gain, the laser pulse must be guided for a long distance in a plasma channel, as for an optical fiber. However, there are still two limitations on the interaction length: the electron dephasing length

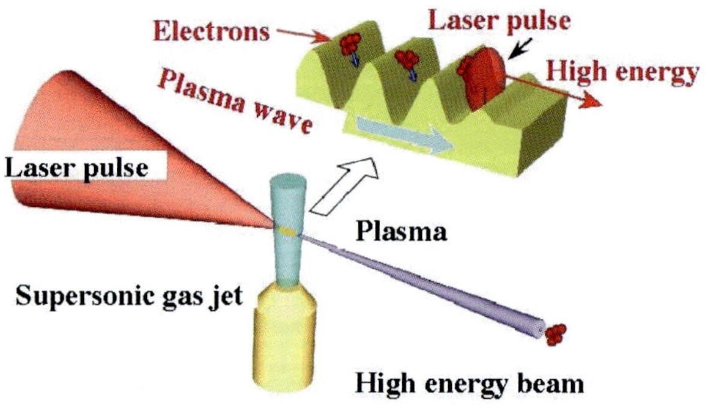

Fig. 24. Schematic drawing of the experimental setup and the wakefield excitation by the laser pulse.

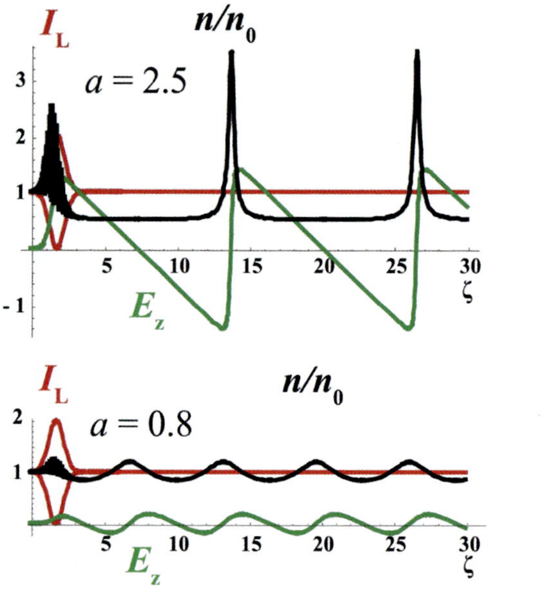

Fig. 25. Shapes of laser pulses I_L, electron densities n/N_0 and electric fields E_z in the nonlinear (upper figure) and linear (lower figure) regimes.

L_d and the pump depletion length L_p. The pump depletion length, $L_p \approx (\lambda_p/a_0^2)\gamma_p^2$, is the distance over which the pump loses a significant fraction of its energy owing to the excitation of the plasma wave. The pump depletion length is only slightly larger than the dephasing length. In a case where the interaction length is limited by the dephasing length, the energy gain of the electrons is given by $W_{\max}(\text{MeV}) \approx 1.3a_0^2 n_c/n_e$ for $a_0 \leq 1$, where n_c is the cutoff density of the plasma, which is close to 500 MeV for $L_d \approx 0.3$ cm. The dephasing limitation could in principle be overcome by staging the LWFA such that, after outrunning the plasma wave, an electron is injected into a new plasma wave at the appropriate phase. Once pump depletion occurs, staging with a fresh pump pulse is necessary.

In the high intensity (or nonlinear) limit, $a_0^2 \gg 1$, the accelerating field is given by $E_{\max}/E_0 = (a_0^2/2)(1 + a_0^2/2)^{-1/2}$. The wave forms of the nonlinear wakefield are similar to the spiky wave shown in the upper picture of Fig. 25. The energy gain in the nonlinear regime is $W_{\max}(\text{MeV}) \approx 1.6a_0^2 n_c/n_e$.

In linear wakefield excitation processes, the accelerating structures have a very good stability, limited only by the shot-to-shot fluctuations in the lasers and the formation of the plasma channels. The electron must be injected at the appropriate phase because the linear wakefield has no ability to pick up electrons from the background plasma. On the other hand, when the laser power exceeds the self-focusing power, the wakefield can reach a very large amplitude, as in an electron bubble in a plasma, and trap plasma electrons directly without external injection. The energy scaling laws in the highly nonlinear regime have been discussed in many articles [54].

6.2. *Experiments with LWFAs*

6.2.1. *Monoenergetic acceleration*

A great number of experiments have been performed in the nonlinear regime with laser powers above 2 TW and laser energies (and durations) in the range of 0.1 J ($\approx$ 30 fs) to 100 J (ps) [56]. Monoenergetic electrons were delivered only in narrow density windows for the given irradiation power of the lasers in which short electron bunches were injected at the appropriate phase of the wakefields.

6.2.2. *Plasma waveguide*

A hollow plasma column which has a parabolic radial electron density distribution, $n_e = n_0 + [n = e(r_0) - n_0]/r_0^2$, is able to guide the weakly relativistic laser pulse ($a_0 \leq 1$) beyond the Rayleigh length, where n_0 is the electron density along the central axis and r_0 is the characteristic radius of the column. The Gaussian beam will be guided at the matched beam size $r_0 = w_0$ provided that the channel depth is $\Delta n_e = (\pi r_e r_0^2)^{-1}$ or $\Delta n_e(\mathrm{cm}^{-3}) = 1.13 \times 10^{20}/r_0^2(\mu\mathrm{m})$, where r_e is the classical electron radius [57].

Several methods of producing plasma channels for laser guiding, such as the laser or electrical breakdown of the inner surface of a vacuum dielectric capillary [58], the explosion of a laser-heated plasma thread [59], the fast z-pinch [60] and the electrical discharge of a gas-filled capillary [61], have been proposed and tested. The laser damping rates of the inner surface breakdown waveguide are 0.13–0.18 cm^{-1}, which are several times higher than those for other methods because the plasma produced by surface erosion contains high-Z atoms such as silicon, aluminum and oxygen. In addition, the lifetime of the inner surface breakdown method is limited to less than a few tens of shots. The gas-filled capillary method presently shows good results in high energy acceleration experiments; multi-GeV electron beams with an energy up to 4.2 GeV, a 6% rms energy spread, a charge of 6 pC, and a 0.3 mrad rms divergence have been produced from a 9-cm-long capillary discharge waveguide with a plasma density of $n_e \approx 7 \times 10^{17}\,\mathrm{cm}^{-3}$ and powered by laser pulses with a peak power up to 0.3 PW [62].

A small conical plasma produced in a strong axial magnetic field > 1 T (plasma micro-optics) is able to deliver electron beams with a good pointing stability. This technique is applicable to the electron injector [63].

6.2.3. *Electron injection*

As presented in the previous section, the electrons must be injected at the appropriate phase in order to deliver a well-controlled electron bunch. Electron injection by an external electron source such as a linac has been attempted. However, no one has succeeded, because the bunch length of the output was much longer than the plasma wavelength [64]. The initial electrons are supplied from the plasma by local acceleration/modulation of the plasma such as the collision of two laser pulses [65], wave breaking at the density down ramp or step [66], and ionization injection of doped high-Z impurity in a target gas [67]. The first demonstration of the controlled injection was performed by the collision of two laser pulses. Recently, the other two injection schemes have been extensively investigated, because they are free from the synchronization of two or three laser pulses.

6.2.4. *Staging acceleration*

In order to accelerate electrons beyond multi-GeV by overcoming the dephasing and/or pump depletion limit, staging acceleration is indispensable. Experimental results demonstrating staging acceleration have been published [62]. Electron bunches produced by ionization injection in the gas jet were focused and injected into the laser-excited wakefield produced in a second acceleration channel, i.e. the gas-filled capillary shown in Fig. 26. The energy gain in the second capillary varied with the injection phase of the wakefield.

6.3. *Applications of LWFAs and problems*

6.3.1. *Applications*

Relativistic electron beams delivered by a laser wakefield are considered to be compact particle

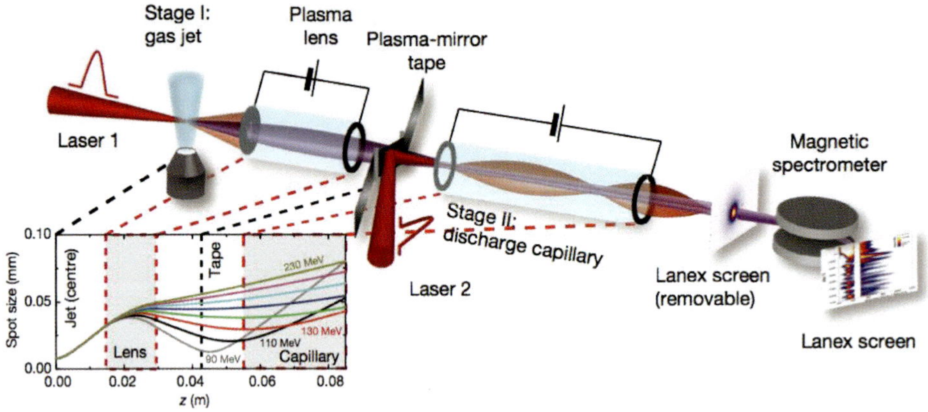

Fig. 26. Setup of the staging acceleration experiment (Steinke, *Nature* (2016) [62]).

accelerators, which inspire a wide range of applications of outstanding particle-beam and radiation sources. Intense THz radiation [68], betatron X-ray radiation [69] and the all-optical inverse Compton gamma ray source [70] have been successfully demonstrated. Moreover, X-ray free electron lasers (XFELs) based on LWFAs have been studied at many institutes [71]. A research program for accelerating high quality electron beam by staging acceleration has been started in Japan for an injector of a mobile XFEL [72], which will be a useful tool for a biochemical test of a lesion. The direct uses of the output beam from an LWFA, such as high energy frontier colliders [73] and cancer therapy machines, have also been considered.

A VHE beam therapy system with electron energies of 150–250 MeV driven by an LWFA shown in

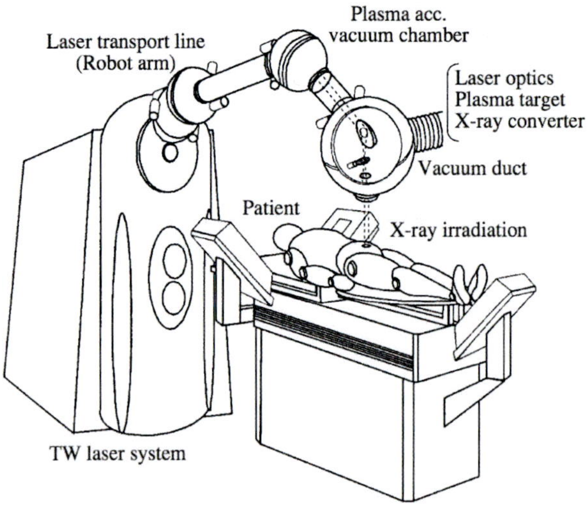

Fig. 27. Setup of the staging acceleration experiment (Kinoshita, *Laser Physics* (2006) [75]).

Fig. 27 is an interesting application of complementary medicine in present cancer treatment systems [74, 75].

For VHEs, the depth–dose curves vary slowly with the depth. Moreover, it is possible to form a dose peak in a deep part of a body by an orthogonal two-beam irradiation [76]. Therefore, the tumor-to-normal tissue dose ratio is greater for VHE beams than for conventional photon beams. However, the VHE beam moderately diffuses, owing to the multiple scattering in a thick body.

6.3.2. *Future problems*

For the applications mentioned above, the crucial requirements are electron beams with a higher stability and higher luminosity, which are predominantly influenced by the stability and repetition rate of the drive lasers and the control of the plasma. The current stability of a-few-percent level at a repetition rate on the order of 1 Hz will be improved to 0.1% by increasing the repetition rate to greater than a few tens of Hz through the introduction of new laser technology such as fiber lasers [77] and thin disk lasers with optical parametric chirped pulse amplification (OPCPA). A deep, stable and long plasma channel is a key element for high energy acceleration beyond 10 GeV.

7. Laser Driven Ion Acceleration

7.1. *High intensity laser*

An ion acceleration driven by an interaction between a laser and a thin foil was known in the 1970s [78]. The proof-of-principle experiments of the laser-driven ion acceleration were performed using nanosecond pulse lasers in the 1990s. A chirped pulse amplification (CPA) method [79] became a turning point of this field in 2000. Namely, the laser pulse of femtoseconds can be realized by the time compression of the chirped pulse. As a corollary, the peak intensity reaches a level above the terawatt. The significant feature of the laser-driven ion acceleration adopting the ultrashort pulse laser is that the effective source size of ion emission is typically several tens of micrometers. Furthermore, the acceleration gradient of the order of MeV/μm, which goes far beyond any definition of the discharge region in conventional RF accelerators, arises in the ultrasmall space [80]. If it is possible to take advantage of this steep acceleration gradient, a compact laser accelerator may be realized [81, 82].

7.2. *Ion acceleration mechanism*

The laser pulse interacts with electrons in the medium at first, and then ions are accelerated by the electromagnetic field induced by the motion of the electrons. For example, in laser pulse irradiation to thin film, as shown in Fig. 28, the electrons are accelerated to the rear surface from the irradiated surface. The fast electrons with the mean free

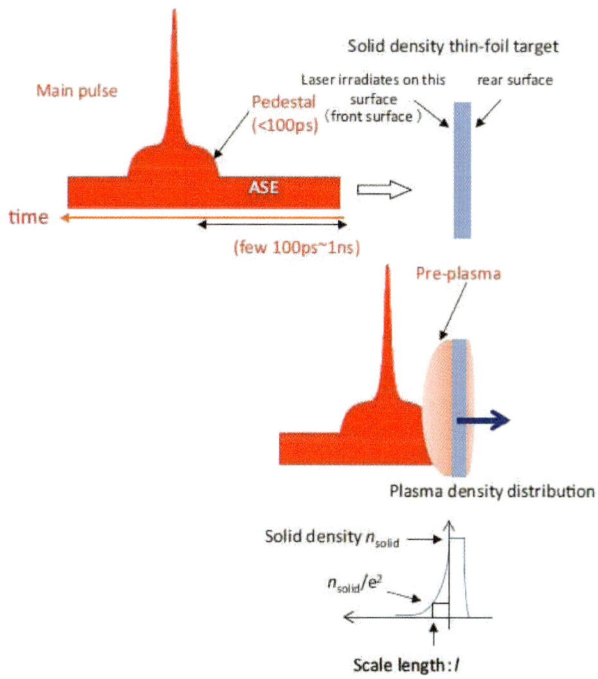

Fig. 28. Sketch of the high intensity laser interacting with the thin foil. (Figure extracted from Ref. 98.)

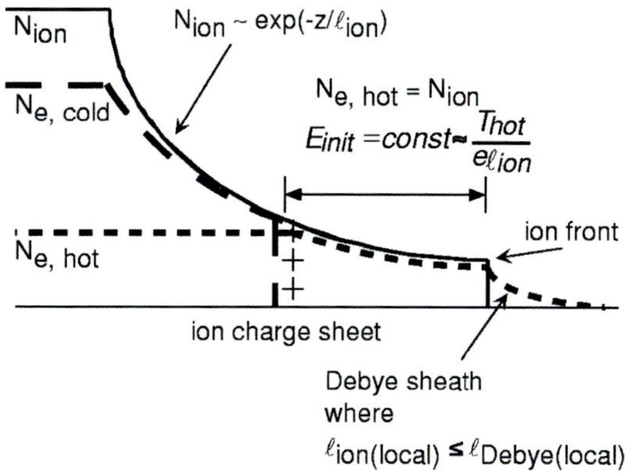

Fig. 29. Schematic view of laser acceleration of ions from the rear surface of the target. The charge separation in the region of the Debye sheath at the front of the expansion plasma induces the acceleration field. (Figure extracted from Ref. 88.)

pass exceeding the thickness of the thin film escape to a vacuum from the rear surface. Therefore, the quasistatic acceleration field is formed by a charge separation occurring on the rear surface of the thin film [83–87]. This acceleration field is referred to as a sheath field. Its spatial size is the Debye length of the plasma, as shown in Fig. 29 [88]. Acceleration models such as the target normal sheath acceleration (TNSA) mechanism [89–92] and the strong charge separation field (SCSF) mechanism [93–97] have been proposed. Furthermore, the radiation-pressure-dominant acceleration (RPDA) mechanism has been theoretically predicted in the region of the laser intensity (10^{23} W/cm^2) beyond the current laser systems.

7.3. *Technology*

Ion acceleration experiments have some issues with improving the conversion efficiency from the laser energy to the ion energy.

A high intensity laser has usually a nanosecond preceding light caused by an amplified spontaneous emission (ASE), which generates an unwanted preformed plasma. In order to reduce the ASE, several devices have been developed: saturable absorber (SA) [99, 100], double CPA laser system [101], optical parametric chirped pulse amplification (OPCPA) combined with SA [102], and cross-polarized wave generation (XPW) [103, 104].

Another issue is to compress the laser pulse spatially and temporally. Large optical components corresponding to the laser beam larger than several centimeters are distorted significantly by their own weight, the laser heating and the fixing stress. Because they cause distortion of the wave front, a spot size is blurred. A deformable mirror combining with a wave front sensor can correct the distortion of the wave front and focus the laser beam to nearly the diffraction limit [105–111].

7.4. *Current experimental situation*

The energy spectra of the ions have usually an exponential profile up to a cutoff energy, as shown in Fig. 30. Figure 31 shows the maximum ion energies as a function of laser irradiance at various facilities. The data are grouped into the pulse duration. The maximum ion energy increases clearly with the laser irradiance. Although not shown here, it has been found that the maximum ion energy has a tendency to increase with a thinner target within the extent that the prepulse does not destroy the target [112, 113]. In this manner, the target development is just as essential as upgrading of the laser system to increase the ion energy.

In such an investigative trend, the records of the maximum energy were recently broken employing thinner targets by F. Wagner *et al.* They achieved the proton acceleration of 85 MeV by the PW laser at GSI [114]. In addition to a thinner target, improving the contrast with a double plasma mirror and controlling the polarization of the laser, the proton acceleration of 93 MeV was recorded by I. J. Kim *et al.* at GIST of Korea [115].

As an alternative approach, the experiments employing cluster gas targets were performed in 2009 and 2013 (Fig. 32) and succeeded in creating subcritical density plasmas, which efficiently absorb laser energy. As a result, the maximum ion energies ten

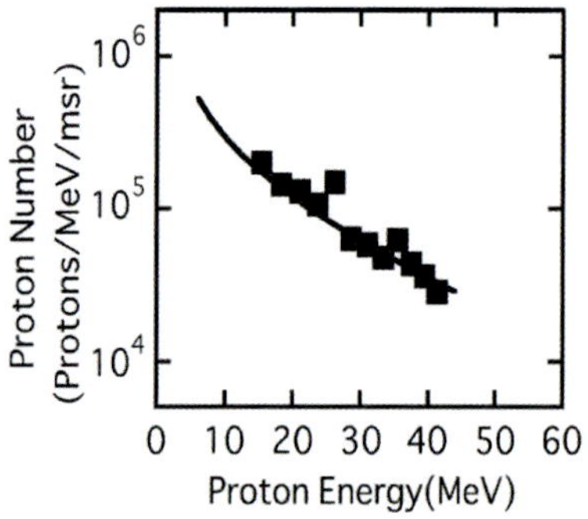

Fig. 30. Typical ion energy spectrum, from Ref. 116.

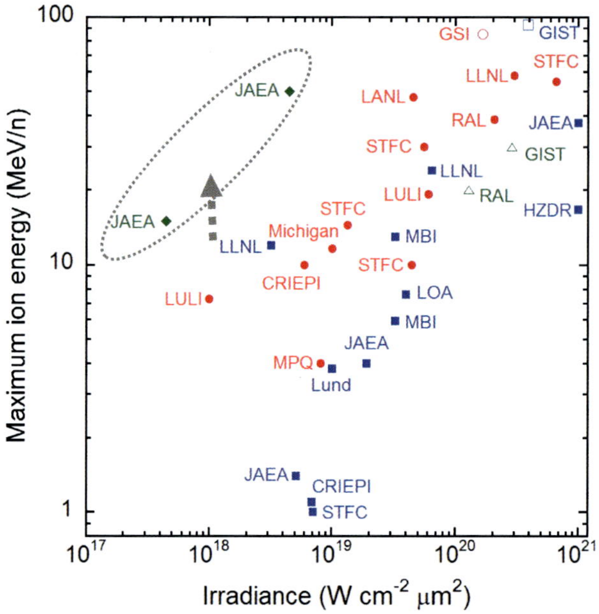

Fig. 31. The circle and square plots are grouped into the pulse duration of more than 150 fs and less than 100 fs, respectively. The open circle [114] and the open square [115] plots are recently recorded results. The diamonds indicate the experiments with the cluster gas targets [117–119]. The triangles indicate the experiments with the near-critical dense carbon nanotube [122] and nanostructure [120] targets on the thin film. The abbreviated names indicate: Central Research Institute of Electric Power Industry (CRIEPI), Gwangju Institute of Science and Technology (GIST), GSI Helmholtzzentrum für Schwerionenforschung GmbH (GSI), Helmholtz-Zentrum Dresden-Rossendorf (HZDR), Japan Atomic Energy Agency (JAEA), Los Alamos National Laboratory (LANL), Lawrence Livermore National Laboratory (LLNL), Laboratoire d'Optique Appliquée (LOA), Laboratoire d'Utilisation des Lasers Intenses (LULI), Lund University (Lund), Max Born Institute (MBI), University of Michigan (Michigan), Max Planck Institute of Quantum Optics (MPQ), Rutherford Appleton Laboratory (RAL), Science and Technology Facilities Council (STFC).

times higher than for the conventional method have been demonstrated as shown in Fig. 31 [117–119]. In a simulation by a plasma particle code, a magnetic vortex grows and generates a strong acceleration electric field in the subcritical density plasma. It suggests that a new acceleration mechanism called "magnetic field ion acceleration" accelerates ions [121]. Meanwhile, near-critical dense carbon nanotube targets [122] or nanostructured targets [120] deposited on the thin film have recently attracted attention as a potential way to increase the acceleration efficiency.

Although the maximum ion energy has been discussed above, we have to point out that absolute

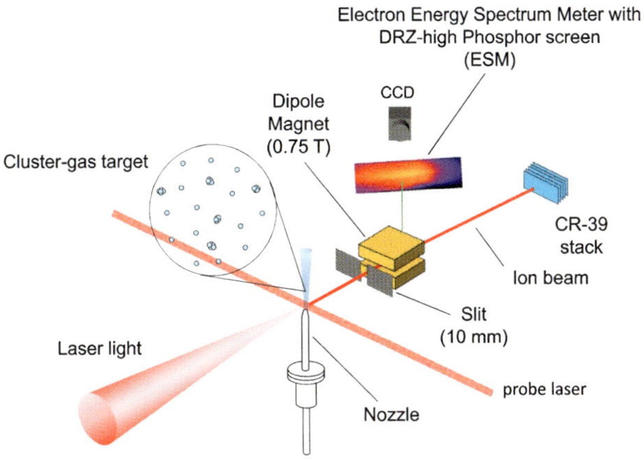

Fig. 32. Experimental setup for the interaction between the laser and the cluster gas target.

differential energy spectra as the number of ions and solid angle should also be evaluated as to whether they are conducive to the practical applications.

7.5. *Application to radiotherapy*

In the case of applying the laser-driven ion acceleration to heavy ion radiotherapy, it could lead to advantages in terms of compactness and costs compared to the conventional accelerators. As application using 10 TW class laser systems, irradiation experiments on living cells have already been carried out to demonstrate the radiobiological effects [123, 124].

In the future, it is quite possible to achieve the maximum ion energy over 100 MeV by upgrading the laser system and optimizing the target condition. The way of controlling the ion beam with large divergence and broad energy has been proposed and demonstrated [125]. The beam optics has also been proposed, supposing that we will reap a laser-based ion accelerator of the hundred MeV class in anticipation of future medical applications [126, 127]. Anyway, for practical use, it is expected to form the ion beam with a quasi-monoenergetic, stable and high repetition rate.

Since the laser technologies are evolving rapidly to meet those needs, realization of the intensity of $10^{23}\,\mathrm{Wcm}^{-2}$ will be possible by focusing to the diffraction limit in the currently running petawatt

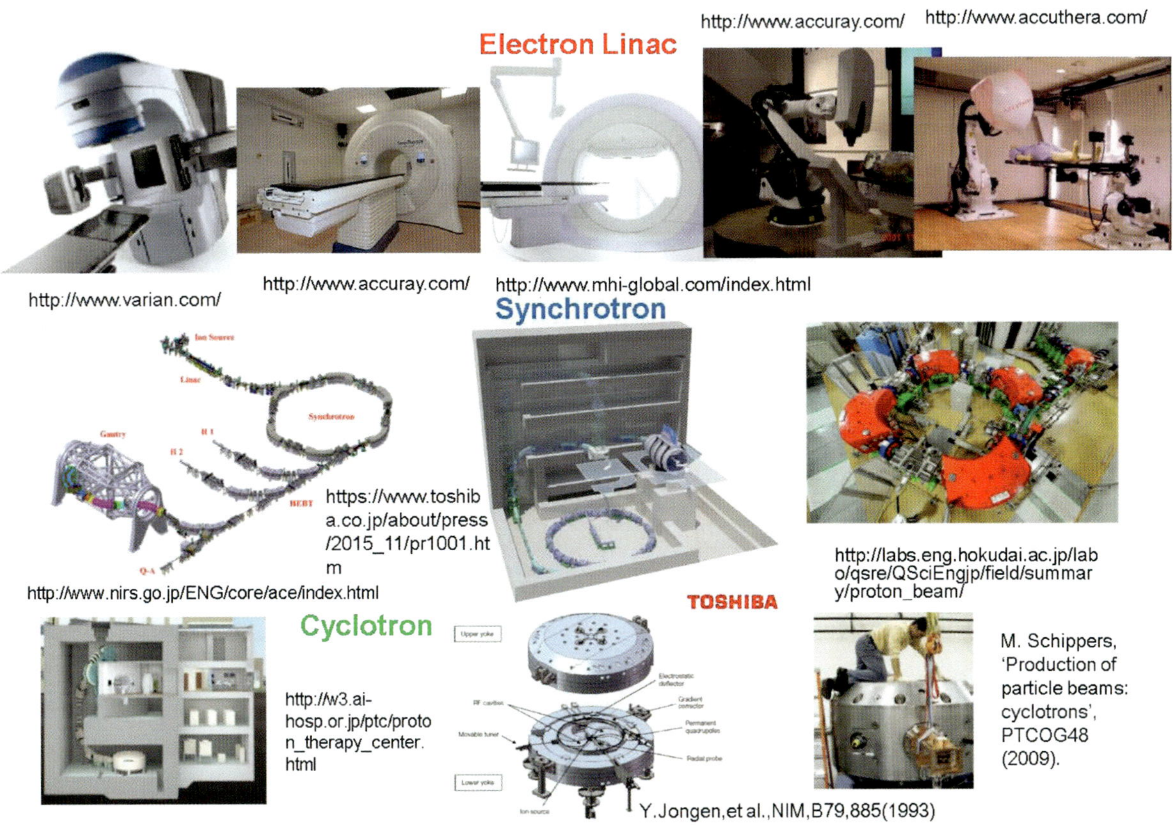

Fig. 33. Downsizing of medical accelerators by advanced technologies.

laser systems. Because the effect of the RPDA mechanism appears in this irradiance range, the laser accelerator would be much closer to practical use.

8. Closing Remarks

We summarize the advanced medical accelerators as shown in Fig. 33. The electron linac X-ray source is downsized by adopting S-, C- and X-band RF. Since the accelerators become more compact, additional functions such as the X-ray CT and robot arm can be combined for realizing image-guided radiation therapy (IGRT) and dynamic tracking therapy. The medical cyclotron and synchrotron are downsized by optimized alignment and superconducting magnet technology.

Further compactness could come true by using advanced laser-based acceleration. By making use of an ~100 THz infrared wave for acceleration, the accelerating structure and beam size become μm-size. Therefore, the application as a μm beam source to radiation biology and therapy can be promising. For the realization of a laser plasma accelerating system for cancer therapy, first the beam parameters of practical medical accelerators, such as the energy, spectrum and intensity, have to be achieved. Then, we can construct a feasible view of the inner contents of Fig. 27.

Acknowledgments

The following individuals contributed to this article: Hitoshi Kobayashi (High Energy Accelerator Research Organization, KEK) for Chap. 2, Takeo Iwai (Department of Radiation Oncology, Yamagata University School of Medicine) for Chap. 3, Masayuki Hirabayashi, Toshiki Tachikawa, Toshinori Mitsumoto, Hiroshi Tsutsui, Toru Ishizuka and Takuya Miyashita (Industrial Equipment Division, Sumitomo Heavy Industries, Ltd.) for Chap. 4, and Satoshi Jinno (Nuclear Professional School, University of Tokyo) for Chap. 7. The principal authors would like to express their deep gratitude to these coauthors.

References

[1] http://www.advancedoncotherapy.com.
[2] E. Fokas *et al.*, *Biochim. Biophys. Acta* **1796**, 216 (2009).
[3] M. Uesaka and E. Tanabe, 6.1 Electron Linear Accelerator, in Comprehensive Biomedical Physics, Vol. 8: *Radiation Source and Detectors* (Elsevier, 2014).
[4] H. Kobayashi, T. Kurihara, H. Matsumoto, M. Yoshioka, H. Kumada, A. Matsumura, H. Sakurai, F. Hiraga, Y. Kiyanagi, T. Nakamura, H. Nakashima, T. Shibata, T. Hashirano, F. Inoue, K. Sennyu, T. Sugano, T. Ohba and S. Tanaka, in *Proc. IPAC2012* (New Orleans, Louisiana, USA), THPPR048.
[5] M. Uesaka and H. Kobayashi, *Reviews of Accelerator Science and Technology* **8**, 181 (2015).
[6] J. R. Alonso and J. R. Castro, Light-ion therapy in the US: From the Bevalac to ?? (LBNL–51559) (USA, 2002).
[7] R. Wilson, *Radiology* **47**, 487 (1946).
[8] J. D. Slater, Clinical proton therapy at Loma Linda University Medical Center. https://three.jsc.nasa.gov/articles/LLUclinical.pdf.
[9] www.ptcog.ch, PTCOG website.
[10] http://labs.eng.hokudai.ac.jp/labo/qsre/QSciEng jp/field/summary/proton_beam.
[11] Y. Kawaguchi, *Mystery on Heavy Ion Cancer Therapy* (Daiwa Shuppan, Japan, 2015), in Japanese.
[12] G. Kraft, History of the heavy ion therapy at GSI. https://three.jsc.nasa.gov/articles/Krafts_GSI.pdf.
[13] T. Furukawa *et al.*, *Med. Phys.* **37**, 5672 (2010).
[14] Y. Iwata *et al.*, *Phys. Rev. ST Accel. Beams* **15**, 044701 (2012).
[15] http://techon.nikkeibp.co.jp/atcl/feature/15/3274 41/011100028/?ST = health&P = 4.
[16] Y. Iwata *et al.*, in *Proc. 12th Annual Meeting of Particle Accelerator Society of Japan* (in Japanese) (2015), p. 850.
[17] D. Ondreka and U. Weinrich, in *Proc. EPAC08* (2008), p. 979.
[18] S. Rossi and CNAO, *Eur. Phys. J. Plus* **126**, 78 (2011).
[19] R. Mayer and S. Vatnitsky, Current status of Medaustron, in *Proc. 2nd Int. Symp. Heavy-Ion Radiotherapy and Advanced Technology* (2016), p. 39.
[20] P. Urschütz, Overview and status of the Austrian particle therapy facility Medaustron. indico.psi.ch/event/Urschuetz.
[21] http://www.hibmc.shingu.hyogo.jp.
[22] http://heavy-ion.showa.gunma-u.ac.jp/en/index.php.
[23] T. Iwai, K. Nemoto, H. Yamashita, I. Kubota, A. Goto and T. Kayama, Status report of carbon ion radiotherapy facility at Yamagata University, in *Proc. 2nd Int. Symp. Heavy-Ion Radiotherapy and Advanced Technology* (2016), pp. 34–37.
[24] Toshiba press release: Toshiba to supply heavy-ion radiotherapy system to Yamagata University

Hospital. https://www.toshiba.co.jp/about/press/2015_11/pr1001.htm.
[25] G. Xiao, Status of heavy-ion therapy at Institute of Modern Physics Chinese Academy of Sciences. http://ptcog.ch/archive/conference_p&t&v/PTCOG53/PresentationsSM/Xiao_Keynote.pdf.
[26] Y. Jongen, A. Laisne, W. Beeckman, *et al.*, *Nucl. Instrum. Methods B* **79**, 885–889 (1993).
[27] W. Beeckman, A. Laisne, Y. Jongen, *et al.*, in *Proc. 14th Int. Conf. Cyclotron and Their Applications* Cape Town, South Africa, 1995, pp. 218–221.
[28] H. Blosser, MSUCL-874 (Michigan State University, 1993).
[29] H. Rocken *et al.*, The Varian 250 MeV Superconducting compact proton cyclotrons: medical operation of the 2nd machine, in *Production and Comissioning Status of Machines No. 3 to 7*.
[30] E. McMillan (1952), US patent 2615129.
[31] T. Antaya, High-field superconducting synchrocyclotron (2007), US patent PCT/US2007/001628.
[32] M. Schippers, *PTCOG 48* (2009).
[33] W. Kleeven *et al.*, *The IBA Superconducting Synchrocyclotron Project* (2013).
[34] J. A. Coderre and G. M. Morris, *Radiat. Res.* **151**, 1 (1999).
[35] H. Tanaka, Y. Sakurai, M. Suzuki, S. Masunaga, T. Mitsumoto, K. Fujita, G. Kashino, Y. Kinashi, Y. Liu, M. Takada, K. Ono and A. Maruhashi, *Appl. Radiat. Isot.* **69**, 1642 (2011).
[36] M. Folkard, B. Vojnovic, S. Gilchrist, K. M. Prise and B. M. Michael, *Nucl. Instrum. Methods B* **210**, 302 (2003).
[37] S. Gerardi, *Radiat. Prot. Dos.* **122**, 285 (2006).
[38] A. W. Bigelow, D. J. Brenner, G. Garty and G. Randers-Pehrson, *IEEE Trans. Plasma Sci.* **36**, 1424 (2008).
[39] T. Funayama, N. Hamada, T. Sakashita and Y. Kobayashi, *IEEE Trans. Plasma Sci.* **36**, 1432 (2008).
[40] K. O. Voss, C. Fournier and G. Taucher-Scholz, *New J. Phys.* **10**, 075011 (2008).
[41] S. Gerardi, *J. Radiat. Res.* **50**, Suppl., A13 (2009).
[42] Y. Kobayashi *et al.*, *J. Radiat. Res.* **50**, Suppl., A29 (2009).
[43] T. Ikeda *et al.*, *J. Phys. Conf. Ser.* **58**, 68 (2007).
[44] K. Kobayashi *et al.*, *Int. Congr. Ser.* **1258**, 207 (2003).
[45] K. Koyama *et al.*, *J. Phys. B* **47**, 234005 (2014).
[46] Y. Takeda and I. Matsui, *Nucl. Instrum. Methods* **62**, 306 (1968).
[47] T. P. Plettner *et al.*, *Phys. Rev. ST Accel. Beams* **9**, 111301 (2006).
[48] X. Lin, *Phys. Rev. ST Accel. Beams* **4**, 051301 (2001).
[49] B. Cowan *et al.*, *Phys. Rev. ST Accel. Beams* **11**, 011301 (2008).
[50] B. C. Stuart *et al.*, *Phys. Rev. Lett.* **74**, 2248 (1995).
[51] E. A. Peralta *et al.*, *Nature* **503**, 91 (2013).
[52] E. A. Breuer and P. Hommelhoff, *Phys. Rev. Lett.* **111**, 134803 (2013).
[53] K. Nakajima *et al.*, *Appl. Sci.* **5**, 1 (2014).
[54] A. Pukov and J. Meyer-ter-vehn, *Appl. Phys. B* **74**, 355 (2002); W. Lu *et al.*, *Phys. Rev. ST Accel. Beams* **10**, 061301 (2007).
[55] T. Tajima and J. M. Dawson, *Phys. Rev. Lett.* **43**, 267 (1979).
[56] S. P. D. Mangle *et al.*, *Nature* **431**, 535 (2004); J. Faure *et al.*, *Nature* **431**, 541 (2004); C. G. R. Geddes *et al.*, *Nature* **431**, 538 (2004); E. Miura *et al.*, *Appl. Phys. Lett.* **86**, 251501 (2006).
[57] P. Sprangle and E. Esarey, *Phys. Fluids B* **4**, 2241 (1992).
[58] F. Dorchies *et al.*, *Phys. Rev. Lett.* **82**, 4655 (1999); B. Cros *et al.*, *Phys. Rev. E* **65**, 026405 (2002); Y. Kitagawa *et al.*, *Phys. Rev. Lett.* **92**, 205002 (2004); A. Zigler *et al.*, *J. Opt. Soc. Am. B-Opt. Phys.* **13**, 68 (1996); Kameshima *et al.*, *Appl. Phys. Exp.* **1**, 066001 (2008).
[59] C. G. Durfee III *et al.*, *Phys. Rev. Lett.* **71**, 2409 (1993); H. M. Milchberg *et al.*, *Phys. Plasmas* **3**, 2145 (1999).
[60] T. Hosokai *et al.*, *Opt. Lett.* **25**, 10 (2000).
[61] D. J. Spence and S. M. Hooker, *Phys. Rev. E* **63**, 015401(R) (2000); D. J. Spence *et al.*, *J. Phys. B* **43**, 4103 (2001); A. Butler *et al.*, *Phys. Rev. Lett.* **89**, 185003 (2002).
[62] S. Steinke *et al.*, *Nature* **530**, 190 (2016).
[63] T. Hoskai *et al.*, *Appl. Phys. Lett.* **96**, 121501 (2010); N. Nakanii *et al.*, *Phys. Rev. ST Accel. Beams* **18**, 021303 (2015).
[64] F. Dorchies *et al.*, *Phys. Plasmas* **6**, 2903 (1999).
[65] H. Kotaki *et al.*, *Phys. Plasmas* **11**, 3296 (2004); J. Faure *et al.*, *Nature* **444**, 737 (2006).
[66] S. Bulanov *et al.*, *Phys. Rev. E* **58**, R5257 (1998); T. Hosokai *et al.*, *Phys. Rev. E* **67**, 036407 (2003); K. Schmidt *et al.*, *Phys. Rev. ST Accel. Beams* **13**, 093101 (2010).
[67] A. Pak *et al.*, *Phys. Rev. Lett.* **104**, 025003 (2010); C. E. Clayton *et al.*, *Phys. Rev. Lett.* **105**, 105003 (2010); J. S. Liu *et al.*, *Phys. Rev. Lett.* **107**, 035001 (2011).
[68] W. Leemans *et al.*, *Phys. Rev. Lett.* **91**, 074802 (2003).
[69] S. Kneip *et al.*, *Nat. Phys.* **6**, 980 (2010).
[70] S. Chen *et al.*, *Phys. Rev. Lett.* **110**, 155003 (2013).
[71] F. Grüner *et al.*, *Appl. Phys. B* **86**, 431 (2007); K. Nakajima, *Nat. Phys.* **4**, 92 (2008).
[72] Y. S. ImPACT. http://www.jst.go.jp/impact/en/program/03.html.
[73] C. B. Schroeder *et al.*, *Phys. Rev. ST Accel. Beams* **13**, 101301 (2010); K. Nakajima *et al.*, *Phys. Rev. ST Accel. Beams* **14**, 091301 (2011).
[74] C. Yeboah, G. A. Sandison and V. Moskvin, *Phys. Med. Biol.* **47**, 1285 (2002); T. Fuchs *et al.*, *Phys. Med. Biol.* **54**, 3315 (2009).

[75] K. Kinoshita *et al.*, *Laser Phys.* **16**, 660 (2006).
[76] C. DesRosiers *et al.*, *Phys. Med. Biol.* **45**, 1781 (2000).
[77] G. Mourou *et al.*, *Nat. Photon.* **7**, 258 (2013).
[78] S. Gitomer *et al.*, *Phys. Fluids* **29**, 2679 (1986).
[79] G. A. Mourou *et al.*, *Phys. Today* **51**, 22 (1998).
[80] L. Romagnani *et al.*, *Phys. Rev. Lett.* **95**, 195001 (2005).
[81] H. Daido, M. Nishiuchi and A. Pirozhkov, *Rep. Prog. Phys.* **75**, 056401 (2012).
[82] A. Macchi, M. Borghesi and M. Passoni, *Rev. Mod. Phys.* **85**, 751 (2013).
[83] A. V. Gurevich *et al.*, *Sov. Phys. JETP* **22**, 449 (1966).
[84] A. V. Gurevich *et al.*, *Sov. Phys. JETP* **36**, 274 (1973).
[85] J. E. Crow *et al.*, *J. Plasma Phys.* **14**, 65 (1975).
[86] A. V. Gurevich and A. P. Meshcherkin, *Sov. Phys. JETP* **53**, 937 (1981).
[87] D. Dorozhkina and V. Semenov, *Phys. Rev. Lett.* **81**, 2691 (1998).
[88] S. Hatchett *et al.*, *Phys. Plasmas* **7**, 2076 (2000).
[89] P. Mora, *Phys. Rev. Lett.* **90**, 185002 (2003).
[90] P. Mora, *Phys. Rev. E* **72**, 056401 (2005).
[91] M. Murakami and M. Basko, *Phys. Plasmas* **13**, 012105 (2006).
[92] J. Fuchs *et al.*, *Nat. Phys.* **2**, 48 (2006).
[93] M. Nishiuchi *et al.*, *Phys. Lett. A* **357**, 339 (2006).
[94] M. Passoni and M. Lontano, *Laser Part. Beams* **22**, 163 (2004).
[95] M. Lontano and M. Passoni, *Phys. Plasmas* **13**, 042102 (2006).
[96] M. Passoni and M. Lontano, *Phys. Rev. Lett.* **101**, 115001 (2008).
[97] M. Passoni *et al.*, *New J. Phys.* **12**, 045012 (2010).
[98] M. Nishiuchi, *Jpn. Soc. Radiat. Chem.* **97**, 13 (2014), in Japanese.
[99] J. Itatani *et al.*, *Opt. Commun.* **148**, 70 (1998).
[100] S. Fourmaux *et al.*, *Opt. Exp.* **19**, 8486 (2011).
[101] M. Kalashnikov *et al.*, *Opt. Lett.* **30**, 923 (2005).
[102] H. Kiriyama *et al.*, *Opt. Lett.* **37**, 3363 (2012).
[103] G. Petrov *et al.*, *Opt. Lett.* **26**, 355 (2001).
[104] V. Chvykov *et al.*, *Opt. Lett.* **31**, 1456 (2006).
[105] F. Druon *et al.*, *Opt. Lett.* **23**, 10435 (1998).
[106] G.-Y. Yoon *et al.*, *Appl. Phys. Lett.* **72**, 27857 (1998).
[107] M. D. Perry *et al.*, *Opt. Lett.* **24**, 1602 (1999).
[108] H. Baumhacker *et al.*, *Opt. Lett.* **27**, 15702 (2002).
[109] S. W. Bahk *et al.*, *Opt. Lett.* **29**, 28379 (2004).
[110] S. W. Bahk *et al.*, *Appl. Phys. B* **80**, 82332 (2005).
[111] C. N. Danson *et al.*, *Nucl. Fus.* **S239**, 46 (2004).
[112] A. Mackinnon *et al.*, *Phys. Rev. Lett.* **88**, 215006 (2002).
[113] Y. Sentoku *et al.*, *Phys. Plasmas* **10**, 2009 (2003).
[114] F. Wagner *et al.*, *Phys. Rev. Lett.* **116**, 205002 (2016).
[115] I. J. Kim *et al.*, *Phys. Plasmas* **23**, 070701 (2016).
[116] K. Ogura *et al.*, *Opt. Lett.* **37**, 2868 (2012).
[117] Y. Fukuda *et al.*, *Phys. Rev. Lett.* **103**, 165002 (2009).
[118] Y. Fukuda *et al.*, *Radiat. Meas.* **50**, 92 (2013).
[119] Y. Fukuda *et al.*, *Proc. SPIE* **8779**, 87790F (2013).
[120] M. Passoni *et al.*, *Phys. Rev. ST Accel. Beams* **19**, 061301 (2016).
[121] T. Nakamura *et al.*, *Phys. Rev. Lett.* **105**, 135002 (2010).
[122] J. Bin *et al.*, *Phys. Rev. Lett.* **115**, 064801 (2015).
[123] A. Yogo *et al.*, *Appl. Phys. Lett.* **94**, 181502 (2009).
[124] J. Bin *et al.*, *Appl. Phys. Lett.* **101**, 243701 (2012).
[125] S. Kar *et al.*, *Nat. Commun.* **7**, 10792 (2016).
[126] U. Masood *et al.*, *Appl. Phys. B* **117**, 41 (2014).
[127] T. Burris-Mog *et al.*, *Phys. Rev. ST Accel. Beams* **14**, 121301 (2011).

Mitsuru Uesaka is a professor of the Nuclear Professional School, University of Tokyo. He obtained his doctoral degree at the Nuclear Engineering Department, University of Tokyo, in 1980. He also studied at the Graduate School of Nuclear Engineering Department, University of Wisconsin–Madison, in 1982–1983. He joined IHI Co. Ltd. in 1985. He became a professor in 1999. Dr. Uesaka was the chair of the ICFA (International Committee for Future Accelerators) panel on advanced and novel accelerators in 2007–2013. His researches focus on compact electron linear accelerator development for cancer therapy and nondestructive testing, laser dielectric accelerators for radiation biology, and medical physics.

Kazuyoshi Koyama is a researcher of KEK. He obtained his doctoral degree at the Nuclear Engineering Department, Tohoku University, in 1976. He worked on the problems of the nuclear fusion and the laser plasma accelerator at ETL and AIST in 1975–2008. He stayed at MPQ in 1983–1984. He was a guest professor of Ustunomiya University in 1997–2008. After leaving AIST, Dr. Koyama was assigned as a project professor of the University of Tokyo in 2008–2012. He has been working at KEK and the University of Tokyo since 2013. His researches focus on laser-driven particle accelerators using plasmas and dielectrics.

Reviews of Accelerator Science and Technology
Vol. 9 (2016) 261–263

DOI: 10.1142/S1793626816800011

Book Review

Engines of Discovery: A Century of Particle Accelerators (Revised and Expanded Edition, by Andrew Sessler and Edmund Wilson, World Scientific, 2007)

William A. Barletta
US Particle Accelerator School, Fermilab,
Department of Physics, MIT, Cambridge, MA, USA
barletta@mit.edu

In the first edition of *Engines of Discovery.* Andrew Sessler and Edmund Wilson presented their compelling case for considering particle accelerators to be the most readily recognizable characteristic of "big science." That book, written in highly readable language, aimed at illuminating the rich and expanding dimensions of an intellectual enterprise that remains far too little known to the general public and even to many academics outside of the scientific fields dominated by accelerators as their most consequential single research tool.

That lack of familiarity is not surprising, as courses about accelerator science and engineering are found on few university campuses, especially in the US. Even if most accelerator scientists can be faulted for paying too little attention for too long to public outreach, Sessler and Wilson have certainly done their part to describe what some of their colleagues boastfully call the "cathedrals of the 20th century."

Since the publication of the first edition, accelerator scientists and engineers have delivered first-of-a-kind advances in research machines to advance the frontiers of physics, chemistry, and biological sciences. For chemists, structural biologists, and materials scientists, the FLASH FEL suggested the potential of a new generation of light source; the world's first hard X-ray FEL user facility was commissioned at SLAC; the first seeded-FEL user facility showed promise of fully phase-coherent soft X-rays at Sincrotrone Trieste. These machines were quickly followed by several large FELs, already completed or still under construction in the US, Europe and Asia. Sessler is too modest in omitting his own very large contribution to the physics of the FEL, namely the first experimental demonstration of high gain operation.

Yet, it was a giant machine for high energy physics that most caught the imagination of the general public and scientists alike. As expected, the operation of the world's largest accelerator, the Large Hadron Collider, which delivered the discovery of the Higgs boson, which completed the Standard Model of the electroweak interaction. The work of accelerator physicists and engineers that preceded these momentous achievements — FELs and hadron colliders — is well-represented in Chaps. 5 and 9, respectively.

The second edition of *Engines of Discovery* does not tamper with the features which made the first edition so attractive: largely self-contained chapters, each of which offers a compelling tale of great insight and invention in the creation of a species of accelerator — linac, cyclotron, synchrotron, collider, light source, neutron spallation source. The authors have chosen lavish illustrations of these stories to bring the reader back in time to share the adventure of technological discovery with the pioneers of accelerator science. Sessler and Wilson do not neglect to remind us of the human side of the story of accelerators, complete with its high hurdles and its eureka moments. The lives of these pioneers — most of whom the authors have known personally — are appealingly yet succinctly described in the numerous sidebar biographies that accompany the words and pictures of their own journeys of discovery.

However, the authors avoid being Pollyannas in their description of the large enterprises of their field. A case in point is the attempt by Ernest Lawrence and Luis Alvarez to build a gigantic, low-frequency RF accelerator, the Material Test Accelerator (MTA), to breed plutonium at the Lawrence Livermore Laboratory, founded by Edward Teller. A sidebar describes the painful period at the Berkeley Rad Lab that accompanied the building of the MTA, a period that coincided with the ostracizing of Teller on the Berkeley campus for his role in the removal of Robert Oppenheimer's security clearance.

The authors have omitted showing a plot of accelerator energy reach energy versus time (Livingston chart) which is usually considered *de rigueur* in histories of the development of accelerator technology. In its place, they have characterized the succession of physical phenomena that temporarily limited the energy reach of particle accelerators until new physics and engineering insights overcame the hurdle to open the path toward new classes of machines. Citing these insights and inventions graphically alongside the accelerator advances and scientific discoveries they enabled would have made for a modified Livingston chart which emphasizes that the history of accelerators is the history of invention.

Individual investigators alone could not have achieved the advances in accelerator performance toward ever-more-challenging beam characteristics. Accelerator building is a "team sport" that generally requires extensive research facilities. Large research laboratories throughout the world have provided the technological and engineering infrastructures that have enabled the heroes of accelerator science to flourish. These laboratories, which have been the homes of machines of ever-higher energy, ever-higher intensity, and often ever-larger size, receive appropriate attention from the authors. As the director of the Radiation Laboratory at Berkeley at the pinnacle of its reputation as a grand center of accelerator science, Sessler was well equipped to convey the intellectual ferment and excitement that have made these national and international laboratories transformative research centers for scientists and engineers to spend their careers.

The second edition of *Engines of Discovery* expands the attention paid beyond the realm of discovery science. With annual sales of accelerators for use in commerce and in medical applications approaching US$5 billion, accelerator technology touches the manufacture of goods with an estimated annual value quoted to be in excess of US$100 billion. The industrial use of small accelerators is essential to modern information and communications technology. Tools for industrial processes include ion implanters, nondestructive test and inspection equipment, polymerization equipment, and facilities for sterilization and disinfestation. Although many think about giant research tools when they hear the word "accelerator," these much smaller devices constitute the overwhelming majority of the world's $\sim$30,000 accelerators. In the first edition Sessler and Wilson had woven the commercial uses of the several classes of accelerators throughout the text. The industrial applications now have a dedicated chapter (XIV) that summarizes the myriad ways ion and electron beams are used to produce a surprisingly wide array of everyday products, from plastic wrap, to automobile tires, to microelectronics, to the ink used on many packages of consumer products. The photograph of an electron beam facility for producing shrink-wrap looks remarkably similar to the klystron gallery of a high-energy linac for discovery science.

A chapter concerning applications for energy and the environment informs the reader about three ways in which accelerators might be used for very-large-scale generation of electrical power. The practicality of any of these approaches is open to debate. Unfortunately, the authors neglect the innovative ways in which accelerator technology has been used at the pilot plant scale to clean the environment.

For their colleagues in the field, the biographical sidebars may be the most intriguing and delightful aspect of this book, even as they are also the most likely to generate numerous nit-picks. This reader[a] found the most significant omission in the first edition to be a description of the role played by Mel Month in founding the US Particle Accelerator School.

Month's persistence and the growth of his vision is well worth noting. In the preface to his book, *Accelerator X-Ray Sources*, Prof. Richard Talman of Cornell writes that the "entire scholarly tradition in US accelerator physics, modest as it is, is due more to Mel Month and his US Particle Accelerator School than to anyone else." While the second edition continues to omit a biographical sidebar on Month, it does offer a sidebar on accelerator schools both in

[a]Disclosure: The author is presently the Director of the USPAS.

the US and at CERN. Happily, this practice of regularly offered, local accelerator schools has now been taken up also in South Korea and Mexico.

The late Andy Sessler was a mentor, a collaborator, and an inspiration to this reader. His collaboration with Ted Wilson is a reminder of the *joie de vivre* that Sessler brought to his work. *Engines of Discovery* will appeal to a broad audience. It offers the general public a fascinating view of a technology that has shaped modern science. For their colleagues involved in accelerator science and technology, Sessler and Wilson offer a happy walk along their own path of discovery in the field that enriches their loves. This book is well worth owning, reading, and rereading.

William A. Barletta is Director of the US Particle Accelerator School at Fermilab. He is also Adjunct Professor of Physics at MIT, UCLA, and Old Dominion University, and Director Emeritus of the Accelerator Division at LBNL, where he was a frequent collaborator with Andy Sessler. In Europe he is Visiting Professor, Faculty of Economics, University of Ljubljana. Prof. Barletta is the coordinating editor-in-chief of *Nuclear Instruments & Methods — Section A*.

Reviews of Accelerator Science and Technology
Vol. 9 (2016) 265–312

DOI: 10.1142/S1793626816300127

Particle Accelerators in China

Chuang Zhang* and Shouxian Fang†
Institute of High Energy Physics, Chinese Academy of Sciences,
Beijing 100049, P. R. China
**Zhangc@ihep.ac.cn*
†Fangsx@ihep.ac.cn

As the special machines that can accelerate charged particle beams to high energy by using electromagnetic fields, particle accelerators have been widely applied in scientific research and various areas of society. The development of particle accelerators in China started in the early 1950s. After a brief review of the history of accelerators, this article describes in the following sections: particle colliders, heavy-ion accelerators, high-intensity proton accelerators, accelerator-based light sources, pulsed power accelerators, small scale accelerators, accelerators for applications, accelerator technology development and advanced accelerator concepts. The prospects of particle accelerators in China are also presented.

Keywords: Particle accelerator; collider; synchrotron radiation source; free electron laser; application.

1. Brief History

The development of particle accelerators in China can be traced back to the early 1950s [1–3]. The history of the past 60 years can be categorized into three periods: the period of initial development, from the early 1950s to the mid-1960s; the period of self-development, from the mid-1960s to the late 1970s; and the period of rapid development, from the late 1970s up to now.

1.1. *The period of initial development*

This period started from the early 1950s, when a number of scientists returned from Western counties to China and threw themselves into scientific research. In the meantime, teams of students and researchers were sent to the former Soviet Union to learn modern science and technology, of which nuclear physics (including its instruments) was one of the main subjects. In about 15 years, some low energy accelerators were constructed and design study of high energy accelerators was also carried out.

In May 1950, the Institute of Modern Physics under the Chinese Academy of Sciences (CAS) was founded. It was renamed the Institute of Physics (IoP) in 1953 and later renamed the China Institute of Atomic Energy (CIAE) in 1958. Design and construction of particle accelerators for nuclear study was one of the main tasks of the institute.

China's first accelerator, a 700 keV electrostatic proton accelerator (Fig. 1, left), was constructed in 1955 by a team led by Prof. Zhongyao Zhao of IoP [4]. In 1957, a 2.5 MeV electrostatic proton accelerator named V2 was completed in 1958 at IoP (Fig. 1, right).

The Lanzhou Division of Physics, the predecessor of the Institute of Modern Physics (IMP) at Lanzhou, was set up in 1957. In the second year, a Cockcroft–Walton accelerator was constructed at IMP; it could accelerate D^+ beams to 400 keV and 150 μA. In the summer of 1958, a 1.2 m cyclotron imported from the former Soviet Union started operation at CIAE. Two 25 MeV betatrons, also imported from the former Soviet Union, were operated at Peking University (PKU) and Tsinghua University (THU) in 1958.

The first heavy-ion cyclotron in China, a 1.5 m cyclotron, was completed and commissioned at IMP in June 1963. The first China-made cyclotron, a 1.2 m cyclotron was completed at the Shanghai Institute of Nuclear Research (SINR), the predecessor of the Shanghai Institute of Applied Physics (SINAP), in 1964. It was updated to a sector-focusing cyclotron of proton energy 10–30 MeV in 1984 (see Fig. 2) [2].

Fig. 1. The 700 keV electrostatic accelerator (*left*) and the 2.5 MeV electrostatic proton accelerator V2 (*right*).

Fig. 2. The first China-made, 1.2 m cyclotron (*left*) and its control room (*right*).

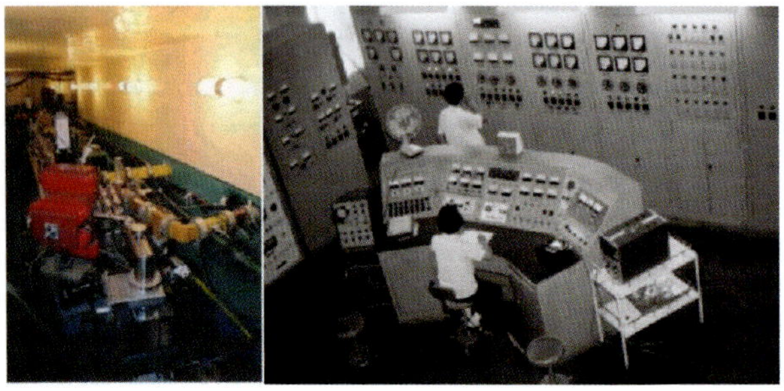

Fig. 3. The 30 MeV electron linac (*left*) and its control room (*right*).

China's first electron linear accelerator, a 30 MeV electron linac (Fig. 3), was constructed by a team led by Prof. Jialin Xie of CIAE in 1964 [1].

Regarding high energy accelerators, the Chinese government started to map out China's experimental high energy physics base in the late 1950s. In 1957, a team led by Prof. Ganchang Wang was sent to the Lebedev Physical Institute, the Joint Institute of Nuclear Research and related institutes of the former Soviet Union for advanced study. Several accelerator projects were proposed and designed during the period from the 1960s to the 1970s. However, due to the economic difficulty in China, all these projects were aborted.

1.2. *The period of self-development*

The collaboration between China and the former Soviet Union was interrupted due to the deterioration of Sino-Soviet relations, while the Cultural Revolution (1966–1976) led to the self-development of particle accelerators being isolated from the world. In this period, the application of particle accelerators in nuclear physics and the national economy was given priority, and a number of low energy accelerators were produced.

By the late 1960s, more than 50 low energy accelerators of various types, such as Cockcroft–Walton accelerators, electrostatic accelerators, electron linacs and microtrons, had been produced by research institutes as well as industry. Ion implanters with beam energy of 20–300 kV were commercially produced and widely used in the semiconduction industry and in material modification.

With the development of electron linacs and their key components, such as klystrons, modulators and accelerating structures, the application of particle accelerators was gradually promoted in China. In 1977, the first two medical electron linacs, BJ-10 of 5–10 MeV (see Fig. 4) and ZJ-10 of 6–10 MeV, were developed by a Beijing medical linac collaboration [Beijing Medical Equipment Institute (BMEI), THU, Beijing Cancer Hospital, etc.] and a Shanghai medical linac collaboration (Shanghai Nuclear Equipment Factory, IHEP, Shanghai Medical Equipment Institute, etc.), respectively [2].

The first irradiation electron linac, BF-5 (see Fig. 5), was developed by the collaboration of THU, BMEI and Beijing Normal University in 1979 [2].

In February 1973, the Institute of High Energy Physics (IHEP) under CAS was founded based on Part 1 of CIAE. Soon after its establishment, IHEP started to work out the design of high energy accelerators and the relevant R&D programs. There were many discussions on the design of high energy accelerators, but none of them was put into construction.

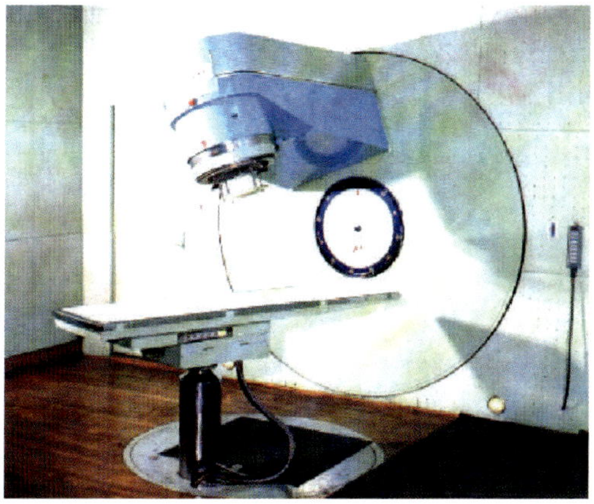

Fig. 4. The first traveling wave medical electron linac BJ-10.

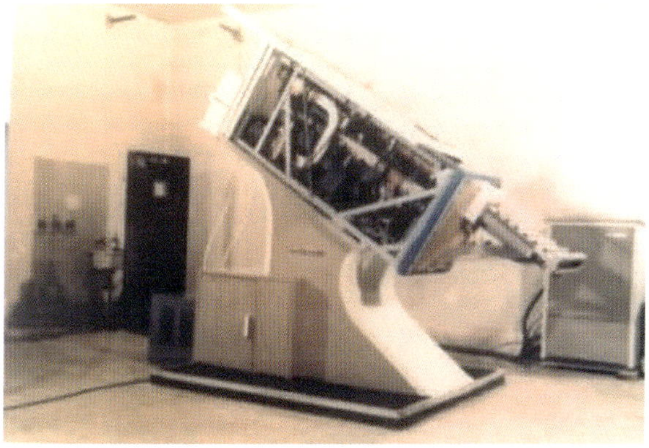

Fig. 5. The first irradiation electron linac, BF-5.

1.3. *The period of rapid development*

The end of the Cultural Revolution and the reform and opening-up of China marked the turning point of China's accelerator development. Many scholars were sent to the accelerator laboratories around the world to learn accelerator physics and technology, and collaboration with institutions of Western countries was vigorously promoted. In the meantime, several accelerator projects were proposed. Three big accelerator projects, namely the Beijing Electron–Positron Collider (BEPC), the Heavy-Ion Research Facility at Lanzhou (HIRFL) and the Hefei Light Source (HLS), were placed in China's eighth five-year plan, and thus started the rapid development period of accelerators in China [5].

In 1975, a plan of design and R&D for a 30 GeV proton synchrotron, Project 753, was approved by the Chinese government. According the plan of the project, an R&D base for high energy accelerators was built on the IHEP campus. It was not until the end of the Cultural Revolution, when another plan called Project 87, aiming at the construction of a high energy physics experimental base, was launched, in 1977. The 50 GeV Beijing Proton Synchrotron (BPS) was its first step [2, 5].

However, Project 87 was suspended in 1980, due to the overall economic situation of the country. After a long and intensive discussion, the construction of BEPC was approved by the Chinese government in December 1983. In the same period, HIRFL at IMP and HLS on the campus of the University of Science of Technology of China (USTC) were also approved by the government.

The construction of BEPC started in October 1984 and was completed in October 1988. In December 1988, HIRFL was completed and it began operation. The construction of HLS was started in November 1984 and completed in 1989 [5].

Several smaller size accelerators were also constructed in that period. The 35 MeV Beijing Proton Linac (BPL), shown in Fig. 6, as a part of the BPS injector, was completed in 1985 at IHEP [6]. BPL operated for nuclear physics study, fast neutron therapy and so on until 2000.

In November 1986, a 4.5 MV electrostatic accelerator was completed at PKU (Fig. 7, left). The 2×6 MeV tandem was jointly constructed in SINR in 1988 (Fig. 7, right) [7].

China's first pulsed power accelerator, Chenguang (see Fig. 8), was jointly constructed by IHEP and the Northwest Institute of Nuclear Technology (NINT) in 1976, and could accelerate electron beams to an energy of 1 MeV with a current of 20 kA and a pulse length of 25 ns. More accelerators of this

Fig. 6. The 35 MeV Beijing Proton Linac.

Fig. 7. The 4.5 MV electrostatic accelerator (*left*) and the 2×6 MeV tandem (*right*).

Fig. 8. The 20 kA, 0.5 MV pulsed accelerator Chenguang.

type were built in the 1980s, such as Flash-I of 100 kA and 8 MV at the China Academy of Engineering Physics (CAEP) and Shanguang-II of 1 MA and 1.5 MV at NINT. China's first induction linac was constructed in 1993 at CAEP. This machine can accelerate 2.3 kA, 60 ns electron beams to 10–12 MeV.

The Beijing Free Electron Laser (BFEL), China's first FEL facility based on the 30 MeV electron linac (shown in Fig. 3), was constructed in early 1990 and achieved saturated lasing in 1993 [8].

Some small scale accelerators were developed in that period. As an example, a heavy-ion integral split ring RFQ was constructed at PKU, in which positive and negative oxygen ion beams can be accelerated simultaneously to 1 MeV [9]. The structure of the RFQ is shown in Fig. 9.

The application of accelerators was also promoted in that period. In 1996, the accelerator mass spectrometry (AMS) program was carried out in PKU using its 6 MeV tandem. The accuracy of the AMS measurement reached 3×10^{-3} for ^{14}C with a sensitivity of 10^{-15} [10]. In the late 1990s, cargo inspection systems for customs applications were

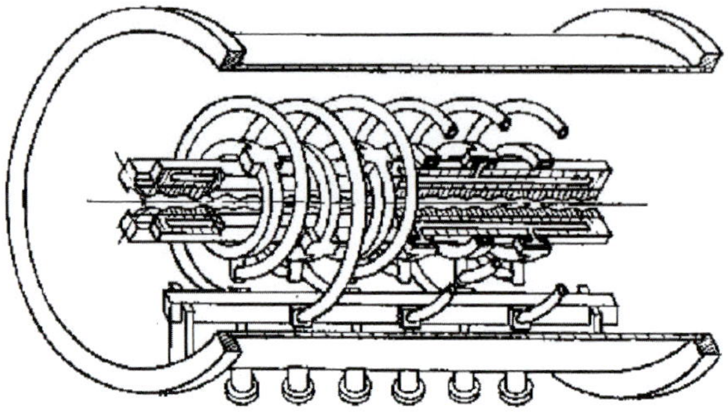

Fig. 9. View of the integral separated ring RFQ.

developed at THU, based on three types of electron linacs: S-band 9 MeV traveling wave linacs, S-band 6 MeV standing wave linacs and X-band 2.5 MeV standing wave linacs [11]. In the meantime, various types of high voltage accelerators, such as Dynamitron, high voltage transformer accelerators, medium frequency transformer accelerators, insulating core transformer accelerators, betatrons, and electron curtain accelerators, were produced by industry for irradiation.

Since the beginning of the new century, the development of China's particle accelerators has further speeded up. The construction of the cooling storage ring HIRFL-CSR started at IMP in 1998 and was completed in 2005. The major upgrade of BEPC, BEPCII, started in 2004 and was completed in 2009. The construction of the Shanghai Synchrotron Radiation Facility (SSRF) started in 2004 and was completed in 2009. In the meantime, an upgrading project of HLS was carried out. The construction of the Beijing Radioactive Ion-Beam Facility (BRIF) started in 2011 and was completed in 2015. A study of the accelerator-driven subcritical system (ADS) was intensively performed in that period. The construction of the China Spallation Neutron Source (CSNS) was started in 2011 and is scheduled to be completed in 2017. The project of the Heavy-Ion Accelerator Facility (HIAF) was approved by the central government at the end of 2015. An X-ray free electron laser test facility, SXFEL, began construction in 2014. An R&D project for a 6 GeV synchrotron radiation (SR) source, the High Energy Photon Source (HEPS), is in progress and its construction is expected in 2018. The Circular Electron–Positron Collider, CEPC, and the future Super Proton–Proton Collider, SPPC, have been also proposed. In the meantime, small-scale accelerators

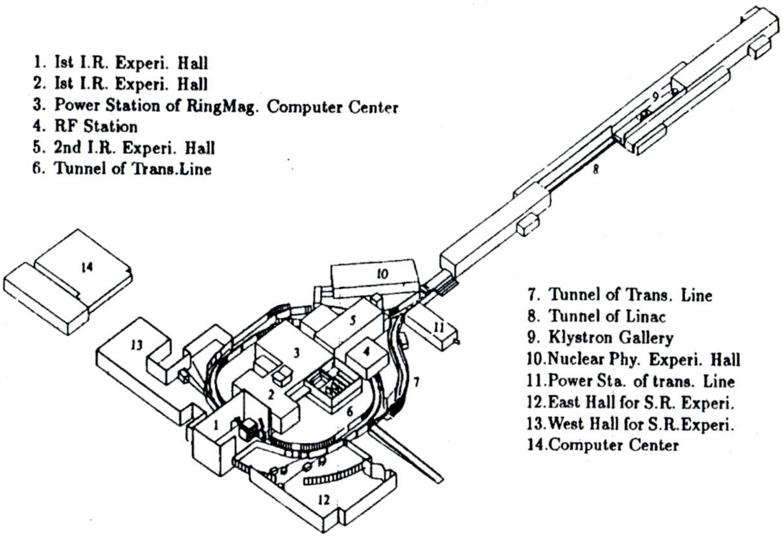

Fig. 10. Schematic layout of BEPC.

for research and application have also continued to thrive [12, 13].

In the following sections, particle colliders, heavy-ion accelerators, high-intensity accelerators, accelerator-based light sources, pulsed power accelerators, small scale accelerators, accelerators for applications, accelerator technology development and advanced acceleration concepts in China will be described in sequence.

2. Particle Colliders

Particle beam colliders have been developed to provide higher center of mass energies E_{cm} for high energy physics research. There are two frontiers in colliders, the high luminosity frontier and the high energy frontier. Over the past 30 years, China has concentrated on the high luminosity colliders operating at the τ and charm energy, i.e. BEPC and BEPCII, and it is now dreaming of the high energy frontier machines, CEPC and SPPC.

2.1. *BEPC*

As the first high energy accelerator in China, BEPC was constructed for both high energy physics and SR researches [14]. The BEPC accelerators consisted of a 202 m-long electron–positron linac injector, a storage ring with a circumference of 240.4 m and, connecting them, a 210 m transport line. There were two interaction points in the storage ring. A general purpose detector, the Beijing Spectrometer (BES), was installed in the southern interaction region (IR). The Beijing Synchrotron Radiation Facility (BSRF), which was equipped with four insertion devices, 12 beam lines and experimental stations, flanked the east and west of the southern areas of the storage ring. Figure 10 shows the layout of BEPC.

The BEPC injector was a traveling wave linac. It provided electron and positron beams for the storage ring, with full energy injection from 1.1–1.55 GeV [15]. Figure 11 is a picture the linac installed in its tunnel.

Fig. 11. The BEPC injector linac in the tunnel. The preinjector was composed of a 4 A, 80 keV electron gun, a prebuncher, a buncher and an accelerating structure with solenoid. There were 16 klystron amplifiers, 56 accelerating tubes and 13 energy doublers (SLED) cavities in the linac. Positron beams were produced by a convertors at 150 MeV.

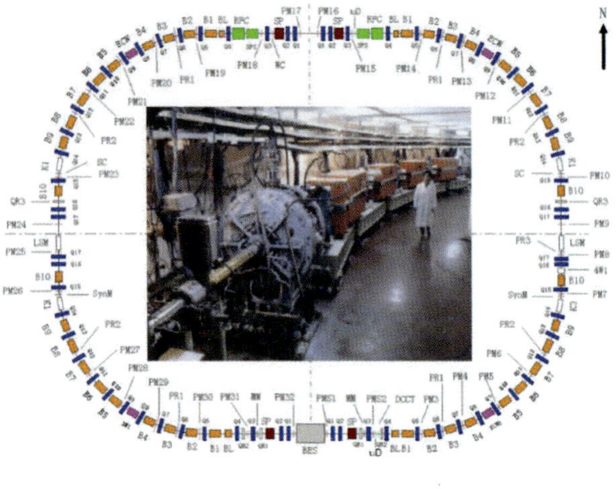

Fig. 12. Schematic layout of the BEPC storage ring.

Electron and positron beams were injected, accumulated and collided in the storage ring. The layout of the BEPC storage ring is shown in Fig. 12.

The BEPC storage ring was a fourfold symmetry, racetrack-shaped machine with an aspect ratio of 82.82 m : 60.02 m. The lattice of each 1/4 ring consisted of a half low beta insertion, an arc and half of an injection section. The arcs were based on the FODO structure. There were 40 bending magnets, 4 low field bending magnets, 60 quadrupoles and 8 insertion quadrupoles in the ring. Four conventional RF cavities were equipped in the dispersion-free sections near the north IP. Table 1 lists the main parameters of BEPC [16–18].

BEPC operated in the τ and charm energy region of 1–2.5 GeV for 17 years, from 1989 to 2004. Beam behaviors were intensively studied, including the lattice and its correction [19], coupling impedance beam instabilities [20, 21], bunch lengthening [22], electron cloud effect [23], ion effect [24], beam-beam effect and luminosity [25]. Figure 13 shows the bunch lengthening measured in BEPC.

Table 1. Main parameters of BEPC.

Operation energy (E)	GeV	1.0–2.5
Injection energy (E_{inj})	GeV	1.3
Circumference (C)	m	240.4
β at IP (β_x^*/β_y^*)	cm	120/5
Tunes ($\nu_x/\nu_y/\nu_z$)		5.8/6.7/0.02 (HEP) 8.72/4.75/0.02 (SR)
Emittance (ε_{x0})	mm · mr	0.4@1.55 GeV (HEP) 0.08@2.2 GeV (SR)
RF (f_{RF})	MHz	199.53
Bunch number (N_b)		1 × 1 (HEP), 60–80 (SR)
Beam current (I_b)	mA	22@1.55 GeV (HEP) 140@2.2 GeV (SR)
Beam–beam param. ξ_y		0.04
Beam lifetime τ	hrs.	8–10 (HEP), 20–30 (SR)
Luminosity	$\text{cm}^{-2}\text{s}^{-1}$	5×10^{30}@1.55 GeV 1×10^{31}@1.89 GeV

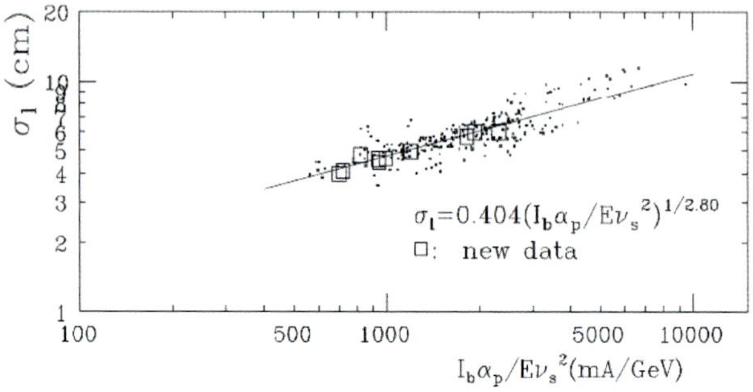

Fig. 13. The bunch lengthening in BEPC.

As a result of its successful operation, the BEPC/BES accurately measured the mass of the τ lepton, carried out on R-scan in the center of mass energy region of 2–5 GeV, and collected the largest data sample of J/Ψ and Ψ' events in the world. A narrow peak was observed near two times the proton energy in the M($p\bar{p}$) distribution for $J/\Psi \to \gamma p\bar{p}$ decays using the BESII 58 million J/Ψ events [26], shown in Fig. 14, although it was still too early to determine whether it was a proton–antiproton bond state or multiple quark state.

The physics opportunity in the charm-τ region called for a collider with significantly higher luminosity. BEPCII, with a luminosity goal of two orders of magnitude higher than for BEPC, was its natural extension.

2.2. *BEPCII*

As with BEPC, the beam energy of BEPCII is 1.0–2.3 GeV, so that its physical potential in the τ and

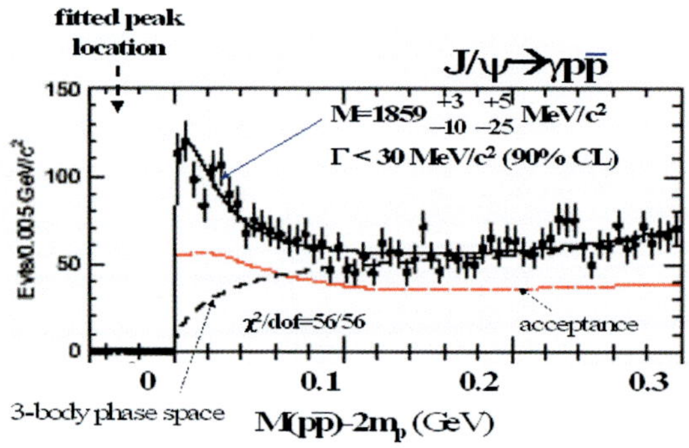

Fig. 14. The new peak observed near 2 M(pp).

Table 2. The luminosity strategy from BEPC to BEPCII.

Parameters	BEPC	BEPCII
Beta function at IP β_y^* (cm)	5.0	1.5
Bunch number k_b	1	93
Beam–beam parameter ξ_y	0.04	0.04
Current per bunch I_b (mA)	35	9.8
Luminosity ratio $L_{\mathrm{BEPCII}}/L_{\mathrm{BEPC}}$	1	~100

charm range is preserved. The collider is optimized at the beam energy of 1.89 GeV, with a design luminosity of $1 \times 10^{33}\,\mathrm{cm^{-2}s^{-1}}$ [27–31].

As a measure of the event production rate, luminosity is one of the most important parameters of colliders. Table 2 describes the strategy of the luminosity upgrade from BEPC to BEPCII.

The multibunch collision scheme in BEPCII requires a double ring structure of the collider. A schematic layout of the BEPCII storage ring is shown in Fig. 15.

The inner ring and the outer ring cross each other in the northern and southern IPs of the BEPCII storage rings. The horizontal crossing angle between two beams at the southern IP, where the detector is located, is 2×11 mrad to meet the requirement of sufficient separation but no significant degradation to the luminosity; while, in the northern crossing region, the two beams cross horizontally and a vertical bump is used to separate the two beams. For the dedicated synchrotron radiation operation of BEPCII, electron beams circulate in the outer ring by means of a pair of horizontal bending coils in the insertion superconducting (SC) magnets in the southern IR and a pair of bending magnets in the northern IR.

The primary challenge of BEPCII is how to deal with effects of the high beam current arising from the required high luminosity. Many key technologies were developed in BEPCII, such as the high intensity electron gun [32], the high efficiency positron convertor [33], SC insertion magnets, the SC RF system [34], bunch-by-bunch feedback [35], the low impedance vacuum chamber, modified strip-type kickers, the septa and two-in-one IR quadrupoles [36]. The west half of the IR during installation of the BESIII detector is shown in Fig. 16.

A number of physical and technical measures were applied to improve the BEPCII performance [37–41]. However, several hardware failures occurred during the high current operation, such as damage to the kicker magnets, SR absorbers, RF couplers, SR monitors, bellows, etc. (shown in Fig. 17).

The designed and reached main parameters of BEPCII are listed in Table 3. As can be seen, the reached parameters of BEPCII agree well with its designed values; while the working points get closer to the half integer, the momentum compaction factor and bunch length are smaller, which is beneficial to luminosity enhancement. The evolution of the BEPCII luminosity since its completion in 2008 is shown in Fig. 18.

Fig. 15. Schematic layout and installed double ring units.

Fig. 16. The BEPCII interaction region, from right to left: the BESIII detector, the SC insertion magnet and valve box, the septum magnet, two-in-one quadupoles Q1a and Q1b, and IR quadupoles Q2(e^-) and Q2(e^+). Electrons move from right to left, and positrons from left to right, and they meet and collide in the center of the detector.

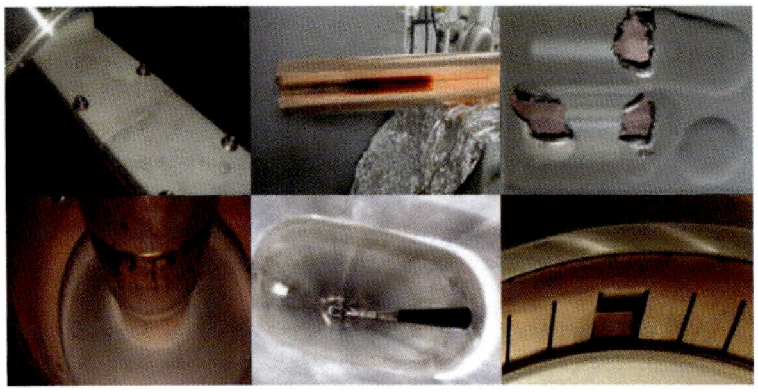

Fig. 17. Some hardware failures, from left to right and up to down: the kicker plate, the SR absorber, the mirror of an SR monitor, the ceramic window of an RF coupler, the connecting bellows and the shielding finger.

Table 3. Main parameters of BEPCII, designed and reached.

Parameters	Unit	Designed	Reached
Optimized energy	GeV	1.89	1.89
RF	MHz	499.8	499.8
Maximum RF voltage	MV	1.5	1.7
Horizontal emittance	nm · rad	144	100
Momentum compaction		0.0235	0.0170
Natural bunch length		1.3	1.15
Working point		6.53/7.58	7.505/5.58
Bunch number		93	90–130
Beam current per ring	mA	910	850
β function at IP	m	1/0.015	1/0.0135
Crossing angle	mrad	11 × 2	11 × 2
Beam–beam parameter		0.04	0.04
Luminosity	$cm^{-2}s^{-1}$	$1.0. \times 10^{33}$	1.0×10^{33}

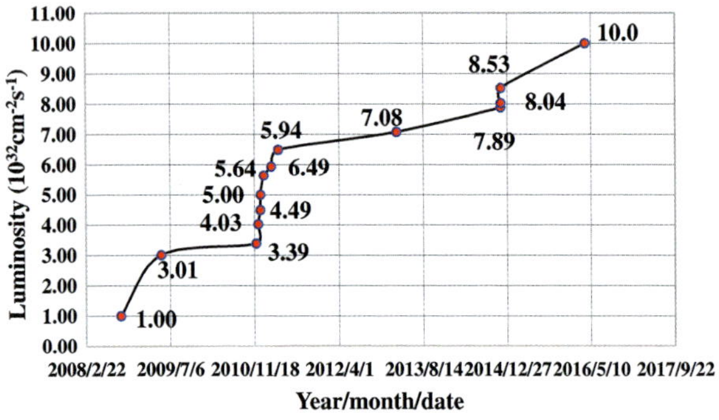

Fig. 18. The evolution of the BEPCII luminosity.

Fruitful experimental results have been obtained by the BESIII collaboration at BEPCII. As an example, a charged charmonium-like structure was observed at a center of mass energy of 4.260 GeV using a $525\,pb^{-1}$ data sample collected with the BESIII detector operating at BEPCII, called Zc(3900) [42] (shown in Fig. 19). The newly

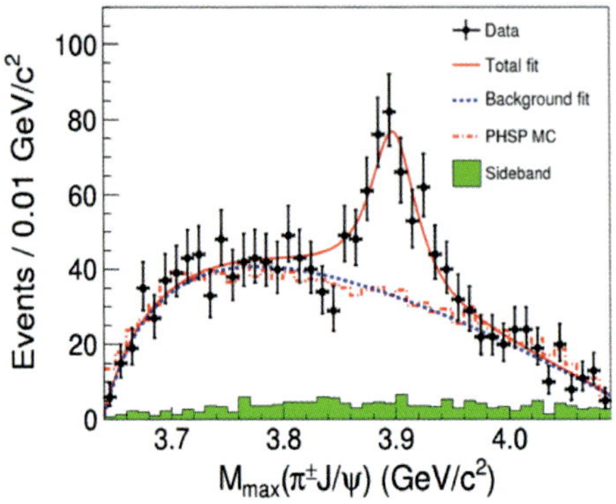

Fig. 19. Observation of Zc(3900).

discovered particle was explained as four-quark combinations. Since then, BESIII has seen a series of other new particles of Zc(4020) and Zc(4025) that appear to contain four quarks.

2.3. *CEPC and SppC*

Soon after the discovery of the Higgs boson at 126 GeV at LHC in 2012, the scheme of CEPC was proposed. SppC could be installed later in the same tunnel as CEPC for pp, ep and ion collisions. In CEPC the electron beam energy is 120 GeV, while in SppC the proton beam energy can reach 25–50 TeV for the different choice of circumference and maximum bending field. CEPC and SppC would be constructed as a global accelerator complex, which might provide a great opportunity to realize collisions of protons or ions with electrons or positrons in an ultra-high center of mass energy range of up to 4.2 TeV. A preliminary conceptual design of CEPC and SppC was carried out [43]. Table 4 gives the main parameters of CEPC and SppC in the preliminary conceptual design stage.

Table 4. The main parameters of CEPC and SppC.

Parameter	Unit	CEPC	SppC
Energy	TeV	0.12	35.6
Circumference	km	54	54
Max. dipole field	T	0.0657	20
Bending radius	km	6.094	5.92
Beam current	mA	16.6	1000
Bunch number		50	5835
SR loss per turn	GeV	3.11	0.0021
SR power per beam	MW	51.7	2.1
β_y^*	mm	1.2	750
Number of IPs		2	2
Luminosity	$cm^{-1}s^{-1}$	2×10^{34}	1.2×10^{35}

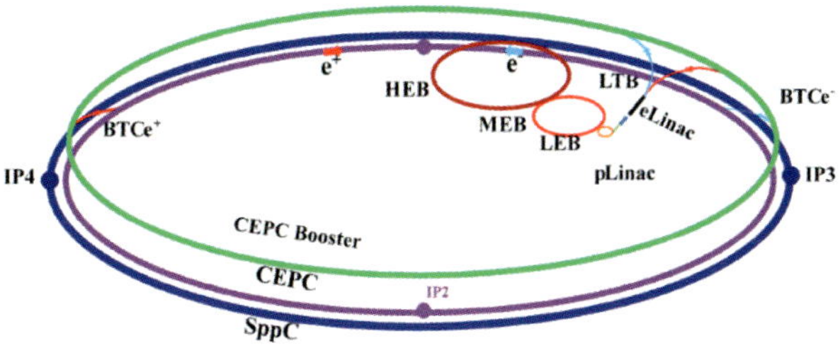

Fig. 20. The accelerator chain of CEPC and the future SppC. Elinac — electron–positron linac; LTB — linac-to-booster transfer line; BTC — booster-to-collider transfer line; plinac: proton linac, LEB — low energy proton booster; MEB — medium energy proton booster; HEB — high energy proton booster.

Figure 20 shows the accelerator chain of CEPC and the future SppC.

A design study of CEPC and SppC is in progress. In the CEPC design, a pretzel scheme is applied in the main ring for the time being, while the double ring and partial double ring schemes are also being considered. Intensive studies of the lattice optimization, dynamic aperture issues, collective effects, beam–beam simulation, beamstrahlung, energy sawtooth effect, lowfield issues of the booster injector, and key technology R&D have been carried out by the design team [44–51].

The proposed CEPC and SppC are going to be a long term program. There is still a long way to go before the dream comes true.

3. Heavy-Ion Accelerators

As described in Sec. 1, heavy-ion accelerators have been developed in China since the 1950s. Many low energy machines, including high voltage accelerators, cyclotrons and RFQs were constructed. This section covers HIRFL and its cooling rings, the HI-13 tandem and its upgrade, and the planned projects HIAF and Beijing-ISOL.

3.1. *HIRFL and CSR*

HIRFL is a major heavy-ion facility in China. It focuses on nuclear physics, atomic physics, heavy-ion related application and interdisciplinary research.

Constructed in the 1980s, HIRFL is a cyclotron system, consisting of a three 33° spiral-sector focusing cyclotron (SFC) with energy constant $K = 69$ as the injector and a four 52° sector separated cyclotron

Fig. 21. The main accelerator (SSC) of HIRFL. SSC is a separated sector isochronous cyclotron. It consists of four sector magnets with a maximum field of 1.6 T. The injection and extraction mean radius of SSC are 1.0 m and 3.21 m, respectively. The RF is 6.5–14 MHz, and the voltage 100–250 kV. The total volume of the vacuum chamber is 100 m^3.

(SSC) of constant $K = 450$ as the main accelerator [52, 53]. Figure 21 shows the main accelerator (SSC) of HIRFL.

The first beam was accelerated in and extracted from SSC by the end of 1988. Many nuclear physics experiments were performed at the sites near SFC and SSC. Some upgrades were carried out in the 1990s to improve the beam intensity, beam quality, and efficiency of the operating cyclotron system and to extend the experimental capability. The major upgrade of HIRFL was performed together with construction of the new project of the cooling storage ring, HIRFL-CSR, in the early 21st century.

HIRFL-CSR is a postaccelerating system of HIRFL. It consists of a main cooling storage ring (CSRm) and an experimental cooling storage ring (CSRe), using the cyclotrons of SFC and SSC as the injector chain [54].

The overall layout of the HIRFL-CSR complex is schematically shown in Fig. 22. In HIRFL-CSR, heavy-ion beams with an energy range of 2–25 MeV/u are injected from the cyclotron system into CSRm, where the beams are accumulated, cooled, accelerated to an energy of 2–25 MeV/u, and then extracted to produce radioactive ion beams (RIBs) or highly charged heavy-ion beams. Those secondary beams will be accepted and stored by CSRe for internal target experiments with the electron cooling system. The beams with an energy range of 100–600 MeV/u can also be extracted from CSRm

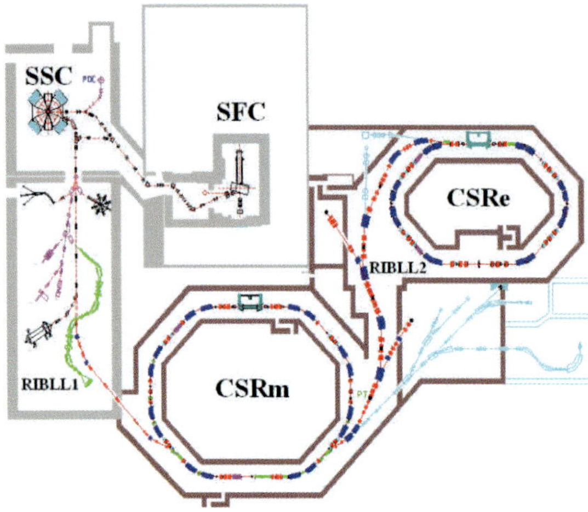

Fig. 22. Overall layout of the HIRFL-CSR complex.

Fig. 23. CSR in its tunnel, CSRm (above) and CSRe (below).

Table 5. The main parameters of HIRFL-CSR.

Parameter	CSRm	CSRe
Circumference (m)	161.0	128.8
Ion species	P-U	P-U
Max. energy (MeV/u)	1100(C^{6+})	500 (U^{90+})
$B\rho_{max}$ (T · m)	12.05	9.40
B_{max} (T)	1.6	1.6
Ramping rate (T/s)	0.1–0.4	0.1–0.2
f_{rf}(MHz)	0.24–1.81	0.4–2.0
Harmonic number	1, 2, 4	1, 2
V_{RF} (kV)	1 × 7.0	2 × 8.0
Vacuum pressure (mbar)	3.0×10^{-11}	3.0×10^{-11}
E-cooler		
Ion energy (MeV/u)	3–50	10–500
Voltage (kV)	2–35	50–300
Length (m)	4.0	4.0

Table 6. The supplied beam parameters in operation.

Beam	$^{12}C^{6+}$	$^{36}Ar^{18+}$	$^{78}Ar^{28+}$	$^{54}Xe^{27+}$
Injector	SFC	SFC + SSC	SFC	SFC
Accumulation	charge exchange	multi-turn	multi-turn	multi-turn
Energy (MeV/u)	600	368–500	300–500	197–235
Extraction	Slow/fast	Fast	Fast	Fast
Intensity (ppp)	2×10^8/ 7×10^9	4×10^8	2×10^8	1×10^8
Exp. terminal	therapy/N Wall	CSRm/ CSRe	CSRe	CSRe

using slow extraction for external target experiments or cancer therapy treatment. Figure 23 shows CSRm in its tunnel.

The construction of HIRFL-CSR was completed in 2005, and its commissioning was performed from 2006 to 2007. Based on the success of the stripping injection, the electron cooling, the multiple multi-turn injection, the varying-harmonic ramping and the cooling stacking and the combination between cooling-stacking, the beam ramping to its final energy and extracting from the ring was realized in 2007 [55–60]. The main parameters of HIRFL-CSR are given in Table 5.

The facility began operation at the end of 2007. During its operation in recent years, HIRFL-CSR supplied heavy-ion beams for experiments at several terminals of both CSRm and CSRe. The experiments have covered high resolution mass measurements, cancer therapy, neutron wall experiments, atomic physics studies using the electron target, and internal gas target experiments using injection beams from the cyclotron injector SFC [61–63]. The beam parameters of HIRFL-CSR operation are listed in Table 6.

Many experimental results have been obtained from HIRFL-CSR. The masses of 12 nuclei were measured for the first time and the mass precision of about 30 nuclei was improved. As an example, direct mass measurements of short-lived nuclides ^{63}Ge, ^{65}As, ^{67}Se, and ^{71}Kr were performed. These results have had an important impact on nucleosynthesis in the rapid proton capture process [64]. Figure 24 shows a histogram of the extracted revolution times for the nuclei, whose masses were determined in the experiments.

3.2. *HI-13 and its upgrade project BRIF*

The HI-13 tandem accelerator produced by HVEC was installed and tested at CIAE in 1986. Proton beams with an average current of 10 μA at 7.5 MV

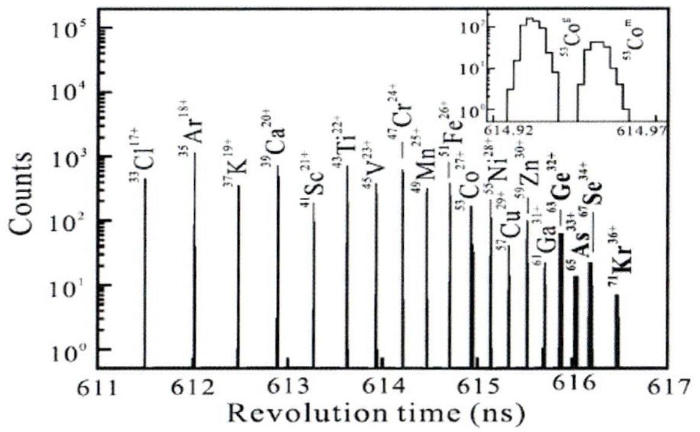

Fig. 24. Histogram of the extracted revolution times for part of the $A = 2Z - 1$ nuclei, whose masses are indicated by bold symbols.

and a peak current of 1.05 mA at a terminal voltage of 13 MV were obtained [65]. In August 1987, the whole facility, including the accelerator and various experiment facilities was completed and passed the acceptance checks [66].

The machine has been improved since it was put into operation. A second stripper consisting of 240 foils was installed in the first dead section on the high energy side, and was brought into operation. Some parts of the accelerator were replaced, such as the laddertron chain, the idler rims, the nylon insulator links, the dividing resistors, some vacuum parts and the accelerator tubes. Figure 25 shows the tandem accelerator.

The HI-13 tandem operated well for 20 years, from 1986. The facility provided more than 40 species of ions from proton to gold for users from 50 institutions with hundreds of research subjects. The research fields covered a wide range of nuclear physics and applications, such as nuclear structures, nuclear astrophysics, nuclear data measurement, AMS and heavy-ion irradiation [67, 68].

Fig. 25. The HI-13 tandem accelerator at CIAE.

As an upgrade project of the HI-13 tandem system, the Beijing Radioactive Ion-Beam Facility (BRIF) was proposed in 1996 [69]. Aiming at research on radioactive ion beams, the BRIF complex consists of a 100 MeV proton compact cyclotron, a two-stage ISOL (isotope separation on-line) system with a mass resolution of 20,000, an SC heavy-ion booster linac with an energy gain of 2 MeV/q, and various experimental terminals. A layout of the BRIF complex is shown in Fig. 26.

The construction and commissioning of the project BRIF are in progress. The commissioning of CYCIAE-100 was launched at the end of 2013 and the first 100 MeV proton beam was extracted on July 4, 2014. The beam current was stably maintained at above 25 μA, and the cyclotron is now ready to provide CW proton beam on the target source for RIB production [70]. As a high intensity proton accelerator, CYCIAE-100 will be described in the next section in detail.

As a part of BRIF, the Isotope Separator On-Line system (BRISOL) was developed at CIAE. The construction and installation of the ISOL system was completed at the end of March 2014, and the stable ion beam test shows that it can reach a mass resolution better than 10,000. It is expected to generate dozens of species of RIBs with 100 MeV proton beams. The commissioning demonstrated the high

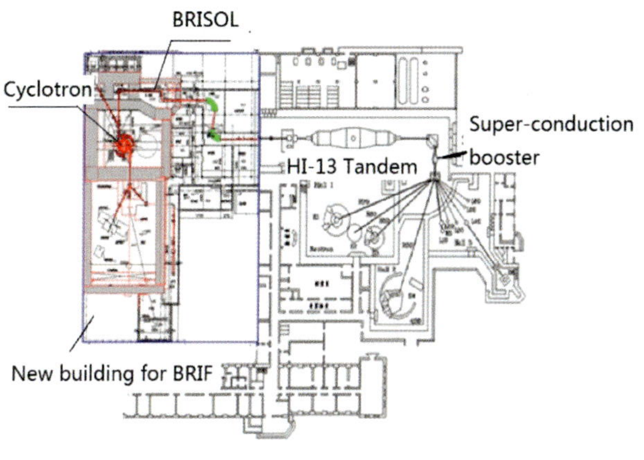

Fig. 26. Layout of the BRIF complex. The radioactive beams are induced in a thick target by a primary proton beam of 100 MeV and 200 μA (CYCIAE-100) from the cyclotron. After diffusion from the thick target materials, these exotic nuclear beams are extracted from the ion source, separated in the online mass separator, accelerated and injected into the tandem and the SC linac to be accelerated, and then used for the experiment.

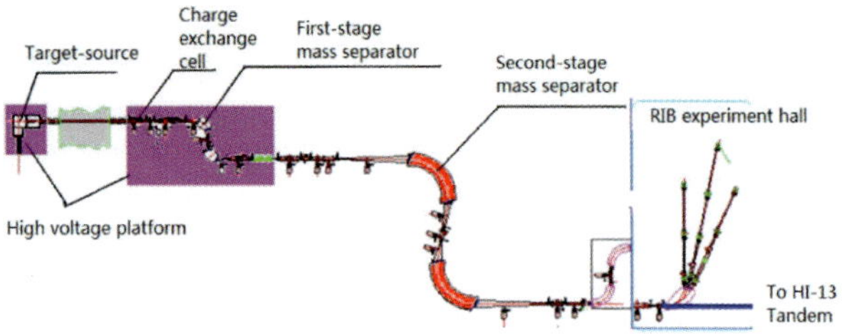

Fig. 27. Layout of the BRISOL facility. To reach a high mass resolving power of 20000, a bienergy system with big and small dipoles is applied to reduce energy dispersion.

reliability of the BRISOL facility [71]. A layout of the BRISOL facility is shown in Fig. 27.

The SC booster system is designed to further increase the energy of heavy-ion beams from the HI-13 tandem accelerator. There are four QWR cavities installed into one cryostat to boost the beam energy by 2 MeV/q. The measured accelerating gradient reached 2.5 MV/m in a low temperature test. The oxygen ion beams were boosted to 1.055 MeV/q by two QWR cavities [70].

The first RIB, K-38, was produced by a 100 MeV proton beam, and separated and accelerated to 300 keV in May 2015. More radioactive beam productions experiments will be carried out with the ISOL system to optimize the charge exchange efficiency and to improve the positive surface ion source. Another target-source is under development to produce more species of radioactive beams other than alkaline ions.

3.3. *Future plans*

The success of the heavy-ion facilities of HIRFL and CSR as well as HI-13 and BRIF, both the accelerators and the experiments, has encouraged scientists to propose more aggressive projects. There are two major heavy-ion accelerator plans in China, i.e. the High Intensity Heavy-Ion Accelerator Facility (HIAF) [72] and the Beijing ISOL facility [73].

Aiming to expand nuclear and related researches into the presently unreachable region and give scientists the possibility of conducting cutting edge research in these fields, IMP proposed the HIAF project in 2009. The purpose of the facility is to provide intense beams of primary and radioactive ions for a wide range of research fields. Radioactive ion beams are used to investigate the structure of exotic nuclei, to learn more about astrophysical nuclear reactions and to measure the masses of

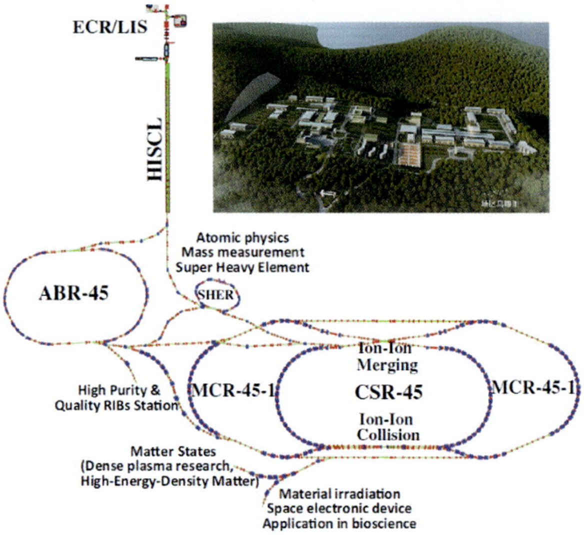

Fig. 28. General layout of the HIAF accelerator complex.

nuclei with high precision. The high energy, highly bunched heavy-ion beams are used to interact with dense plasma in order to probe the physics of nuclear fusion. Highly charged ions are used for atomic physics and applied science research. A general layout of the HIAF complex is shown in Fig. 28.

As seen in Fig. 28, the HIAF accelerator complex consists of ion sources, a high intensity heavy-ion SC linac HISCL, an accumulation ring and a booster ring ABR-45, and a multifunction storage ring system. The injector of HIAF includes the ion sources and the high current linac HISCL. Two types of ion sources, i.e. the electron cyclotron resonance (ECR) source and the laser ion source (LIS), are used in the injector. The heavy-ion beams with the energy of 25–100 MeV/u from HISCL are accumulated, cooled and accelerated to the energy range of 1.2–3.4 GeV/u in ABR-45. These energetic heavy-ion beams will be extracted directly to produce radioactive ion beams or highly charged heavy-ion beams for internal and external target experiments. The beams can also be delivered to the cooling storage ring CSR-45 or the colliding rings MCR-45-1 and MCR-45-2 for ion–ion collisions and ion–ion merging researches, or further stacking to increase their intensity [72].

The design study of HIAF is in progress [74–76]. The main design parameters of the typical U^{34+} ion beams in HIAF are listed in Table 7. The lower and upper parts of the intensity range represent the

Table 7. The main design parameters of U^{34+} ion beams in HIAF.

Machine	Ion	Energy	Intensity
ECR source	U^{34+}	60 kV/q	0.02 pmA
LIS source	U^{34+}	60 kV/q	0.1 pmA
HISCL	U^{34+}	25 MeV/u	0.015–0.075 pmA
	U^{76+}	40 MeV/u	0.003–0.015 pmA
ABR-45	U^{34+}	1.2 GeV/u	0.25–1.3 × 10^{11} ppp
	U^{76+}	3.4 GeV/u	0.5–2.5 × 10^{10} ppp
CSR-45	U^{34+}	1.2 GeV/u	1.0–5.0 × 10^{11} ppp
	U^{76+}	3.4 GeV/u	0.2–1.0 × 10^{11} ppp
	U^{92+}	4.4 GeV/u	0.2–1.0 × 10^{11} ppp
MCR-45-1(2)	U^{34+}	1.2 GeV/u	1.0–5.0 × 10^{11} ppp
	U^{76+}	3.4 GeV/u	0.2–1.0 × 10^{11} ppp
	U^{92+}	4.4 GeV/u	0.2–1.0 × 10^{11} ppp

design intensity of different machines in the cases of ECR or LIS being used.

The HIAF project was approved by the Chinese government at the end of 2015, with a total budget of US$350 million and a site was selected by the sea near Huizhou city of Guangdong province.

The Beijing ISOL is a project jointly proposed by PKU and CIAE. This facility aims at both basic science research and application. A schematic layout is shown in Fig. 29.

As shown in Fig. 29, there are two driven modes in the Beijing ISOL; CARR (Chinese Advanced Research Reactor) and the deuteron linear accelerator (D-linac). The large thermal neutron fission cross section for ^{235}U and the high neutron flux in CARR allow the delivery of very high fission yields and the 20–50 MeV, 5–10 mA D-linac makes use of the (d, n) reaction to produce fast neutrons and ions. The Beijing ISOL facility plans to use both the ISOL and PF (projectile fragmentation) techniques. The primary beam coming from the ISOL target will be highly charged and accelerated up to 150 MeV/u by the postaccelerator. The radioactive beams can further generate most neutron-rich ion beams under in-flight fragmentation.

With the unprecedented neutron-rich beam intensities, unique physics opportunities can be anticipated with the Beijing ISOL, such as mapping the neutron drip line, discovering new magic numbers, synthesizing superheavy elements using neutron-rich beams, exploring cluster and halo exotic structures, and performing reaction and decay measurements for nuclei across the astrophysical r-process. The double magic ^{78}Ni nuclei may reach an intensity of about 250 pps. In the aspect of application, the intense neutron beam can be used to test and evaluate the material properties, and the ion beams can also be used to investigate the irradiation mechanism of materials [73].

The Beijing ISOL is in its proposal stage. It is expected that domestic and international cooperation will be promoted during the whole process of its design, construction and operation.

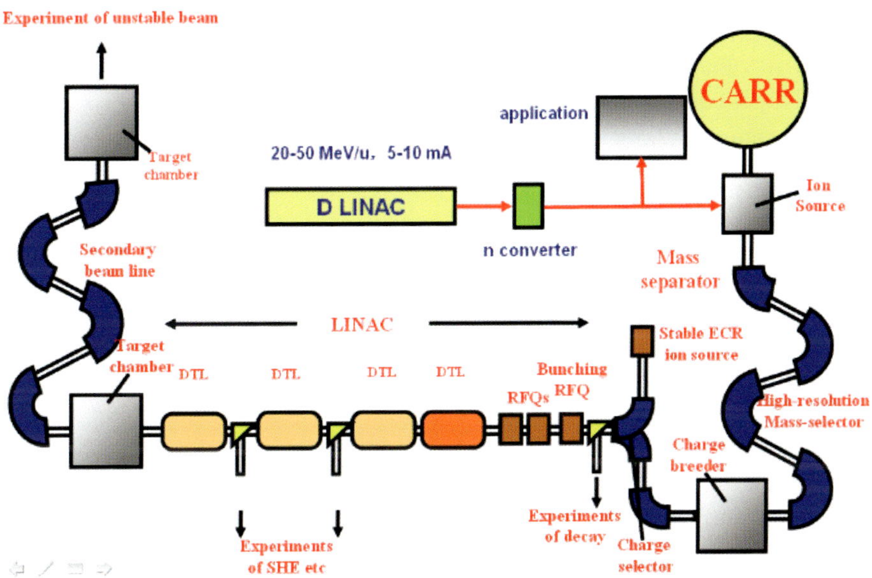

Fig. 29. Schematic layout of the proposed Beijing ISOL.

4. High Intensity Proton Accelerators

High intensity proton accelerators have developed rapidly in the world, because of their wide application in high energy physics research, RIB generation, spallation neutron production, material irradiation, isotope production, ADS and other fields. High intensity accelerators of the machine type can be high voltage accelerators, proton linacs, cyclotrons, rapid-cycling synchrotrons, high energy storage rings or others. As described in the previous sections, many accelerators of that type have already been built in China. In this section, the China Spallation Neutron Source, the high intensity cyclotrons and China's ADS program are emphasized.

4.1. *CSNS*

The investigation into the construction of a spallation neutron source in China started in 2000. Finally, it was determined that this machine should be world-class but modest-size within China's financial capability, and would provide a multidisciplinary platform of scientific research for both institutions and industry, and the beam power was chosen to be 100 kW in the first phase with potential for further upgrading to 500 kW in the second phase. Since then, intensive design study, key technology R&D, and detailed reviews of the project have been performed [77–79]. The CSNS project was approved by the Chinese government in 2008.

CSNS is a proton-accelerator-based facility for delivering spallation neutrons to users. It consists mainly of an 80 MeV H^- linac, a 1.6 GeV rapid cycling synchrotron (RCS), a target, system and neutron instruments. Two beam transport lines, LRBT (linac-to-ring-beam transport) and RTBT (ring-to-target-beam transport) — connect the linac to the RCS and the RCS to the target, respectively. The spallation target is made of tungsten, with heavy water as the cooling medium. CSNS may accommodate about 20 neutron scattering instruments, and 3 will be equipped on day one. Figure 30 shows a schematic layout of CSNS.

The designed average beam power of CSNS phase I is 100 kW. For the phase II upgrade, the SC spoke cavity sections would be installed in the reserved 85 m space following to the DTL tanks to increase the linac energy from 80 MeV to 250 MeV. With the higher injection energy, more protons can

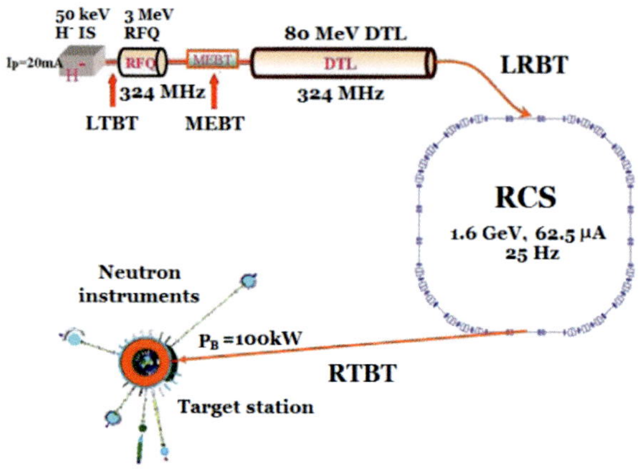

Fig. 30. Schematic layout of CSNS. 50 keV H^- beams from the ion source are accelerated to 3 MeV by the RFQ, and matched into the drift tube linac (DTL) through the medium energy beam transport line (MEBT). The beams are then accelerated to 80 MeV with four DTLs and injected into the RCS through the LRBT line. H^- ions are converted into protons using stripping painting and accumulated in the RCS. The proton beams are accelerated to 1.6 GeV with a repetition rate of 25 Hz, extracted, and transported through the RTBT line, then bombard the target to generate neutrons.

Table 8. The main parameters of CSNS phase I and phase II.

Parameter	Phase I	Phase II
Beam power on target (kW)	100	500
Linac energy (GeV)	80	250
RCS energy (GeV)	1.6	1.6
RCS circumference (m)	228	228
Average beam current (μA)	62.5	312.5
RCS repetition rate (Hz)	25	25
Linac RF (MHz)	324	324
RCS RF (MHz)	1.02–2.44	1.61–2.44
Target material	Tungsten	Tungsten

be accumulated in the RCS. The beam power of CSNS phase II is 500 kW [80, 81]. The main parameters of CSNS are given in Table 8.

The construction of the CSNS project started in September 2011 in Dongguan, Guangdong province. After that, the physical design was optimized [82]; space charge effect was studied [83] and the collimator was developed [84]; linac construction was in progress [85], electron cloud instability in the RCS was studied [86], and the magnets and their power supplies were manufactured [87, 88]; the ferrite-loaded RF system was promoted [89]; the ceramic vacuum ducts were produced, and the injection and extraction and many other systems were developed [90–92]. The construction of the buildings and

Fig. 31. The construction site of CSNS.

accelerator components is well in progress. Figure 31 shows the site of CSNS under construction.

By the time of writing, most of the accelerator components have been fabricated, tested and installed in the CSNS tunnels.

The ion source and the LEBT were installed in the tunnel in 2014. The H^- beam current of 40 mA at the front end of the linac was achieved. The first DTL, DTL-1, was installed in the tunnel in 2015 and the beam commissioning was successfully completed. The low power RF tuning of the other three tanks is in progress. Figure 32 shows the MEBT and DTL-1 in the linac tunnel.

Regarding the RCS, all the 24 dipoles, 48 quadrupoles and many other devices were installed in the ring tunnel. Figure 33 shows the dipoles and quadrupoles installed in the RCS tunnel.

According to the schedule of the CSNS project, the beam commissioning of the DTL linac will be completed at the end of 2016; the installation of the RCS will be finished in August, and the tuning of the RCS components will be completed at the end of November 2016. The RCS beam commissioning

Fig. 32. The MEBT and DTL-1 in the linac tunnel.

Fig. 33. Dipoles and quadrupoles in the RCS tunnel.

is scheduled to start at the beginning of 2017. It is expected that the first beam will be on the neutron target in June 2017 [92].

4.2. *CYCIAE series cyclotrons*

Cyclotrons have the advantage of providing a high average beam current. CIAE has been working on cyclotrons of different type for more than 50 years. In the last 20 years, it has developed the CYCIAE series cyclotrons. The first high intensity proton cyclotron, CYCIAE-30, was constructed and applied for medical and industrial isotope production at CIAE in the 1990s [93]. Proton beams with an energy of 15–30 MeV and intensity of 370 μA were obtained. The 10 MeV central region model cyclotron, CYCIAE-CRM, was successfully built in 2006, and an internal beam current of up to 430 μA was obtained in 2009 [94]. The upgrading of CYCIAE-CRM is in progress, in order to increase the beam current to the mA level. Two cyclotrons were completed at CIAE: the 14 MeV CYCIAE-14 in 2012 [95] and the 100 MeV CYCIAE-100 in 2014 [96].

As mentioned in the Subsec. 3.1, CYCIAE-100 serves as a driving machine for RIB production in the BRIF project [97]. A schematic structure of CYCIAE-100 is drawn in Fig. 34.

CYCIAE-100 is a compact proton cyclotron with a design beam current of 200 μA. It consists of a multicusp ion source, a spiral inflector for beam injection, a 44.37 MHz RF system and a magnet with four sectors. Table 9 gives the main parameters of CYCIAE-100.

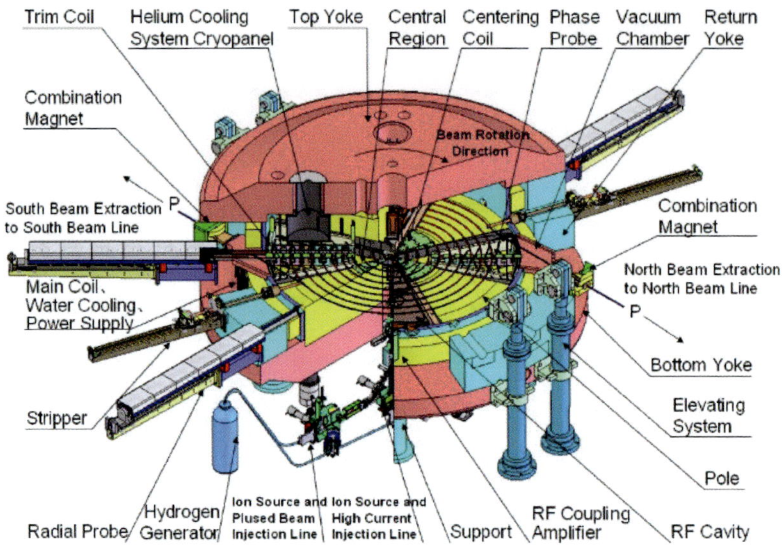

Fig. 34. Schematic structure of CYCIAE-100.

Table 9. The main parameters of CYCIAE-100.

Source		Injection	
Type	Multi–cusp	Energy	∼40 keV
Current	>5 mA	Inflector	Spiral
RF system			
No. of dees	2	Dee voltage	60–120 kV
Dee angle	36°	Harmonic mode	4
Frequency		44.37 MHz	
Magnet			
Number of sectors		4	
Sector angle		∼ 47°	
Field in hill		1.35 T	
Field in valley		0.15 T	
Radius of pole		2000 mm	
Outer radius of yoke		3080 mm	
Gap between hills		50–60 mm	
Total weight of iron		435 t	

Fig. 35. Photo of CYCIAE-100 on its site.

The design and construction of CYCIAE-100 started in 2006. During the construction period, an intensive beam dynamic study was performed [98]; all the systems, such as the RF system [99], the magnet [100] and the power supplies, were well produced, tested and integrated into the cyclotron [97]. Figure 35 shows the completed cyclotron of CYCIAE-100.

The beam commissioning was performed step by step, starting from a test of the ion source. Finally, on July 4, 2014, the first 100 MeV proton beam was obtained at the extraction beam line. The beam current reached 25 μA. It is expected that CYCIAE-100 will be capable of providing 200–500 μA proton beams in the coming years [101].

4.3. *ADS*

Along with the rapid development of nuclear power, better use of the uranium resource and safe disposal of long-lived radioactive nuclear waste become two of the most important issues. ADS is an effective way to solve these problems. In an ADS facility, proton beams are utilized to drive a subcritical reactor with the spallation neutrons generated by protons bombarding a metal target, and atoms of nuclear waste are transmuted into stable or less radioactive elements in the reactor and so generate nuclear energy.

China's study of ADS started in the 1990s. A project entitled "Basic Research on the Physics and Technology of ADS on a Clean Nuclear Power System" was sponsored by the Ministry of Science and Technology (MOST). The first five-year program was performed from 2000 to 2005, when the study of ADS physics and related technology was carried out. The second five-year program started in 2007. In this period, a subcritical experimental reactor assembly, Venus-1, was constructed and tested at CIAE. Regarding the accelerators, a high current proton injector, consisting of an ECR ion source, an LEBT and an RFQ accelerator, was constructed at IHEP. 3.5 MeV proton beams with a peak current of 46 mA and a duty factor of 12% were obtained [102–106].

Later, the priority of the ADS strategy in China shifted to the transmutation of nuclear waste from nuclear energy production, while fast reactors focus on nuclear fuel breeding [107].

In 2011, CAS launched a program entitled "Accelerator-Driven Nuclear Waste Transmutation (C-ADS)" as a strategic pilot science project. The research activities were focused on high power proton accelerators, high power spallation targets, and subcritical reactors [108]. A desired roadmap for longterm development of ADS in China was proposed by CAS, and it is sketched out in Fig. 36.

According to this roadmap, the first budget of about US$260 million was launched for an ADS R&D and test setup construction, including the accelerators, the targets and the subcritical reactor [108, 109].

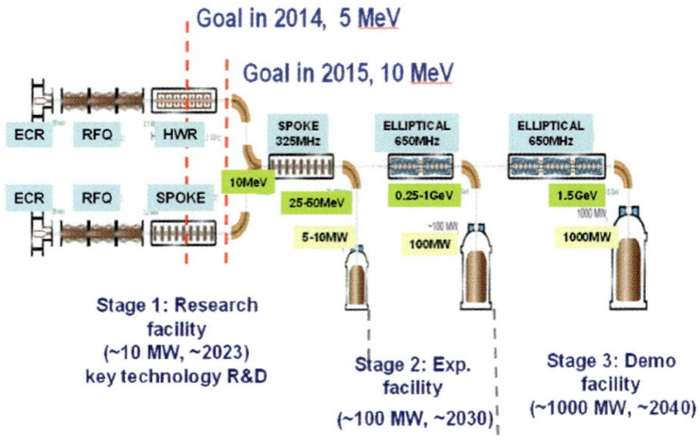

Fig. 36. The desired roadmap of ADS in China. There are three stages of development, i.e. a research facility for key technology R&D with the 25 MeV, 10 mA linac and 10 MW reactor, an experimental facility with the 250 MeV, 10 mA linac and 100 MW reactor and a demo facility with the 1.5 GeV, 10 mA linac and 1 GW reactor.

A block diagram of the C-ADS accelerator is shown in Fig. 37.

As can be seen, there are two injectors that are being developed in the C-ADS project: injector I consists of an ECR ion source, a 325 MHz RFQ and a low-β single spoke resonator named Spoke012; injector II is composed of an ECR ion source, a 162.5 MHz RFQ and a low β half-wave resonator named HWR010. The former being developed at IHEP and the latter at IMP. The 25 MeV injector test facility is being jointly constructed by IMP and IHEP. The technical decision for the injector on RF and cavity types will be made according to the R&D and experiment results. There will be two identical injectors in the demo facility.

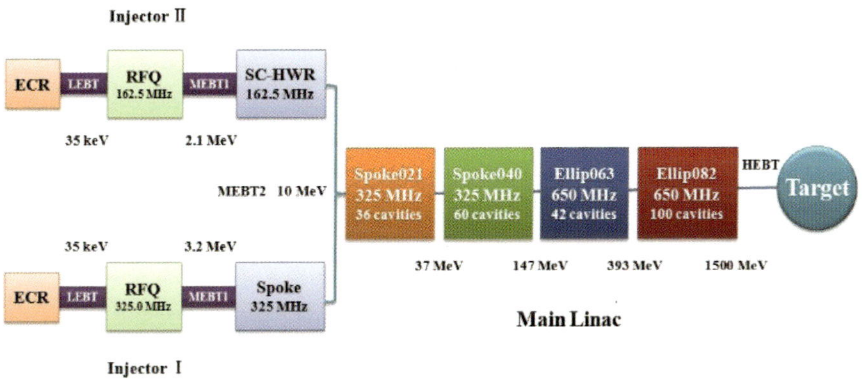

Fig. 37. Block diagram of the C-ADS accelerator. It consists of three distinct regions: (1) an injector (0–10 MeV) based on the 325 MHz RFQ + spoke resonator or 162.5 MHz RFQ + HWR (half-wave resonator); (2) a low β accelerator (10–147 MeV) with two types of single spoke resonators operating at 325 MHz, i.e. $\beta = 0.21$ (Spoke021) and $\beta = 0.40$ (Spoke040); (3) a high-β accelerator (147–1500 MeV) with two types of elliptical cavities operating at 650 MHz, i.e. $\beta = 0.63$ (Ellip063) and $\beta = 0.82$ (Ellip082). In total, 23 cavities are required.

The design and construction of both the 325 MHz and 162.5 MHz ADS injectors have been well processed, including the physical design study [110–114], ECR ion sources, high intensity CW RFQs [115, 116], SC cavities [117, 118], high power couplers [119–121], the interface to the high power target and many other accelerator components [122, 123].

The 10 MeV test facilities of injector I and injector II were successfully constructed. The beam energy of 10 MeV and the CW or pulsed current of 10 mA were reached in their beam commissioning [109, 124]. Figures 38 and 39 show the test facilities mounted in the accelerator tunnels.

The success of the injector test facilities and key technology R&D for C-ADS encourage the research to go ahead. The project CIADS (China Initiative Accelerator-Driven System), as the second stage of the ADS development in the roadmap, was proposed by CAS and approved by the Chinese government at the end of 2015. A schematic layout of CIADS is shown in Fig. 40.

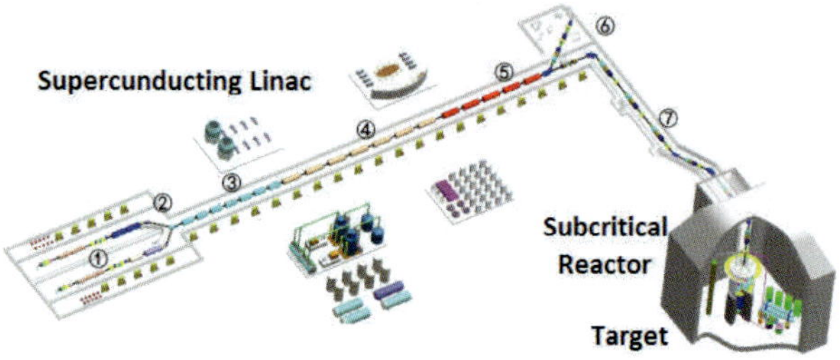

Fig. 40. Schematic layout of CIADS: (1) two mutual backup injectors; (2) medium energy transport line; (3) Spoke021 section; (4) Spoke040 section; (5) Spoke 63 section; (6) experimental area; (7) high energy transport line.

Table 10. The characteristic parameters of CIADS in comparison with the demo ADS facility.

Parameter	Unit	CIADS	Demo
Beam average current	mA	10	>10
Beam energy	MeV	250	~1500
Beam power on target	MW	2.5	c ≥15
Reactor heat power (MW_t)	10	10	1000

Fig. 38. The injector I test facility in the tunnel.

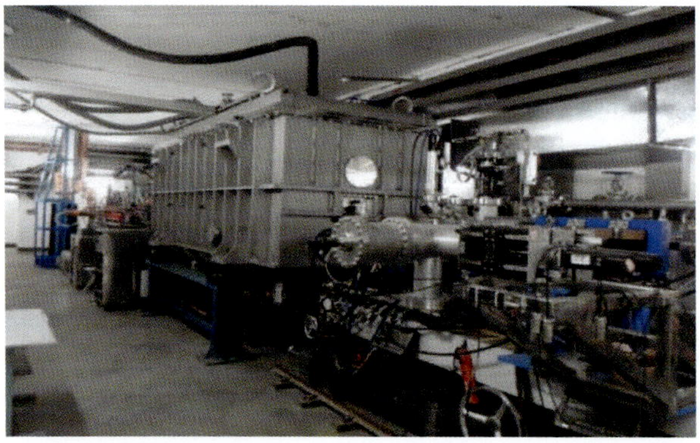

Fig. 39. The injector II test facility in the tunnel.

The characteristic parameters of CIADS in comparison with the demo ADS facility are given in Table 10 [109].

It is clearly seen from the table that the ongoing project CIADS is an early but important step toward the industrial demonstration of the full scale ADS facility.

5. Accelerator Based Light Sources

Accelerator-based light sources are among the most successful scientific facilities in the world; they can provide high brightness radiation from tera hertz to γ-rays. Starting from the 1980s, synchrotron radiation sources and free electron lasers (FELs) have been rapidly developed in China. In this section, the development of China's SR sources and FELs is emphasized, and energy recovery linacs (ERLs) are mentioned. Compton back-scattering X-rays will be discussed in Sec. 7.

5.1. *Synchrotron radiation sources*

There are three operational SR sources in China, i.e. BSRF, HLS and SSRF. The High Energy Photon Source (HEPS) is in its design and R&D stage. The main parameters are listed in Table 11.

Table 11. The main parameters of SR sources in China.

Name	BSRF	HLS	SSRF	HEPS
Generation	I	II	III	III
Location	Beijing	Hefei	Shanghai	Beijing
Energy (GeV)	2.5	0.8	3.5	5–6
Circ. (m)	240	66	432	1296
Lattice type	FODO	DBA	DBA	7BA
Emit. (nm)	120	20–36	3.9	0.06
Current (mA)	250	300	300	200
Completion	1989	1991	2009	—

5.1.1. *BSRF*

As a first generation SR source, BSRF is attached to BEPC, operating in either parasitic or dedicated mode. As mentioned in Sec. 2, in the case of the dedicated synchrotron radiation operation of BEPCII, electron beams circulate in the outer ring by means of two pairs of horizontal bending magnets in the southern and northern IRs. Surrounding the BEPCII storage ring are 3 experimental halls, 14 beam lines, (5 of them from wiggler magnets), and 15 experimental stations in BSRF. Ten of the stations can be operated in parasitic mode during the particle physics experiments. The machine operates in dedicated SR mode for about two months each year, providing SR beams for more than 200 user groups [29, 30]. Figure 41 shows a layout of BSRF.

5.1.2. *HLS and HLS-II*

HLS is a second generation dedicated VUV and soft X-ray light source, comprising an 800 MeV electron storage ring, an electron linac as the injector and a beam transfer line. It was designed and constructed in the late 1980s [125] and the phase II project proceeded in early 2000s [126, 127]. To enhance the performance of HLS, an upgrade project called HLS-II was carried out during 2010–2014 [128].

In HLS-II,the magnet lattice of the storage ring was reconstructed with four DBA cells in order to reduce beam emittance and to provide more straight sections for insertion devices. The beam energy of the injector linac was increased from 200 MeV in HLS to the full energy of 800 MeV for HLS-II. Most machine systems, including the injector, injection system, RF system, magnets, power supplies, vacuum, beam diagnostics, transverse and longitudinal multibunch feedbacks, and control system were upgraded [128–134]. Figure 42 shows the new storage ring of HLS-II.

HLS-II came into operation at the beginning of 2015. Table 12 lists the main parameters of HLS-II in comparison with HLS [128, 135].

As a future plan, the Hefei Advanced Light Source (HALS) was proposed, whose purpose is to produce VUV and soft X-ray synchrotron radiation with high brilliance and lateral coherence, with beam

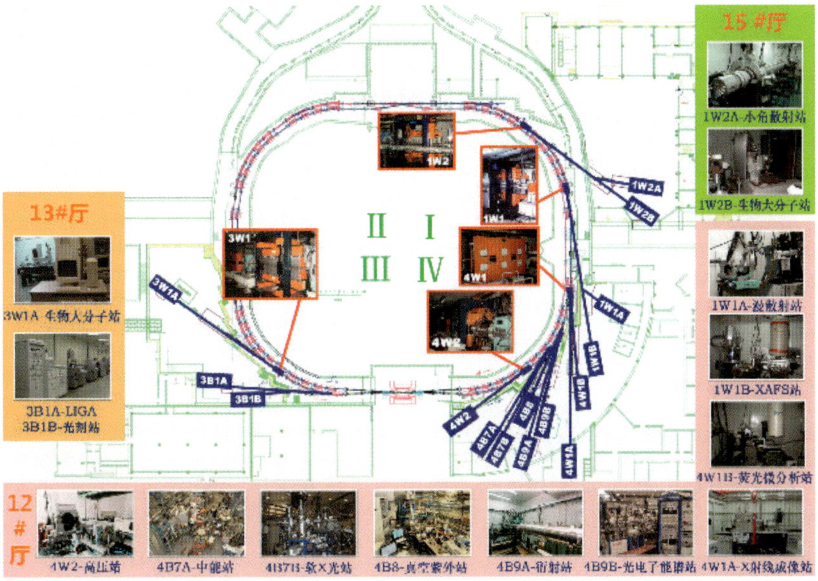

Fig. 41. Layout of BSRF, the storage ring, the beam lines and the experimental stations.

Fig. 42. The HLS-II storage ring.

Table 12. The main parameters of HLS-II in comparison with HLS.

Parameter	HLS	HLS-II
Injection energy (MeV)	200	800
Operation energy (MeV)	800	800
Circumference	66.13	66.13
Magnet lattice	4 × TBA	4 × DBA
RF	204	204
Transverse tunes	3.54/3.60	4.41/3.21
Beam emittance (nm)	∼160	20–36
Straight sections (m)	4 × 3.36	4 × (4.0 + 2.3)
Beam current (mA)	250–300	>300

energy of 1.5 GeV and ultralow emittance of 0.2 nm · rad [136].

5.1.3. *SSRF*

SSRF is a third generation light source based on a 3.5 GeV storage ring [137–142]. The project was approved by the central government in January 2004. The groundbreaking took place in the Shanghai Institute of Applied Physics (SINAP) at the Zhang jiang Hi-Tech Park in December 2004. The construction was carried out well on schedule and many advanced technologies were developed [143–149]. An electron beam was stored and accumulated in the SSRF storage ring, and the user operation started in April 2009, on schedule. A layout of the SSRF complex is shown in Figs. 43 and 44 is the photo of the SSRF storage ring.

SSRF has achieved high performance and has been and kept steady, improving since it was put into operation in April 2009. The machine switched to top-up operation for routine user experiments in December 2012 [153]. The fast orbit feedback was successfully commissioned in 2013, and has been

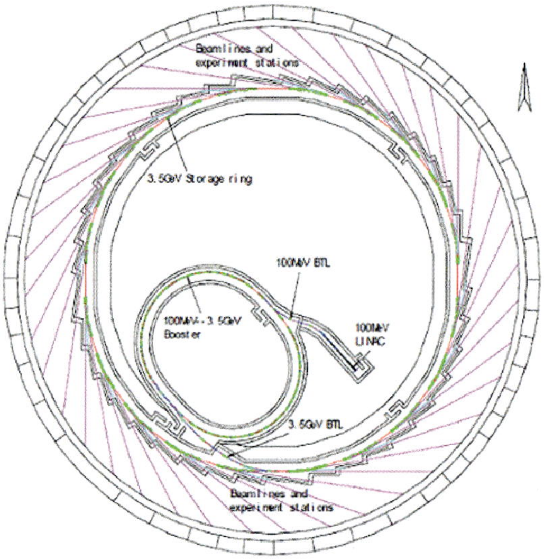

Fig. 43. Layout of the SSRF complex, which is composed of a 150 MeV linear accelerator [150], a full energy booster [151], a 3.5 GeV storage ring [152], beam lines and experimental stations.

Fig. 44. The SSRF storage ring.

implemented in users' operation together with the slow orbit feedback system controlling the long term orbit stability to 0.26 μm and 0.25 μm (RMS) in the horizontal and vertical planes, respectively. The beam current was increased from 200 mA to 250 mA during 2015. The major efforts at increasing the beam current in SSRF were made on conditioning the SRF cavities, suppressing beam instabilities, etc. [154, 155].

SSRF is currently equipped with 13 beam lines and relevant experimental stations, and has served 2000 user groups with 14,000 individual users from all over the country. Taking advantage of the high efficiency operation, many results have been obtained

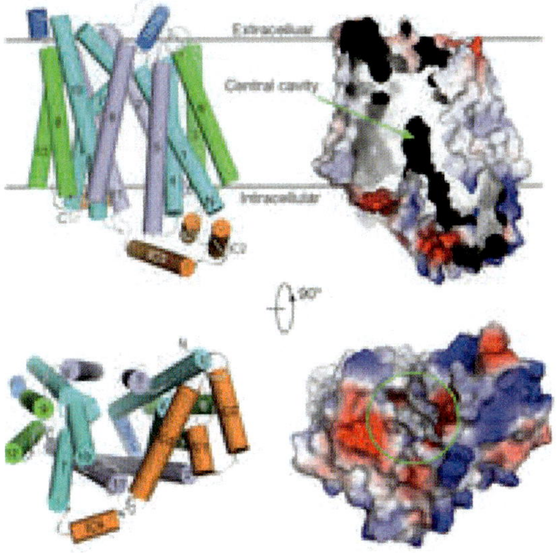

Fig. 45. Overall structure of the human glucose transporter GLUT1 resolved in an SSRF user experiment.

at SSRF. As an example, the overall structure of the human glucose transporter GLUT1 [156] resolved at beam line BL17U of SSRF is shown in Fig. 45.

The SSRF Phase II beam line project was approved by the central government in March 2015. The main task is to build 16 more beam lines to fulfill the pressing demands of users. In the storage ring, 13 new insertion devices will be installed, including 4 cryogenic permanent undulators and an SC wiggler with a magnetic field of 4.2 T. The storage ring lattice will also be upgraded with two superbend cells, in order to extend the photon energy of the X-rays radiated from bending magnets and add more SR source points.

As the future plan beyond the Phase II beam line project, a major upgrade toward a diffraction-limited storage ring is under consideration. This upgrade will aim at reducing the natural emittance of the SSRF storage ring by a factor of 10, therefore increasing the brightness by about two orders of magnitude. Table 13 lists the main parameters of the current SSRF and the proposed SSRF-U [157, 158].

5.1.4. *HEPS*

HEPS, a storage-ring-based light source with a beam energy of 6 GeV and transverse emittances of a few tens of pm · rad, is to be built in Beijing, and is now under design and R&D. Figure 46 shows a computer graphic of the HEPS facility.

HEPS is mainly composed of a 300 MeV electron linac, a full energy booster and an ultralow emittance storage ring of 5–6 GeV. The main parameters of the HEPS storage ring are listed in Table 14 [159].

The ultralow emittance of the high energy storage ring is very challenging for both beam physics and machine hardware, such as lattice design, dynamic aperture, beam instabilities and their suppression, high gradient magnets, combining function bending magnets, on-axis injection, high power RF system, ultralow vacuum pressure, high performance insertion devices, high accuracy power supplies, fast orbit feedback and high stability tunnel environment. An R&D project called HEPS-TF was launched to study all these beam physics and hardware issues [159–167]. It is expected that the construction of HEPS will start in 2018.

5.2. *FELs*

As fourth generation light sources, FELs possess high brightness, full coherency and ultrashort pulse duration, operating in a wide frequency range, from terahertz radiation and infrared to the visible spectrum, ultraviolet and X-ray. Therefore, FEL facilities have wide application in frontier researches.

FEL R&D has taken place in China since the mid-1980s, and several FEL facilities were built in the 1990s [168–170]. In CIAE, an RF-linac-driven infrared FEL was constructed for FEL fundamental research and linac technology development [171].

Table 13. The main parameters of SSRF and SSRF-U.

Parameter	Current	Phase II	SSRF-U
Beam energy (GeV)	3.5	3.5	3.0
Circumference (m)	432	432	432
Lattice type	DBA	DBA	7 BA
Emittance (nm · rad)	3.89	4.22 (with SW)	0.202
Momentum compaction	4.27×10^{-4}	4.2×10^{-4}	2.0×10^{-4}
Straight sections (m)	$4 \times 12.0 + 16 \times 6.5$	$4 \times 12.0 + 16 \times 6.5$	$4 \times 10.35 + 16 \times 5.6$

Fig. 46. Computer graphic of the HEPS facility.

Table 14. The main design parameters of HEPS.

Parameter	Unit	Value
Beam energy	GeV	6.0
Beam current	mA	200
Circumference	m	1295.6
Lattice type		7BA
Natural emittance	pm · rad	59.4
Transverse tunes (H/V)		11.16/41.12
Energy loss per turn	MeV	1.995
Momentum compaction		3.74×10^{-5}
Length of straight sections	m	48×6.0

An induction-linac-based FEL amplifier was built in CAEP and spontaneous emission, constant and variable parameter undulator amplifier experiments were carried out [169]. The Shanghai Institute of Optics and Fine Mechanics (SIOM) generated radiative emission of 8 mm wavelength and 1 MW power with a 0.5 MeV pulse line accelerator. Similar experiments were carried out in the Southwest Institute of Applied Electronics, in cooperation with the University of Electronic Science and Technology of China, and output laser power of 7.6 MW at 37 GHz was observed. Cherenkov FELs were also pursued at several of China's institutes. At USTC, a Nd glass laser of wavelength 353.3 nm was used as input to produce a third harmonic of 117.8 nm VUV coherent radiation for application [170]. The Beijing FEL, BFEL, was constructed in the late 1980s based on a 30 MeV S-band electron linac. The first lasing of BFEL was observed in 1993 [172], while the saturation was realized in 1996 [173]. Figure 47 shows the BFEL facility and its lasing signal. The characteristics of the IR-FEL in BFEL is summarized in Table 15 [170].

The early research on FELs, mainly in the infrared range using low energy electron linacs, has laid a foundation for FEL development in China. In

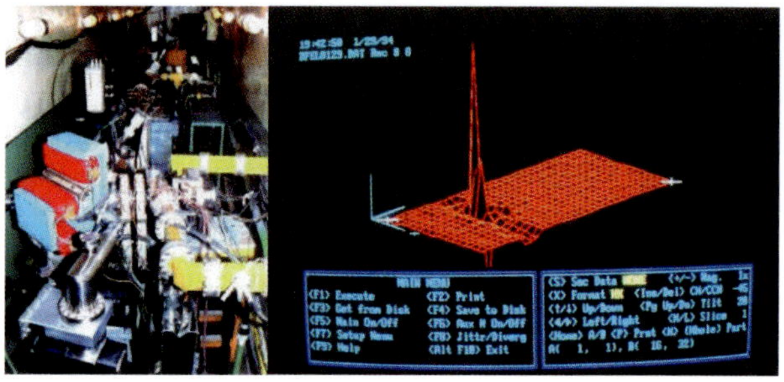

Fig. 47. Photo of BFEL and its lasing signal.

Table 15. The laser characteristics of BFEL.

Parameter	Value	Unit
Total spectral range	9–11	μm
Output energy in optical macropulse	2–10	mJ
Average power in macropulse	200	kW
Peak power	200	MW
Spectral width	0.3–2	%
Small signal gain	32	%
Optical extraction efficiency	48	%
Beam divergence	2	mrad

recent years, remarkable advances have been made in high gain FELs.

The Shanghai deep ultraviolet FEL (SDUV-FEL) is a single pass FEL facility designed for generating coherent radiation at wavelengths down to the deep ultraviolet region. It consists of a 160 MeV linac to provide high brightness electron beams, various short undulators to modulate the electron energy, and a long undulator section of about 10 m for FEL amplification. The linac comprises a low emittance electron injector with a photocathode RF gun, S-band main linac sections, and a bunch compressor [174]. The SDUV linac is shown in Fig. 48.

Fig. 48. The SDUV linac in the tunnel.

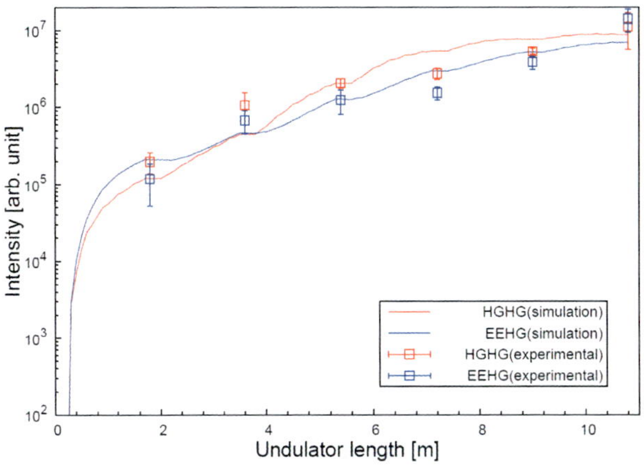

Fig. 49. Gain curves of the EEHG-FEL and HGHG-FEL at SDUV-FEL.

As an FEL research platform completed in 2009, SDUV-FEL has operated for a series of successful FEL experiments since then, such as continuous wavelength tuning and two-stage cascading of the high gain harmonic generation (HGHG) FEL [175], experimental demonstration of a longitudinal phase space linearizer by corrugated structures [176] and of polarization switching by a crossed planar undulator [177], and the first lasing of an echo-enabled harmonic generation (EEHG) FEL [178]. The gain curves of the EEHG-FEL and the HGHG-FEL at SDUV-FEL are displayed in Fig. 49, which shows that the experimental results agree well with the simulation.

In the meantime, theoretical research on phase-merging enhanced harmonic generation (PEHG) [179], the self-seeding FELs, generating mode-locked multi-chromatic radiation in the seeded HGHG-FEL [180], noise propagation in the seeded FEL, the emittance-exchange-based HHG-FEL, the seeded FEL using a tapered undulator, the coherent hard X-ray FEL based on an echo-enabled staged harmonic generation scheme [181] and the XFEL oscillator with intermediate energy electron beam [182] has been carried out at SINAP.

There are now two high gain FEL projects under construction in China, i.e. the Dalian Coherent Light Source (DCLS) [183, 184] and the Soft X-ray FEL test facility (SXFEL) [185], which cover the spectra from VUV to soft X-ray. Also, an infrared FEL facility operating with wavelengths from 5 μm to 200 μm is at the engineering design stage, jointly performed by USTC and Xiamen University [186].

DCLS is an FEL user facility based on the HGHG scheme. It is being constructed on the new campus of Dalian Institute of Chemical Physics, CAS, in the city of Dalian, northeast of China. A schematic layout of DCLS and its linac is shown in Fig. 50. The main design parameters of DCLS are given in Table 16.

The construction of DCLS started in October 2014. The key technologies have been tested with prototype R&D, the building and the infrastructure are ready for installation, and mass production of the

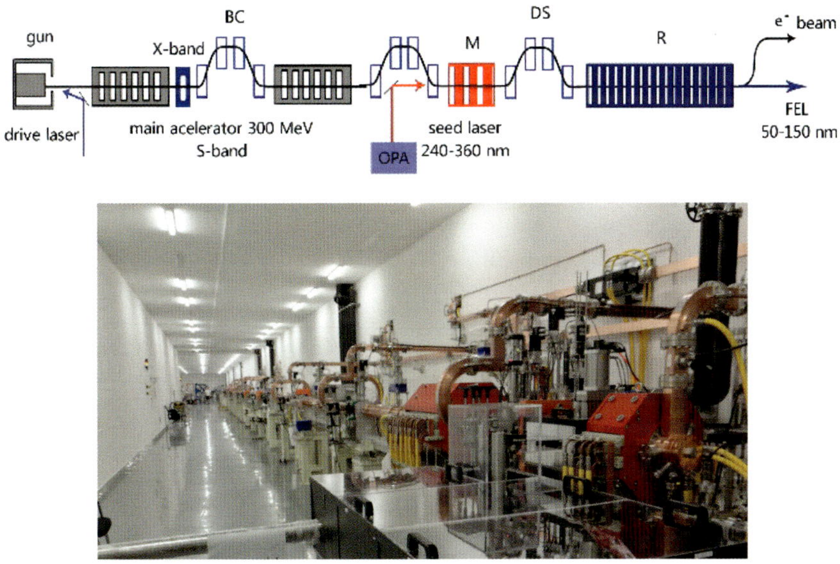

Fig. 50. Schematic layout of DCLS and its linac.

Table 16. The main design parameters of DCLS.

Parameter	Value	Unit
Beam energy	300	MeV
Energy spread	1×10^{-4}	
Normalized emittance	1–2	μm · rad
Peak current	300	Å
Seed laser wavelength	240–360	nm
Seed laser width (FWHM)	1.0	ps
Radiator period length	30	mm
FEL wavelength	50–150	nm
FEL pulse energy	≥100	μJ
Photon flax per pulse	10^{12}	

Table 17. The main design parameters of SXFEL.

Parameter	Value	Unit
Beam energy	840	MeV
Peak current	>500	A
Repetition rate	1–10	Hz
Normalized emittance	<2.0	mm · mrad
Energy spread	<0.15%	
Bunch length (FWHM)	<1.0	ps
Bunch charge	0.5	nC
FEL wavelength	8.8	nm
FEL power	>100	MW

machine components is in progress. The first lasing of DCLS is expected at end 2016.

As an important step toward a hard X-ray FEL in China, SXFEL is being constructed on the SSRF campus by a joint team of SINAP and THU. The test facility is designed to be a multistage seeded FEL with the baseline goal of generating fully coherent FEL radiation at a wavelength of around 8.8 nm. SXFEL consists of a 130 MeV S-band photocathode injector, a C-band linac with an energy of 840 MeV, an undulator section with two HGHG stages, and a diagnostic beam line. A layout of SXFEL is shown in Fig. 51. Table 17 gives the main design parameters of SXFEL.

The groundbreaking of SXFEL took place at the end of 2014. The construction of the accelerator tunnel and the klystron gallery has been completed. The production of key components, such as the photoinjector, C-band accelerating structures, movable chicanes, cavity beam position monitors, three types of undulators, is in progress [187]. The project is scheduled to realize the first lasing in 2017.

Based on the test facility of SXFEL, a user facility is under development by upgrading its linac energy to 1.5 GeV to serve the users with FEL radiation at a wavelength ranging from 2.2 to 10 nm. The groundbreaking for the SXFEL user facility project is expected to be performed in 2016.

As a future development, a hard X-ray FEL is being planned at CAEP and SINAP. The proposed hard X-ray FEL by SINAP is based on a 6 GeV linac to be constructed in the same tunnel as SXFEL. The unique feature of this facility is that an innovative scheme of two-beam cascaded EEHG will be adopted and provide fully coherent hard X-ray FEL output [188]. Table 18 lists its main design parameters, and Fig. 52 sketches the proposed XFEL.

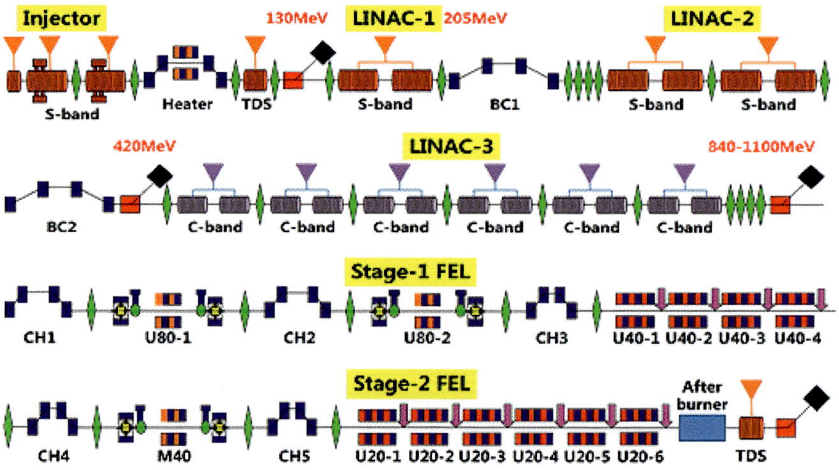

Fig. 51. Layout of SXFEL. The main linac comprises three accelerating sections and two bunch compressors, BC1 at 205 MeV and BC2 at 420 MeV. Linac-1 is operated at S-band frequency, and linac-2 and linac-3 at C-band. In the lasing section, there are two HGHG stages, 5 chicanes, 13 undulators and 5 beam spectrometers.

Table 18. The main design parameters of SXFEL.

Beam parameters	
Beam energy (GeV)	6.0
Normalized slice emittance (μm · rad]	0.4
Bunch charge (nC)	0.2
Peak current (Å)	>3000
FWHM bunch length (fs)	~100
Undulator parameter	
Period (mm)	15
Peak magnetic field [T]	0.924
Section length (m)	4.8
Total length (m)	100
FEL parameter	
Wavelength (Å)	1.0
Saturation power (GW)	6.5
Output power with taper (GW)	~100
Spectral bandwidth (%)	0.002
Peak brilliance (photons/mm^2/mrad^2/s/0.1%BW)	1.3×10^{33}

5.3. *ERLs*

ERLs are also known as one of the technology options for the fourth generation light sources, due to their high repetition rate, short pulse, high spatial coherence and high brilliance. They are usually designed based on state-of-the-art low emittance electron guns and SC linacs.

The study of ERLs started relatively late in China. Some key technologies were developed and ERL test facilities were proposed. A SC ERL test facility (SETF) was initiated in 2006 by PKU [189]. SETF consists mainly of a 5 MeV DC-SRF injector and a cryomodule with a nine-cell cavity for accelerating electrons to 15–20 MeV. A beam transport loop with two arcs was designed to match to the main accelerator [190]. An undulator and a chicane are in the loop to produce 4–8 micron lasers. A layout of SETF is shown in Fig. 53.

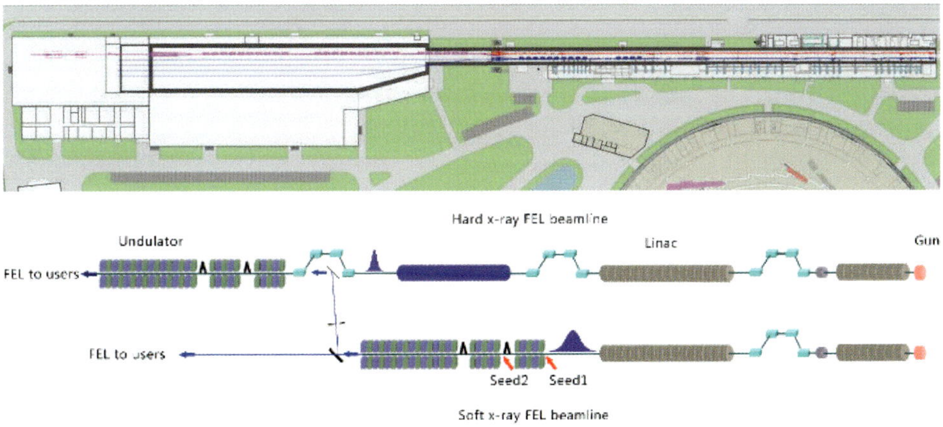

Fig. 52. The proposed hard XFEL facility: location (above) and schematic layout: (below).

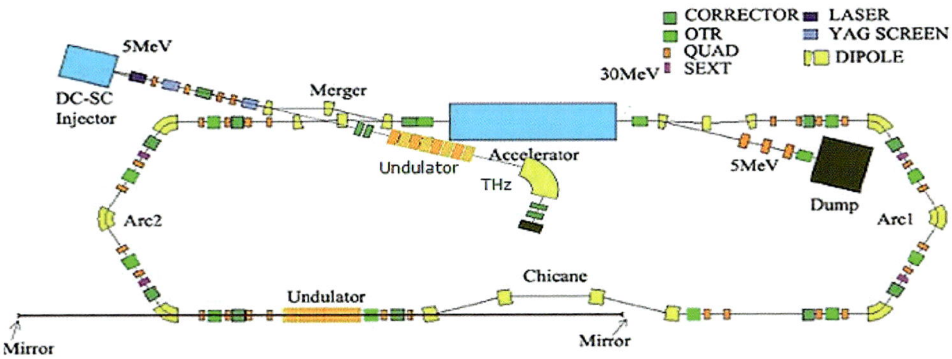

Fig. 53. Schematic layout of the ERL test facility at PKU.

As a step toward the ERL test facility, a compact high repetition rate THz radiation source was developed at PKU based on a DC-SRF photoinjector through velocity bunching [191, 192]. This compact THz source can generate approximately 1 W of super-radiant THz radiation, with a repetition rate of 16.25 MHz and a frequency that can be tuned from 0.24 THz to 0.42 THz by varying the electron beam energy from 2.4 MeV to 3.1 MeV [193].

A compact ERL test facility was proposed by IHEP in 2011, aimed at studying the ERL's key technology, such as the photocathode high voltage DC gun, the low emittance injector, the merger system, the CW multicell SC cavity, and some beam physics problems, including CSR, BBU effects and so on [194, 195]. A 1.3 GHz SC unit and a photocathode DC gun are being constructed at IHEP [196]. SINAP has been developing a VHF photocathode RF gun and the superconducting technology for the ERL. As the future development, an X-ray ERL and FEL facility based on HEPS is being considered in IHEP [197].

6. Pulsed Power Accelerators

Pulsed power accelerators (PPAs) can provide electron, proton and ion beams with high pulsed current to generate X-rays, γ-rays, high power microwaves and lasers for various applications. They have been developed in China since the early 1960s. Over these five decades, the accelerated beam energy has increased from hundreds of kilovolts to tens of megavolts, the pulsed current from tens of kiloamperes to a few megaamperes, and the pulsed power to the trillion-watt scale. PPAs have been widely applied in China, to flash radiography, Z-pinches, inertial confinement fusion, high power microwaves, FELs, environmental protection and imploding liner, [12, 13, 198].

As mentioned in Sec. 1, the first PPAs in China, Chenguang, was produced in 1976 with beams of 1 MeV and 20 kA. In 1979, a high impedance electron PPA, Flash-I, was constructed at CAEP; it can provide pulsed electron beams of 8 MeV, 100 kA and 80 ns. In the 1980s, several water medium forming line accelerators were developed at CIAE, the Institute of Electronics, the National University of Defense Technology and other institutions. A low impedance PPA, Flash-II, was completed by NINT in 1989. This machine can provide 0.9 MeV, 0.9 MA and 70 ns pulsed electron beams for X-ray flash radiography and other applications. Figure 54 shows the Flash-II accelerator.

Fig. 54. The pulsed power accelerator Flash-II at NINT.

In the 1990s, a variety of new PPA technologies were developed in China, in collaboration with foreign institutions, such as inductive energy storage, the Tesla transformer, the linear transformer driver and the repetition frequency technique. Based on these new technologies, several accelerators, such as Qiangguang-I, PHOTO, SINU8 and PS-20, were produced.

Qiangguang-I is a multifunctional electron PPA with a water–oil and capacitive–inductive structure. Constructed by NINT in 2000, this accelerator can generate 6 MeV, 2 MA electron beams with pulse lengths of 20–200 ns for the study of X-ray effects and the Z-pinch. PHOTO is an electron-beam-pumped excimer laser facility, where an electron PPA serves as the final amplifier excited by a linear transformer driver (LTD). SINUS8 is a repetition frequency PPA based on a Tesla transformer with the electron beam parameters of 800 keV, 8 kA, 100 Hz and 30 ns. RS-20 is a repetition frequency electron PPA of 3 MeV, 20 kA and 100 ns for irradiation application. Also, a Tesla type compact repetition frequency electron PPA of 1 MeV, 20 kA, 40 ns and 100 Hz was constructed at CAEP in 2006 for high power microwave research [13].

The development of linear induction accelerators (LIAs) began in the late 1970s at CAEP, when the R&D program on key technologies and components was carried out. Prototypes of the induction module and other devices were designed and tested in the 1980s. The first LIA in China was built in 1988, with the electron beam parameters of 1.5 MeV,

3 kA and 60 ns. The SG-I linear induction accelerator of 3.5 MeV, 3 kA and 60 ns was constructed in the 1990s for FEL research [199]. An LIA X-ray facility (LIAXF) was built in 1993, with the beam properties of 10 MeV, 2 kA and 60 ns and the spot size of 6 mm. In 1995, LIAXF was upgraded to LIAXFU (12 MeV, 2.5 kA and 60 ns, and a spot size of 4 mm) [200]. An advanced LIA, Dragon-I (20 MeV, 2.5 kA and 70 ns, and a spot size of less than 1.5 mm), was completed in 2006 [201–203]. An innovative triple pulse LIA, Dragon-II, was completed in 2014. Dragon-II is based on the technique of controlled-pulse-adder multipulse with the electron beam parameters of 20 MeV, 2 kA and 3 × 70 ns, and the spot size of less than 2 mm [204]. Figure 55 shows a photo of Dragon-II.

High density physics studies such as the Z-pinch and quasi-isentropic compression of materials require a pulse current at mega-ampere levels to reach the ultrastate of high temperature, high pressure and high magnetic field. The Z-pinch is one of the main approaches to inertial confinement fusion. Due to the increasing experimental demands, the primary test stand (PTS), named Julong-I in Chinese, was developed at CAEP in 2013. PTS is a multiterawatt facility with a current of 8–10 MA and a rise time of 90 ns [205]. Figure 56 gives a side view of PTS.

Fig. 55. The linear induction accelerator Dragon-II.

Fig. 56. Side view of PTS (Julong-I). It consists of 24 modules connected in parallel. Each module consists of a 6 MV Marx generator, an intermediate storage capacitor, a laser-triggered spark gap switch, a pulse-forming line, a self-breakdown water switch and a triplet transmission line.

7. Small Scale Accelerators

Small scale accelerators are smaller than the large facilities described in the previous sections. With the characteristics of special design, dedicated purpose, compact structure and relatively low cost, they are widely applied in scientific research. In contrast to accelerators for general applications discussed in the next section, they are built for certain scientific purposes.

A great number of small scale accelerators were constructed by research institutes and universities in China [5, 13].

In addition to the early constructed 2.5 MeV electrostatic proton accelerator V2, the 30 MeV electron linac and the 35 MeV proton linac, IHEP set up a test beam facility and slow positron beams, taking advantage of the BEPC injector.

At IMP, a 320 kV, high charge state ion platform was set up based on an ECR ion source. With a low energy of 0.1–10 MeV, this high intensity heavy ion research device, consisting mainly of an ECR ion source and an RFQ, is being built with funding from the Natural Science Foundation of China (NSFC).

A 2 × 6 MV tandem was constructed by SINR in 1988. SINR also developed a unique compact cyclotron based on a supersensitive AMS facility devoted to carbon-14 analysis (see Fig. 57).

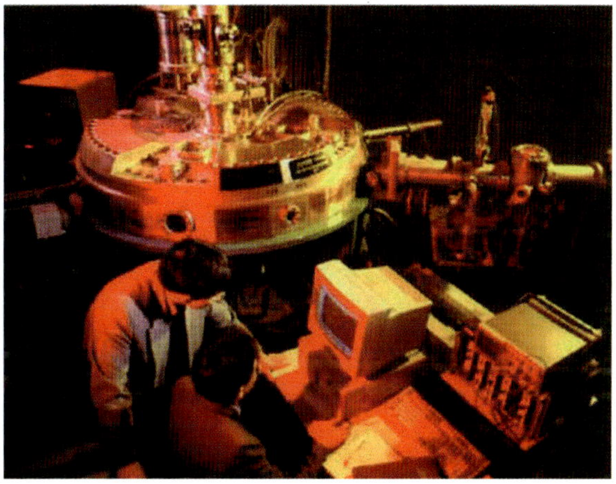

Fig. 57. The minicyclotron mass spectrometer.

At SINAP, two small electron linacs with shared RF power and a shared tunnel were constructed around 2003. The 30 MeV linac delivers 250 fs bunch trains to the undulator, and thus produces coherent terahertz radiation with 1 W average power, while the 10 MeV linac with a 10 ps bunch length is dedicated for pulse radiolysis. A 15 MeV electron linac with 1.5 kW average beam power was constructed for nuclear data analysis of the thorium–uranium cycle.

At CIAE, compact cyclotrons were developed for nuclear and medical research. A heavy-ion accelerator is under construction at IMP for nuclear astrophysics experiments supported by NSFC. A 30 MeV electron linac was built at CAEP, with a photocathode RF gun as its injector.

At Sichuan University, a 1.2 M cyclotron, a J-2.5 electrostatic accelerator, a 2 MeV electron accelerator, a 600 kV electron–ion accelerator and a 280 keV ion implanter were equipped for nuclear physics and applied research.

A number of small accelerators were built at PKU, including heavy-ion RFQs, AMS facilities, an RFQ-based neutron source and others, among which is the 4.5 MV electrostatic accelerator, designed and constructed in the mid-1980s for nuclear study.

Intensive studies of electron linacs have been carried out at THU, where the photocathode installation, the Tsinghua Thomson backscattering X-ray source (TTX), and the MeV ultrafast electron diffraction device were built and a compact pulsed hadron source (CPHS) is under construction.

Supported by NSFC, an in situ telescope–accelerator interface facility was set up at Wuhan University in 2008; it consists of a 200 kV ion implanter, a 2×1.7 MV tandem and a transmission electron microscope [206], shown in Fig. 58.

Two GIC 4117 tandems were operated at Beijing Normal University and Lanzhou University respectively for PIXE analysis, RBS study and nuclear reaction research.

The construction of a MeV ultrafast electron diffraction facility funded by NSFC is in progress at Shanghai Jiao Tong University (SJTU).

In the following subsections, three typical small scale accelerators are described, i.e. compact neutron sources, Thomson scattering facilities and MeV ultra-fast electron diffraction devices.

7.1. *Compact neutron sources*

Two compact neutron sources are reported in this subsection: the PKU Neutron Imaging Facility (PKUNIFTY) and the Compact Pulsed Hadron Source (CPHS) of THU.

PKUNIFTY is a thermal neutron imaging facility based on an RFQ accelerator-driven compact neutron source [207, 208]. Figure 59 shows a systematic diagram and a photo of it.

Deuteron beams of 50 keV generated by a 2.45 GHz ECR ion source are injected into the RFQ cavity through the low energy beam line containing two magnetic focusing solenoids. The RFQ,

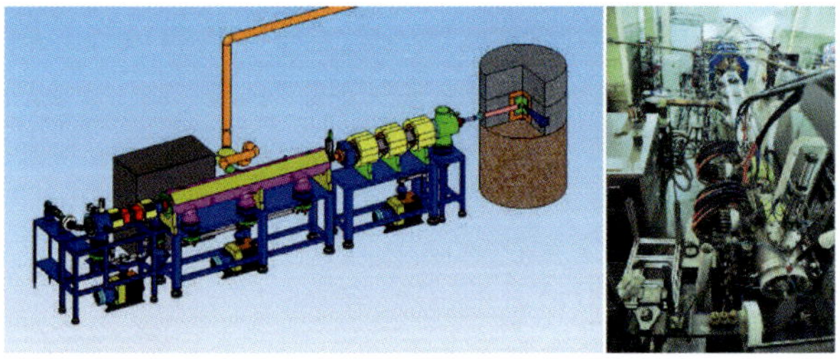

Fig. 59. Systematic diagram (*left*) and photo (*right*) of PKUNIFTY.

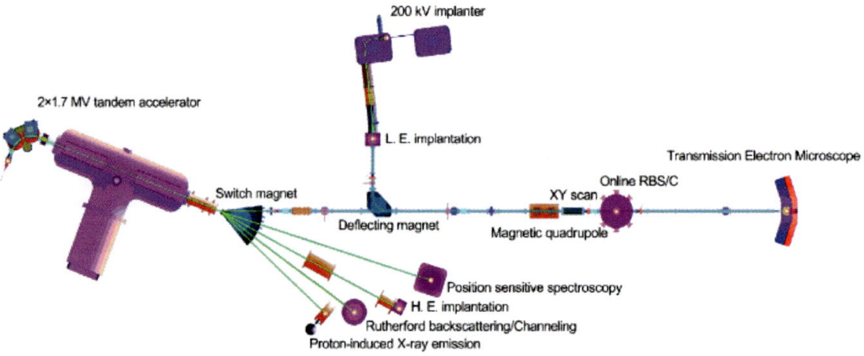

Fig. 58. The *in situ* TEM-implanter/accelerator interface facility.

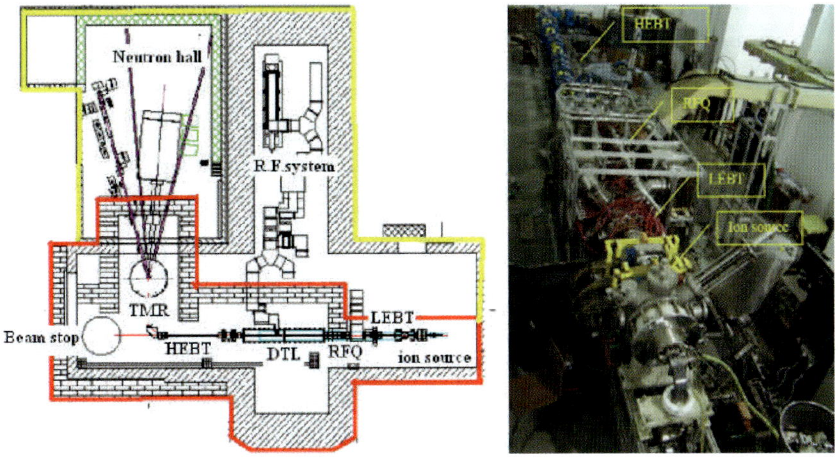

Fig. 60. Schematic layout (*left*) and photo (*right*) of CPHS.

operating at 201.5 MHz frequency, accelerates the deuteron beams to 2 MeV with a peak current of 40 mA and a duty factor of 10%. The intense deuteron beams then bombard the beryllium target to generate fast neutrons. The target system consists of the target body, a moderator, a reflector and a collimator. It can provide a thermal neutron flux of 5×10^5 n/cm^2/s on the imaging plane [209].

The construction of PKUNIFTY was completed in 2012. Since then the facility has been operated for pedagogical training, scientific research and technique development of neutron radiography [210].

CPHS was proposed, designed and is now under construction at THU [211]. A schematic layout and a photo of CHPS are shown in Fig. 60, and its main parameters are given in Table 19.

As a proton-linac-based neutron source, CPHS consists of a proton linac, a neutron target station, a small angle neutron scattering instrument, a neutron imaging/radiology station and a proton irradiation station. The proton linac is composed of an ECR ion source, a low energy beam line, an RFQ accelerator, a DTL linac and a high energy beam line connecting to the target. The construction of CPHS has been in progress since 2009 [212–214]. Most of the accelerator components have been produced, and the commissioning of the 3 GeV accelerator, composed of an ion source, a low energy beam line and an RFQ, has been carried out [215, 216].

Table 19. The main design parameters of CHPS.

Parameter	Value	Unit
Beam energy	13	MeV
Average current	1.25	mA
Repetition frequency	50	Hz
Particle per pulse	1.56×10^{14}	Protons
Duration of pulse	500	μs
Peak current	50	mA
RF	325	MHz
Target material	Beryllium	

7.2. *Compton backscattering facilities*

Compton backscattering facilities can generate ultrafast, high flux for X-rays and γ-rays by means of scattering between free charged particle beams and photon beams. Several facilities have been built or planned in China. TTX was built and operated for experiments. In the Compton backscattering regime, the BEPCII-injector-based γ-ray source and the Shanghai laser electron gamma source (SLEGS) are planned.

The design and construction of the TTX source were carried out starting from 2001 [217–219]. The TTX source is based on Thomson scattering, to generate a femtosecond laser pulse with a relativistic electron beam, aiming for a compact, ultrafast, high-flux, monochromatic and tunable X-ray source for scientific, medical and industrial applications. A systematic diagram and a photo of TTX are shown in Fig. 61.

TTX comprises a photocathode RF electron injector and a femtosecond terawatt laser system. The linac system consists of an S-band photocathode RF gun, an S-band traveling wave tube and two

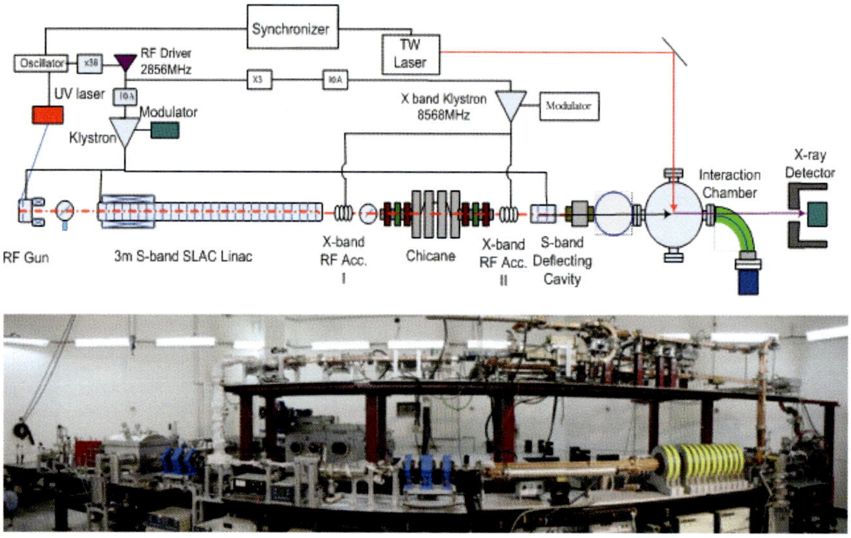

Fig. 61. Systematic diagram and photo of the TTX facility.

Table 20. The main parameters of the laser and electron beams in TTX.

Parameter	Laser	Electron
Energy	800 nm (1.55 eV)	46.7 MeV
Intensity	300 mJ	0.2 nC
Beam size (rms)	~100 μm	~35 μm × 40 μm
Pulse length	~70 fm (FWHM)	~2.7 ps

Table 21. The preliminary design parameters of the BEPC γ-ray source.

Parameter	Laser	Electron
Energy	1064 nm (1.17 eV)	0.8–2.5 GeV
Intensity	~2 J	2 nC
Beam size	~0.75 mm	~0.5 × 0.5 mm^2
Pulse length	~10 ns (FWHM)	10 ps

X-band structures, which can generate ultrashort, high brightness electron pulses to scatter with the laser pulses. Table 20 lists the main parameters of the laser and electron beams in TTX.

The first X-ray beam was generated by TTX in 2011. Since then the intensity of the hard X-rays measured in photons per pulse has been steadily increased from 2×10^5 in 2011 and 2×10^6 in 2012 to 1×10^7 in 2013 and 2×10^7 in March 2016.

To extend the capability of TTX with more stable X-ray output and higher average photon flux, a compact electron-storage-ring-based X-ray source, called TTX-II, was proposed. The design study of TTX-II is underway. The designed parameters of TTX-II are as following: electron beam energy of 500 MeV, bunch charge of 1 nC, ring circumference of 4.8 m, RF of 500 MHz and X-ray flux of 10^{10} photos per pulse [220, 221].

In the high energy regime, a beam line was built for accurate beam energy measurement. IHEP also has plans to set up a Compton backscattering facility based on the BEPC injector; the designed γ-ray flux is 1×10^6 photons per pulse. Table 21 lists the preliminary design parameters of the BEPC γ-ray source.

SLEGS is a high intensity γ-ray beam line, based on the 3.5 GeV storage ring of SSRF. It can generate γ-rays up to the maximum energy of 22 MeV by Compton backscattering between a CO_2 laser of hundreds of watts and 3.5 GeV electron beams. The flux of noncollimated γ-rays is estimated to be 10^9–10^{10} [222]. A layout of SLEGS is shown in Fig. 62.

Recently, Compton backscattering γ-rays have been generated using laser-plasma-accelerated electron beams, as detailed in Sec. 9.2.

7.3. *MeV ultrafast electron diffraction facilities*

Time-resolved ultrafast electron diffraction (UED) is a powerful tool for studying ultrafast structural dynamics at the atomic level with a-few-hundred-femtosecond temporal resolution [223, 224]. Due to the strong space charge effects and velocity mismatch, the temporal resolution of UED at the energy

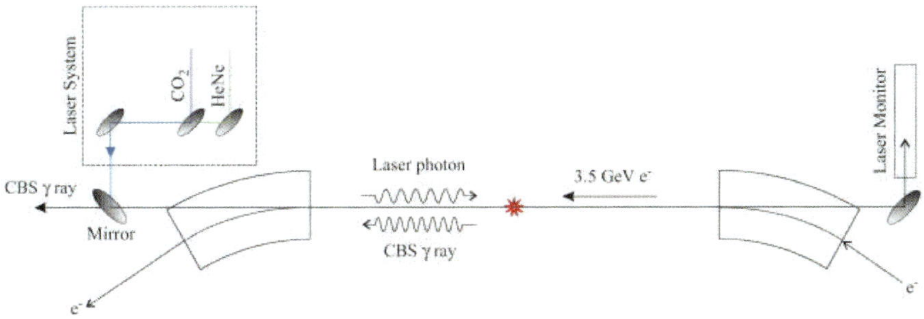

Fig. 62. Layout of SLEGS.

range of kilo-electron-volts (keV UED) is limited to about 1 ps. In the energy range of mega-electron-volts, the pulse charge density can be increased by 3–4 orders of magnitude, and a high quality single shot diffraction pattern with a sub-100 fs electron pulse can be achieved with MeV UED.

In China, MeV UED has been studied since 2008, mainly at THU, SJTU and PKU.

THU began MeV UED research in 2008. A prototype of continuously time-resolved (CTR) MeV UED was set up at the TTX facility, and a high quality diffraction pattern was obtained in 2009. A dedicated UED MeV platform was built in 2014 at THU, and a pump–probe experiment was successfully carried out on a gold monocrystal sample [225–228]. A schematic layout of the THU MeV UED is shown in Fig. 63.

Figure 64 shows the Tsinghua MeV UED system, and Table 22 lists its main parameters.

Based on the research on keV UED, SJTU has established an MeV UED experimental research platform [229]. A schematic layout of the RF-gun-driven UED is drawn in Fig. 65, and a photo of the facility is shown in Fig. 66.

The femtosecond laser system can operate at 1 kHz and the maximal energy for each laser pulse

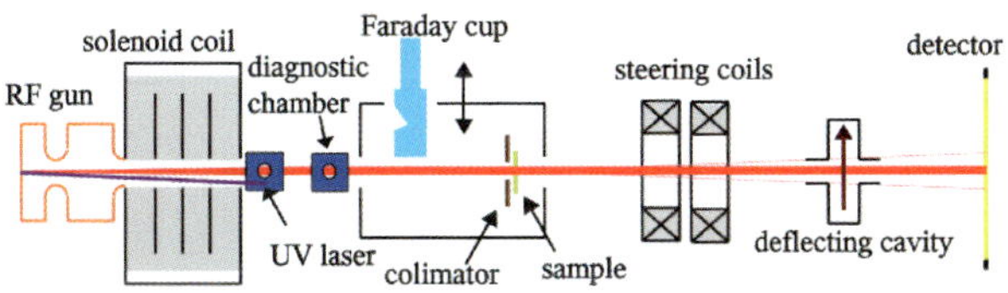

Fig. 63. Systematic layout of the MeV UED system at THU. The system is composed of a 1.6-cell photocathode RF gun, a UV laser system, a collimator, steering coils, an RF deflecting cavity and a detector screen. Samples are placed at the back of the collimator, and the diffraction pattern can be seen on the screen.

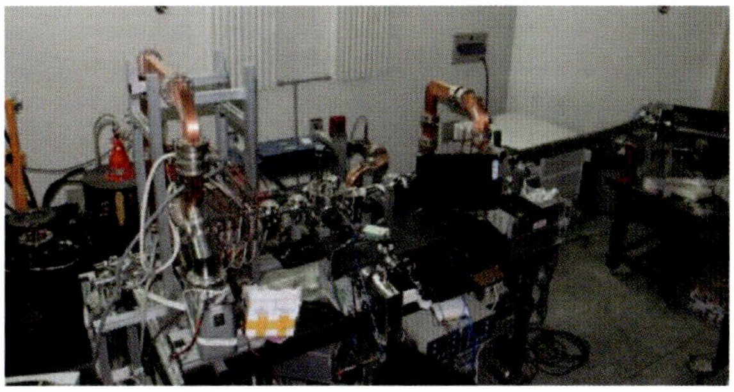

Fig. 64. Photo of the Tsinghua MeV UED system.

Table 22. Parameters of the MeV UED system at THU.

Parameter	Value
Beam energy	1.5–3.0 MeV
Initial charge at cathode	1.0 pC
Gun RF field	60 MV/m
Electron launching phase	20°
Bunch length (rms) at cathode	0.3 ps
Bunch radius (uniform) at cathode	0.20 mm
Normalized emittance at cathode	0.1 μm

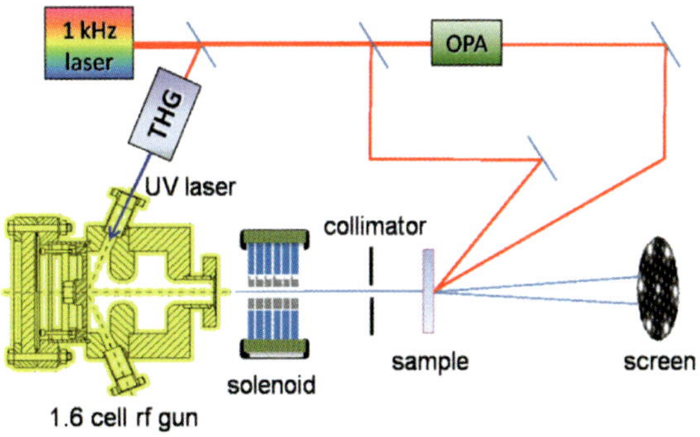

Fig. 65. Systematic layout of MeV UED at SJTU. The electron beam is produced in a 1.6-cell S-band photocathode RF gun with a femtosecond UV laser. The beam is focused by a solenoid to form a convergent electron diffraction pattern on a fluorescent screen about 3.4 m downstream of the sample.

Fig. 66. Photo of MeV UED at SJTU.

Table 23. The main parameters of the MeV UED facility at SJTU.

Beam energy	∼2.5	MeV
Peak field at cathode	∼65	MV/m
Laser injection phase	∼30	deg.
Beam charge	0.1–10	pC
Laser spot size (rms)	∼300	μm
Aperture hole diameter	3.0	mm
Bunch length at sample	100–300	fs
Sample to cathode distance	60	cm
Detector to sample distance	340	cm

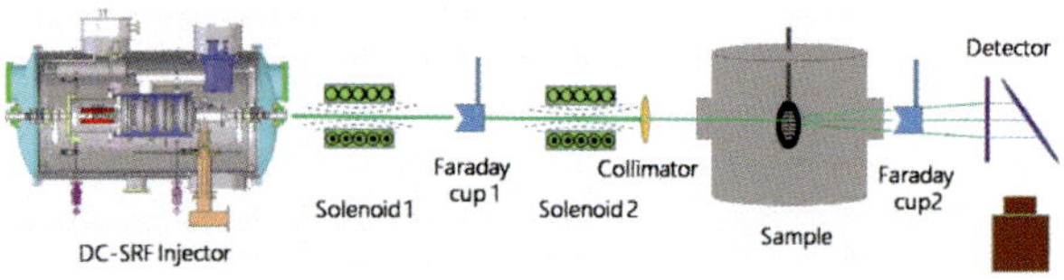

Fig. 67. Schematic layout of the MeV MHz UED beam line. Solenoids 1 and 2 are 1.2 m and 3.5 m downstream of the photocathode, respectively; the sample is 7.0 m and the detector is 9.6 m downstream of the photocathode.

is about 18 mJ [230]. The main parameters of the MeV UED facility at SJTU are given in Table 23.

In the meantime, an advanced MeV UED instrument with a high temporal resolution of 50 fs, a high spatial resolution of a few nanometers and a high repetition rate of 1 kHz is under construction at SJTU, supported by NSFC.

A proof-of-principle experiment of MeV UED operating at the megahertz repetition rate using an SC RF photoinjector was set up jointly by PKU and SJTU [231]. A schematic layout of the experiment is shown in Fig. 67.

8. Accelerators for Applications

Particle accelerators have wide application in society. In China, application accelerators are independently or jointly produced by research institutions and companies. Usually, the institutions and industry collaborate closely on the construction of large accelerator projects. The technology has been transferred to industry smoothly in various ways, such as joint R&D activities, direct technical transfer and joint companies [232, 233]. In this section, different kinds of accelerators are described, such as RF electron linacs, DC high voltage accelerators, cyclotrons and synchrotrons, applied in the radiotherapy, irradiation and nondestructive tests (NDTs) [234–237].

8.1. *Accelerators for radiotherapy*

Currently, about 2000 accelerators are operated for radiotherapy in China. About 30% of them were made in China. Most accelerators for radiotherapy are linacs. Proton and heavy-ion accelerators are also being developed in the country for oncology treatment. As for radioisotope production, cyclotrons are playing the major role.

There are several medical linac manufacturers in China, and now more and more investment is attracted to the medical accelerator industry. BMEI was the biggest medical linac manufacturer in China before merged with Ekita (Beijing), which produced 6 MeV, 14 MeV and 20 MeV radiotherapy sets. A 16 MeV traveling wave linac (HM-J16) with a microwave power feedback system for medical application was produced by Yangzhou Marine Electric and Medical Instrument Company. The main product of IHEP-Daheng medical company is the KBA1800 medical linac with an electron beam energy of 18 MeV. Neusoft Medical of Liaoning province produces 6 MeV sets with an intensity-modulated radiation therapy system. SHINVA Medical of Shandong province produced its first medical accelerator in 2000, and developed a series of radiotherapy sets based on 6 MeV linacs of XHA600, XHA600C and XHA600D (see Fig. 68), as well as the medium energy set of XHA1400. The XHA series machines are equipped with multileaf collimators (MLCS) and support intensity-modulated radiotherapy (IMR) [238].

The trend of developing electron linacs for radiotherapy in China is toward high accelerating gradient and compact structure, as well as high beam quality and high dose rate. At the same time, advanced techniques of radiotherapy such as intensity-modulated therapy, image-guided radiation therapy (IGRT) and 4D-IGRT, are being incorporated into the treatment delivery system.

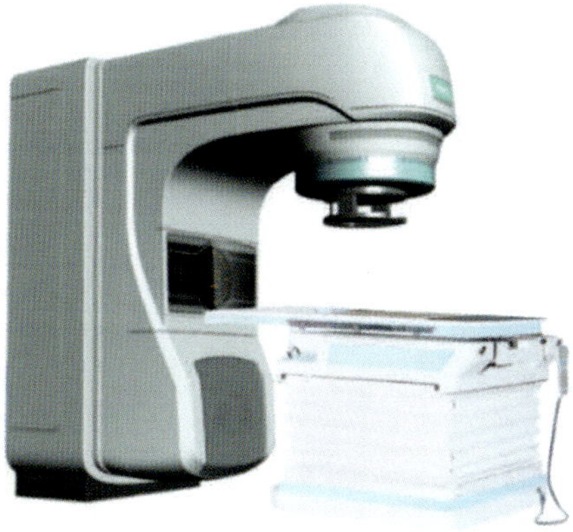

Fig. 68. A radiotherapy set, XHA600D, produced by SHINVA.

Due to the Bragg peak in the absorption curves, proton and heavy-ion beam therapy has advantages over X-ray and electron beam therapy. A heavy-ion beam therapy facility was set up at IMP based on the HIRFL beam lines. Hundreds of patients were treated using this facility, with good results. Two heavy-ion medical machines (HIMMs) with beam energy of 250–400 MeV/u have been constructed in the hospitals of Gansu province by IMP [239, 240]. Figure 69 shows an HIMM synchrotron constructed in the hospital of Wuwei, Gansu Province.

A 70–235 MeV proton accelerator facility, Shanghai Advanced Proton Treatment (SAPT), has been developed by SINAP [241]. Also, a 250 MeV SC cyclotron was designed at Huazhong University of Science and Technology (HUST) for radiotherapy.

Regarding radioisotope production, a 30 MeV cyclotron (CYCIAE 30) designed for producing radioactive isotopes for medical use was completed at the end of 1994 by CIAE (see Fig. 70). It can accelerate 350 μA proton beams to 30 MeV and

Fig. 69. The synchrotron of HIMM in commissioning.

Fig. 70. The cyclotron CYCIAE 30 for radioisotope production.

produce more than 60 kinds of isotopes [93, 242]. A small cyclotron to produce short-lived isotopes for positron emission tomography (PET) was developed by CAEP and 11 MeV, 50 μA beams were obtained. The 14 MeV high current compact cyclotron of CIAE, CYCIAE-14, can produce a variety of isotopes for PET systems. The designed proton beam current is 400 μA [96, 237].

8.2. *Accelerators for irradiation*

Accelerators for irradiation are applied in the fields of material modification, polymers, healthcare, food, environment and others. Along with the rapid development of the economy in China, the demands of accelerators for irradiation are steadily on the increase. There are mainly two kinds of accelerators that have been developed for radiation processing in China, i.e. high power electron linacs and high voltage accelerators [13, 232]. In contrast to the medical accelerators, the irradiation accelerators in China are mostly produced by domestic enterprises.

IHEP produces a series of irradiation accelerators of 10 MeV, 15 kW standing wave linacs and 10 MeV, 20–40 kW traveling wave linacs based on the technology developed during the construction of BEPC. The L-band 10 MeV, 40 kW linacs for irradiation were jointly developed and produced by IHEP and the Wuxi EL-PONT company, as shown in Fig. 71 [243].

THU and NUCTECH jointly developed return-type S-band traveling wave linacs for irradiation. CIAE has produced various electric linac systems, including 14 MeV, 15 kW and 10 MeV, 15 kW irradiation facilities. NUCTECH have also developed a

Fig. 71. The L-band 10 MeV, 40 kW linac for irradiation.

Fig. 72. Electron linacs for irradiation. *Left*: The 10 MeV, 15 kW linac produced by CIAE. *Right*: The 4.5 MeV, 2 kW linac for sterilization produced by NUCTECH.

mobile irradiation system. Shandong Vanform, Wuxi EL-PONT, CGN Zhongke HIWITS and other high tech companies have made a series of irradiation products with electron beam power from 5 kW to 20 kW, in cooperation with the research institutions of IHEP, SINAP, Nanjing University and USTC. Figure 72 shows the 10 MeV, 15 kW linac produced by CIAE and the 4.5 MeV, 2 kW linac for sterilization produced by NUCTECH [232].

A variety of high voltage accelerators have been developed and applied in industry and other fields in China, including Dynamitrons, high voltage transformer accelerators, Marx generators, medium frequency transformer accelerators, insulating core transformer accelerators, ion implanters, electron curtain accelerators and Rhodotrons [13].

SINAP has developed a series of Dynamitrons with electron energy from 1.5 MeV to 5.0 MeV and

Fig. 73. High voltage accelerators developed by SINAP. *Left*: A Dynamitron under installation on site. *Right*: The 1.2 MeV, 50 mA accelerator for EB cleaning flue gases.

intensity from 10 mA to 40 mA. It has also produced electron curtain accelerators of 0.3 MeV and 50 mA. Figure 73 shows the high voltage accelerators developed by SINAP.

CGN Zhongke HIWITS, Jiansu Dasheng and Wuxi EL-PONT have manufactured Dynamitrons from 0.5 MeV and 100 mA to 10 MeV and 2 mA as industrial products. Based on the IMP technology, Lanzhou Jintai has developed DG series accelerators with beam energy 0.2–2.5 MeV and beam current of 40–50 mA for irradiation. Beijing Research Institute of Automation for Machinery Industry (RIAMB) developed a series of insulating core transformer accelerators of 0.3 MeV and 40 mA, 0.6 MeV and 40 mA, and 1.0 MeV and 20 mA. HUST has been working on insulating core transformer accelerators and has achieved a high power conversion efficiency of 85%. Ningbo Chaoneng has been working on Rhodotrons of beam energy 5–10 MeV [13].

8.3. *Accelerators for nondestructive tests*

Accelerators, usually 1–20 MeV electron linacs, can act as high energy X-ray sources generated by beam–material interactions for nondestructive tests (NDTs). The main applications of the accelerator-based NDTs are high energy X-ray industry computer tomography (ICT) and X-ray inspection, especially for such large objects as cargos.

Based on the technology developed at BEPC, IHEP has built ICT systems with 6 MeV (or 4 MeV, 9 MeV or 14 MeV) electron linacs. NUCTECH has manufactured ICT products based on 15 MeV accelerators. RIAMB produces a series of NDT linacs of

Fig. 74. Photo of the ICT produced by NUCTECH.

2 MeV, 4 MeV, 6 MeV, 9 MeV and 12 MeV. Granpect and SPRI supply a series of NDT products with linac energy from 2 MeV to 15 MeV [13, 232]. Figure 74 shows the ICT produced by NUCTECH.

Regarding X-ray inspection, THU and NUCTECH developed several series of cargo inspection systems, including the fixed system (9 MeV traveling wave or standing wave linac), the mobile system (S-band or X-band 2.5 MeV linac), the relocatable system (6 MeV standing wave linac), the fast scan system, the portal system, the rail system, and the air cargo system for different inspection usages [244, 245]. The series of cargo inspection systems produced by NUCTECH are shown in Fig. 75.

New technologies based on dual energy X-rays, neutrons and monoenergetic X-rays have been studied to achieve the required material identification.

Fig. 75. The series of cargo inspection systems produced by NUCTECH.

Fig. 76. Customer distribution of cargo inspection systems produced by NUCTECH (updated on Jan. 29, 2016).

The dual energy system uses an S-band standing wave linac with electron energy of 9 MeV and 6 MeV, and the movable dual energy system uses a linac of 6 MeV and 3 MeV. NUCTECH has combined the dual energy and fast scan technologies to produce material identification images with a truck moving at a speed of 15 km/h. As one of the main vendors, NUCTECH has provided around 1100 sets of cargo inspection systems to more than 140 countries [246]. The customer distribution of cargo inspection systems produced by NUCTECH is shown in Fig. 76.

9. Accelerator Technology and Advanced Accelerator Concepts

The progress of particle accelerators in China is closely related to the development of accelerator physics and technology, as shown in the previous sections. In this section, some examples of technology development in accelerator projects are discussed and the study of advanced accelerator concepts is described.

9.1. *Accelerator technology*

Particle accelerators involve a great deal of advanced technologies, such as high power microwave and RF systems, accurate magnets and their power supplies, permanent undulators and wigglers, magnet ion sources, photocathode electron guns, positron sources, ultrahigh vacuum, beam diagnosis, computer control, bunch-by-bunch feedback, SC cavities, cryogenics, insertion devices, accurate mechanics, survey and alignment. Along with the development of China's economy, a number of accelerator projects have been designed and constructed; and

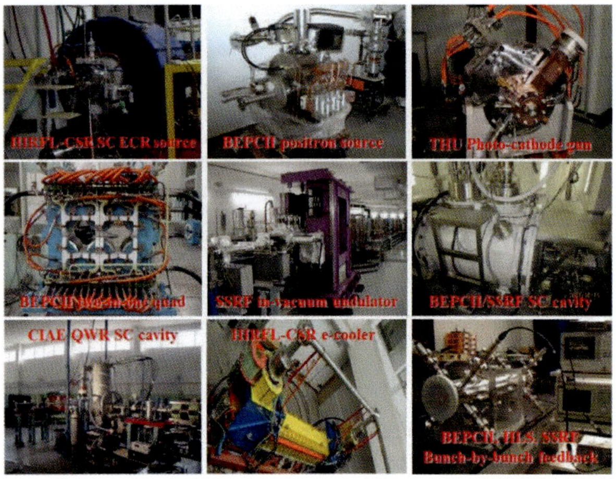

Fig. 77. Examples of accelerator technology developed during the accelerator projects.

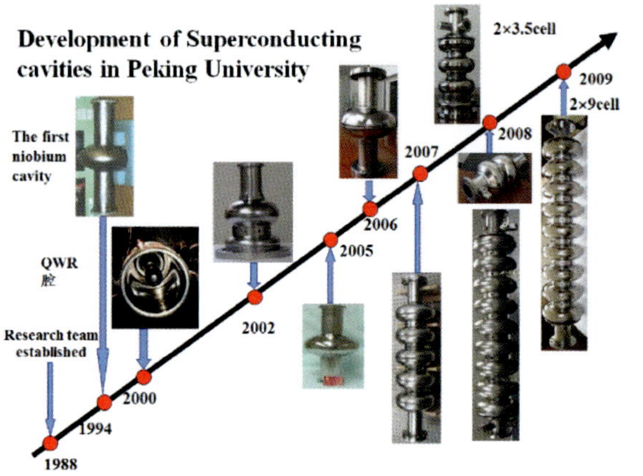

Fig. 78. Twenty-year development of SRF technology at PKU.

the corresponding technology has made progress by leaps and bounds. As examples, Fig. 77 shows some advanced devices constructed and applied in China's accelerators [233].

Among them, the development of superconducting RF (SRF) technology is a typical example.

SRF technology was first explored in China in the early 1970s, at IHEP. Unfortunately, the research was suspended for about 10 years, until an SRF laboratory was founded at PKU at the end of the 1980s. The early efforts of the PKU SRF group were made in the design and fabrication of L-band high-beta cavities using homemade niobium sheets, and in the study of Nb–Cu sputtering technology to construct SC QWR for CIAE. In 2003, the "SRF-accelerator-based FEL light source" project proposed by PKU and joined in by IHEP and SINAP was approved by MOST. A series of R&D around the project was carried out at PKU, including the construction, processing and testing of the DC-SC photoinjector, the R&D of multicell SC cavities, the study of the large grain niobium cavities, and the development of a spoke cavity for ion acceleration. The maximum accelerating gradient of the nine-cell TESLA type SC cavities reached 34.6 MV/m [247–251]. Based on SRF technology, PKU constructed an SC l with beam current above 1 mA and a compact THz source using the DC-SC photoinjector [252]. Figure 78 illustrates the 20-year development of SRF technology in PKU.

Efforts have also been made by the Chinese ILC team in the development of SC cavities. At IHEP, single and nine-cell cavities with fine and large grains were manufactured, processed and tested, and an SRF accelerating unit is undergoing tests to be used as the booster for an SRF linac [226].

Regarding SRF in electron storage rings, two KEKB type cavities and three CESR type SRF cavities of 500 MHz were installed in BEPCII and the SSRF storage rings respectively, and have operated successfully since then.

SRF technology will be applied in the ongoing project CIADS. As its R&D program, two alternative types of SRF cavities have been developed: one is a low β single spoke resonator type (Spoke012, 021 and 040) developed by IHEP; the other is a 162.5 MHz low-β half-wave resonator type (HWR010 and HWR015) developed by IMP. All of the Spoke type and HWR type SRF cavities of the two injectors have been manufactured, tested and operated with beams, with satisfactory results [109, 253–256]. The vertical test results of Spoke012 and HRW010 SRF cavities are given in Figs. 79 and 80, respectively. It can be seen from the figures that the measured accelerating gradients of all the cavities are better than the designed values.

At the same time, the construction of the QWR cavities of the BRIF SC booster is in progress. CIAE has also been developing thin film SRF cavities for the BRIF upgrading.

9.2. *Advanced accelerator concepts*

Studies of advanced accelerator concepts began in the late 1970s, when the idea of phase-adjusted focusing laser acceleration was presented [257]. A scheme of electron bunch acceleration by intense laser pulse

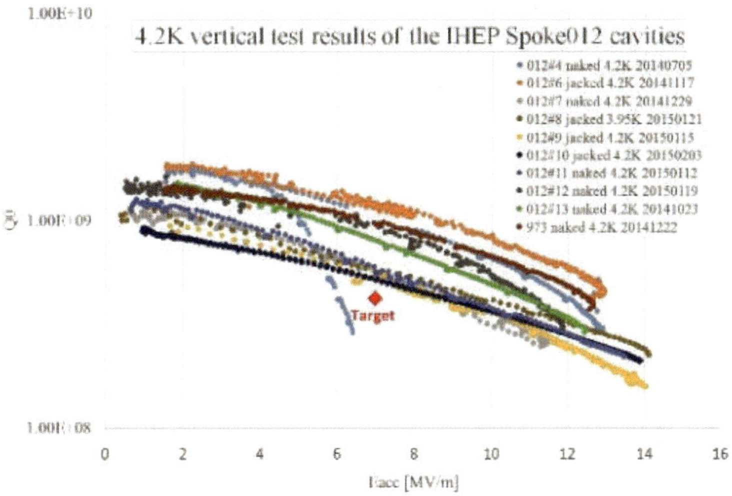

Fig. 79. Vertical test result of Spoke012 SRF cavities.

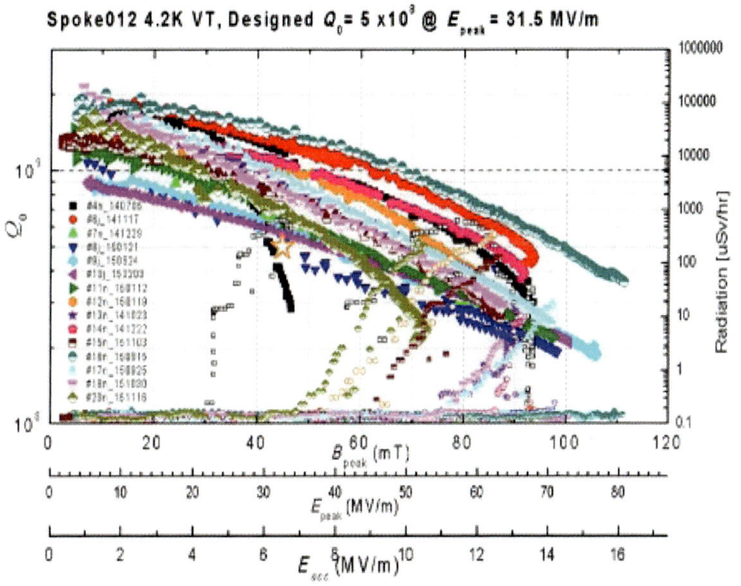

Fig. 80. Vertical test result of HWR010 SRF cavities.

in vacuum with the capture and acceleration scenario was proposed in 2003 [258, 259].

In the past 15 years, studies of advanced concepts in China have concentrated on laser wakefield accelerators (LWFAs). Intensive studies, both theoretical and experimental, have been carried out by the teams of the Institute of Physics (IoP), SIOM, PKU, SJTU, THU, IHEP, SINAP, CAEP and other institutions, and significant progress has been made in acceleration mechanism and new schemes, cascaded acceleration, accelerated beam stability and repeatability, control of the beam energy spectrum, possible applications, and others [260–270]. Near 1 GeV electron beams and 60 MeV proton beams were obtained in the experiments with domestic and international collaboration.

In 2010, SIOM experimentally demonstrated a near-GeV cascaded LWFA using ionization-induced injection [271], and later in 2013, a two-stage LWFA with seeding phase control and gradient injection was realized and electrons of peak energy above 1 GeV were obtained [272, 273]. Figure 81 shows a schematic experimental setup of the cascaded LWFA at SIOM.

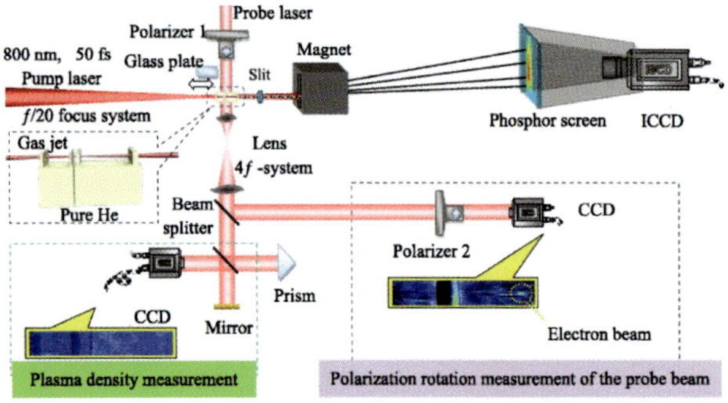

Fig. 81. Schematic experimental setup of the cascaded LWFA at SIOM. The laser beam was focused by a parabolic mirror into the entrance of an ablative capillary.

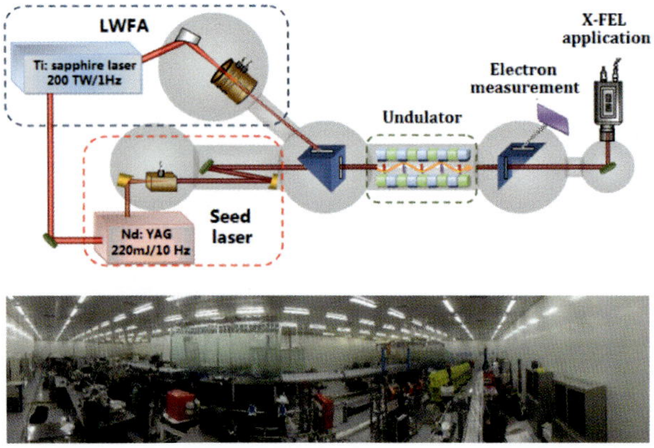

Fig. 82. Schematic layout of the LWFA-based XFEL facility at SIOM and photo of the setup. The amplified chirped T: sapphire laser beam is split into two beams to drive LWFA of electrons and the HGHG X-FEL, respectively.

As a key instrument project supported by NSFC, an LWFA-based seeded HGHG X-ray FEL setup is being developed at SIOM [274]. Figure 82 shows a schematic layout of the facility. The facility has been in commissioning and the undulator emission was observed with the LWFA-accelerated 400 MeV electron beams. Based on the high quality monoenergetic electron beams generated by the LWFA system, a Compton backscattering experiment was performed, and ultrahigh brilliance MeV γ-rays with a peak brilliance of $\sim 3 \times 10^{22}$ photons/(s$\cdot$mm$^2\cdot$mrad^2 0.1% BW) at 1 MeV was obtained [275].

Regarding the ion acceleration, PKU has proposed a concept of generating high current monoenergetic proton beams by a circularly polarized laser pulse in the phase-stable acceleration regime [276]. 60 MeV proton and 600 MeV carbon beams were obtained by a collaboration of PKU, LANL, MPQ and other institutions on the experiments [277–279].

Experimental research on laser-driven proton acceleration was also carried out at SIOM. Proton acceleration by high contrast 5 TW and 100 TW lasers was performed in 2012–2013 [280, 281]. In 2014, the effects of a deforming mirror on proton acceleration were studied, the proton radiography of a dragonfly was obtained, and ion acceleration by an 800 nm femtosecond laser-driven collisionless shock was observed [282].

In the same time, a compact laser plasma accelerator (CLAPA) is being constructed at PKU as a key instrument project funded by MOST. The system has been designed to demonstrate the feasibility of laser–plasma accelerators for medical and biological applications, with protons accelerated to energies from 1 to 15 MeV. It will also be employed to generate high quality electron beams using the ionization injection schemes and dual laser beams for better control of the beam quality [283]. A schematic layout of CLAPA is shown in Fig. 83, and photos of CLAPA under construction are shown in Fig. 84.

The study of advanced accelerator concepts is still in its early stage in China. The accelerated beam quality should be improved, the laser and target technology should be developed further, and the application of laser–plasma accelerators in compact light sources, particle beam injectors, medical treatment

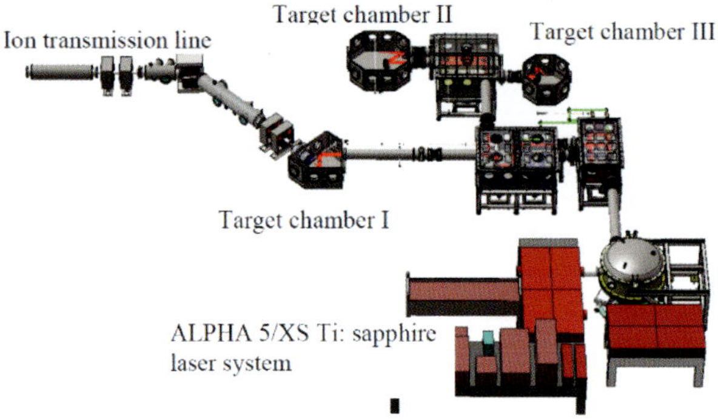

Fig. 83. Schematic layout of CLAPA.

Fig. 84. Photos of CLAPA under construction.

and other fields should be promoted. There is still a long way to go before the advanced concepts are converted into advanced accelerators.

We have come a long way from the 1950s. In the past 60 years, particle accelerator research has grown out of nothing and, by the efforts of several generations in China, has gradually caught up with the advanced levels in world research. Looking forward to the future, we are full of confidence that China's accelerators will achieve higher energy, higher intensity, higher efficiency and better performance, in order to serve the purpose of scientific research and technology development, as well as extensive applications in society.

Acknowledgments

The authors would like to thank Prof. Jiawen Xia of IMP, Prof. Tianjue Zhang of CIAE, Profs. Shinian Fu, Weimin Pan and Qing Qin of IHEP, Prof. Zhentang Zhao of SINAP, Prof. Jianjun Deng of CAEP, Profs. Chuanxiang Tang, Huaibi Chen and Wei Lu of THU, Profs. Kexin Liu and Xueqing Yan of PKU, and Prof. Ruxin Li of SIOM for kindly providing the detailed and updated information on the relevant accelerators. Furthermore, we are particularly grateful to our colleagues of generations who have devoted themselves to the development of particle accelerators in China. They are practical authors of this article.

Appendix A: The Institutions Mentioned in This Paper

Name	Location
China Institute of Atomic Energy, CIAE	Beijing
Institute of Modern Physics, IMP	Lanzhou
Peking University, PKU	Beijing
Tsinghua University, THU	Beijing
Shanghai Institute of Applied Physics, SINAP	Shanghai
Beijing Medical Equipment Institute, BMEI	Beijing
Institute of High Energy Physics, IHEP	Beijing
University of Science of Technology of China, USTC	Hefei
Shanghai Institute of Optics and Fine Mechanics, SIOM	Shanghai
Wuhan University	Wuhan
Beijing Normal University	Beijing
Lanzhou University	Lanzhou
Shanghai Jiao Tong University, SJTU	Shanghai
University of Science & Technology, HUST	Wuhan
Institute of Physics, IoP	Beijing

Appendix B: The Accelerator Facilities Mentioned in This Paper

Name	City	Year
2.5 MeV electrostatic proton accelerator, V2	Beijing	1957
IMP 1.5 m heavy-ion cyclotron	Lanzhou	1963
SINR 1.2 m proton cyclotron	Shanghai	1964
CIAE 30 MeV electron linac	Beijing	1964
Beijing medical electron linac, BJ-10	Beijing	1977
Shanghai medical electron linac, ZJ-10	Shanghai	1977
Beijing Proton Linac, BPL	Beijing	1985
4.5 MV electrostatic accelerator	Beijing	1986
13 MeV tandem, HI-13	Beijing	1986
Beijing Electron–Positron Collider, BEPC	Beijing	1988
Heavy-Ion Research Facility at Lanzhou, HIRFL	Lanzhou	1988
Hefei Light Source, HLS	Hefei	1989
Beijing Free Electron Laser, BFEL	Beijing	1993
SINR 2 × 6 MV tandem	Shanghai	1993

Name	City	Year
30 MeV compact cyclotron at CIAE, CYCIAE-30	Beijing	1995
HIRFL Cooling Storage Ring, HIRFL-CSR	Lanzhou	2006
In situ TEM-accelerator interface facility	Wuhan	2008
Shanghai Synchrotron Radiation Facility, SSRF	Shanghai	2010
Shanghai deep ultraviolet free electron laser, SDUV-FEL	Shanghai	2011
Tsinghua Thomson backscattering X-ray source, TTX	Beijing	2011
PKU Neutron Imaging Facility, PKUNIFTY	Beijing	2011
14 MeV compact cyclotron at CIAE, CYCIAE-14	Beijing	2012
100 MeV compact cyclotron at CIAE, CYCIAE-100	Beijing	2014
MeV UED facility at THU	Beijing	2014
Beijing Radioactive Ion Beam Facility, BRIF	Beijing	2015
China's accelerator-driven system, CADS	Beijing/ Lanzhou	2016
Dalian Coherent Light Source, DCLS	Dalian	Constr
Soft X-ray FEL test facility, SXFEL	Shanghai	Constr.
MeV UED facility at SJTU	Shanghai	Constr.
Heavy-Ion Medical Machines, HIMM	Wuwei	Constr.
Shanghai Advanced Proton Treatment, SAPT	Shanghai	Constr.
China Spallation Neutron Source, CSNS	Dongguan	Constr.
Campact Pulsed Hadron Source, CPHS	Beijing	Constr.
Heavy-Ion Accelerator Facility, HIAF	Huizhou	Planned
Beijing ISOL	Beijing	Planned
China Initiative Accelerator-Driven System, CIADS	Huizhou	Planned
High Energy Light Source, HELS	Beijing	Planned
Hard X-ray FEL	—	Planned
Circular Electron–Positron Collider, CEPC	—	Planned

References

[1] J. L. Xie, Development of linear accelerators in China, in *Proc. 1984 Linear Accelerators Conference* (1984), pp. 14–18.

[2] C. E. Chen, W. J. Zhao and X. R. Liang, Development of low energy accelerators in China, *Physics* **18**, 581 (1989), in Chinese.

[3] S. X. Fang, The development of China's accelerators I have experienced, in *Proc. IPAC 2013* (2013), pp. 3144–3148.

[4] M. H. Yie, *Acta Phys. Sin.* **60**, 19 (1963), in Chinese.

[5] S. X. Fang, Brief overview of particle accelerators in China, in *Proc. APAC 1998*, (1998), p. 3–7.

[6] Q. Y. Zhou, in *Proc. Third Japan–China Joint Symposium on Accelerators for Nuclear Science and Application* (1987), pp. 53–57.

[7] C. E. Chen, in *Proc. First Japan–China Joint Symposium on Accelerators for Nuclear Science and Application* (1987), pp. 19–23.

[8] J. L. Xie *et al.*, First lasing of Beijing FEL, NIMA **341**, p. 43.

[9] C. E. Chen *et al.*, Experimental studies on the acceleration of positive & negative ions with a heavy ion ISR RFQ, in *Proc. EPAC 1996* (1996).

[10] C. E. Chen *et al.*, The test run of the AMS system at Peking University, in *Proc. PAC 2001* (2001), pp. 2616–2618.

[11] Y. Z. Lin *et al.*, Three types of linacs for customs large container inspection application, in *Proc. PAC 2001*, pp. 2808–2810.

[12] Chinese Nuclear Society, *Report on Advances in Nuclear Science and Technology, 2007–2008* (in

Chinese), Science and Technology of China Press (2008).

[13] Chinese Nuclear Society, *Report on Advances in Nuclear Science and Technology, 2014–2015* (in Chinese), Science and Technology of China Press (2016).

[14] S. X. Fang and S. Y. Chen, The Beijing electron–positron collider, in *Proc. 14th Int. Conf. High-Energy Accelerators* (1989), pp. 51–55.

[15] J. Wang, Q. Han and D. Ling, The status of BEPC 1.3 GeV linac, in *Proc. 1994 Int. Linac Conf.* (1994), pp. 202–204.

[16] Y. Z. Wu, (Presented for the BEPC storage ring commissioning group), The BEPC storage ring — status and prospects, in *Proc. PAC91* (1991), pp. 180–182.

[17] S. H. Wang, For the BEPC team, BEPC status and plans, in *Proc. PAC 95* (1995), pp. 506–519.

[18] C. Zhang, For the BEPC team, status of BEPC and plans of BEPCII, in *Proc. APAC 2001* (2001), pp. 6–10.

[19] S. Y. Chen *et al.*, *HEP & NP* (formerly *Chin. Phys. C*) **10**(2), 208–214 (1986), in Chinese.

[20] Z. Y. Guo *et al.*, Single bunch phenomena in BEPC, in *Proc. PAC 1993* (1993), pp. 3237–3239.

[21] Z. Y. Guo *et al.*, Beam instability studies in the BEPC, in *Proc. EPAC 2000* (2000), pp. 1129–1129.

[22] Z. Y. Guo, Q. Qin and C. Zhang, *HEP & NP* (formerly *Chin. Phys. C*) **20**(11), 1034–1040 (1996), in Chinese.

[23] Z. Y. Guo *et al.*, *HEP & NP* (formerly *Chin. Phys. C*) **25**(03), 251–257 (2001), in Chinese.

[24] C. Zhang, *HEP & NP* (formerly *Chin. Phys. C*) **22**(11), 1050–1056 (1998).

[25] S. Wang, S. X. Fang and C. Zhang, Beam–beam effect in BEPC miniI-β scheme, in *Proc. PAC 1999* (1999), pp. 1695–1697.

[26] BES Collaboration, *Phys. Rev. Lett.* **91**, 022001, 1–5 (2003).

[27] BEPCII team, BEPCII design report IHEP-BEPCII-SB-03-03 (2003), in Chinese.

[28] C. Zhang, L. Ma *et al.*, *Design and construction of BEPCII* (Shanghai Scientific & Technical Publisher, 2015), in Chinese.

[29] C. Zhang, For the BEPC & BEPCII team, Performance of the BEPC and progress of the BEPCII, in *Proc. APAC 2004* (2004), pp. 15–19.

[30] J. Q. Wang *et al.*, For the BEPCII team, status and performance of BEPCII, in *Proc. IPAC'10* (2010), pp. 2359–2363.

[31] Q. Qin, For the accelerator team of BEPCII, Performance and prospects of BEPCII, in *Proc. IPAC'12* (2010), pp. 1030–1034.

[32] B. Liu *et al.*, Experiment and operation of BEPCII electron gun system, *Chin. Phys. C* **30**(05), 466–469.

[33] G. X. Pei, BEPII positron source, *Chin. Phys. C* **30**(01), 66–70 (2006).

[34] G. W. Wang, The construction of the BEPC II RF system, *Chin. Phys. C* **32**(S1), 169–171 (2008).

[35] J. H. Yue *et al.*, The experiments of transverse feedback system on the BEPC, *Chin. Phys. C* **30**(02), 163–165 (2006).

[36] Z. S. Yin *et al.*, Design and field measurement of the BEPC-II interaction region dual-aperture quadrupoles, *NIMA* **573**, 323–328 (2007).

[37] S. H. Wang *et al.*, BEPCII linac beam commissioning and operation, *Chin. Phys. C* **30**(S1), 60–62 (2006).

[38] C. Zhang, BEPCII: Construction and commissioning, *Chin. Phys. C* **33**(S2), 60–64 (2009).

[39] G. Xu *et al.*, Study on the accelerator physics issues in the storage ring of BEPCII, *Chin. Phys. C* **30**(S1), 69–70 (2006).

[40] Y. Y. Wei and Q. Qin, Application of response matrix at the BEPC storage ring, *Chin. Phys. C* **30**(S1), 43–45 (2006).

[41] C. H. Yu *et al.* BEPCII performance and beam dynamics studies on luminosity, in *Proc. IPAC 2016* (2016), pp. 1014–1018.

[42] BESIII Collaboration, *Phys. Rev. Lett.* **110**, 252001, 1–7 (2013).

[43] The CEPC-SPPC Study Group, CEPC-SPPC Preliminary Conceptual Design Report, IHEP-CEPC CEPC-DR-2015-01 (Mar. 2015).

[44] Q. Qin, Overview of the CEPC accelerator, in *Proc. HP 2014* (2014), pp. 16–21.

[45] H. P. Geng, Lattice and its related beam dynamics issues in the CEPC storage ring, in *Proc. IPAC 2015* (2015), pp. 2469–2471.

[46] A preliminary design of the CEPC interaction region, in *Proc. IPAC 2015* (2015), pp. 2019–2021.

[47] D. Wang *et al.*, CEPC parameter choice and partial double ring design, in *Proc. IPAC 2016* (2016), pp. 3788–3790.

[48] S. Bai *et al.*, CEPC parameter choice and partial double ring design, in *Proc. IPAC 2016* (2016), pp. 3802–3804.

[49] T. J. Bian *et al.*, MDI design in CEPC partial double ring, in *Proc. IPAC 2016* (2016), pp. 3791–3794.

[50] J. Y. Zhai *et al.*, Preliminary conceptual design of the CEPC SRF system, in *Proc. SRF 2015* (2015), pp. 272–275.

[51] H. J. Liu, J. Gao and Y. C. Liu, Cavity and HOM coupler design for CEPC, *Chin. Phys. C* **40**(5), 057001 (2016).

[52] B. W. Wei, Present status of HIRFL, in *Proc. 9th Int. Conf. Cycl. Appl.* (1981), pp. 23–32.

[53] B. W. Wei, Commissioning of HIRFL, in *Proc. 11th Int. Conf. Cycl. Appl.* (1989), pp. 9–12.

[54] J. W. Xia *et al.*, The heavy ion cooler-storage-ring project (HIRFL-CSR), *NIMA* **488**, 11–25 (2002).

[55] W. L. Zhan, *NIMA* **805**, 533–540 (2008).
[56] W. L. Zhan, H. S. Xu, G. Q. Xiao *et al.*, *NIMA* **834**, 694–700 (2010).
[57] J. C. Yang, J. W. Xia and W. L. Zhan, HIRFL-CSR complex, *Chin. Phys. C* **33**(09), 804–810 (2009).
[58] Y. J. Yuan, J. W. Xia and M. J. Li, Charge stripping accumulation of light heavy ions in HIRFL-CSR main ring, *Chin. Phys. C* **32**(09), 754–757 (2008).
[59] L. J. Jia *et al.* Intrabeam scattering of heavy ions at HIRFL-CSR, *Chin. Phys. C* **32**(09), 754–757 (2008).
[60] L. T. Sun *et al.*, Brief review of multiple charge state ECR ion, *NIMB* **235**, 524–529 (2005).
[61] Y. J. Yuan *et al.*, Status of the HIRFL-CSR complex, *NIMB* **317**, 214–217 (2013).
[62] L. J. Mao *et al.*, Electron cooling experiments at HIRFL-CSR, in *Proc. SAP 2014*, pp. 113–115.
[63] Y. J. Yuan *et al.*, Status and upgrades of HIRFL, in *Proc. HIAT 2015*, pp. 110–114.
[64] X. L. Tu *et al.*, *Phys. Rev. Lett.* **106**, 112501 (2012).
[65] J. Qin *et al.* Test and commissioning of the HI-13 tandem accelerator, *NIMA* **268**, 316–318 (1988).
[66] X. L. Guan *et al.*, The first five year experience with the HI-13 tandem accelerator at CIAE, *NIMA* **328**, 34–38 (1993).
[67] S. S. Jiang *et al.* Accelerator mass spectrometry with the Beijing HI-13 tandem, *NIMB* **52**, 285–289 (1990).
[68] W. P. Liu *et al.*, Current progress of nuclear astrophysics study and BRNBF at CIAE, *NIMB* **204**, (2003), pp. 62–70.
[69] X. L. Guan, M. W. Fan and W. S. Jiang, Proposal for a radioactive nuclear beam facility at CIAE, *NIMA* **382**, (1996), pp. 241–244.
[70] T. J. Zhang, J. J Yang *et al.*, The beam commissioning of BRIF and future cyclotron development at CIAE, *NIMB* **376**, 434–439 (2016).
[71] B. Tang *et al.*, The commissioning of the BRISOL facility, *NIMB* **376**, 429–433 (2016).
[72] J. C. Yang *et al.*, High Intensity Heavy Ion Accelerator Facility (HIAF) in China, *NIMB* **317**, 263–265 (2013).
[73] B. Q. Cui *et al.*, The Beijing ISOL initial conceptual design report, *NIMB* **317**, 263–265 (2013).
[74] Z. J. Wang *et al.*, Conceptual design of superconducting heavy ion linear injector for HIAF, in *Proc. LINAC 2012*, pp. 561–563.
[75] X. Gao *et al.*, The SHER-HIAF ring lattice design, in *Proc. IPAC 2013* (2013), pp. 2561–2563.
[76] L. J. Mao *et al.*, Electron cooling system in the booster synchrotron of the HIAF project, *NIMA* **786**, 91–96 (2015).
[77] CSNS team, Preliminary design report of the China Spallation Source (CSNS), IHEP-CSNS-Report/2004-01E (2004).
[78] S. X. Fang *et al.*, *J. Kor. Phys.* **48**(4), 697 (2006).
[79] J. Wei, S. N. Fu and S. X. Fang, China spallation neutron source accelerators: design, research, and development, in *Proc. EPAC 2006* (2006), pp. 366–368.
[80] J. Y. Tang, Rapid cycling synchrotrons and accumulator rings for high-intensity hadron beams, *RAST* **06**, 143–169 (2013).
[81] S. N. Fu *et al.*, Status of CSNS project, in *Proc. IPAC 2013* (2013), pp. 3995–3999.
[82] S. Wang, Introduction to the overall physics design of CSNS accelerators, *Chin. Phys. C* **33**(S2), 1–3 (2009).
[83] S. Y. Xu and S. Wang, Study on space charge effects of the CSNS/RCS, **35**(12), 1152–1158 (2011).
[84] Thermal analysis and cooling structure design of the primary collimator in CSNS/RCS, *Chin. Phys. C.*
[85] S. N. Fu *et al.*, Linac construction for China spallation neutron source, in *Proc. LINAC 2012* (2012), pp. 807–809.
[86] N. Wang, Q. Qin and Y. D. Liu, The electron proton instability in the CSNS ring, *Chin. Phys. C* **33**(S2), 8–11 (2009).
[87] C. D. Deng *et al.*, CSNS magnet system and prototypes fabrication, *Chin. Phys. C* **32**(S1), 71–73 (2008).
[88] X. Qi, Progress on the CSNS power supply system, in *Proc. IPAC 2015* (2015), pp. 2021–2913.
[89] H. L. Huang *et al.*, Measurements on different ferrite materials, *Chin. Phys. C* **32**(S1), 212–214 (2008).
[90] J. Qiu *et al.*, Studies of transverse phase space painting for the CSNS RCS injection, *Chin. Phys. C* **31**(10), 942–946 (2007).
[91] L. Shen *et al.*, Design of injection painting bumper magnets pulse power supply for CSNS/RCS, *Chin. Phys. C* **32**(S1), 22–24 (2008).
[92] S. Wang, S. N. Fu and J. Zhang, Construction and beam commissioning of CSNS accelerators, in *Proc. IPAC 2016* (2016), pp. 47–51.
[93] M. Fan, X. Zhang and Z. Li, The construction of high intensity cyclotron CYCIAE-30, *Chin. Sci. Bull.* **40**(20), 1825 (1995).
[94] T. Zhang *et al.*, Test stand design and construction for high intensity cyclotron development, *Chin. Phys. C* **32**(S1), 237–240 (2008).
[95] T. J. Zhang *et al.*, Overall design of CYCIAE-14, a 14 MeV PET cyclotron, *NIMB* **269**, 2950–2954 (2011).
[96] T. Zhang *et al.*, CYCIAE-100, a 100 MeV H^- cyclotron for RIB production, *NIMB* **261**, 1027–1031 (2007).

[97] T. J. Zhang *et al.*, Design & construction status of CYCIAE-100, *NIMB* **266**, 4117–4122 (2008).

[98] T. J. Zhang *et al.*, The beam dynamics study for the CYCIAE-100 cyclotron, *NIMA* **676**, 90–95 (2012).

[99] B. Ji *et al.*, Theoretical and practical study on RF model cavity of 100 MeV H^- cyclotron, *Chin. Phys. C* **32**(S1), 160–162 (2008).

[100] T. J. Zhang *et al.*, Magnet design and engineering technique study on a 100 MeV compact H^- cyclotron, *Chin. Phys. C* **32**(S1), 1–3 (2008).

[101] T. J. Zhang *et al.*, The beam commissioning of BRIF and future cyclotron development at CIAE, *NIMB* **376**, 434–439 (2016).

[102] S. X. Fang *et al.*, High power proton accelerator for nuclear power plant, in *Proc. CNIC-01468 CNNC-0002* (2000), pp. 95–107.

[103] H. H. Xia and Z. X. Zhao, The summary of researches on ADS in China, frontiers of energy and power engineering in China, *Front. Energy Power Eng. China* **1**(2), 71–79 (2007).

[104] D. Z. Ding, The conceptual study of accelerator driven radiological clean nuclear power system (AD-RCNPS) in China, in *Proc. 3rd Workshop on Neutron Science Project* (1998), pp. 132–140.

[105] X. L. Guan *et al.*, The activities of HPPA technology related to ADS in China, in *Proc. APAC 2001* (2001), pp. 138–142.

[106] S. N. Fu *et al.*, A high-current RFQ accelerator for the ADS system in China, *J. Kor. Phys. Soc.* **48**(4), 806–809 (2006).

[107] S. X. Fang *et al.*, Suggestion on the accelerator driven subcritical reactor system (ADS), *Bull. Chin. Acad. Sci.* **24**(06), 641–644 (2009).

[108] W. L. Zhan and H. S. Xu, Advanced fission energy program — ADS transmutation system, *Bull. Chinese Acad. Sci.* **27**(03), 375–381 (2012).

[109] W. M. Pan and J. P. Dai, ADS based on linear accelerators, *RAST* **8**, 55–76 (2015).

[110] Z. H. Li *et al.*, Physics design of an accelerator for an accelerator-driven subcritical system, *PRSTAB* **16**, 080101 (2013).

[111] F. Yan *et al.*, Physics design for the C-ADS main linac basing on two different injector designs, *Chin. Phys. C* **38**(2), 27004 (2014).

[112] C. Meng, Z. H. Li and J. Y. Tang, Error analysis and lattice improvement for the C-ADS Injector-I, *Chin. Phys. C* **38**(6), 67008 (2014).

[113] Z. Gao *et al.*, The study on transient beam loading compensation for China ADS proton linac injector II, *Chin. Phys. C* **40**(5), 057005 (2016).

[114] Z. J. Wang *et al.*, End-to-end simulation of the C-ADS Injector-II with a 3-D field map, *Chin. Phys. C* **37**(4), 47003 (2013).

[115] C. Li *et al.*, Beam dynamics study of RFQ for CADS with a 3D space-charge-effect, *Chin. Phys. C* **38**(3), 37005 (2014).

[116] C. Meng *et al.*, Beam commissioning of C-ADS Injector-I RFQ accelerator, in *Proc. IPAC 2015* (2015), pp. 3827–2829.

[117] H. Li *et al.*, Development of a 325 MHz $\beta = 0.12$ superconducting single spoke cavity for China-ADS, *Chin. Phys. C* **38**(7), 77008 (2014).

[118] Y. He *et al.*, SRF cavities for ADS project in China, in *Proc. SRF 2013* (2013), pp. 868–872.

[119] X. Chen *et al.*, Development of power coupler for superconducting spoke cavities for China ADS proton linac, in *Proc. SRF 2013* (2013), pp. 1024–1026.

[120] F. B. Meng *et al.*, Study and design of RF coupler for Chinese ADS HWR superconducting cavity, *Chin. Phys. C* **38**(11), 117003 (2014).

[121] A. M. Shi *et al.*, Design of the new couplers for C-ADS RFQ, *Chin. Phys. C* **39**(4), 47004 (2015).

[122] W. Wu *et al.*, Development of a superconducting solenoid for CADS, *Chin. Phys. C* **38**(1), 17003 (2014).

[123] H. Jia *et al.*, Design and construction of the MEBT1 for CADS injector scheme II, *Chin. Phys. C* **39**(10), 107003 (2015).

[124] F. Yan *et al.*, Commissioning of the China-ADS Injector-I Testing Facility, in *Proc. IPAC 2016* (2016), pp. 2048–2051.

[125] D. H. He *et al.*, Hefei national synchrotron radiation light source, *At. Energ. Sci. Technol.* **19**, 735–742 (1985), in Chinese.

[126] Z. P. Liu and W. M. Li, Progress of the NSRL phase II Project, in *Proc. APAC 2001* (2001), pp. 235–238.

[127] H. L. Xu *et al.*, The adjustment of Phase II upgraded program of Hefei electron storage ring, *Chin. Phys. C* **30**(S1), 153–155 (2006).

[128] L. Wang *et al.*, The upgrade project of Hefei light source (HLS), in *Proc. IPAC 2010* (2010), pp. 2588–2590.

[129] Z. H. Bai *et al.*, Lattice study for the HLS-II storage ring, *Chin. Phys. C* **37**(4), 47004 (2013).

[130] L. Shang *et al.*, The new injection system of the HLS II, in *Proc. IPAC 2013* (2013), pp. 687–689.

[131] Q. Luo *et al.*, The magnet subsystem of HLS II, in *Proc. IPAC 2013* (2013), pp. 3537–3539.

[132] C. F. Xu *et al.*, Preliminary study of the higher-harmonic cavity for HLS-II, *Chin. Phys. C* **36**(9), 891–894 (2012).

[133] Q. Luo *et al.*, New beam diagnostics and related study on HLS photo-injector and HLS II, in *Proc. IPAC 2015* (2015), pp. 3578–3580.

[134] Y. Wang *et al.*, Vacuum system of the storage ring of HLS-II, in *Proc. IPAC 2015* (2015), pp. 2976–2977.

[135] J. Y. Li *el al.*, Operational status of HLS-II, in *Proc. IPAC 2016* (2016), pp. 4155–4158.

[136] L. Wang *et al.*, Our future light source at NSRL, in *Proc. PAC09* (2009), pp. 2321–2323.

[137] S. Y. Chen, H. J. Xu and Z. T. Zhao, The SSRF accelerator complex, in *Proc. APAC 2001* (2001), pp. 239–242.
[138] Z. T. Zhao and H. J. Xu, For the SSRF project team, progress of the Shanghai Synchrotron Radiation Facility, in *Proc. APAC 2004* (2004), pp. 1–3.
[139] Z. T. Zhao, H. J. Xu and H. Ding, For the SSRF project team, construction of the Shanghai Synchrotron Radiation Facility, in *Proc. APAC 2007* (2007), pp. 593–596.
[140] Z. T. Zhao, H. J. Xu and H. Ding, For the SSRF project team, commissioning status of the Shanghai Synchrotron Radiation Facility, in *Proc. EPAC 2008* (2008), pp. 998–1000.
[141] Z. T. Zhao and H. J. Xu, Operational status of the Shanghai Synchrotron Radiation Facility, in *Proc. IPAC 2010* (2010), pp. 2421–2413.
[142] Z. T. Zhao *et al.*, Performance optimization and upgrade of the SSRF storage ring, in *Proc. IPAC 2013* (2013), pp. 178–180.
[143] B. C. Jiang *et al.*, Investigation of fast ion instability in SSRF, *NIMA* **614**, 331–334 (2010).
[144] M. Liu *et al.*, A new timing system: The real time synchronized data bus, in *Proc. IPAC 2010* (2010), pp. 1396–1398.
[145] M. Z. Zhang *et al.*, Study of a new injection scheme for the SSRF storage ring, in *Proc. IPAC 2012* (2012), pp. 685–687.
[146] J. Liu *et al.*, RF system for SSRF storage ring, in *Proc. PAC 2009* (2009), pp. 2116–2118.
[147] T. J. Shen *et al.*, Construction of SSRF magnet power supply system, in *Proc. APAC 2007* (2007), pp. 664–665.
[148] D. K. Jiang *et al.*, The vacuum system for the SSRF storage ring, in *Proc. APAC 2007* (2007), pp. 708–710.
[149] L. R. Shen and D. K. Liu, The SSRF control system, in *Proc. IPAC 2010* (2010), pp. 2704–2706.
[150] M. H. Zhao *et al.*, Design and commissioning of a 150 MeV linac for SSRF, *Chin. Phys. C* **32**(S1), 244–246 (2008).
[151] H. H. Li *et al.*, Commissioning of the SSRF booster, in *Proc. EPAC 2008* (2008), pp. 2073–2075.
[152] Z. M. Dai *et al.*, Status of the SSRF storage ring, in *Proc. EPAC 2008* (2008), pp. 1998–2000.
[153] Z. T. Zhao *et al.*, Progress towards top-up operation at SSRF, in *Proc. IPAC 2010* (2010), pp. 3008–3010.
[154] M. Z. Li *et al.*, Beam-based alignment of SSRF storage ring, *Chin. Phys. C* **33**(04), 301–305 (2009).
[155] J. Hou *et al.* Studies of closed orbit correction and slow orbit feedback for the SSRF storage ring, *Chin. Phys. C* **33**(02), 145–150 (2009).
[156] D. Deng *et al.* Crystal structure of the human glucose transporter GLUT1 (131), *Nature* **510**, 121–125 (2014).
[157] Z. T. Zhao *et al.*, Consideration on the future major upgrades of the SSRF storage ring, in *Proc. IPAC 2015* (2015), pp. 1672–1674.
[158] S. Q. Tian *et al.*, Lattice design of the SSRF-U storage ring, in *Proc. IPAC 2015* (2015), pp. 304–306.
[159] X. Jiang *et al.*, BAPS preliminary design report. IHEP internal report (2012).
[160] G. Xu *et al.*, Recent physical studies for the HEPS project, in *Proc. IPAC 2016* (2016), pp. 2886–2888.
[161] G. Xu, Y. Jiao and Y. M. Peng, ESRF-type lattice design and optimization for the high energy photon source, *Chin. Phys. C* **40**(2), 27001 (2016).
[162] Y. Jiao and G. Xu, Exploring the ultimate linear and nonlinear performance of the HEPS hybrid 7BA design, in *Proc. IPAC 2016* (2016), pp. 2886–2888.
[163] G. Xu *et al.*, On-axis beam accumulation enabled by phase adjustment of a double-frequency RF system for diffraction limited storage rings, in *Proc. IPAC 2016* (2016), pp. 2032–2035.
[164] Y. M. Peng *et al.*, Candidate booster design for the HEPS project, in *Proc. IPAC 2016* (2016), pp. 3263–3265.
[165] S. K. Tian *et al.*, Intra-beam scattering studies for low emittance at BAPS, *Chin. Phys. C* **37**(6), 067001 (2013).
[166] F. L. Long *et al.*, Conceptual design for the HEPS power supply system, in *Proc. IPAC 2016* (2016), pp. 2540–2542.
[167] H. J. Wang *et al.*, Overall design of magnet girder system for HEPS-TF, in *Proc. IPAC 2016* (2016), pp. 2383–2385.
[168] J. L. Xie, Linac for FEL in China, in *Proc. LINAC 1994* (1994), pp. 950–953.
[169] Z. X. Hui, FEL studies in China, in *Proc. EPAC 1990* (1990), pp. 121–124.
[170] J. L. Xie, Free electron laser research in China, in *Proc. PAC 1996*, pp. 162–166.
[171] W. Z. Zhou *et al.*, Progress report of the FEL injector at CIAE, *NIMA* **341**, ABS37–ABS38 (1994).
[172] J. L. Xie *et al.*, First lasing of the Beijing FEL, *NIMA* **341**, 34–38 (1994).
[173] J. L. Xie *et al.*, The saturation of the Beijing FEL, *NIMA* **358**, 256–259 (1995).
[174] Z. T. Zhao *et al.*, The Shanghai high-gain harmonic generation DUV free-electron laser, *NIMA* **528**, 591–594 (2004).
[175] B. Liu *et al.*, Demonstration of a widely-tunable and fully-coherent high-gain harmonic-generation free-electron laser, *PRSTAB* **16**, 020704 (2013).
[176] H. X. Deng *et al.*, Experimental demonstration of longitudinal beam phase-space linearizer in a

free-electron laser facility by corrugated structure, *Phys. Rev. Lett.* **113**, 254802 (2014).
[177] H. X. Deng *et al.*, Polarization switching demonstration using crossed-planar undulators in a seeded free-electron laser, *PRSTAB* **17**, 020704 (2013).
[178] Z. T. Zhao *et al.*, First lasing of an echo-enabled harmonic generation free-electron laser, *Nat. Photon.* **6**(6), 360–363 (2012).
[179] H. Deng and C. Feng, Using off-resonance laser modulation for beam-energy-spread cooling in generation of short-wavelength radiation, *Phys. Rev. Lett.* **111**, 084801 (2013).
[180] Z. Wang, D. Xiang and Z. T. Zhao, Generation of coherent mode-locked radiation in a seeded free electron laser, in *Proc. IPAC 2016* (2016), pp. 777–780.
[181] C. Feng and Z. T. Zhao, Coherent hard X-ray free-electron laser based on echo-enabled staged harmonic generation scheme, in *Proc. IPAC 2010* (2010), pp. 2120–2122.
[182] J. H. Dai, H. X. Deng and Z. M. Dai, Proposal for an X-ray free electron laser oscillator with intermediate energy electron beam, *Phys. Rev. Lett.* **108**, 034802 (2012).
[183] T. Zhang *et al.*, Design of a wavelength continuously tunable ultraviolet coherent light source, in *Proc. IPAC 2012* (2012), pp. 1727–1729.
[184] G. L. Wang *et al.*, Generating polarization controllable FELs at Dalian coherent light source, in *Proc. IPAC 2016* (2016), pp. 770–773.
[185] Z. T. Zhao *et al.*, Shanghai soft X-ray free electron laser test facility, in *Proc. IPAC 2011* (2011), pp. 3011–3013.
[186] H. Li *et al.*, Linac design of the IR-FEL project in China, in *Proc. FEL 2015* (2015), pp. 427–429.
[187] D. Wang, For the SXFEL team, soft X-ray free electron laser at SINAP, in *Proc. IPAC 2016* (2016), pp. 1028–1031.
[188] Z. T. Zhao, C. Feng, J. Chen *et al.*, Two-beam based two-stage EEHG-FEL for coherent hard X-ray generation, *Sci. Bull.* **61**(9), 720–727 (2016).
[189] Z. C. Liu *et al.*, Optical design of the energy recovery linac FEL at Peking University, in *Proc. FEL 2006* (2006), pp. 277–280.
[190] S. L. Huang *et al.*, Optics layout for the ERL test facility at Peking University, in *Proc. ERL 2011* (2011), pp. 49–91.
[191] F. Zhu *et al.*, A compact high-average current photo-injector, *Chin. Phys. C* **30**(10), 1006–1009 (2007).
[192] S. W. Quan *et al.*, Stable operation of the DC-SRF photoinjector, *NIMA* **798**, 117–120 (2015).
[193] X. D. Wen *et al.*, Superradiant THz undulator radiation source based on a superconducting photo-injector, *NIMA* **810**, 75–79 (2016).
[194] S. H. Wang *et al.*, Design study on the ERL test facility at IHEP-Beijing, in *Proc. ERL 2011* (2011), pp. 52–55.
[195] Y. Jiao and O. Z. Xiao, Beam dynamics studies of the photo-injector in low-charge operation mode for the ERL test facility at IHEP, *Chin. Phys. C* **38**(6), 67003 (2014).
[196] S. G. Liu, T. M. Huang and J. Q. Xu, BXERL photo-injector based on a 217 MHz normal conducting RF gun, *Chin. Phys. C* **35**(9), 865–869 (2011).
[197] S. Y. Chen, S. H. Wang and X. W. Zhu, Towards one machine, two purposes: using a common SC linac for XFEL and ERL simultaneously, *Chin. Phys. C* **34**(01), 112–114 (2010).
[198] J. J. Deng *et al.*, Overview of pulsed power research at CAEP, *IEEE Trans. Plasma Sci.* **43**(8), 2660–2765 (2015).
[199] J. Deng, B. Ding and J. Shi, Beam conditioning studies of SG-1 FEL, *High Power Laser Part. Beams* **7**(4), 488–492 (1995).
[200] J. Shui, B. Ding and J. Deng, 12 MeV linear induction electron accelerator, *High Power Laser Part. Beams* **9**(4), 599–604 (1997).
[201] J. J. Deng *et al.*, Design of the Dragon-I linear induction accelerator, in *Proc. 21st Int. Conf.* (2002), pp. 40–42.
[202] J. Deng *et al.*, Dragon-I linear induction accelerator, in *Proc. Int. Lin. Accel. Conf.* (2006), pp. 49–51.
[203] J. Li *et al.*, Commissioning of Dragon-I linear induction accelerator, *Chin. Phys. C* (S1), 247–249 (2008).
[204] J. J. Deng *et al.*, The Dragon-II, a triple pulse high power LIA, in *Proc. 5th Eur. Asian Pulsed Power Conf.* (2014).
[205] J. Deng *et al.*, Initial performance of the primary test stand, *IEEE Trans. Plasma Sci.* **41**(10), 2580–2583 (2013).
[206] L. P. Guo *et al.*, Establishment of *in situ* TEM–implanter/accelerator interface facility at Wuhan University, *NIMA* **586**, 143–147 (2008).
[207] Y. B. Zhou *et al.*, PKUNIFTY: A neutron imaging facility based on an RFQ, *NIMA* **651**, 62–66 (2011).
[208] Progress of PKUNIFTY — a RFQ accelerator-based neutron imaging facility at Peking University, *Phys. Procedia* **43**, 79–85 (2013).
[209] W. W. Wen *et al.*, Neutronic design and simulated performance of Peking University Neutron Imaging Facility (PKUNIFTY), *NIMA* **651**, 67–72 (2011).
[210] H. Wang *et al.*, Preliminary energy-filtering neutron imaging with time-of-flight method on PKUNIFTY — a compact accelerator-based neutron imaging facility at Peking University, *NIMA* **823**, 65–71 (2016).

[211] J. Wei *et al.*, Compact pulsed hadron source — a university based accelerator platform for neutron and proton application, in *Proc. PAC 2009* (2009), pp. 1360–1362.
[212] J. Wei *et al.*, The compact pulsed hadron source construction status, in *Proc. IPAC 2010* (2010), pp. 633–635.
[213] X. L. Guan *et al.*, The compact pulsed hadron source status, in *Proc. HIAT 2012* (2012), pp. 112–114.
[214] S. X. Zheng *et al.*, Status of CPHS project, in *Proc. IPAC 2013* (2013), pp. 3887–3889.
[215] Q. Z. Xing *et al.*, High power test and beam commissioning of the CPHS RFQ accelerator, in *Proc. IPAC 2013* (2013), pp. 3305–3307.
[216] Q. Z. Xing *et al.*, Present status of the high current proton linac at Tsinghua University and its beam measurements and applications, in *Proc. HB 2014* (2014), pp. 208–210.
[217] C. X. Tang *et al.*, Tsinghua Thomson scattering X-ray source, *NIMA* **608**, S70–S74 (2009).
[218] C. X. Tang *et al.*, A simulation study of Tsinghua Thomson scattering X-ray source, *Chin. Phys. C* **33**(S2), 146–150 (2009).
[219] Y. C. Du *et al.*, Soft X-ray generation experiment at the Tsinghua Thomson scattering X-ray source, *NIMA* **637**, S168–S171 (2011).
[220] H. S. Xu *et al.*, Design of a compact storage ring for the TTX, in *Proc. IPAC 2011* (2011), pp. 3005–3007.
[221] H. S. Xu *et al.*, Laser electron storage ring for TTX, in *Proc. IPAC 2013* (2013), pp. 2171–2173.
[222] W. Guo *et al.*, A high intensity beam line of γ-rays up to 22 MeV energy based on Compton backscattering, *NIMA* **578**, 457–462 (2007).
[223] A. H. Zewail and J. M. Thomas, *4D Electron Microscopy*, Imperial College Press, London (2009).
[224] G. Sciaini and R. J. D. Miller, *Rep. Prog. Phys.* **74**, 096101 (2011).
[225] R. K. Li *et al.*, The megaelectron-volt ultrafast electron diffraction experiment at Tsinghua University, in *Proc. PAC 2009* (2009), pp. 590–592.
[226] R. K. Li and C. X. Tang, Temporal resolution of MeV ultrafast electron diffraction based on a photocathode RF gun, *NIMA* **605**, 243–248 (2009).
[227] R. K. Li *et al.*, Single-shot continuously time-resolved MeV ultrafast electron diffraction, *Rev. Sci. Instrum.* **81**(3), 036110 (2010).
[228] R. K. Li *et al.*, Recent Progress of MeV ultrafast electron diffraction at Tsinghua University, in *Proc. IPAC 2010* (2010), pp. 229–231.
[229] D. Xiang *et al.*, Accelerator-based single- shot ultrafast transmission electron microscope with picosecond temporal resolution and nanometer spatial resolution, *NIMA* **759**, 74–82 (2014).
[230] F. C. Fu, High quality single shot ultrafast MeV electron diffraction from a photocathode radio-frequency gun, *Rev. Sci. Instrum.* **85**, 083701 (2014).
[231] L. W. Feng *et al.*, Ultrafast electron diffraction with megahertz MeV electron pulses from a superconducting radio-frequency photoinjector, *Appl. Phys. Lett.* **107**, 224101 (2015).
[232] C. X. Tang, Present status of the accelerator industry in Asia, in *Proc. IPAC 2010*, pp. 2447–2451.
[233] C. Zhang, Accelerator-based mega-science projects in China and their impact on economy, in *Proc. IPAC 2011* (2011), pp. 1986–1990.
[234] Y. Z. Lin, Applications of low energy accelerators in China, in *Proc. APAC 2001* (2001), pp. 852–857.
[235] C. X. Tang, The development of accelerator applications in China, in *Proc. APAC 2004* (2004), pp. 528–532.
[236] Y. Z. Lin, Applications of low energy linacs in China, in *Proc. LINAC 2002* (2002), pp. 279–283.
[237] M. W. Fan, T. J. Zhang and J. W. Xia, Medical cyclotron and development in China, in *Proc. C 2010* (2010), pp. 425–429.
[238] http://www.shinva.com/rod_index/columnsId=36.html.
[239] J. C. Yang *et al.*, Design of a compact structure cancer therapy synchrotron, *NIMA* **756**, 19–22 (2014).
[240] J. Shi *et al.*, Slow extraction design in HIMM, in *Proc. IPAC 2013* (2013), pp. 2567–2569.
[241] M. Liu *et al.*, A scheme design of collimator for gantry in proton therapy facility, *NIMA* **791**, 47–53 (2015).
[242] T. J. Zhang and M. W. Fan, Initial operation of CIAE medically used cyclotron, in *Proc. PAC 1997* (1997), pp. 3834–3836.
[243] G. X. Pei *et al.*, A 10 MeV L-band linac for irradiation applications in China, in *Proc. LINACL 2012* (2012), pp. 147–149.
[244] C. X. Tang *et al.*, Low-Energy linacs and their applications in Tsinghua University, in *Proc. LINAC 2006* (2006), pp. 256–258.
[245] C. X. Tang, Low energy accelerators for cargo inspection, *RAST* **8**, 143–163 (2015).
[246] http://www.nuctech.com/en/SitePages/HomePage.aspx.
[247] J. E. Chen and K. Zhao, The status of RF superconductivity RD in China, in *Proc. SRF 1997* (1997), pp. 224–232.
[248] J. E. Chen and K. Zhao, The growth of SRF in China, in *Proc. SRF 2007* (2007), pp. 1–5.
[249] J. E. Chen, K. X. Liu and K. Zhao, TESLA technology in China, in *Proc. SRF 2009* (2009), pp. 42–45.
[250] J. Gao *et al.*, IHEP 1.3 GHz SRF technology R&D status, in *Proc. SRF 2011* (2011), pp. 87–89.
[251] F. Zhu *et al.*, A 1.3 GHz cryomodule with 2×9-cell cavity for SETF at Peking University, in *Proc. SRF 2015* (2015), pp. 1443–1445.

[252] J. K. Hao *et al.*, Development of DC-SRF injector at Peking University, in *Proc. IPAC 2015* (2015), pp. 2944–2947.

[253] H. Li *et al.*, Development of a 325 MHz $\beta = 0.12$ superconducting single spoke cavity for China-ADS, *Chin. Phys. C* **38**(7), 77008 (2014).

[254] H. Li *et al.*, Development of a very low beta superconducting single spoke cavity for China-ADS linac, in *Proc. SRF 2013* (2013), pp. 942–945.

[255] Z. J. Wang *et al.*, The comparison of ADS injector II with HWR cavity and CH cavity, in *Proc. IPAC 2011* (2011), pp. 2616–2618.

[256] W. M. Yue *et al.*, R&D of IMP superconducting HWR for China ADS, in *Proc. LINAC 2012* (2012), pp. 600–602.

[257] H. W. Peng and J. J. Zhuang, A phase-adjusted focusing laser accelerator, *Sci. China* **22**(10), 953–963 (1979).

[258] P. X. Wang *et al.*, The output properties of laser-driven GeV electron bunches, *NIMA* **481**, 581–586 (2002).

[259] X. P. Zhang *et al.*, Field properties and vacuum electron acceleration in a laser beam of high-order Laguerre–Gaussian mode, *NIMA* **281**, 4103–4108 (2008).

[260] Y. Tian *et al.* Electron emission at locked phases from the laser-driven surface plasma wave, *Phys. Rev. Lett.* **109**, 115002 (2012).

[261] J. F. Hua *et al.*, Generating 10–40 MeV high quality monoenergetic electron beams using a 5 TW 60 fs laser at Tsinghua University, *Chin. Phys. C* **39**(1), 017001 (2015).

[262] B. Liu *et al.*, Generating overcritical dense relativistic electron beams via self-matching resonance acceleration, *Phys. Rev. Lett.* **110**, 045002 (2013).

[263] W. T. Li *et al.*, A controllable electron injector based on ponderomotive acceleration of electrons inside the wakefield, *Plasma Phys. Control. Fus.* **55**, 095013 (2013).

[264] X. Y. Zhao *et al.*, Electron injection and acceleration in the plasma bubble regime driven by an ultraintense laser pulse combined with using dense-plasma wall and block, *Phys. Plasma* **19**, 033108 (2012).

[265] G. Q. Liao *et al.*, Bursts of terahertz radiation from large-scale plasmas irradiated by relativistic picosecond laser pulses, *Phys. Rev. Lett.* **114**, 255001 (2015).

[266] J. L. Liu *et al.*, Two-stage acceleration of protons from relativistic laser–solid interaction, *Phys. Rev. STAB* **15**, 101301 (2012).

[267] M. Chen *et al.*, Tunable synchrotron-like radiation from centimeter scale plasma channels, *Light: Sci. Appl.* **5**, e16015 (2016).

[268] M. Zeng *et al.*, High quality electron beam acceleration by ionization injection, *Phys. Plasmas* **23**, 063113 (2016).

[269] F. Li *et al.*, Low-energy-spread laser wakefield acceleration using ionization injection with a tightly focused laser in a mismatched plasma channel, *Plasma Phys. Control. Fus.* **58**, 034004 (2016).

[270] W. M. Wang *et al.*, Towards sub-TeV electron beams driven by ultra-short, ultra-intense laser pulses, *J. Plasma Phys.* **78**(4), 461–468 (2012).

[271] J. S. Liu *et al.*, All-optical cascaded laser wakefield accelerator using ionization-induced injection, *Phys. Rev. Lett.* **107**, 035001 (2011).

[272] W. T. Wang *et al.*, Control of seeding phase for a cascaded laser wakefield accelerator with gradient injection, *Appl. Phys. Lett.* **103**, 243501 (2013).

[273] H. Lu *et al.*, Laser wakefield acceleration of electron beams beyond 1 GeV from an ablative capillary discharge waveguide, *Appl. Phys. Lett.* **99**, 091502 (2011).

[274] Y. Xu *et al.*, A stable 200 TW/1 Hz Ti:sapphire laser for driving full coherent XFEL, *Opt. Laser Technol.* **79**, 141–145 (2016).

[275] C. H. Yu *et al.*, Ultrahigh brilliance quasi-monochromatic MeV γ-rays based on self-synchronized all-optical Compton scattering, *Sci. Rep.* **6**, 29518 (2016).

[276] X. Q. Yan *et al.*, Generating high current monoenergetic proton beams by a circularly polarized laser pulse in the phase-stable acceleration regime, *Phys. Rev. Lett.* **100**, 135003 (2008).

[277] A. Henig *et al.*, Radiation pressure acceleration of ion beams driven by circularly polarized laser pulses, *Phys. Rev. Lett.* **103**, 245003 (2009).

[278] H. Y. Wang *et al.*, Laser-driven three-stage heavy-ion acceleration from relativistic laser–plasma interaction, *Phys. Rev. E* **89**, 013107 (2014).

[279] J. H. Bin, Ion acceleration using relativistic pulse shaping in near-critical-density plasmas, *Phys. Rev. Lett.* **115**, 064801 (2015).

[280] W. P. Wang *et al.*, Generation of low-divergence mega-electron-volt ion beams from thin foil irradiated with an ultrahigh-contrast laser, *Appl. Phys. Lett.* **101**, 214103 (2012).

[281] W. P. Wang *et al.*, Effects of nanosecond-scale prepulse on generation of high-energy protons in target normal sheath acceleration, *Appl. Phys. Lett.* **102**, 224101 (2013).

[282] H. Zhang *et al.*, Collisionless shocks driven by 800 nm laser pulses generate high-energy carbon ions, *Phys. Plasmas* **22**, 013113 (2015).

[283] R. F. Li *et al.*, Compact laser plasma, accelerator at PKU, in *Proc. IPAC 2016* (2016), pp. 2069–2071.

Chuang Zhang is a senior research scientist and professor of the Institute of High Energy Physics (IHEP), Chinese Academy of Sciences. He joined IHEP in 1975, and took part in the design, construction, commissioning and operation of the Beijing Electron–Positron Collider (BEPC) and its upgrade project, BEPCII. He was also involved in design study of the China Spallation Source (CSNS) and the proposed Circular Electron–Positron Collider (CEPC). He was Deputy Director of IHEP from 1998 to 2005, Chair of the Asian Committee for Future Accelerators from 2007 to 2009, a member of the CERN Machine Advisory Committee from 2010 to 2013, SPC Chair of IPAC 2013 and Vice President of the Chinese Physical Society from 2007 to 2015.

Shouxian Fang graduated from the Physics Department of Fudan University in 1955. He worked at the Institute of Modern Physics and the Institute of Atomic Energy before 1968 and is currently working at the Institute of High Energy Physics as a professor of physics. He studied accelerator physics and particle dynamics in 1957–1960 at the USSR's Lebedev Physics Institute, and served as Project Director of the BEPC and Director of the Institute of High Energy Physics (IHEP), with responsibility for the BEPC Project in 1988–1992. In 1991, Prof. Fang was elected a member of Chinese Academia Science (CAS), and in 2004–2008 he was a member of the presidium of the Chinese Academy of Sciences. From 1992 to 2012, he was Director of BEPC National Laboratory. From 1999 to 2003, he was a member of the expert group of the accelerator-driven clean nuclear energy project, and led the high current proton linac and proton superconducting cavity R&D team and joined in the design and construction of the first 3.5 MeV RFQ in China. In 2007, he started to promote and lead proton therapy accelerator research in Shanghai and then became a general consultant to this project, involving himself deeply in the detailed design of the project. Since 2010, Prof. Fang has been Chairman of the Science and Technology Committee for the Chinese Spallation Neutron Source (CSNS) project, and since 2011 a general advisor to the Chinese ADS pilot scientific project entitled "Study of the Key Technologies for Accelerator-Driven Nuclear Waste Transmutation," which was approved by the CAS in 2010.

Reviews of Accelerator Science and Technology
Vol. 9 (2016) 313–335

DOI: 10.1142/S1793626816300139

Jerome Lewis Duggan: A Nuclear Physicist and a Well-Known, Six-Decade Accelerator Application Conference (CAARI) Organizer

Floyd Del McDaniel
Department of Physics, University of North Texas,
Denton, TX 76203, USA
mcdaniel@unt.edu

Barney L. Doyle
Sandia National Laboratories, P. O. Box 5800,
Albuquerque, NM 87185-1056, USA

Jerry Duggan was an experimental MeV-accelerator-based nuclear and atomic physicist who, over the past few decades, played a key role in the important transition of this field from basic to applied physics. His fascination for and application of particle accelerators spanned almost 60 years, and led to important discoveries in the following fields: accelerator-based analysis (accelerator mass spectrometry, ion beam techniques, nuclear-based analysis, nuclear microprobes, neutron techniques); accelerator facilities, stewardship, and technology development; accelerator applications (industrial, medical, security and defense, and teaching with accelerators); applied research with accelerators (advanced synthesis and modification, radiation effects, nanosciences and technology); physics research (atomic and molecular physics, and nuclear physics); and many other areas and applications. Here we describe Jerry's physics education at the University of North Texas (B. S. and M. S.) and Louisiana State University (Ph.D.). We also discuss his research at UNT, LSU, and Oak Ridge National Laboratory, his involvement with the industrial aspects of accelerators, and his impact on many graduate students, colleagues at UNT and other universities, national laboratories, and industry and acquaintances around the world. Along the way, we found it hard not to also talk about his love of family, sports, fishing, and other recreational activities. While these were significant accomplishments in his life, Jerry will be most remembered for his insight in starting and his industry in maintaining and growing what became one of the most diverse accelerator conferences in the world — the International Conference on the Application of Accelerators in Research and Industry, or what we all know as CAARI. Through this conference, which he ran almost single-handed for decades, Jerry came to know, and became well known by, literally thousands of atomic and nuclear physicists, accelerator engineers and vendors, medical doctors, cultural heritage experts... the list goes on and on. While thousands of his acquaintances already miss Jerry, this is being felt most by his family and us (B.D. and F.D.M).

Keywords: Accelerator research; Jerry Duggan; nuclear physics; atomic physics; X-ray.

1. Introduction

This article is about the life and career of Dr. Jerry Duggan, who was Regents Professor of Physics at the University of North Texas (UNT) before he retired in 2012 after 39 years of service at UNT and 10 years of service at the Oak Ridge Associated Universities and the Oak Ridge Institute of Nuclear Studies. Jerry died on Sunday, August 31, 2014, in Denton, Texas at the age of 81. He had been hospitalized since June 10, 2014, after accidentally falling and hitting his head, which resulted in brain swelling and bleeding. He spent the summer of 2014 in the hospital and in rehabilitation. Jerry was a great husband, father, brother, uncle, teacher, colleague, and friend to everyone that was fortunate enough to know him. Besides his family, he loved two things — fishing and sharing his love of physics. It brought him tremendous joy to watch his students become physicists, professors, doctors and, most importantly, productive scientists. One of his greatest pleasures was following the careers of the students from the Accelerator Laboratory in Denton and knowing in some way he had been an influence

Fig. 1. Jerome Lewis Duggan, Ph.D., Regents Professor of Physics, University of North Texas, Denton, Texas.

on the students' success. Physics was both his career choice and his passion. He was once quoted as saying, "I enjoy research because I get to study the nucleus of the atom. It's really the last frontier to study." Figure 1 is a recent picture of Jerry.

Jerry's Early Life

Jerome Lewis Duggan was born on August 4, 1933, in Columbus, Ohio. He spent his early years in Columbus, Wilkes-Barra, Pennsylvania, and Syracuse, New York, and moved with his family to Denison, Texas in 1944. Jerry was an all-district running back for the Denison High School Yellow Jackets football team. He had won a full scholarship in football to the Colorado School of Mines, until he tore ligaments in his leg. Figure 2 shows Jerry in a high school picture. Figure 3 shows Jerry and wife, Nancy, at the 1952 Denison High School Class Reunion, which was held in 1981.

Jerry's Education

Jerry developed a love for accelerators as early as 1955. After injuring his leg, he transferred from the Colorado School of Mines to North Texas State University in Pre-engineering. He received both his bachelor's degree in Physics and in Mathematics in 1955 and his master's degree in Physics in 1956 from North Texas State University (NTSU), which was renamed the University of North Texas (UNT) in 1988. For his master's thesis project, he built a 150 keV Cockcroft–Walton accelerator using an X-ray transformer power supply and the rf ion source that was developed by the Physics Division at Oak Ridge National Laboratory (ORNL). After graduating from NTSU with a master's degree in 1956, Jerry taught one year at NTSU and then attended graduate school at Louisiana State University (LSU) from 1957 to 1961. He received a Ph.D. in Physics from LSU in 1961. His dissertation research was done at ORNL under an Atomic Energy Commission Graduate Research Fellowship. He was awarded the fellowship to do his Ph.D. research in the tandem accelerator laboratory at ORNL under the stewardship of Philip D. Miller, Director of the 6.5 MV EN Tandem Accelerator Laboratory at ORNL. This gave Jerry a great opportunity to work with some of the world's outstanding accelerator and nuclear physicists.

Jerry the Family Man

Jerry was a loved member of his family. He had four children: Dane Duggan, Dedra Kalmbach, Denise Duggan, and Christy Wallis. In Fig. 4, Jerry, on the far right, is shown with his immediate and extended families. Figure 5 shows Jerry with older brother Joe Duggan and sister Patti Duggan Davey. Figure 6 shows Jerry and his wife, Nancy Bach Duggan.

Jerry's Gadgets

Jerry loved watches that had features tracking the heart rate, he had a stopwatch so he could time his running, and many other things. He would have loved the Apple watch. He bought about a dozen GPSs over the years. He had several portable DVD players, two or three speakers for the TV and, of course, his smartphone. He was fascinated with electronics and would buy the newest gadget and spend hours figuring out how to work it. He had blood pressure machines and cholesterol-checking devices, and a heart monitor. He did love his gadgets.

Jerry was also a tinkerer and repairer of almost any aspect of the four accelerators that were in the

Fig. 2. Jerry at 15 years old in high school in Denison, Texas. He is first from the left in the front row.

Ion Beam Modification and Analysis Laboratory in the Physics Department of UNT. If he did not know exactly how to fix a problem with the accelerator or detector or vacuum system or electronics, Jerry would contact his extensive list of friends, colleagues, and contacts for advice. He knew everyone in the world that was involved with particle accelerators and associated components. Figure 7 shows Jerry sitting below and working on the 2.5 MV Van de Graaff accelerator. Figure 8 shows Jerry working on a scattering chamber at the end of a beam line.

Jerry the Fisherman

Jerry had a love of fishing and would go to Ray Roberts Lake near Denton or to the Gulf Coast at Port Aransas, Texas to fish. Jerry and Nancy liked the Coast so much that they bought a place near the water. Figure 9 is a picture of Port Aransas on the Gulf Coast. Figure 10 is a picture of Jerry fishing at Port Aransas. Figure 11 shows Jerry catching a hammerhead shark at the Gulf Coast. Figure 12 shows Jerry and Del fishing at Port Aransas during spring break. Figure 13 is a photo of a houseboat that was co-owned by Jerry, Del, and Jim Sybert and over the years enjoyed by friends and colleagues. The houseboat was used for many social gatherings and cookouts.

Jerry, a Trained Athlete and a Long-Distance Runner

Jerry was a serious runner and entered many 5 K and 10 K races. He won many medals for his running. Jerry and Del used to run around the track at the Oak Ridge High School every time we went to Oak Ridge for experiments. Figure 14 shows Jerry and Del at a 5 K fun run in Muenster, Texas.

Jerry's Ph.D. Work and Early Career

Jerry received his Ph.D. in 1961 from Louisiana State University, but did his dissertation research at Oak Ridge National Laboratory (ORNL) under the mentorship of Philip Miller, who was Director of the

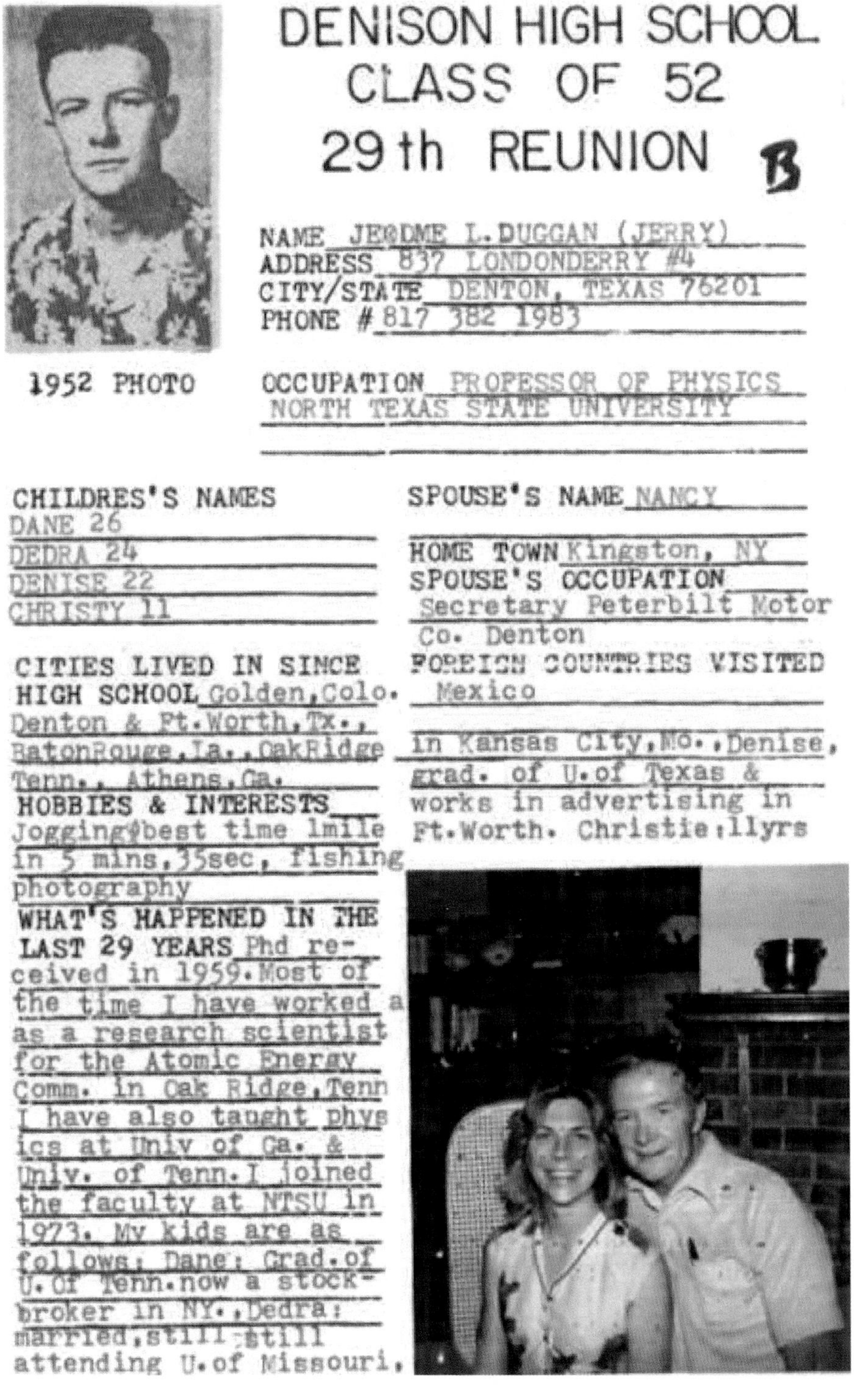

DENISON HIGH SCHOOL
CLASS OF 52
29th REUNION B

NAME JEROME L. DUGGAN (JERRY)
ADDRESS 837 LONDONDERRY #4
CITY/STATE DENTON, TEXAS 76201
PHONE # 817 382 1983

1952 PHOTO

OCCUPATION PROFESSOR OF PHYSICS NORTH TEXAS STATE UNIVERSITY

CHILDRES'S NAMES
DANE 26
DEDRA 24
DENISE 22
CHRISTY 11

SPOUSE'S NAME NANCY
HOME TOWN Kingston, NY
SPOUSE'S OCCUPATION Secretary Peterbilt Motor Co. Denton

CITIES LIVED IN SINCE HIGH SCHOOL Golden, Colo. Denton & Ft. Worth, Tx., BatonRouge, La., OakRidge Tenn., Athens, Ga.

FOREIGN COUNTRIES VISITED Mexico

HOBBIES & INTERESTS Jogging(best time 1mile in 5 mins, 35sec, fishing photography

WHAT'S HAPPENED IN THE LAST 29 YEARS Phd received in 1959. Most of the time I have worked a as a research scientist for the Atomic Energy Comm. in Oak Ridge, Tenn I have also taught physics at Univ of Ga. & Univ. of Tenn. I joined the faculty at NTSU in 1973. My kids are as follows: Dane: Grad. of U. of Tenn. now a stockbroker in NY., Dedra: married, still still attending U. of Missouri, in Kansas City, Mo., Denise, grad. of U. of Texas & works in advertising in Ft. Worth. Christie: 11yrs

Fig. 3. Jerry's high school class reunion in 1981. He and his wife, Nancy, are shown in the lower picture.

6.5 MV EN Tandem Accelerator Laboratory. This led to very strong ties with ORNL, which Jerry maintained for most of his career. Figure 15 is a picture of part of the campus of the ORNL.

After graduation, Jerry became an Assistant Professor at the University of Georgia from 1961 to 1963 and then took a permanent position as a Senior Staff Member at the Oak Ridge

Fig. 4. Jerry's immediate and extended families, in 2011.

Fig. 5. Jerry with older brother Joe Duggan and sister Patti Duggan Davey, in 1981.

Associated Universities (ORAU) and the Oak Ridge Institute of Nuclear Studies (ORINS), where he performed research at ORNL and coordinated a program that brought university students and faculty from around the country to ORNL to do research. While at ORAU, he taught classes in nuclear science and accelerator physics for professors and students from many universities throughout the country. During this time, Jerry developed a large series of high-quality nuclear science experiments to train both students and faculty members in nuclear science experiments and instrumentation. This led to his love of teaching and his ultimate move to a university faculty position. While at ORINS, he was also

Fig. 6. Jerry and wife Nancy Bach Duggan, in 1982.

Fig. 7. Jerry working on the 2.5 MV Van de Graaff accelerator in the Ion Beam Modification and Analysis Laboratory at UNT.

a Resident Graduate Faculty Member at the University of Tennessee from 1967 to 1973.

In 1968–1969, Jerry returned to the University of Georgia for a year. Del met Jerry in 1968, while he was a Ph.D. graduate student at that university working on neutron scattering, polarization, and spin flip with major professor Malcolm Steuer. Jerry took a sabbatical leave from ORAU and worked in the Accelerator Laboratory of the Department of Physics and Astronomy at the University of Georgia, and that is where Del and Jerry got to know each other well and probably led to Del's being asked to come to UNT as an Assistant Professor after his postdoctoral appointment at the University of Kentucky.

As a side note, at the physics department picnic that summer, as we approached the football season in Georgia, the faculty and students played football. As we chose teams, Jerry was selected as the quarterback for one team and Del as the quarterback for the other. Needless to say, no one told Del or

Fig. 8. Jerry working on a scattering chamber at the end of a beam line.

Fig. 9. The Gulf Coast at Port Aransas, Texas.

any other person on his team that Jerry had been a star running back at Dennison High School and had started college at the Colorado School of Mines on a football scholarship. As Del remembers, most on Del's team were just trying to be nerdy physicists.

Fig. 10. Jerry fishing at Port Aransas, Texas. We are not sure where all these black drum fish came from. Jerry claimed he caught them.

Of course, this was the time when we used pen pocket protectors and carried slide rules on our belts.

Barney met Jerry at a CAARI conference in 1976 while he was a postdoc at Kansas State University. He did his Ph.D. with Steve Shafroth at UNC/Chapel Hill on multiple inner shell ionization by light and heavy MeV ions. This was a topic being studied by Jerry and his colleagues at UNT at that time (then North Texas State University), so obviously that resonated with Barney. As testimony to his view on the importance of CAARI, which Jerry started, Barney has attended every biennial CAARI conference since 1976 except one, and that one was just a week before a conference on nuclear microprobes that he chaired in Santa Fe, New Mexico the following week. While Jerry's technical accomplishments are truly outstanding, he will be most remembered for his vision in starting CAARI, a conference on accelerator invention, development, engineering, physics, and applications.

Fig. 11. Jerry catching a hammerhead shark at the Gulf Coast of Texas.

Fig. 12. Jerry and Del fishing at Port Aransas, Texas. Jerry had pushed a number of the fish over to Del's pile, and that is why we were laughing.

Jerry, Nuclear Science Training Courses and Laboratory Manuals

From 1963 to 1973, Jerry was a Senior Scientist at the Oak Ridge Institute of Nuclear Studies, which later became the Oak Ridge Associated Universities (ORAU). In this capacity he lectured and conducted the training courses on radioisotopes, health physics, and nuclear instrumentation and measurements. Jerry had also written training materials and application manuals over the last 20 years for major nuclear instrumentation vendors, such as ORTEC, Tennelec, and The Nucleus. For ORTEC, he wrote *AN 34 Manual on Nuclear Science Experiments for College Curricula and Training Schools.* In 2006, ORTEC put this book of experiments on their website [1] to provide wider availability. He also completed a similar manual called *Laboratory Investigations in Nuclear Science* for Tennelec and The Nucleus [2]. He continued writing manuals on advanced experiments in nuclear science at North Texas State University, with the support of the National Science Foundation [3]. As can be seen from Jerry's list of publications, he had considerable experience in the measurements and instrumentation associated with virtually all kinds of radiation measurements and in the use of small accelerators in teaching [4–17].

Fig. 13. The houseboat on Lewisville Lake, halfway between Dallas and Denton.

Fig. 14. Jerry and Del at a 5 K fun run in Muenster, Texas, about one hour northwest of Denton.

Jerry's Research

From 1961 to 1970, most of Jerry's research work concerned nuclear physics with particle accelerators. Most of the experiments focused on ^{3}He-induced nuclear stripping reactions on light nuclei (^{3}He, n), (^{3}He, p), and (^{3}He, d) [18–26]. At the time, ORNL had one of the best neutron time-of-flight spectrometers in the world [18]. Jerry and his colleagues were one of the first groups to assign the spins and parities of many light nuclei using the distorted wave Born approximation to fit the angular distributions over a wide variety of incident ^{3}He energies [18–27].

In 1970, a paper by Tom Johansson of the Lund Institute in Sweden changed Jerry's research direction to atomic physics [28]. Jerry, like Barney and Del, and others, noticed that the atomic cross sections were much larger than the nuclear cross sections and that experiments were completed much faster. Johansson's paper illustrated the incredible sensitivity that could ultimately be obtained by particle-induced X-ray emission (PIXE) [28]. In order to understand the mechanisms involved in these interactions, Jerry started a comprehensive study of K-, L-, and M-shell X-ray production cross sections for a wide variety of targets, projectiles, and energies [29–67]. These measurements have proven to be very important in understanding the complex mechanisms that are involved in ion–atom

Fig. 15. Oak Ridge National Laboratory, Oak Ridge, Tennessee.

interactions in the MeV region of energy. In addition, Jerry found that, for heavier ions, electron capture from the target atom to the heavy ion became very important [68–79]. The experimental measurements were compared with the perturbed stationary state theoretical calculations with energy loss, Coulomb deflection, and relativistic effects (ECPSSR), in collaboration with George Basbas and Gregory Lapicki [80–82].

In 1973, Jim Sybert, Chair of Physics and a long-time friend of Jerry, convinced him to come back to UNT as a Professor of Physics, and Jerry convinced Floyd "Del" McDaniel to come to UNT in 1974. Jim Sybert, Jerry Duggan, and Jim Mackey, who also became a Professor of Physics at UNT, all attended LSU for their Ph.D. degrees. For many years, Jerry and Del and their graduate students made three or four trips a year to ORNL to conduct experiments for a week or two at a time. During this period, a number of students completed their experimental work for their master's or Ph.D. degrees at ORNL.

Early in his career, Jerry realized the many practical and useful applications of particle accelerators in industry and even in medicine [83–86]. In the 1980s, he began devoting a significant fraction of his research effort to helping solve materials and industrial problems using particle accelerators. He began a new program at North Texas State University (NTSU) to study surface or near-surface phenomena by using a number of ion beam techniques to address these problems. Research proposals were written to a number of federal agencies, State of Texas agencies, and industrial collaborators to purchase a new National Electrostatics Corporation (NEC) Model 9SDH-2 3 MV Tandem Pelletron Accelerator and associated beam lines, scattering chambers, electronics, and particle, X-ray, and gamma ray detectors. In 1985, the development of the Ion Beam Modification and Analysis Laboratory (IBMAL) for materials research at NTSU became Jerry's primary project [87] (www.phys.unt.edu/ibmal). In 1987, a new, 1750-square-foot laboratory was constructed and 7000 more square feet of laboratory space was modified for this new materials research facility.

The new materials modification and characterization laboratory at UNT now houses equipment for trace element accelerator mass spectrometry (TEAMS) [88–97], nuclear reaction analysis (NRA), charged particle activation analysis (CPAA), Rutherford backscattering spectrometry (RBS), hydrogen forward scattering (HFS), ion channeling (IC), particle-induced X-ray emission (PIXE), ion implantation (II), sputter-initiated resonance ionization spectroscopy (SIRIS), and heavy ion microprobe analysis (HIMA) [98, 99]. Many of these accelerator-based materials analysis techniques were developed in collaboration with Texas Instruments Inc., and other industrial partners who were members of the NSF-sponsored Industry–University Cooperative Research Center at UNT. Jerry was a driving force in building IBMAL into the largest research laboratory on the UNT campus [100] and the largest accelerator research laboratory in Texas. Figure 16 shows a plan view of IBMAL in the Department of Physics at UNT.

In 1993, Jerry and the laboratory staff began a collaborative research project with Texas Instruments to study single event upsets (SEUs) in semiconductor circuits. Jerry was involved with the development of a nuclear microprobe which would provide a 0.5-micron beam that could be scanned over a 100 microns $\times$ 100 microns area to produce an elemental surface map of the sample for most elements [101]. The microprobe was developed jointly by UNT and Texas Instruments. Again, proposals were written to support the purchase of a focused ion microprobe from the Microanalytical Research Center in Melbourne, Australia. A microprobe facility was built for SEUs, ion-beam-induced charge (IBIC) collection studies, micro-PIXE, micro-RBS, and biomedical applications [102].

In 1997, Jerry and staff began collaborating with Sandia National Laboratories in Albuquerque, New Mexico on the development and applications of radiation effects microscopy. Jerry and Del started taking students to Sandia National Laboratories for research experiments with Barney Doyle and his research colleagues using the 6.5 MV EN Tandem Accelerator and the heavy ion microprobe [103–106]. During this time, a number of students completed their experimental work for their master's or Ph.D. degrees at Sandia National Laboratories.

Over the years, Jerry was involved in a number of accelerator-based research projects, including hafnium and zirconium diffusion studies of silicon [107, 108], ion photon emission microscopy of luminescent layers [109], Monte Carlo simulations and experimental studies for medical radioisotope

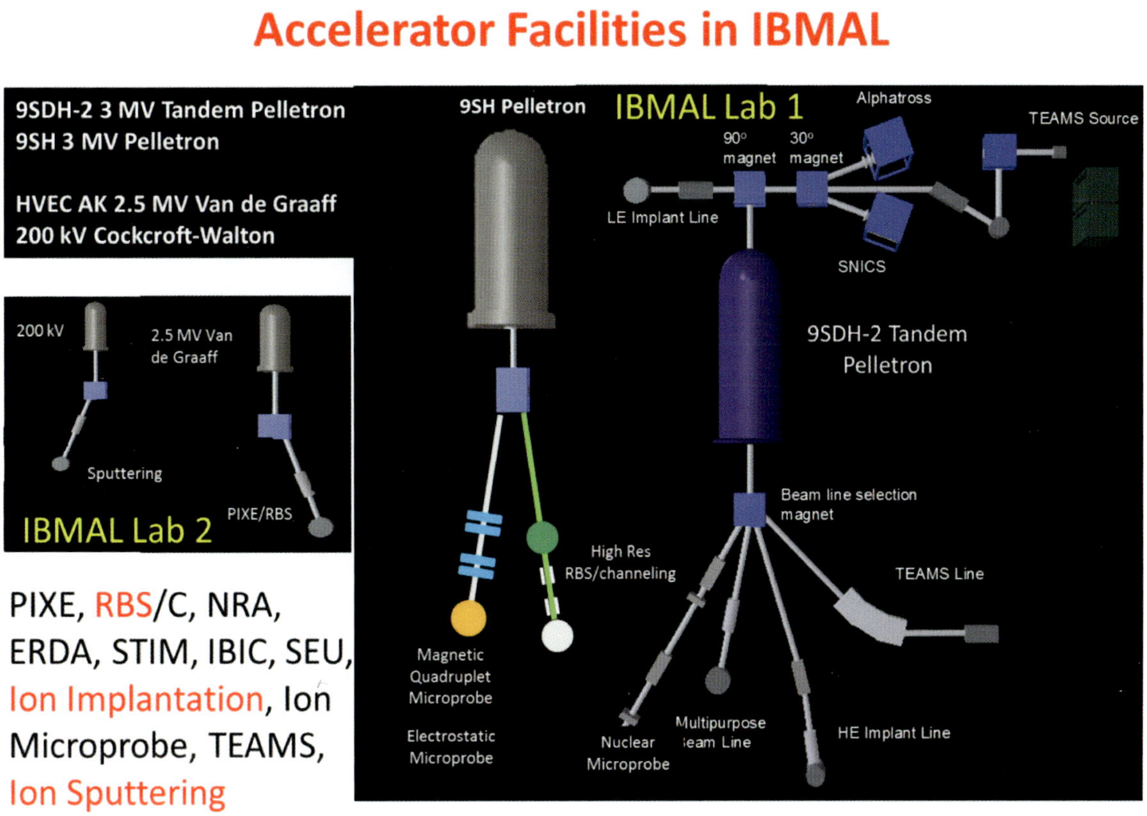

Fig. 16. IBMAL laboratory 1: the NEC 9-SDH-2 3 MV tandem and the NEC 9SH 3 MV single-ended accelerators and associated beam lines and experimental end stations. IBMAL laboratory 2: the Cockcroft–Walton accelerator and the 2.5 MV Van de Graaff accelerator and associated beam lines and apparatus.

yttrium-90 production using a 33 MeV linac [110], the role of impurities in carbon nanotubes [111, 112], stopping power studies of GaAs films [113], formation of optically active metalsilicides using ion implantation [114, 115], studies of the accuracy of several methods for determining impurity concentrations [116], the effects of heavy ion irradiation on the shallow dopants [117], hydrogen absorption in carbon nanotubes by elastic recoil detection [118, 119], microprobe analysis of Post oak seedlings [120], formation of Ru_2Si_3 nanocrystals in silica using ion implantation and thermal oxidation [121], unique methods of forming Ge_xSi_{1-x} thin films on insulators [122], and numerous others. To summarize Jerry's research activities over the years, he was able to diagnose the problems and find the solutions in many different and sometimes unrelated research areas. Part of this skill set came from his excellent training and work ethic, but some undoubtedly came from his many interactions with other colleagues and experts at the accelerator conferences.

Jerry, a Mentor of Students

Jerry will be remembered as a mentor of students. We cannot tell you how many international students he taught to drive a car in the mall parking lot and then helped them get a driver's license, because they did not drive in their home countries. If students came here and needed dental care, Jerry helped them get to a dentist. If their spouses needed a job, Jerry helped them find a job. After they graduated, Jerry helped them find a job. Figure 17 is a picture of Jerry with a number of graduate students in IBMAL.

While reading through the comments written by previous students, we were struck by their admiration for Jerry. We will include parts of two short comments: "I learned so much from him, not only about physics but life in general. He truly cared about us, all of us." And the second comment: "I remember Jerry was such a great physicist with a gentle heart toward his friends and his students, full of passion in his work." These two comments in a sense summarize Jerry's professional life.

Fig. 17. Jerry with a number of students in the Ion Beam Modification and Analysis Laboratory.

Jerry obtained the NSF funding that started the Student–Faculty Accelerator Laboratory Visitation Program at UNT. This program provided funding for students and their faculty members mainly from four-year colleges and master's programs to come to UNT for advanced accelerator laboratory experiments. Here, they could have access to unique research equipment that was not available to them at their home institution. The students would return to their universities and complete Advanced Laboratory Reports for class credits at their universities. Jerry and the staff and graduate students at IBMAL ran the concentrated courses over weekends at the average of 10–15 times per year. Figure 18 is a picture of one group that came to UNT for the concentrated training in advanced accelerator laboratory experiments. Figure 19 is a picture of a UNT graduate student giving instructions to another student–faculty visitation group. The students always had a good time, worked hard, and learned a lot about accelerator-based analysis. Because of this introduction to the physics department at UNT and some of the faculty and students, a number of students applied to graduate school and became graduate students at UNT. So, this program that Jerry started became a very good graduate student recruiting tool.

The International Conference on the Application of Accelerators in Research and Industry (CAARI)

One of the many things for which Jerry will be remembered is the International Conference on the Application of Accelerators in Research and Industry, called the "Denton Conference" by our physics colleagues around the world, and known as just CAARI today.

After accepting a senior staff position at the Oak Ridge Institute of Nuclear Studies (ORINS) in 1963, Jerry did research with his earlier research group from ORNL and taught classes. Along with Jim Lambert and Paul Treado from Georgetown University, Jerry also conducted summer Institutes in Experimental Nuclear Physics for college teachers. As part of the Institute of Nuclear Studies, he used a 400 keV Van de Graaff accelerator to perform Rutherford backscattering spectrometry (RBS) and nuclear reaction analysis (NRA).

From these early experiences, Jerry saw a need and decided to have a small conference which was focused on teaching faculty members how to take advantage of the accelerator as a teaching tool, and to outline some modest research activities that were being done with accelerators. In 1968, the first conference was held in the Oak Ridge Playhouse. The

Fig. 18. Jerry with a group of students and their faculty members who came to the accelerator lab for the NSF-sponsored Student–Faculty Accelerator Laboratory Visitation Program.

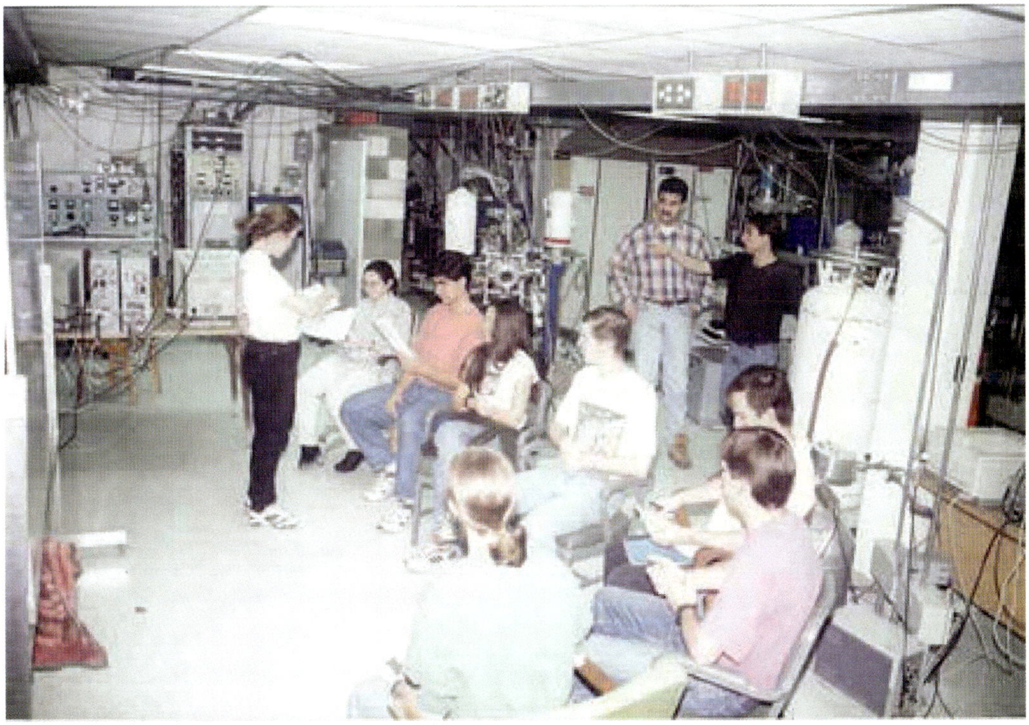

Fig. 19. A UNT graduate student giving instructions to a student–faculty visitation group.

conference was called "The Use of Small Accelerators in Teaching and Research," and it consisted of 23 invited talks [123].

The second accelerator conference was again held in Oak Ridge (in 1970), with 110 participants [124]. In 1973, Jerry moved to UNT as Professor of Physics and brought the conference with him. Ira Lon Morgan joined Jerry as co-organizer, and some industrial applications were added to the 1974 meeting in Denton, which had about 240 participants [125].

Lon convinced Jerry that the conference was missing an opportunity by not having more industrial applications at the conference. For the 1976 conference, quite a few industrial applications were added [126]. It is interesting to note that the attendance jumped from 240 participants in 1974 to over 400 in 1976. Lon was the cochairman of this conference series from 1976 to 2002 [126–139]. After Morgan joined the accelerator conference, the name was changed to the International Conference on the Application of Accelerators in Research and Industry (CAARI). Figure 20 is a picture of Lon Morgan. The growth of the CAARI conference series, the locations, and the various publishers are shown in Table 1.

Fig. 20. Ira Lon Morgan.

Table 1. Growth, locations, and publishers of CAARI over the years.

Year	Attendees	Location	Publisher
1968	110	Oak Ridge	DOE
1970	125	Oak Ridge	DOE
1974	220	Denton	DOE
1976	375	Denton	IEEE
1978	400	Denton	IEEE
1980	550	Denton	IEEE
1982	600	Denton	IEEE
1984	500	Denton	NIMB
1986	720	Denton	NIMB
1988	960	Denton	NIMB
1990	790	Denton	NIMB
1992	950	Denton	NIMB
1994	1080	Denton	NIMB
1996	815	Denton	AIP
1998	640	Denton	AIP
2000	675	Denton	AIP
2002	510	Denton	AIP
2004	495	Fort Worth	NIMB
2006	550	Fort Worth	NIMB
2008	619	Fort Worth	AIP
2010	513	Fort Worth	AIP
2012	534	Fort Worth	AIP
2014	550	San Antonio	NIMB/Procedia

The first three conferences were totally supported by DOE and ORTEC. From that point on, DOE and NSF were the major sponsors and a number of vendors were minor sponsors. During the years of this conference series, all of the national laboratories that conduct accelerator research have contributed with speakers and financial support. Through the years, virtually all of the accelerator manufacturers, nuclear electronics companies, and vacuum component companies have also supported the industrial exhibits, which helps support students and foreign scientists.

At the opening ceremony of CAARI 2004, Jerry handed the conference duties over to Floyd "Del" McDaniel (UNT) and Barney L. Doyle of Sandia National Laboratories (SNL). In 2010, Gary Glass (UNT) and Yongqiang Wang of Los Alamos National Laboratory (LANL) were added as cochairs of CAARI. In 2015, Arlyn Antolak (SNL–Livermore) was added as a cochair. From 2004 to 2012, the conference was held in Fort Worth, Texas every even year, as in the past [140–144]. In 2014, the conference was moved to San Antonio, Texas [145]. The next conference in this series will be held back in Fort Worth, at the Renaissance Worthington Hotel, from October 30 to November 4, 2016.

Jerry started this conference series in Oak Ridge, Tennessee in 1968 primarily because there was an obvious need for an applications-oriented accelerator conference. In the early 1970 s, after the conference was moved to Denton, all the talks were held in just one room, an auditorium of the UNT Student Union. For those reading this who have attended CAARI recently, that's hard to imagine, as there were about ten parallel sessions last time! In the early days, *all* the attendees sat through *all* of the presentations which covered *all* of the subtopics at the meeting. You had basic nuclear physicists learning about applied nuclear physics, and vice versa. The same was true of accelerator-based atomic physics, which was the research that Jerry, Del, and Barney were mainly doing at that time. And then there were all the industrial applications that were being born at that time, like ion implantation for doping semiconductors and strengthening the near-surface of metals used for prostheses. Basic experimental and theoretical ion–solid interactions were topics that were then coupled with other talks on the development of plasma-facing components for tokamaks and even medical applications such as proton radiotherapy of cancer. Nuclear and atomic inner shell reaction cross sections were reported by the basic physicists, and then other scientists or engineers would show how they were being used for materials analysis, and development of isotopes for medical imaging and treatment. Radiation-generating devices and techniques were discussed not only concerning the first, second, third, and fourth generations of synchrotrons, but also for CT, PET, and MRI imaging of diseases. Some of the first experiments that generated and used radioactive isotopic beams for fundamental nuclear physics were presented at CAARI, as were the invention and use of accelerator mass spectrometry to date the Shroud of Turin by measuring C-14/C-12 isotopic ratios... and so on. Jerry's insight in developing and industry in continuing and growing CAARI were truly remarkable. This meeting still stands alone for its breadth of topics that are of interest to the accelerator community, coupled with its strong focus on industrial applications.

The conference has grown and is now one of the largest of the American Physical Society (APS) Topical Conferences. Through Jerry's activities with this conference series, he worked with individuals in most of the major universities, all of the government research laboratories, and most of the industries involved with atomic and nuclear applications. This accelerator conference is now a major publication outlet for the accelerator community.

A colleague and friend, George Vourvopoulos, met Jerry in the early 1960 s at ORNL. He had arrived from Athens, Greece as a young scientist eager to learn about science and life in the US. Jerry, although only a few years older, took George under his wing, helping him and guiding him through physics and life in the US. Their paths soon diverged, with Jerry starting his long association with Denton and North Texas State University (NTSU). He did not, however, use NTSU as his first and only connection with science. Soon, we learned about his other big love and success: the Denton conferences.

Physicists, chemists, biologists, materials scientists, medical scientists, and physicians from university faculty, researchers from companies, and national laboratories from many countries would come to the Denton conferences to present the results of their research, and listen and exchange freely views with scientists from many other countries. Sometimes these professors would wear their national costumes, and although English was the official conference language the accents were varied indeed.

It did not take long to put Denton and NTSU on the scientific world map. The expression "We heard this in Denton" started becoming familiar. The exchange of information, ideas, experimental results, theoretical calculations, etc., all ingredients of research, extended to many parts of the world. Soon, other scientists in several countries realized that the form of exchanging of ideas and results that Jerry had started in Denton was helping the communication among scientists, and avoided the unnecessary duplication of expensive research.

We believe that the Denton conferences are one of the most important inheritance pieces that Jerry left to us. He served nuclear and accelerator science and promoted the free exchange of scientific information — a true gift to all of us who benefitted from his dedication. Figure 21 is a picture of Jerry standing in the lobby of the Renaissance Worthington Hotel in Fort Worth, with the logo of the CAARI conference series at his feet.

Jerry received a number of honors during his illustrious career. In 1987, he was awarded the UNT President's Award, which honors the person or group

Fig. 21. Jerry standing with the CAARI logo in the lobby of the Renaissance Worthington Hotel in Fort Worth during a CAARI meeting.

that has brought the most favorable recognition to the University. In 1995, he was given a special medal at the PIXE meeting, for "lifetime contributions to the field of accelerators." In 1996, Jerry was made a Fellow in the American Physical Society, with the citation reading, "For outstanding contributions in the application of low energy nuclear technology for analysis in the semiconductor, metals, and geophysical industries, and for initiating an international conference as a forum for the interaction of Industrial and Academic Physicists." In 2000, he was promoted to Regents Professor of Physics at the University. One of the things that Jerry was most proud of was the 15 Ph.D. students and 16 M.S. students he had supervised over the years. During his career, he published over 200 refereed papers.

As a small token of the heartfelt appreciation of the entire accelerator community for Jerry's scientific and conference organizational skills, *the Proceedings of the 19th Conference on the Application of Accelerators in Research and Industry* was dedicated to Jerry L. Duggan. The 23rd conference in this series was held in May 2014 in San Antonio, and Jerry attended as Conference Chair Emeritus. He continued to interact with colleagues and students, discussing the latest research on the applications of particle accelerators. It was a great honor for him to see the conference continue with 500+ attendees from more than 50 countries.

For Barney and Del and many others around the world, Jerry has been our friend and colleague for most of our lives. We have never had a friend as kind and generous as this man, who has had such a profound influence on particle accelerator physics and its applications around the world.

Acknowledgments

We would like to thank the many families, friends, and colleagues who have contributed to this article, especially Jerry's wife, Nancy Duggan; Jerry's daughter, Denise Duggan; and colleague and friend George Vourvopoulos, from Athens, Greece.

References

[1] J. L. Duggan, *Experiments in Nuclear Science*, In Aug. 2006, ORTEC put this book of experiments on their website, www.ortec-online.com, Application Note AN-34. The book has 25 experiments and 170 pages. This manual of nuclear educational experiments has been reprinted 5 times, with over 10,000 copies in print worldwide. ORTEC's decision to put the complete book on their website will make the experiments available to the largest possible audience.

[2] J. L. Duggan, *Laboratory Investigations in Nuclear Science* (Tennelec–The Nucleus, Oak Ridge, TN, 1988).

[3] J. L. Duggan, F. D. McDaniel and J. Hehn, Project supported by the National Science Foundation. *Advanced Experiments in Nuclear Science* (North Texas State University Press, 1976).

[4] J. L. Duggan, A scattering chamber experiment for the modern physics laboratory, *Am. J. Phys.* **33**(4), 312–318 (1965).

[5] J. L. Duggan, J. F. Yegge and K. Ford, Rutherford scattering with a low-energy Cockcroft–Walton accelerator, *Am. J. Phys.* **35**(8), 765–770 (1967).

[6] J. L. Duggan, W. D. Adams, R. J. Scroggs and L. S. Anthony, Charged particle detector experiments for the modern physics laboratory, *Am. J. Phys.* **35**(7), 631–637 (1967).

[7] W. K. Robinson, W. D. Adams and J. L. Duggan, Some X-ray experiments for the modern physics laboratory, *Am. J. Phys.* **36**(8), 683–689 (1968).

[8] J. L. Duggan and L. S. Anthony, ^{6}Li + d experiment, *Am. J. Phys.* **35**(3), 194–200 (1968).

[9] J. L. Duggan, A study of the ^{7}Li(p, α) α reaction as a student exercise, *J. Nucl. Energy* **22**, 611–619 (1968).

[10] J. L. Duggan and J. F. Yegge, A Rutherford elastic scattering 'experiment', *J. Chem. Educ.* **45**, 85 (1968).

[11] W. K. Robinson, J. L. Nagi and J. L. Duggan, A time-of-flight neutron experiment for the undergraduate laboratory, *Am. J. Phys.* **37**(5), 482–488 (1969).

[12] J. L. Duggan, The use of small accelerators in teaching, [invited paper] in *1970 Symposium on the Application of Small Accelerators*, American Nuclear Society (U.S. Department of Health, Education, and Welfare, Rockville, MD, 1970).

[13] D. Desmarais and J. L. Duggan, An undergraduate alpha particle time-of-flight experiment for determining the mean excitation energy for electronic stopping power of Al, Cu, Ag, and Au, *Am. J. Phys.* **52**(5), 408–411 (1984).

[14] D. Desmarais and J. L. Duggan, Alpha particle induced inner-shell ionization measurements for the undergraduate laboratory, *Am. J. Phys.* **52**(6), 507–513 (1984).

[15] D. Desmarais and J. L. Duggan, The study of alpha-particle decay schemes of heavy nuclei, *Am. J. Phys.* **11**, 1079–1085 (1990).

[16] D. Desmarais and J. L. Duggan, Electron conversion of selected nuclides, *Am. J. Phys.* **59**(1), 56–62 (1991).

[17] J. E. Noakes and J. L. Duggan, A high resolution 4π alpha spectrometer with silicon surface barrier detectors, *J. Radio-analyt. Chem.* **43**, 399–409 (1978).

[18] J. L. Duggan, P. D. Miller and R. F. Gabbard, The Be^{9}(He^{3}, n)C^{11}, Li^{7}(He^{3}, n)B^{9} and C^{13}(He^{3}, n)O^{15} reactions by time-of-flight techniques, *Nucl. Phys.* **46**, 366 (1964).

[19] P. D. Miller, J. L. Duggan and M. M. Duncan, Absence of states in Be^{9} at 7.56, 13.91, and 18.00 MeV, *Nucl. Phys.* **54**, 155 (1964).

[20] M. M. Duncan, J. L. Duggan and R. D. Purrington, Direct interactions with three particle final states, *Phys. Rev.* **134**, 164 (1964).

[21] W. R. Coker, M. M. Duncan, J. L. Duggan and P. D. Miller, An investigation of the ^{9}Be(^{3}He, p)^{11}B reaction at low energies, *Nucl. Phys.* **91**, 97–111 (1967).

[22] P. D. Miller, J. L. Duggan, M. M. Duncan, R. L. Dangle, W. R. Coker and J. Lin, ^{11}B(^{3}He, d)^{12}C spectroscopic factors, *Nucl. Phys. A* **136**, 229 (1969).

[23] J. Y. Park, J. L. Duggan and P. D. Miller, Elastic scattering of ^{3}He particles by ^{9}Be and ^{11}B between 4.0 and 18.0 MeV, *Nucl. Phys. A* **134**, 277 (1969).

[24] J. L. Duggan, J. Y. Park, S. D. Danielopoulos, P. D. Miller, J. Lin, M. M. Duncan and R. L. Dangle, ^{3}He elastic scattering from ^{10}B and ^{14}C in the range 4 to 18 MeV, *Nucl. Phys. A* **151**(1), 107–119 (1970).

[25] W. R. Coker, P. D. Miller, J. L. Duggan, M. M. Duncan, R. L. Dangle and J. Lin, The ^{14}C(^{3}He, α_o)^{13}C reaction at 6.0, 8.0, and 10 MeV, *Nucl. Phys. A* **168**, 307–313 (1971).

[26] R. R. Wu, R. L. Dangle, M. M. Duncan, J. L. Duggan, P. D. Miller and J. Lin, A study of the ^{11}B(^{3}He, p)C^{13} reaction at 12, 10, 8, 6, and 4 MeV, *Nucl. Phys. A* **199**, 23–30 (1973).

[27] W. R. Coker, J. Lin, J. L. Duggan and P. D. Miller, Multistep contributions to ^{11}B(^{3}He, A)^{10}B from 8.0 to 12.0 MeV, *Phys. Lett. B* **45**, 321 (1973).

[28] T. B. Johansson, R. Akselsson and S. A. E. Johansson, X-ray analysis: Elemental trace analysis at the 10^{-12} g level, *Nucl. Instrum. Methods* **84**, 141–143 (1070).

[29] J. L. Duggan and W. M. Beck, Charged particle produced X-ray fluorescence, in *Advances in X-Ray Analysis*, Vol. 15 (Plenum, New York, 1972), p. 407.

[30] R. P. Chaturvedi, T. J. Gray, J. L. Duggan, C. C. Sachtleben and J. Lin, X-ray Production cross-sections for Ti, Co, Ge, Rb and Sn by 8–44 MeV oxygen ion bombardment, in *Advances in X-Ray Analysis*, Vol. 17 (Plenum, 1974), p. 445.

[31] S. Datz, J. L. Duggan *et al.*, Cross sections for L-subshell ionization in Au by the collision of protons and helium ions, *Phys. Rev.* **9**, 192 (1974).

[32] F. D. McDaniel, T. J. Gray, R. K. Gardner, G. M. Light, J. L. Duggan, H. A. Van Rinsvelt, R. D. Lear, G. H. Pepper, J. W. Nelson and A. R. Zander, K-shell X-ray production cross sections of selected elements Ti to Sb for 1.0 to 5.0 MeV/amu -Li ions, *Phys. Rev. A* **12**(4), 1271–1280 (1975).

[33] R. M. Wheeler, R. P. Chaturvedi, J. L. Duggan, J. Tricomi and P. D. Miller, X-ray production cross sections for fourteen elements from calcium to palladium for incident carbon ions, *Phys. Rev. A* **13**(3), 958–964 (1976).

[34] J. Tricomi, J. L. Duggan, F. D. McDaniel, P. D. Miller, R. P. Chaturvedi, R. M. Wheeler, J. Lin, K. A. Kuenhold, L. A. Rayburn and S. J. Cipolla, K-shell X-ray production in Ge, Rb, Y, Zr, and Ag by ^{14}N ion impact, *Phys. Rev. A* **15**, 2269 (1977).

[35] F. D. McDaniel, J. L. Duggan, P. D. Miller and G. D. Alton, K-shell ionization of elements Ca to Zn for 0.5 to 2.5 MeV/amu ^{14}N ion bombardment, *Phys. Rev. A* **15**, 846–855 (1977).

[36] S. R. Wilson, F. D. McDaniel, J. R. Rowe and J. L. Duggan, K-shell X-rays of selected elements from Nb through Gd for incident protons and alpha particles from 0.6 to 2.4 MeV, *Phys. Rev. A* **16**, 903–912 (1977).

[37] G. Monigold, F. D. McDaniel, J. L. Duggan, R. Rice, A. Toten, R. Mehta and P. D. Miller, K-shell ionization of elements $_{15}$P to $_{28}$Ni for 0.4 to 3.8 MeV/amu $_5$B-ion bombardment, *Phys. Rev. A* **18**, 380–391 (1978).

[38] R. Mehta, F. D. McDaniel, J. L. Duggan and P. D. Miller, K-shell ionization of elements $_{29}$Cu to $_{37}$Rb for 0.4–3.8 MeV/amu ^{10}B ion bombardment, *Phys. Rev. A* **19**, 1363–1366 (1979).

[39] R. K. Rice, F. D. McDaniel, G. Basbas and J. L. Duggan, K-shell ionization by low velocity ions, *Phys. Rev. A* **24**, 758 (1981).

[40] R. Mehta, J. L. Duggan, J. L. Price, F. D. McDaniel and G. Lapicki, M-shell X-ray production cross sections in thin targets of $_{79}$Au, $_{82}$Pb, $_{83}$Bi, and $_{92}$U by 0.3–2.6 MeV H^+ and He^+ ions, *Phys. Rev. A* **26**, 1883–1891 (1982).

[41] R. Mehta, J. L. Duggan, J. L. Price, G. Lapicki, P. M. Kocur and F. D. McDaniel, $^1_1H^+$ and $^2_4He^+$ induced M-shell X-ray-production cross sections for selected elements in the rare-earth region, *Phys. Rev. A* **28**, 3217–3222 (1983).

[42] J. L. Duggan, P. M. Kocur, J. L. Price, F. D. McDaniel, R. Mehta and G. Lapicki, L-shell X-ray production cross sections of Ni, Cu, Ge, As, Rb, Sr, Y, Zr and Pd by (0.25–25) MeV protons, *Phys. Rev. A* **32**(4), 2088–2092 (1985).

[43] J. L. Price, J. L. Duggan, F. D. McDaniel, G. Lapicki and R. Mehta, K-shell X-ray production by 0.5–2.5 MeV $^9_4Be^+$ ions incident upon selected elements from fluorine to potassium, *Phys. Rev. A* **34**(4), 2830–2833 (1986).

[44] G. Lapicki, R. Mehta, J. L. Duggan, P. M. Kocur, J. L. Price and F. D. McDaniel, Multiple outer-shell ionization effect in inner-shell ionization by light ions, *Phys. Rev. A* **34**, 3813–3821 (1986).

[45] M. C. Andrews, F. D. McDaniel, J. L. Duggan, P. D. Miller, P. L. Pepmiller, H. F. Krause, T. M. Rosseel, L. A. Rayburn, R. Mehta and G. Lapicki, L- and M-shell X-ray production cross sections of Nd, Gd, Ho, Yb, Au, and Pb by 25-MeV carbon and 32-MeV oxygen ions, *Phys. Rev. A* **36**(8), 3699–3706 (1987).

[46] J. L. Price, J. L. Duggan, F. D. McDaniel, G. Lapicki and R. Mehta, M-shell X-ray production cross sections for 0.5–2.5-MeV Be^+ ions incident upon selected elements from praseodymium to bismuth, *Phys. Rev. A* **37**(2), 365–369 (1988).

[47] Y. C. Yu, M. R. McNeir, D. L. Weathers, J. L. Duggan, F. D. McDaniel and G. Lapicki, K-shell X-ray production cross sections in $_6$C, $_8$O, $_9$F, $_{11}$Na, $_{12}$Mg, and $_{13}$Al for 0.75 to 4.5 MeV protons, *Phys. Rev. A* **44**(9), 5702–5706 (1991).

[48] M. R. McNeir, Y. C. Yu, D. L. Weathers, J. L. Duggan, F. D. McDaniel and G. Lapicki, L-shell X-ray production cross sections in $_{26}$Fe, $_{28}$Ni, $_{29}$Cu, $_{30}$Zn, $_{31}$Ga, and $_{32}$Ge by 0.5- to 5.0-MeV protons, *Phys. Rev. A* **44**(7), 4372–4378 (1991).

[49] Y. C. Yu, M. R. McNeir, D. L. Weathers, J. L. Duggan, F. D. McDaniel, D. K. Marble, Z. Y. Zhao and G. Lapicki, K-shell X-ray production cross sections in $_6$C, $_8$O, $_9$F, $_{11}$Na, $_{12}$Mg, and $_{13}$Al by 0.5 to 8.0 MeV helium ions, *Phys. Rev. A* **44**(9), 7252–7257 (1991).

[50] M. R. McNeir, Y. C. Yu, D. L. Weathers, D. K. Marble, Z. Zhao, J. L. Duggan, F. D. McDaniel and G. Lapicki, L-shell X-ray production cross sections in $_{26}$Fe, $_{28}$Ni, $_{29}$Cu, $_{30}$Zn, $_{31}$Ga, and $_{32}$Ge by 0.5 to 8.0 MeV helium ions, *J. Phys. B* **27**, 5295–5308 (1994).

[51] R. Mehta, H. L. Sun, D. K. Marble, J. L. Duggan, F. D. McDaniel and G. Lapicki, L-shell X-ray production by 2–12 MeV carbon ions in fifteen selected elements from copper to lead, *J. Phys. B* **28**, 1187–1200 (1995).

[52] J. D. McCoy, J. L. Duggan, E. D. Robinson and S. J. Cipolla, Proton and alpha particle induces X-ray emission as a proposed surface analysis tool, in *Proc. Int. Symp. on the Use of Nuclear Techniques in the Basic Metals Industries* (Helsinki, Finland, 1972).

[53] J. Lin, R. F. Carlton and J. L. Duggan, X-ray production cross sections via proton and alpha particle bombardment, in *Proc. Conf. on the Inner Shell Ionization Phenomena and Future Applications*, eds. R. W. Fink, USAEC Conf. No. 720404 (1973).

[54] G. Monigold, F. D. McDaniel, J. L. Duggan, R. Mehta, R. Rice and P. D. Miller, K-shell X-ray production cross sections of selected elements Al to Ni for 4.0 to 38.0 MeV ^{10}B ions, in *Proc. Conf. on the Application of Small Accelerators*, Oct. 1976. Published by IEEE Nuclear and Plasma Sciences Society, IEEE Conf. No. 76CH1175-9 NPS, pp. 70–74.

[55] S. R. Wilson, J. R. Rowe, J. L. Duggan and F. D. McDaniel, K-shell X-ray production cross sections of selected elements from Nb through Gd for incident protons and alphas from 0.6 to 2.4 MeV, in *Proc. Conf. on the Application of Small Accelerators*, Oct. 1976. Published by IEEE Nuclear and Plasma Sciences Society, IEEE Conf. No. 76CH1175-9 NPS, pp. 186–190.

[56] K. A. Kuenhold, J. L. Duggan, F. D. McDaniel, A. D. Ray, A. Zander and P. D. Miller, Production of L-X-rays by 9.5 to 41.8 MeV fluorine ions incident on six elements from Pr to Bi, in *Proc. Conf. on the Application of Small Accelerators*, Oct. 1976. Published by IEEE Nuclear and Plasma Sciences Society, IEEE Conf. No. 76CH1175-9 NPS, pp. 334–337.

[57] F. D McDaniel and J. L. Duggan, K-shell X-rays for Ti, Ni, Ge, and Rb for incident ions from ^{1}H to ^{35}Cl in the energy range from 1–3 MeV/amu, in *Proc. Fourth Int. Conf. on Beam Foil Spectroscopy*

and Heavy-Ion Atomic Physics (Gatlinburg, TN, Sep. 15–19, 1975), eds. Sellin and D. Pegg (Plenum, New York, 1976), Vol. 2, pp. 519–537.

[58] T. M. Button, R. K. Rice, J. L. Duggan and F. D. McDaniel, L-shell X-ray production of Ga, As, Se, Br, and Cs by 0.4 to 2.4 $^1_1$H and $^4_2$He ions (*IEEE Trans. Nucl. Sci.*, New York), Vol. NS-26, No. 1 (1979), p. 1139.

[59] R. S. Peterson, F. D. McDaniel, J. L. Duggan, R. S. Thoe and P. D. Miller, Mo X-ray production in collisions of $_9$F with $_{10}$Ne (*IEEE Trans. Nucl. Sci.*, New York), Vol. NS-26, No. 1 (1979), p. 1159.

[60] T. M. Button, R. S. Peterson, F. D. McDaniel, A. Toten, M. C. Andrews, III, R. Rice, J. D. Gressett, J. L. Duggan, R. S. Thoe, G. D. Alton and P. D. Miller, Radiative electron capture X-ray profiles from fast collisions of F^{+q} with Ne (*IEEE Trans. Nucl. Sci.*, New York, 1979), Vol. NS-26, No. 1, p. 1157.

[61] R. K. Rice, T. M. Button, J. L. Duggan and F. D. McDaniel, K-shell X-ray production of Na, Mg, Al, P, S and Cl by 0.4–2.4 MeV $^1_1$H and $^4_2$He ions (*IEEE Trans. Nucl. Sci.*, New York), Vol. NS-226, No. 1 (1979), p. 1150.

[62] R. Mehta, J. L. Duggan, F. D. McDaniel, P. M. Kocur, J. L. Price and G. Lapicki, Survey of M-shell X-ray production for 19 elements by H^+ and $HE^{+,++}$ ions, in *Proc. XIII Int. Conf. on Physics of Electronic and Atomic Collisions* (Berlin, Germany, July 27–Aug. 2, 1983), Vol. 13 (1983), p. 431.

[63] R. Mehta, J. L. Duggan, F. D. McDaniel, P. M. Kocur and J. L. Price, M-shell X-ray production cross sections for 19 elements in the range $Z = 54$–92 for incident H and He ions from 15 keV to 40 MeV (*IEEE Trans. Nucl. Sci.*, New York), Vol. NS-30, No. 2 (1983), p. 950.

[64] M. C. Andrews, F. D. McDaniel, J. L. Duggan, R. Mehta, P. D. Miller, H. Krause, P. L. Pepmiller, T. M. Rosseel, L. A. Rayburn and G. Lapicki, L-shell X-ray production cross sections for selected elements with $Z = 60$ to 82 incident 25 MeV carbon and 32 MeV oxygen ions, *Nucl. Instrum. Methods* **10**, 181 (1985).

[65] M. C. Andrews, F. D. McDaniel, J. L. Duggan, R. Mehta, G. Lapicki, P. D. Miller, P. L. Pepmiller, H. F. Krause, T. M. Rosseel and L. A. Rayburn, L-shell X-ray production cross sections in Nd, Gd, Ho, Yb, Au and Pb for 25-MeV carbon and 32-MeV oxygen ions, *Nucl. Instrum. Methods B* **10/11**, 181–185 (1985).

[66] J. L. Price, J. L. Duggan, F. D. McDaniel, G. Lapicki and R. Mehta, L-shell X-ray production cross sections for 0.5–2.5 MeV Be^+ ions incident upon selected elements from $Z = 29$ to 47, *Nucl. Instrum. Methods B* **24/25**, 85 (1987).

[67] F. D. McDaniel, J. L. Duggan, J. D. Gressett, D. K. Marble, J. F. Culwell and G. Lapicki, L-shell and M-shell X-ray production cross sections for selected elements for 0.1 to 2.5 MeV protons, *Abstracts for International Conference on Spectroscopy and Collisions of Few Electron Ions* (Bucharest, Romania; Aug. 29–Sep. 2, 1988) (The Central Institute of Physics, Bucharest, 1988), p. 44.

[68] F. D. McDaniel, J. L. Duggan, G. Basbas, P. D. Miller and G. Lapicki, Projectile charge state dependence of K-shell ionization by silicon ions: A comparison of Coulomb ionization theories for direct ionization and electron capture with X-ray production data, *Phys. Rev. A* **16**, 1375–1378 (1977).

[69] F. D. McDaniel, A. Toten, R. S. Peterson, J. L. Duggan, S. R. Wilson, J. D. Gressett, P. D. Miller and G. Lapicki, Projectile charge state dependence of target L-shell ionization by 1.86 MeV/amu fluorine and silicon ions and 1.8 MeV/amu chlorine ions, *Phys. Rev. A* **19**, 1517 (1979).

[70] R. Mehta, J. L. Duggan, F. D. McDaniel, M. C. Andrews, G. Lapicki, P. D. Miller, L. A. Rayburn and A. R. Zander, Direct ionization and electron capture in M-shell X-ray production by fluorine ions, *Phys. Rev. A* **28**(5), 2722–2726 (1983).

[71] Y. C. Yu, H. L. Sun, J. L. Duggan, F. D. McDaniel, J. Y. Yin and G. Lapicki, Charge-state dependence of M-shell X-ray production in $_{67}$Ho by 2–12 MeV carbon ions, *Phys. Rev. A* **52**(5), 3836–3846 (1995).

[72] H. L. Sun, Y. C. Yu, E. K. Lin, C. W. Wang, J. L. Duggan, A. R. Azordegan, F. D. McDaniel and G. Lapicki, Charge-state dependence of K-shell X-ray production in aluminum by 1–12-MeV carbon ions, *Phys. Rev. A* **53**(6), 4190–4197 (1996).

[73] A. Toten, F. D. McDaniel, R. S. Peterson, J. L. Duggan, S. R. Wilson, J. D. Gressett, P. D. Miller and G. Lapicki, Projectile charge state dependence of target L-shell X-ray production (*IEEE Trans. Nucl. Sci.*, New York), Vol. NS-26, No. 1 (1979), p. 1143.

[74] R. Mehta, J. L. Duggan, F. D. McDaniel, M. C. Andrews, R. M. Wheeler, R. P. Chaturvedi, P. D. Miller and G. Lapicki, Projectile charge state dependence of target M-shell ionization by 1.42 MeV/amu fluorine ions (*IEEE Trans. Nucl. Sci.*, New York), Vol. NS-28, No. 2 (1981), p. 1112.

[75] R. Mehta, F. D. McDaniel, J. L. Duggan, M. C. Andrews, P. D. Miller, L. A. Rayburn and A. R. Zander, Projectile charge state dependence in M-shell ionization of Au, Pb, Bi, and U by 25 and 35 MeV fluorine ions, in *Proc. XII Int. Conf. on the Physics of Electronic and Atomic Collisions* (Gatlinburg, TN; July 15–21, 1981), Vol. 12 (1981), p. 830.

[76] F. D. McDaniel, J. L. Duggan, R. Mehta, M. C. Andrews, A. Toten, J. D. Gressett, D. Johnson,

S. R. Wilson, P. D. Miller, G. Lapicki, G. Basbas, L. A. Rayburn, A. R. Zander, R. M. Wheeler, R. P. Chaturvedi and R. S. Peterson, Ionization and electron transfer for the K-, L-, and M-shells for 1.86 MeV/u ions on selected target systems, in *Proc. XIII Int. Conf. on Physics of Electronic and Atomic Collisions* (July 27–Aug. 2, 1983, Berlin, Germany), Vol. 13 (1983), p. 428.

[77] R. Mehta, J. L. Duggan, F. D. McDaniel, M. C. Andrews, G. Lapicki, P. D. Miller and L. A. Rayburn, Electron capture in M-shell X-ray production from heavy elements by incident 25 and 35 MeV fluorine ions, *IEEE Trans. Nucl. Sci.* **30**(2), 906 (1983).

[78] M. C. Andrews, F. D. McDaniel, J. L. Duggan, R. Mehta, P. D. Miller, H. Krause, P. L. Pepmiller, T. M. Rosseel, L. A. Rayburn and G. Lapicki, M-shell electron capture and direction ionization of selected elements with $Z = 60$ to 82 by incident 25 MeV carbon and 32 MeV oxygen ions, *Nucl. Instrum. Methods B* **10**, 186 (1985).

[79] M. C. Andrews, F. D. McDaniel, J. L. Duggan, P. D. Miller, P. L. Pepmiller, H. F. Krause, T. M. Rosseel, L. A. Rayburn, R. Mehta and G. Lapicki, M-shell electron capture and direct ionization of gold by 25-MeV carbon and 32-MeV oxygen ions, *Nucl. Instrum. Methods B* **10/11**, 186–189 (1985).

[80] R. Rice, G. Basbas and F. D. McDaniel, Extended tables for plane-wave Born approximation calculation of direct Coulomb ionization of the K-shell by charged particles, *At. Data Nucl. Data Tables* **20**, 503 (1977).

[81] D. E. Johnson, G. Basbas and F. D. McDaniel, Nonrelativistic plane-wave Born-approximation calculations of direct Coulomb M-subshell ionization by charged particles, *At. Data Nucl. Data Tables* **24**, 1 (1979).

[82] G. Lapicki and F. D. McDaniel, Electron capture from K-shells by fully stripped ions, *Phys. Rev. A* **22**, 1896 (1980); **23**, 975 (1981).

[83] J. L. Duggan and I. L. Morgan, Industrial applications for accelerators, *IEEE Trans. Nucl. Sci.* **22**(3), 1216 (1975).

[84] J. L. Duggan, Novel uses of small accelerators in medicine and industry, *IEEE Trans. Nucl. Sci.* **30**(3), 3039 (1983).

[85] J. L. Duggan, Industrial and medical applications of accelerators with energies less than 20 MeV, *IEEE Trans. Nucl. Sci.* **30**(4), 3039 (1983).

[86] G. Norton and J. L. Duggan, Industrial applications of electrostatic accelerators, in *Handbook of Accelerator Physics and Engineering*, eds. A. W. Chao and M. Tigner (World Scientific, 1998), p. 24.

[87] J. L. Duggan, F. D. McDaniel, S. Matteson, D. E. Golden, J. M. Anthony, B. Gnade and J. A. Keenan, The University of North Texas Ion Beam Modification and Analysis Laboratory, in *Proc. Tenth Int. Conf. on the Application of Accelerators in Research and Industry* (Denton, Texas; Nov. 7–9, 1988), *Nucl. Instrum. Methods B* **40/41**, 709 (1989).

[88] J. M. Anthony, S. Matteson, F. D. McDaniel and J. L. Duggan, Accelerator mass spectrometry at the University of North Texas, *Nucl. Instrum. Methods B* **40/41**, 731–733 (1989).

[89] S. Matteson, F. D. McDaniel, J. L. Duggan, J. M. Anthony, T. J. Bennett and R. L. Beavers, A high resolution electrostatic analyzer for accelerator mass spectrometry, *Nucl. Instrum. Methods B* **40/41**, 759–761 (1989).

[90] J. D. Gressett, D. L. Maxson, S. Matteson, F. D. McDaniel, J. L. Duggan and J. M. Anthony, Computer automation of an accelerator mass spectrometry system, *Nucl. Instrum. Methods B* **40/41**, 755–758 (1989).

[91] F. D. McDaniel, S. Matteson, D. K. Marble, L. S. Hodges, A. M. Arrale, J. Y. Hajsaleh, M. R. McNeir, J. L. Duggan and J. M. Anthony, Molecular dissociation in accelerator mass spectrometry for trace impurity characterization in electronic materials, in *Proc. High Energy and Heavy Ion Beams in Materials Analysis Workshop* (Albuquerque, NM; June 14–16, 1989), eds. J. R. Tesmer, C. J. Maggiore, M. Nastasi, J. C. Barbour and J. W. Mayer (Materials Research Society, Pittsburgh, PA, 1990), p. 269.

[92] S. Matteson, D. K. Marble, L. S. Hodges, J. Y. Hajsaleh, A. M. Arrale, M. R. McNeir, J. L. Duggan, F. D. McDaniel and J. M. Anthony, Molecular-interference-free accelerator mass spectrometry, *Nucl. Instrum. Methods B* **45**, 575–579 (1990).

[93] J. M. Anthony, S. Matteson, D. K. Marble, J. L. Duggan, F. D. McDaniel and D. J. Donahue, Applications of accelerator mass spectrometry to electronic materials, in *Proc. First European Conference on the Application of Accelerators in Applied Research and Technology* (Frankfurt, Germany; Sep. 4–9, 1989), *Nucl. Instrum. Methods B* **50**, 262 (1990).

[94] J. M. Anthony, S. Matteson, J. L. Duggan, P. S. Elliott, D. K. Marble, F. D. McDaniel and D. L. Weathers, Atomic mass spectrometry of materials, *Nucl. Instrum. Methods B* **52**, 493 (1990).

[95] S. Matteson, J. L. Duggan, D. K. Marble, F. D. McDaniel, D. L. Weathers, D. K. Wilson, J. M. Anthony and R. L. Beavers, An ultra-clean micro probe ion source for atomic mass spectrometry, *Nucl. Instrum. Methods B* **52**, 327 (1990).

[96] F. D. McDaniel, S. Matteson, D. L. Weathers, D. K. Marble, J. L. Duggan, P. S. Elliott, D. K. Wilson and J. M. Anthony, The University of North Texas Atomic Mass Spectrometry Facility for Detection of Impurities in Electronic Materials, *Nucl. Instrum. Methods B* **52**, 310 (1990).

[97] J. M. Anthony, S. Matteson, D. K. Marble, J. L. Duggan, F. D. McDaniel and D. J. Donahue, Applications of accelerator mass spectrometry to electronic materials, *Nucl. Instrum. Methods B* **50**, 262–266 (1990).

[98] F. D. McDaniel, D. L. Weathers, J. L. Duggan, S. Matteson and D. K. Marble, Ion beam techniques for materials modification and analysis, in *Proc. Second Int. Conf. on Applications of Nuclear Techniques* (Heraklio, Crete, Greece; June 24–30, 1990), eds. G. Vourvopoulos and T. Paradellis (World Scientific, 1991), p. 307.

[99] J. L. Debrun, K. Bethge and J. L. Duggan, Applied research and technology with accelerators state of the art, 1993, *Nucl. Instrum. Methods B* **89**, 447–450 (1994).

[100] B. Rout, M. S. Dhoubhadel, P. R. Poudel, V. C. Kummari, B. Pandey, N. T. Deoli, W. J. Lakshantha, S. J. Mulware, J. Baxley, J. E. Manuel, J. L. Pacheco, S. Szilasi, D. L. Weathers, T. Reinert, G. A. Glass, J. L. Duggan and F. D. McDaniel, An overview of the facilities, activities, and developments at the University of North Texas Ion Beam Modification and Analysis Laboratory (IBMAL), in *IX Int. Symp. Radiation Physics* (Puebla, Mexico; Apr. 14–17, 2013), *AIP Conf. Proc.* **1544**, 11–18 (2013). DOI: 10.1063/1.4813454.

[101] F. D. McDaniel, J. L. Duggan, C. Yang, B. N. Guo, M. El Bouanani and M. Nigam, The high-energy heavy ion nuclear microprobe at the University of North Texas, in *Proc. 7th Int. Conf. on Nuclear Microprobe Technology and Applications* (Bordeaux, France; Sep. 10–15, 2001), *Nucl. Instrum. Methods B* **181**, 315–319 (2001).

[102] C. Yang, B. N. Guo, M. El Bouanani, M. Nigam, J. L. Duggan and F. D. McDaniel, The recent progress of the high-energy heavy ion nuclear microprobe at the University of North Texas, *AIP Conf. Proc.* **576**, 539–542 (2000).

[103] B. N. Guo, S. N. Renfrow, B. L. Doyle, D. S. Walsh, T. J. Aton, M. El Bouanani, J. L. Duggan and F. D. McDaniel, Ion beam induced charge collection (IBICC) from integrated circuit test structures using a 10 MeV carbon microbeam, in *Fifteenth Int. Conf. on the Application of Accelerators in Research and Industry* (Denton, TX; Nov. 4–7, 1998), *AIP Conf. Proc. CP475* (AIP Press), eds. J. L. Duggan and I. L. Morgan (American Institute of Physics, 1999), pp. 318–321.

[104] B. N. Guo, S. N. Renfrow, B. L. Doyle, D. S. Walsh, T. J. Aton, M. El Bouanani, J. L. Duggan and F. D. McDaniel, Ion beam induced charge collection (IBICC) from integrated circuit test structures using a 10 MeV carbon microbeam (American Institute of Physics, Conference Series CP475, 1999), p. 1121.

[105] B. N. Guo, M. El Bouanani, S. M. Renfrow, M. Nigam, D. S. Walsh, B. L. Doyle, J. L. Duggan and F. D. McDaniel, Diffusion time resolved ion beam induced charge collection from stripe-like test junctions induced by heavy-ion microbeams, *Nucl. Instrum. Methods B* **181**, 315–319 (2001).

[106] B. N. Guo, M. El Bouanani, S. N. Renfrow, D. S. Walsh, B. L. Doyle, J. L. Duggan and F. D. McDaniel, Heavy ion microbeam studies of diffusion time resolved charge collection from p–n junctions, *AIP Conf. Proc.* **576**, 531–534 (2000).

[107] M. Quevedo-Lopez, M. El Bouanani, S. Addepalli, C. Huang, J. L. Duggan, B. E. Gnade, R. M. Wallace, I. Colombo, M. Douglans and M. Visokay, Hafnium diffusion studies of hafnium silicate into silicon, *Appl. Phys. Lett.* **79**, 4192 (2001).

[108] M. Quevedo-Lopez, M. El Bouanani, S. Addepalli, J. L. Duggan, B. E. Gnade, R. M. Wallace, M. N. Visokay, M. Douglas, M. J. Bevan and L. Colombo, Thermally induced Zr incorporation into Si from zirconium silicate thin films, *Appl. Phys. Lett.* **79**, 2958 (2001).

[109] C. Yang, B. L. Doyle, P. Rossi, M. Nigam, M. El Bouanani, J. L. Duggan and F. D. McDaniel, Luminescent layers for ion-photon emission microscopy, in *Proc. 7th Int. Conf. on Nuclear Microprobe Technology Applications* (Bordeaux, France; Sep. 10–15, 2000), *Nucl. Instrum. Methods B* **181**, 535–538 (2001).

[110] D. Necsoiu, I. L. Morgan, H. Hupf, W. J. Courtney, J. Kinross-Wright, M. El Bouanani, J. L. Duggan and F. D. McDaniel, Monte Carlo simulation & experimental studies of yttrium-90 production using a 33 Mev linac, *Appl. Radiat. Isot.* **57**, 509–515 (2002).

[111] F. Naab, O. W. Holland, T. Imholt, F. D. McDaniel, J. L. Duggan and J. Roberts, The role of impurities in the interaction of carbon nanotubes with microwave radiation, in *Proc. 10th Int. Conf. on Particle Induced X-Ray Emission and Its Analytical Applications, PIXE 2004* (Portoroz, Slovenia; June 4–8, 2004), http://pixe2004.ijs.si/proceedings.

[112] F. Naab, O. W. Holland, J. L. Duggan and F. D. McDaniel, Ion beam analysis of carbon nanotubes, *J. Phys. Chem. B* **109**, 1415–1419 (2005).

[113] M. Nigam, J. L. Duggan, M. El Bouanani, C. Yang, G. V. Ravi Prasad, E. D. Sosa, S. Matteson and F. D. McDaniel, Stopping power of thin GaAs films for Si & P ions, *Nucl. Instrum. Methods B* **219-220**, 273–277 (2004).

[114] L. J. Mitchell, O. W. Holland, K. Hossain, E. B. Smith, J. L. Duggan and F. D. McDaniel, Formation of optically-active, metal-silicides using ion implantation and or oxidation, *Nucl. Instrum. Methods B* **241**, 548–552 (2005).

[115] L. J. Mitchell, O. W. Holland, F. Naab, J. L. Duggan and F. D. McDaniel, Ion beam synthesis of C based optically based nanoclusters, *J. Non-cryst. Solids* **352**(23–25), 2562–2564 (2005).

[116] J. L. Duggan, F. Naab, O. W. Holland, F. D. McDaniel, J. J. Xu, Z. Y. Zhao, B. N. Guo, J. Lin, K. H. Shin and U. Jeong, Analysis and accuracy of several methods of determining the concentration of ^{11}B implanted silicon, *Nucl. Instrum. Methods B* **243**, 205–210 (2006).
[117] P. Pelicon, M. El Bouanani, G. V. Ravi Prasad, A. Razpet, J. Simcic, B. N. Guo, D. Birt, J. L. Duggan and F. D. McDaniel, The loss of boron in ultra-shallow boron implanted Si under heavy ion irradiation, *Radiat. Eff. Defects S.* **161**(8), 487–494 (2006).
[118] F. U. Naab, M. Doubhadel, J. R. Gilbert, L. K. Savage, O. W. Holland, J. L. Duggan and F. D. McDaniel, Direct measurement of hydrogen absorption in carbon nanotubes/nanofibers by elastic recoil deflection, *Phys. Lett. A* **356**, 152–155 (2006).
[119] F. D. McDaniel, O. W. Holland, F. Naab, L. J. Mitchell, M. Doubhadel and J. L. Duggan, Ion beam analysis of hydrogen retained in carbon nanotubes and carbon films, *Nucl. Instrum. Methods B* **249**, 330–334 (2006).
[120] B. C. Boling, F. U. Naab, D. Smith, J. L. Duggan and F. D. McDaniel, Leaf elemental analysis in mycorrhizal post oak seedlings, *Nucl. Instrum. Methods B* **251**, 182–190 (2006).
[121] L. J. Mitchell, L. C. Phinney, O. W. Holland, K. Hossain, F. Naab, E. B. Smith, P. Paudel, M. Doubhadel, J. L. Duggan and F. D. McDaniel, An investigation into the formation of Ru_2Si_3 nanocrystals in silica using ion implantation and thermal oxidation, *Nucl. Instrum. Methods B* **261**(1–2), 669–673 (2007).
[122] K. Hossain, O. W. Holland, T. D. Mishima, M. B. Santos, J. L. Duggan and F. D. McDaniel, A unique method of forming Ge_xSi_{1-x} thin-films on insulator, *Mater. Res. Soc. Symp. Proc.* **1036E** (2008).
[123] J. L. Duggan, The First Conference on the Use of Small Accelerators in Teaching and Research, AEC Conf. No. 680411 (1968).
[124] J. L. Duggan, The Second Conference on the Use of Small Accelerators in Teaching and Research, AEC Conf. No. 700322 (1970).
[125] J. L. Duggan and I. L. Morgan, The Third Atomic Energy Commission Conference on the Use of Small Accelerators (eds.), Vol. II: *The Use of Small Accelerators in Industrial Applications*, AEC Conf. Vols. 741040-P1 and -P2 (Oct. 1974).
[126] J. L. Duggan and I. L. Morgan, *Scientific and Industrial Applications of Small Accelerators* (eds.), IEEE, New York, 76CH 1175-9 NPS (Apr. 1976).
[127] J. L. Duggan and I. L. Morgan, *IEEE Trans. in Nuclear Science; 1978 Conference on the Applications of Small Accelerators in Research and Industry*, Vol. NS-26, No. 1, Part II (eds), (Feb. 1979).
[128] J. L. Duggan and I. L. Morgan, *IEEE Transactions in Nuclear Science; 1980 Conference on the Application of Accelerators in Research and Industry*, Vol. NS-28, No. 2 (eds.), (Apr. 1981).
[129] J. L. Duggan and I. L. Morgan, *IEEE Transactions in Nuclear Science; 1982 Conference on the Application of Accelerators in Research and Industry*, Vol. NS-30, No. 2 (eds.), (Apr. 1983).
[130] J. L. Duggan, I. L. Morgan and J. A. Martin, *Eighth Conference on the Application of Accelerators in Research and Industry* (eds.), *Nucl. Instrum. Methods B* **10/11** (1985).
[131] J. L. Duggan and I. L. Morgan, *Ninth Conference on the Application of Accelerators in Research and Industry* (eds.), *Nucl. Instrum. Methods B* **24/25**, (1987).
[132] J. L. Duggan and I. L. Morgan, *Application of Accelerators in Research and Industry 88* (eds.), *Nucl. Instrum. Methods B* **40/41** (1989).
[133] J. L. Duggan and I. L. Morgan, *Application of Accelerators in Research and Industry 90* (eds.), *Nucl. Instrum. Methods B* **56/57** (1990).
[134] J. L. Duggan and I. L. Morgan, *Application of Accelerators in Research and Industry 92* (eds.), *Nucl. Instrum. Methods B* **79** (1993).
[135] J. L. Duggan and I. L. Morgan, *Application of Accelerators in Research and Industry 94* (eds.), *Nucl. Instrum. Method Phys. Res. B* **99** (1995).
[136] J. L. Duggan and I. L. Morgan, *Application of Accelerators in Research and Industry 96* (eds.), American Institute of Physics Conf. Series, Vol. CP392 (1997).
[137] J. L. Duggan and I. L. Morgan, *Application of Accelerators in Research and Industry 98* (eds.), American Institute of Physics Conf. Series, Vol. CP475 (1999).
[138] J. D. Duggan and I. L. Morgan, *Application of Accelerators in Research and Industry: CAARI 2000* (eds.), American Institute of Physics Conf. Series, Vol. CP576 (2001).
[139] J. L. Duggan and I. L. Morgan, in *Proc. 17th Conf. on the Application of Accelerators in Research & Industry* (eds.), American Institute of Physics Conf. Series, Vol. 680 (2003).
[140] *Proc. Eighteenth Int. Conf. on the Application of Accelerators in Research and Industry* (Fort Worth, TX; Oct. 10–15, 2004). Published as *Nucl. Instrum. Methods B* **241**, 1–1038 (2005), eds. F. Del McDaniel, B. L. Doyle, G. Vizkelethy, B. M. Johnson and J. M. Sisterson (Elsevier, The Netherlands, 2005).
[141] *Proc. Nineteenth Int. Conf. on the Application of Accelerators in Research and Industry* (Fort Worth, TX; Aug. 20–25, 2006). Published as *Nucl. Instrum. Methods B* **261**, 1–1224 (2007), eds. F. Del McDaniel, B. L. Doyle, I. Ben-Itzhak, R. W. Hamm, B. M. Johnson, J. M. Sisterson, D. R. Slaughter, S. Thevuthasan, G. Vizkelethy, Y.-Q.

Wang and D. P. Wells (Elsevier, The Netherlands, 2007).

[142] *Proc. Twentieth Int. Conf. on the Application of Accelerators in Research and Industry* (Fort Worth, TX; Aug. 10–15, 2008). Published as American Institute of Physics Conference Proceedings CP#1099, pp. 1–1076, eds. F. Del McDaniel and B. L. Doyle (American Institute of Physics, 2009).

[143] *Proc. Twenty-First Int. Conf. on the Application of Accelerators in Research and Industry* (Fort Worth, TX; Aug. 8–13, 2010). Published as American Institute of Physics Conference Proceedings CP#1336, pp. 1–781, eds. F. Del McDaniel and B. L. Doyle (American Institute of Physics, 2011).

[144] *Proc. Twenty-Second Int. Conf. on the Application of Accelerators in Research and Industry* (Fort Worth, TX; Aug. 5–10, 2012). Published as American Institute of Physics Conference Proceedings CP#1525, pp. 1–762, eds. F. Del McDaniel, B. L. Doyle, G. Glass and Y.-Q. Wang (American Institute of Physics, 2013).

[145] *Proc. Twenty-Third Int. Conf. on the Application of Accelerators in Research and Industry* (San Antonio, TX; May 5–10, 2014). Published as *Nucl. Instrum. Methods B* (2015) and in *Physics Procedia* (2015), eds. F. Del McDaniel, B. L. Doyle, G. Glass and Y.-Q. Wang (Elsevier, The Netherlands, 2015).

Floyd Del McDaniel is Emeritus Professor of Physics and Materials Science and Engineering and Emeritus Associate Dean for Research in the College of Arts and Sciences at the University of North Texas in Denton, TX. He became an assistant professor in 1975, associate professor in 1979, full professor in 1984, and Regents Professor in 2002. He was chair of the Department of Physics from 2003–2007, and was Associate Dean for Research for the College from 2009–2015, when he retired. He received a B.S. in Physics and Mathematics from the University of Memphis (1966) and an M.S. (1968) and PhD (1971) in Nuclear Physics from the University of Georgia. He has been Director of the Accelerator Laboratory at UNT (Ion Beam Modification and Analysis Laboratory (IBMAL)) since 1992. Dr. McDaniel is the co-chair of the biennial International Conference on the Application of Accelerators in Research and Industry, (http://www.CAARI.com) and has been on the organizing committee since 1974. His research has been recognized by selection as a Regents Professor and a Toulouse Scholar at the University of North Texas; and an R&D 100 award.

Barney L. Doyle is a Distinguished Member of the Technical Staff at Sandia National Laboratories in Albuquerque, New Mexico, USA. He received a B.S. in Physics at Kansas State University in 1971, and a PhD in Physics from University of North Carolina at Chapel Hill, NC in 1976 where he performed his research at the Triangle Universities Nuclear Laboratory (TUNL). He joined the Solid State Physics Center at Sandia National Laboratories in Albuquerque, NM in 1977. After 9 years in the technical staff, Doyle was promoted to manage the Ion Beam Laboratory (IBL), and was in that position for 25 years until he returned to the staff in 2011. His research has been recognized with several national awards including three R&D-100 Awards and a United States Congressional Commendation, all for various developments with accelerators. Five techniques he invented are patented, and he is a Fellow of the American Physical Society.